# RATIONAL HOMOTOPY THEORY II

# RATIONAL HOMOTOPY THEORY II

Yves Félix
Catholic University of Louvain, Belgium

Steve Halperin
University of Maryland, USA

Jean-Claude Thomas
Université d'Angers, France

World Scientific

NEW JERSEY • LONDON • SINGAPORE • BEIJING • SHANGHAI • HONG KONG • TAIPEI • CHENNAI

*Published by*

World Scientific Publishing Co. Pte. Ltd.
5 Toh Tuck Link, Singapore 596224
*USA office:* 27 Warren Street, Suite 401-402, Hackensack, NJ 07601
*UK office:* 57 Shelton Street, Covent Garden, London WC2H 9HE

**Library of Congress Cataloging-in-Publication Data**
Félix, Y. (Yves)
Rational homotopy theory II / by Yves Félix (Catholic University of Louvain, Belgium), Steve Halperin (University of Maryland, USA), Jean-Claude Thomas (Universite d'Angers, France).
pages cm
Sequel to Rational homotopy theory (2001), but self contained--Introduction, following table of contents.
Includes bibliographical references and index.
ISBN 978-9814651424 (hardcover : alk. paper)
1. Homotopy theory. I. Halperin, Stephen. II. Thomas, J.-C. (Jean-Claude) III. Title. IV. Title: Rational homotopy theory two.
QA612.7.F47 2015
514'.24--dc23
2014046652

**British Library Cataloguing-in-Publication Data**
A catalogue record for this book is available from the British Library.

# Preface

Sullivan's seminal paper, *Infinitesimal Computations in Topology*, includes the application of his techniques to non-simply connected spaces, and these ideas have been used frequently by other authors. Our objective in this sequel to [18] is to provide a complete description with detailed proofs of this material. This then provides the basis for new results, also included, and which we complement with recent advances for simply-connected spaces. There do remain many interesting unanswered questions in the field, which hopefully this text will make it easier for others to resolve.

This monograph may also be seen as an exploration of the minimal models introduced by Sullivan, to whom this work is gratefully dedicated.

# Contents

# Introduction

Rational homotopy theory assigns to topological spaces invariants which are preserved by continuous maps $f$ for which $H_*(f;\mathbb{Q})$ is an isomorphism. The two standard approaches of the theory are due respectively to Quillen [58] and Sullivan [61], and [62]. Each constructs from a class of CW complexes $X$ an algebraic model $\mathcal{M}_X$, and then constructs from $\mathcal{M}_X$ a CW complex $X_{\mathbb{Q}}$, together with a map $\varphi_X : X \to X_{\mathbb{Q}}$. Both $H_*(X_{\mathbb{Q}};\mathbb{Z})$ and $\pi_n(X_{\mathbb{Q}})$ are rational vector spaces, and with appropriate hypotheses

$$H_*(\varphi_X) : H_*(X)\otimes\mathbb{Q} \to H_*(X_{\mathbb{Q}};\mathbb{Z})\,, \quad \text{and}$$

$$\pi_n(\varphi_X) : \pi_n(X)\otimes\mathbb{Q} \to \pi_n(X_{\mathbb{Q}}), \quad n\geq 2,$$

are isomorphisms.

In each case the model $\mathcal{M}_X$ belongs to an algebraic homotopy category, and a homotopy class of maps $f : X \to Y$ induces a homotopy class of morphisms $\mathcal{M}_f : \mathcal{M}_X \to \mathcal{M}_Y$ (in Quillen approach) and a homotopy class of morphisms $\mathcal{M}_f : \mathcal{M}_Y \to \mathcal{M}_X$ (in Sullivan's approach). These are referred to as *representatives* of $f$.

In Quillen's approach, $X$ is required to be simply connected and $\mathcal{M}_X$ is a rational differential graded Lie algebra which is free as a graded Lie algebra. In this case $H_*(\varphi_X)$ and $\pi_{\geq 2}(\varphi_X)$ are always isomorphisms.

Here, as in [18], we adopt Sullivan's approach, and in this Introduction provide an overview of the material in the monograph, together with brief summaries of the individual Chapters.

Sullivan's approach associates to each path connected space $X$ a cochain algebra $\mathcal{M}_X$ of the form $(\wedge V, d)$ in which the free commutative graded algebra $\wedge V$ is generated by $V = V^{\geq 1}$, and $V = \oplus_m \wedge^m V$ with $\wedge^m V =$

$V\wedge\cdots\wedge V$ ($m$ factors). Additionally, each $\wedge V^{\leq k}$ is preserved by $d$, and $d$ also satisfies a "nilpotence" condition: $(\wedge V, d)$ is called a *minimal Sullivan algebra*. A minimal Sullivan algebra determines a simplicial set $\langle \wedge V, d\rangle$ with spatial realization $|\wedge V, d|$, and when $(\wedge V, d)$ is the model of the CW complex $X$, this determines (up to homotopy) the map

$$\varphi_X : X \to X_{\mathbb{Q}} = |\wedge V, d|\,.$$

This approach makes non-simply connected spaces accessible to rational homotopy theory. For example, if $H_1(X;\mathbb{Q})$ is finite dimensional then $\pi_1(X_{\mathbb{Q}})$ is the Malcev completion of $\pi_1(X)$: $\pi_1(\varphi_X)$ induces an isomorphism

$$\varprojlim_n \pi_1(X)/\pi_1^n(X)\otimes\mathbb{Q} \xrightarrow{\cong} \pi_1(X_{\mathbb{Q}})\,,$$

where $(\pi_1^n(X))$ denotes the lower central series of $\pi_1(X)$. On the other hand, this approach also comes at the cost of a finiteness condition: If $X$ is simply connected then $H_*(\varphi_X)$ and $\pi_{\geq 2}(\varphi_X)$ are isomorphisms if and only if $H_*(X;\mathbb{Q})$ is a graded vector space of finite type.

In the case of non-simply connected CW complexes $X$, two additional ingredients are required for rational homotopy theory: first, the action by covering transformations of $\pi_1(X)$ on the cohomology $H^*(\widetilde{X};\mathbb{Q})$ of the universal covering space of $X$; second, a Sullivan representative $\psi$ for a *classifying map* mapping $X$ to the classifying space for $\pi_1(X)$ and inducing an isomorphism of fundamental groups. If $H^*(\widetilde{X};\mathbb{Q})$ has finite type, then the groups $\pi_{\geq 2}(X)\otimes\mathbb{Q}$ can be computed from a minimal Sullivan model of $\widetilde{X}$, and if the action of $\pi_1(X)$ on $H^*(\widetilde{X};\mathbb{Q})$ is nilpotent, then this Sullivan model can be computed from $\psi$.

Minimal Sullivan algebras $(\wedge V, d)$ are equipped with a homotopy theory and a range of invariants analogous to those which arise in topology. Key among these are the graded *homotopy Lie algebra* $L = \{L_p\}_{p\geq 0}$ and, when $\dim H^1(\wedge V, d) < \infty$, the *group* $G_L$. Here $L_p = \mathrm{Hom}(V^{p+1},\mathbb{Q})$, the Lie bracket is dual to the component $d_1 : V \to \wedge^2 V$ of $d$, and an exponential map converts $L_0$ to $G_L$. The group $G_L$ acts by conjugation in $L$ and also in $H(\wedge V^{\geq 2},\overline{d})$ where $(\wedge V^{\geq 2},\overline{d})$ is obtained by dividing by $V^1\wedge\wedge V$. A third key invariant is the *Lusternik-Schnirelmann category*, $\mathrm{cat}\,(\wedge V, d)$, defined as the least $m$ for which $(\wedge V, d)$ is a homotopy retract of $(\wedge V/\wedge^{>m} V,\overline{d})$.

When $(\wedge V, d)$ is the minimal Sullivan model of a CW complex $X$ such that $\dim H^1(X;\mathbb{Q}) < \infty$, then $G_L = \pi_1(X_{\mathbb{Q}})$, $L_{\geq 1} = \pi_{\geq 1}(\Omega X_{\mathbb{Q}})$, and there is a natural map $H(\wedge V^{\geq 2},\overline{d}) \to H(\widetilde{X};\mathbb{Q})$ equivariant via $\pi_1(\varphi_X)$ with respect to the actions of $G_L$ and the action by covering transformations of $\pi_1(X)$. Finally, as in the simply connected case ( [18]) $\mathrm{cat}\,(\wedge V, d)\leq \mathrm{cat}\,X$.

Sullivan models $(\wedge V, d)$ are constructed via a functor, $A_{PL}$ (inspired by the differential forms on a manifold) from spaces to rational cochain algebras, for which $H(A_{PL}(X))$ and $H^*(X; \mathbb{Q})$ are naturally isomorphic algebras. Then $(\wedge V, d)$ is the unique (up to isomorphism) minimal Sullivan algebra admitting a morphism $m : (\wedge V, d) \to A_{PL}(X)$ for which $H(m)$ is an isomorphism. Moreover, any morphism $\varphi : (\wedge W, d) \to A_{PL}(X)$ from an arbitrary minimal Sullivan algebra determines by adjunction a homotopy class of maps $|\varphi| : X \to |\wedge V, d|$; in particular, $|m|$ is homotopic to the map $\varphi_X$ above.

This second step can be applied to construct a minimal Sullivan model $(\wedge V, d) \to (A, d)$ for any commutative cochain algebra satisfying $H^0(A, d) = \Bbbk$. While these may not be Sullivan models of a topological space, the homotopy machinery of minimal Sullivan algebras is established independently of topology, and so can be applied in this more general context.

In particular, minimal Sullivan algebras become a valuable tool in the study of graded Lie algebras $E = E_{\geq 0}$ with lower central series denoted by $(E^n)$, provided that $E_0$ acts nilpotently by the adjoint representation in each $E_i$, $i \geq 1$, and that

$$\dim E_0/[E_0, E_0] < \infty\,, \quad \dim E_i < \infty\,, \quad i \geq 1\,, \quad \text{and } \cap_n E_0^n = 0\,.$$

Such Lie algebras are called *Sullivan Lie algebras*. Note that we may have $\dim E_0 = \infty$.

For a Sullivan Lie algebra, $E$, $\varinjlim_n C^*(E/E^n)$ is a minimal Sullivan algebra, called the *associated Sullivan algebra* for $E$, and $\varprojlim_n E/E^n$ is its homotopy Lie algebra. (Here $C^*(-)$ is the classical Cartan-Chevalley-Eilenberg cochain algebra construction.) In the reverse direction, if $(\wedge V, d)$ is any minimal Sullivan algebra for which $\dim H^1(\wedge V, d) < \infty$ then its homotopy Lie algebra is a Sullivan Lie algebra whose associated Sullivan algebra is $(\wedge V, d_1)$ with $d_1$ the component of $d$ mapping $V$ to $\wedge^2 V$.

In summary, the interplay between spaces, minimal Sullivan algebras and graded Lie algebras is illustrated by the diagram

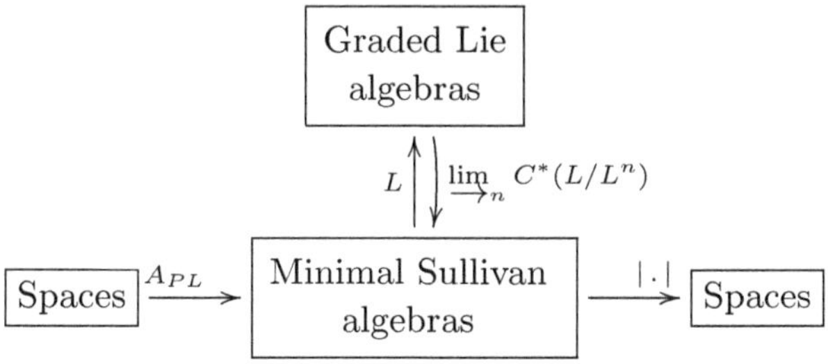

A crucial technical tool in the theory of minimal Sullivan algebras is the conversion of cochain algebra morphisms to $\Lambda$-*extensions* $(\wedge V, d) \to (\wedge V \otimes \wedge Z, d)$ in which $(\wedge V, d)$ is a minimal Sullivan algebra and $d : Z^p \to \wedge V \otimes \wedge Z^{\leq p}$ satisfies a "nilpotence" condition; when $V^1 \neq 0$ it may happen that $Z^0 \neq 0$. Division by $\wedge^+ V \otimes \wedge Z$ gives a quotient cochain algebra $(\wedge Z, \overline{d})$ and the $\Lambda$-extension determines *holonomy representations* of $L$ and, if $\dim H^1(\wedge V, d) < \infty$, of $G_L$ in $H(\wedge Z, \overline{d})$.

These $\Lambda$-extensions are the Sullivan analogues of fibrations, and $(\wedge Z, \overline{d})$ is the Sullivan analogue of the fibre. The analogy is not merely abstract: suppose

$$\begin{array}{ccc} (\wedge V, d) & \longrightarrow & (\wedge V \otimes \wedge Z, d) \\ \downarrow{\scriptstyle \alpha} & & \downarrow{\scriptstyle \beta} \\ A_{PL}(Y) & \xrightarrow{A_{PL}(p)} & A_{PL}(X) \end{array}$$

is a commutative diagram in which $Y$ is a CW complex and $p$ is the projection of a fibration with fibre $F$. Then $\beta$ factors to give a morphism $\gamma : (\wedge Z, \overline{d}) \to A_{PL}(F)$, and, in this setting $H(\gamma) : H(\wedge Z, \overline{d}) \to H^*(F; \mathbb{Q})$ is equivariant via $\pi_1(|\psi|)$ with respect to the holonomy representation of $G_L$ in $H(\wedge Z, \overline{d})$ and $\pi_1(Y)$ in $H(F; \mathbb{Q})$.

There are two important examples of $\Lambda$-extensions. First, if $(\wedge V, d)$ is a minimal Sullivan algebra then $(\wedge V^1, d) \to (\wedge V, d)$ is a $\Lambda$-extension, the Sullivan analogue of a classifying map for a CW complex $X$. If this morphism is a Sullivan representative for the classifying map, and if $H^*(X; \mathbb{Q})$ has finite type and the covering space action of $\pi_1(X)$ is nilpotent, then $X$ is a *Sullivan space* and the quotient $(\wedge V^{\geq 2}, \overline{d})$ is a minimal Sullivan model for $\widetilde{X}$. Many classical examples are Sullivan spaces, including all closed orientable Riemann surfaces.

Second, if $(\wedge V, d)$ is any minimal Sullivan algebra, converting the augmentation $(\wedge V, d) \to \Bbbk$ yields a $\Lambda$-extension $(\wedge V, d) \to (\wedge V \otimes \wedge U, d)$ with $H(\wedge V \otimes \wedge U, d) = \Bbbk$; this is the *acyclic closure* of $(\wedge V, d)$. Here the differential in the quotient $\wedge U$ is zero and so the holonomy representation is a representation of the homotopy Lie algebra $L$ of $\wedge V$ in $\wedge U$. The $\Lambda$-extension also determines a diagonal

$$\Delta : \wedge U \to \wedge U \otimes \wedge U$$

which makes $\wedge U$ into a commutative graded Hopf algebra, and there is a natural homomorphism $\eta_L : UL \to \mathrm{Hom}\,(\wedge U, \Bbbk)$ of graded algebras which converts right multiplication by $L$ to the dual of the holonomy representation.

In particular, in the case of the acyclic closure of the associated Sullivan algebra of a Sullivan Lie algebra, $E$, this yields a morphism $UE \to \mathrm{Hom}(\wedge U, \Bbbk)$ which identifies $\wedge U$ as a sort of "predual" of $UE$. In this setting we define

$$\operatorname{depth} E = \text{least } p \text{ (or } \infty) \text{ such that } \mathrm{Tor}_p^{UE}(\Bbbk, \wedge U) \neq 0\,.$$

This generalizes the definition in [18] for Lie algebras $E = E_{\geq 1}$ of finite type, because in this case $UE = (\wedge U)^\#$ and $\mathrm{Ext}_{UE}^*(\Bbbk, UE)$ is the dual of $\mathrm{Tor}_*^{UE}(\Bbbk, \wedge U)$.

The invariant $\operatorname{depth} E$ plays an important role in the growth and structure theorems for the homotopy Lie algebra of a simply connected space of finite LS category. These were established after [18] appeared, and so are included here. The extent to which they may be generalized to non-simply connected spaces remains an open question.

Although the present volume is a sequel to [18] it can be read independently, since all the definitions, conventions and results are stated here, whether or not they appear in [18], although we do quote proofs from [18] whenever this is possible. As in [18], we work where possible over an arbitrary field $\Bbbk$ of characteristic zero, and with rare exceptions, definitions and notation are unchanged from [18]; in particular, $V^\#$ denotes the dual of a graded vector space $V$. Also, for simplicity, the cohomology algebra $H^*(X;\Bbbk)$ of a space $X$ is denoted by $H(X)$.

That said, by and large the material in this monograph either is a non-trivial extension of, or is in addition to, the content of [18]. In particular, it includes:

- the extension of Sullivan models from simply connected spaces to path connected spaces with general (not necessarily nilpotent) fundamental group $G$.
- an analysis of $L_0$, the *fundamental Lie algebra of* $(\wedge V, d)$.
- a description of the holonomy action of $\pi_1(B)$ on $H^*(F)$ in terms of Sullivan models.
- a complete proof that under the most general possible hypotheses the Sullivan fibre associated with a fibration $B \longleftarrow E$ is the Sullivan model of the fibre $F$ of $p$, even when $B$ is not simply connected.
- an analysis of the minimal Sullivan model of a classifying space and the introduction of Sullivan spaces.
- the definition of the depth of $L$ for any Sullivan algebra and a homological analysis of its properties extending those provided in [18] when $L_0 = 0$.

- complete proofs of the growth and structure theorems for the higher rational homotopy groups of a connected CW complex.

We now proceed to a more detailed Chapter by Chapter description.

## Chapter 1: Basic definitions and constructions

This Chapter provides the definitions and constructions which serve as the base for the theory. The focus is on three categories : topological spaces, simplicial sets, and commutative cochain algebras, and specifically in the last category on the cochain algebras $(\wedge V, d)$ described above. Consistent with [18] these are called *Sullivan algebras* if $V = V^{\geq 1}$. In general they are called $\Lambda$-*algebras*, and $\Lambda$-algebras with $V^0 \neq 0$ appear naturally. A $\Lambda$-algebra is *minimal* if $d(V^n) \subset \wedge V^{\leq n}$.

Homotopy and based homotopy are defined for continuous maps, morphisms of simplicial sets, and for cochain algebra morphisms from a $\Lambda$-algebra; the various constructions are shown to preserve homotopy and based homotopy (Theorem 1.1, p. 33). In particular we show that a morphism from a $\Lambda$-algebra lifts up to (based) homotopy through a quasi-isomorphism.

The principal constructions in Chapter 1 then include the cochain algebra $A_{PL}(X)$ of polynomial differential forms on the singular simplices of a topological space $X$, and the minimal Sullivan model of a commutative cochain algebra, which is shown to be unique up to isomorphism.

On the other hand, the homotopy groups $\pi_n(\wedge V, d)$ of a minimal Sullivan algebra are defined as follows:

$$\pi_n(\wedge V, d) = \begin{cases} (V^n)^\# , & n \geq 2 \\ G_L , & n = 1 , \end{cases}$$

where $G_L$ is the exponential group of the fundamental Lie algebra $L_0 = (V^1)^\#$. It is defined in Chapter 2, but only when dim $H^1(\wedge V, d) < \infty$.

A morphism $\varphi : (\wedge V, d) \to A_{PL}(Y)$ induces homomorphisms

$$\pi_n(\varphi) : \pi_n(Y) \to \pi_n(\wedge V, d), \; n \geq 1$$

(Theorem 1.4, p. 37 for $n \geq 2$ and Theorem 2.4, p. 75, for $n = 1$) with $\pi_1(\varphi)$ defined only if dim $H^1(\wedge V, d) < \infty$.

In particular, associated with a $\Lambda$-algebra $(\wedge V, d)$ is the spatial realization $| \wedge V, d|$ of a simplicial set $\langle \wedge V, d \rangle$, and the adjoint to the identity of $\langle \wedge V, d \rangle$ determines a morphism $m_{|\wedge V, d|} : (\wedge V, d) \to A_{PL}(| \wedge V, d)|)$. When

$(\wedge V, d)$ is a minimal Sullivan algebra, Theorems 1.3, 1.4 (p. 37) and 2.4 (p. 75) show that

$$\pi_n(m_{|\wedge V,d|}) : \pi_n(|\wedge V, d|) \xrightarrow{\cong} \pi_n(\wedge V, d), \quad n \geq 1$$

are isomorphisms of groups. Thus the minimal Sullivan model $(\wedge V, d)$ of a space $X$ computes the cohomology of $X$ and the homotopy groups of $|\wedge V, d|$.

Finally, the spatial realization of the adjoint of $m_X$ is a continuous map $|m_X| : |Sing\, X| \to |\wedge V, d|$, which, if $X$ is a CW complex, is homotopy equivalent to the map $\varphi_X : X \to |\wedge V, d|$ described earlier. If $X$ is simply connected then $H^*(\varphi_X; \mathbb{Q})$ is an isomorphism if and only if $H^*(X; \mathbb{Q})$ is a graded vector space of finite type (Theorem 1.5, p. 42).

## Chapter 2: Homotopy Lie algebras and Sullivan Lie algebras

This Chapter introduces the other major category central to the monograph: that of graded Lie algebras $L = L_{\geq 0}$. Associated with a graded Lie algebra are the following constructions:

- The *lower central series* of the ideals $L^{n+1} = [L, L^n]$, and the completion $\widehat{L} = \varprojlim_n L/L^n$. We say $L$ is *pronilpotent* if $L \xrightarrow{\cong} \widehat{L}$.
- The universal enveloping algebra $UL$, the ideal $I_L \subset UL$ generated by $L$, and the completion $\widehat{UL} = \varprojlim_n UL/I_L^n$, where $I_L^n$ is the $n^{\text{th}}$ power of $I_L$. The completion $\widehat{I}_L = \varprojlim_n I_L/I_L^n$, is the augmentation ideal of $\widehat{UL}$.
- The Cartan-Chevalley-Eilenberg construction $C_*(L) = (\wedge sL, \partial)$, where $\partial(sx \wedge sy) = (-1)^{\deg x} s[x, y]$, and the dual commutative cochain algebra $C^*(L) = (\otimes_q (\wedge^q sL)^\#, \partial^\#)$.

With a minimal Sullivan algebra $(\wedge V, d)$ is associated the *quadratic Sullivan algebra* $(\wedge V, d_1)$ defined by $d_1 : V \to \wedge^2 V$ and $d - d_1 : V \to \wedge^{\geq 3} V$, and its homotopy Lie algebra $L = L_{\geq 0}$, defined by

$$L_k = (V^{k+1})^\# \quad \text{and} \quad \langle v, [x, y] \rangle = (-1)^{\deg y + 1} \langle d_1 v, x, y \rangle.$$

The Sullivan condition implies that $\cap_n L^n = 0$. Here the finiteness conditions

$$\dim H^1(\Lambda V, d) < \infty \quad \text{and} \quad \dim V^i < \infty, \quad i \geq 1$$

are respectively equivalent to the conditions

$$\dim L_0/[L_0, L_0] < \infty \quad \text{and} \quad \dim L_i < \infty, \quad i \geq 1,$$

and when they hold $L$ is a *Sullivan Lie algebra.*

Sullivan Lie algebras are important for much of the theory and for the applications: in particular, the homotopy Lie algebra of a minimal Sullivan algebra $(\wedge V, d)$ satisfying the finiteness conditions above is a pronilpotent Sullivan Lie algebra (Theorem 2.1, p. 50). In this case the surjections $L \to L/L^{n+1}$ are the duals of the inclusions $V_n \to V$ defined by

$$V_0 = V \cap \ker d_1 \quad \text{and} \quad V_{n+1} = V \cap d_1^{-1}(\Lambda^2 V_n).$$

If $L$ is a Sullivan Lie algebra then $\varprojlim_n C^*(L/L^n)$ is a quadratic Sullivan algebra, the *associated quadratic Sullivan algebra* for $L$, and $\widehat{L}$ is the homotopy Lie algebra for this Sullivan algebra.

Now consider an arbitrary graded Lie algebra $L = L_{\geq 0}$. The diagonal $\Delta : L \to UL \otimes UL$, $x \mapsto x \otimes 1 + 1 \otimes x$ extends to a morphism $\widehat{\Delta} = \widehat{UL} \to \widehat{UL \otimes UL}$. A sub Lie algebra $P_L \subset \widehat{I}_L$ and a subgroup $G_L \subset \widehat{UL_0}$ are then defined by

$$P_L = \{x \in \widehat{I}_L \mid \widehat{\Delta} x = x \hat{\otimes} 1 + 1 \hat{\otimes} x\}, \quad \text{and}$$

$$G_L = \{1 + y \in 1 + \widehat{I}_L \mid \widehat{\Delta}(1+y) = (1+y)\hat{\otimes}(1+y)\}.$$

The exponential $\widehat{I}_L \to 1 + \widehat{I}_L$ and the logarithm $1 + \widehat{I}_L \to \widehat{I}_L$ restrict to inverse bijections

$$P_{L_0} \xrightarrow{\cong} G_L \quad \text{and} \quad G_L \xrightarrow{\cong} P_{L_0}.$$

When $L$ is a pronilpotent Sullivan Lie algebra, then the inclusion $L_0 \to UL$ extends to an isomorphism

$$L_0 \xrightarrow{\cong} P_{L_0}\,.$$

In particular, if $L$ is the homotopy Lie algebra of a minimal Sullivan algebra for which $\dim H^1(\wedge V, d) < \infty$, then $L_0 = (V^1)^\#$ and Theorem 2.4 (p. 75) establishes that

$$\pi_1(|\Lambda V, d|) \xrightarrow{\cong} G_L$$

is an isomorphism of groups. On the other hand, by definition,

$$\pi_{n+1}(|\wedge V, d|) \cong L_n\,, \qquad n \geq 1\,.$$

Moreover, the Whitehead product defines an action of $\pi_1$ on $\pi_n$ for any topological space, reducing to conjugation when $n = 1$. On the other hand, the adjoint representation of $L_0$ in $L$ gives, via the exponential map, an adjoint representation of $G_L$ in $L$ and the isomorphisms above are equivariant with respect to the adjoint representations of $G_L$ and the Whitehead product in $\pi_*(|\wedge V, d|)$.

For any group $G$ and subsets $S, T \subset G$, $[S,T]$ is the subgroup generated by the commutators $[a,b] = aba^{-1}b^{-1}$, $a \in S$, $b \in T$. The *lower central series* for $G$ is the sequence of normal subgroups defined by $G^{n+1} = [G, G^n]$. In the case $G = G_L$ where $L = L_0$ is a pronilpotent Sullivan Lie algebra, the exponential map restricts to bijections $L^n \xrightarrow{\cong} G_L^n, n \geq 1$ and thus it follows that $G_L = \varprojlim_n G_L/G_L^n$ (Theorem 2.2, p. 64).

## Chapter 3: Fibrations and $\Lambda$-extensions

The relative analogue of a $\Lambda$-algebra is an inclusion of commutative cochain algebras of the form

$$(B,d) \to (B \otimes \wedge Z, d)$$

in which the restriction of the differential to $1 \otimes Z$ satisfies a relative version of the Sullivan condition defined in Chapter 1, with the additional condition that $d(1 \otimes Z) \subset (k\!\!k \otimes \wedge Z) \oplus (B^+ \otimes \wedge Z)$. These are called $\Lambda$-*extensions* and are *minimal* if $d(Z^n) \subset B \otimes \wedge Z^{\leq n}$. The quotient cochain algebra $(\wedge Z, \bar{d})$ is a $\Lambda$-algebra, the *Sullivan fibre* of the $\Lambda$-extension.

The $\Lambda$-extensions are the Sullivan analogues of fibrations, and satisfy analogous properties. In particular homotopy rel $B$ of morphisms from $(B \otimes \wedge Z, d)$ is defined, and given a commutative diagram of commutative cochain algebras

$$\begin{array}{ccc} (B,d) & \xrightarrow{\alpha} & (A,d) \\ \downarrow & & \downarrow \\ (B \otimes \wedge Z, d) & \xrightarrow{\psi} & (C,d)\,, \end{array}$$

there is a morphism $\varphi : (B \otimes \wedge Z, d) \to (A,d)$ extending $\alpha$ and such that $\eta \circ \varphi \sim \psi$ *rel* $B$; moreover, $\varphi$ is unique up to homotopy rel $B$.

Every continuous map is homotopy equivalent to a fibration, which in turn is unique up to fibre homotopy equivalence. The analogue for commutative cochain algebras asserts that any morphism $\varphi : (B,d) \to (C,d)$ with $H^\circ(B) = H^\circ(C) = k\!\!k$ factors through a minimal $\Lambda$-extension as

$$(B,d) \to (B \otimes \wedge Z, d) \xrightarrow{\simeq} (C,d)$$

(Theorem 3.1, p. 102) and that the $\Lambda$-extension is unique up to isomorphism (Theorem 3.2, p. 110). This result is significantly more difficult than in the case considered in [18] when $H^1(\varphi)$ is injective. With that assumption one has $Z = Z^{\geq 1}$ whereas in the general case it can happen that $Z^0 \neq 0$.

Thus $\wedge Z$ will not have finite type as a vector space and the usual degree arguments do not apply.

Two applications of this result are relied on throughout the rest of this monograph. On the one hand, suppose $\epsilon_B : (B, d) \to \Bbbk$ is an augmented commutative cochain algebra. If $H^0(B) = \Bbbk$ then there is a $\Lambda$-extension

$$(B \otimes \wedge U, d) \xrightarrow{\cong} \Bbbk,$$

the *acyclic closure* of $(B, d)$. When $(B, d)$ is a minimal Sullivan algebra then $d$ determines the isomorphism of degree 1, $\alpha : U \xrightarrow{\cong} V$ satisfying $d(1 \otimes u) - \alpha(u) \otimes 1 \in V \wedge \wedge^+ (V \otimes U)$. In particular, the differential in the fibre is 0.

The second application converts the analogy with fibrations into a genuine comparison. Suppose $F \xrightarrow{j} X \xrightarrow{p} Y$ is a fibration with $X$ and $Y$ path connected, and let $m_Y : (\wedge V, d) \xrightarrow{\cong} A_{PL}(Y)$ be a minimal Sullivan model. Here we obtain a commutative diagram.

$$\begin{array}{ccccc}
(\wedge V, d) & \longrightarrow & (\wedge V \otimes \wedge Z, d) & \xrightarrow{\Bbbk \otimes_{\wedge V} -} & (\wedge Z, \bar{d}) \\
{\scriptstyle m_Y}\downarrow\simeq & & {\scriptstyle m_X}\downarrow\simeq & & \downarrow{\scriptstyle m_F} \\
A_{PL}(Y) & \xrightarrow[A_{PL}(p)]{} & A_{PL}(X) & \xrightarrow[A_{PL}(j)]{} & A_{PL}(F)\,.
\end{array}$$

Chapter 5 will show that $m_F$ is a quasi-isomorphism under the weakest possible hypotheses; in general this is not the case. For example in the case of the covering projection $\mathbb{R} \to S^1$, $F$ is the discrete set of integers, and $(\wedge v \otimes \wedge Z, d)$ will be the acyclic closure of $(\wedge v, 0)$ with $\deg v = 1$. Thus $\wedge Z = \wedge u$ with $\deg u = 0$, which is not isomorphic to $H(F)$.

## Chapter 4: Holonomy

Suppose $F \xrightarrow{j} X \xrightarrow{p} Y$ is a fibration defined over a based space $(Y, *)$. Then for any loop $\sigma : (I, \partial I) \to (Y, *)$ the inclusion of $F$ extends to a map $I \times F \to X$ covering $\sigma$, which then restricts to a map $\lambda_\sigma : \{1\} \times F \to F \subset X$. The correspondence $[\sigma] \to [\lambda_\sigma]$ is an antihomomorphism to the group of homotopy classes of homotopy equivalences of $F$, written $\alpha \mapsto \mathrm{hol}\, \alpha$. This is the classical *holonomy action* of $\pi_1(Y, *)$ on $F$.

Analogously, suppose $(\wedge V, d) \to (\wedge V \otimes \wedge Z, d) \to (\wedge Z, \bar{d})$ is a minimal $\Lambda$-extension of a minimal Sullivan algebra $(\Lambda V, d)$ with homotopy Lie algebra

$L$, and for which $\dim H^1(\wedge V, d) < \infty$. If $x \in L$ then a derivation $\theta(x)$ in $(\wedge Z, \bar{d})$ is defined by

$$\theta(x)\Phi = -\Sigma \langle v_i, x\rangle \theta_i \Phi ,$$

where $v_i$ is a basis of $V^1$ and

$$d(1 \otimes \Phi) - 1 \otimes \bar{d}\,\Phi - \Sigma v_i \otimes \theta_i \Phi \in (\wedge V)^{\geq 2} \otimes \wedge Z , \qquad \Phi \in \wedge Z .$$

The Sullivan condition for the $\Lambda$-extension implies that if $x \in L_0$ then $\theta(x)$ acts locally nilpotently. It follows that $\exp \theta(x)$ is defined, and is an automorphism of $(\wedge Z, \bar{d})$.

Since $L_0$ is a Sullivan Lie algebra, $\exp : L_0 \to G_L$ is a bijection, and a homomorphism, *hol*, from $G_L$ to the group, $\mathcal{G}_{(\wedge Z, \bar{d})}$ of homotopy classes of automorphisms of $(\wedge Z, \bar{d})$ is defined by $\mathrm{hol}(\exp_{L_0} x) = [\exp \theta(x)]$ (Theorem 4.1, p. 126). This is the *holonomy action* of $G_L$ in $(\wedge Z, \bar{d})$.

These two definitions of holonomy actions are related as follows. Suppose that

$$\begin{array}{ccccc} (\wedge V, d) & \longrightarrow & (\wedge V \otimes \wedge Z, d) & \longrightarrow & (\wedge Z, \bar{d}) \\ \downarrow{\scriptstyle \varphi_Y} & & \downarrow{\scriptstyle \varphi_X} & & \downarrow{\scriptstyle \varphi_F} \\ A_{PL}(Y) & \xrightarrow{A_{PL}(p)} & A_{PL}(X) & \xrightarrow{A_{PL}(j)} & A_{PL}(F) , \end{array}$$

is a commutative diagram connecting the $\Lambda$-extension above to the polynomial forms on the fibration $Y \longleftarrow X \longleftarrow F$. Then (Theorem 4.2, p. 137) for $\alpha \in \pi_1(Y)$,

$$\varphi_F \circ hol(\, \pi_1(\varphi_Y)\alpha) \sim A_{PL}(hol\, \alpha) \circ \varphi_F ,$$

where $\mathrm{hol}\,\alpha$ means any continuous map $F \to F$ representing $\mathrm{hol}\,\alpha$.

Finally, recall that the derivations $\theta(x)$ of $(\wedge Z, \bar{d})$ were defined for all $x \in L$. In particular they induce derivations $\bar{\theta}(x) = H(\theta(x))$ in $H(\wedge Z, \bar{d})$ and it turns out that this is a representation of $L$ in $H(\wedge Z, \bar{d})$ : the *holonomy representation.* In particular, when $\bar{d} = 0$ the holonomy representation is just the correspondence $x \mapsto \theta(x)$. In general, it follows from the Sullivan condition for the $\Lambda$-extension that $\theta$ extends to a representation of $\widehat{UL}$ in $H(\wedge Z, \bar{d})$.

## Chapter 5: The model of the fibre is the fibre of the model

As in Chapter 3, we consider a commutative diagram

$$\begin{array}{ccccc} (\wedge V, d) & \longrightarrow & (\wedge V \otimes \wedge Z, d) & \longrightarrow & (\wedge Z, \overline{d}) \\ {\scriptstyle m_Y} \downarrow \simeq & & {\scriptstyle m_X} \downarrow \simeq & & {\scriptstyle m_F} \downarrow \\ A_{PL}(Y) & \longrightarrow & A_{PL}(X) & \longrightarrow & A_{PL}(F) \end{array}$$

connecting a minimal $\Lambda$-extension of a minimal Sullivan algebra $(\wedge V, d)$ to the polynomial forms on a fibration. We suppose $X$ and $Y$ are path connected, $Y$ is a CW complex, and $m_Y$ and $m_X$ are quasi-isomorphisms. If the holonomy action of $\pi_1(Y,*)$ is locally nilpotent, and if one of $H(Y)$ or $H(F)$ is a vector space of finite type, then $m_F$ is a quasi-isomorphism (Theorem 5.1, p. 145).

It follows from the Theorems of Chapter 4 that if $\dim H^1(\wedge V, d) < \infty$ and if $m_F$ is a quasi-isomorphism, then the holonomy action of $\pi_1(Y,*)$ is locally nilpotent. Thus this hypothesis is necessary. Moreover, the Theorem fails for the trivial fibration $Y \times F$ if for some $p$ and $q$ $\dim H^p(Y)$ and $\dim H^q(F)$ are infinite. Thus a finiteness condition is also necessary.

A proof of this Theorem was first published in [27] in 1977. The current proof is simpler, and more intuitive.

## Chapter 6: Loop spaces and loop space actions

Suppose $(\wedge V \otimes \wedge U, d)$ is the acyclic closure of a minimal Sullivan algebra satisfying our two standard conditions

$$\dim H^1(\wedge V, d) < \infty \qquad \text{and} \quad \dim V^i < \infty\,, \quad i \geq 2\,.$$

This then determines a co-associative morphism of graded algebras $\Delta : \wedge U \to \wedge U \otimes \wedge U$, which satisfies $\Delta \circ \overline{\theta}(x) = (id \otimes \overline{\theta}(x)) \circ \Delta$ where $\overline{\theta} : L \to \mathrm{Der} \wedge U$ denotes the holonomy representation for the acyclic closure.

The morphism $\Delta$ dualizes to

$$\Delta^\# : (\wedge U)^\# \otimes (\wedge U)^\# \to (\wedge U)^\#\,,$$

making $(\wedge U)^\#$ into a graded algebra. The augmentation $\varepsilon_{\wedge U} : \wedge U \to \Bbbk$ defined by $\varepsilon_{\wedge U}(U) = 0$ dualizes to $\varepsilon^\#_{\wedge U} : \Bbbk \to (\wedge U)^\#$ and $\varepsilon^\#_{\wedge U}$ is the identity for the graded algebra $(\wedge U)^\#$.

Moreover, $\Delta$ realizes to a continuous map

$$|\Delta| : |\wedge U| \times |\wedge U| \to |\wedge U|\,,$$

which makes $|\wedge U|$ into a topological group. The natural map $m_{|\wedge U|} : \wedge U \to A_{PL}(|\wedge U|)$ then induces a homomorphism $H(m_{|\wedge U|}) : \wedge U \to H(|\wedge U|)$ which dualizes to a morphism

$$H_*(m_{|\wedge U|}) : H_*(|\wedge U|) \to (\wedge U)^\#$$

of graded algebras.

On the other hand, let $a \otimes \Phi \mapsto a \cdot \Phi$, $a \in \widehat{UL}, \Phi \in \wedge U$, denote the holonomy representation, $\widehat{U\overline{\theta}}$ of $\widehat{UL}$ in $\wedge U$. Then (Theorem 6.1, p. 180) an isomorphism

$$\eta_L : \widehat{UL} \xrightarrow{\cong} (\wedge U)^\#$$

of graded algebras is defined by $\langle \eta_L(a), \Phi \rangle = \varepsilon_{\wedge U}(a \cdot \Phi)$. As observed in Chapter 2, $\Bbbk[G_L] \otimes UL_{\geq 1}$ is a subalgebra of $\widehat{UL}$. When $\Bbbk = \mathbb{Q}$ there is a commutative diagram (Theorem 6.2, p. 184) of graded Hopf algebras,

$$\begin{array}{ccc} \mathbb{Q}[G_L] \otimes UL_{\geq 1} & \xrightarrow{\quad \eta_L \quad} & (\wedge U)^\# \\ & \underset{\rho_U}{\overset{\cong}{\searrow}} & \Big\uparrow H_*(m_{|\wedge U|}) \\ & & H_*(|\wedge U|)\,. \end{array}$$

Thus $H_*(m_{|\wedge U|})$ maps $H_*(|\wedge U|)$ isomorphically onto $\eta_L(\mathbb{Q}[G_L] \otimes UL_{\geq 1})$.

Next suppose $(\wedge V \otimes \wedge Z, d)$ is a $\Lambda$-extension of $(\wedge V, d)$. As with the acyclic closure, this determines a homotopy class of morphisms of $\Lambda$-algebras,

$$\Delta_{\wedge Z} : (\wedge Z, \overline{d}) \to (\wedge Z, \overline{d}) \otimes \wedge U$$

and thereby a homomorphism $H(\Delta_{\wedge Z}) : H(\wedge Z, \overline{d}) \to H(\wedge Z, \overline{d}) \otimes \wedge U$. Composition of $H(\Delta_{\wedge Z})$ with evaluation with $(\wedge U)^\#$ then gives a map

$$H(\wedge Z, \overline{d}) \otimes (\wedge U)^\# \to H(\wedge Z, \overline{d})\,,$$

and (proof of Theorem 6.3, p. 185) the isomorphism $\eta_L$ identifies this with the holonomy representation of $\widehat{UL}$ in $H(\wedge Z, \overline{d})$.

When $\Bbbk = \mathbb{Q}$ the homomorphism $H_*(m_{|\wedge U|})$ converts this map to a representation

$$H(\wedge Z, \overline{d}) \otimes H_*(|\wedge U|) \to H(\wedge Z, \overline{d})$$

of the algebra $H_*(|\wedge U|)$, which $\eta_L$ identifies with the holonomy representation of $\mathbb{Q}[G_L] \otimes UL_{\geq 1}$.

Finally, these constructions for Sullivan models are connected to topological fibrations. For this we suppose a continuous map $h : (X, *) \to (Y, *)$

has been converted to a fibration $Y \stackrel{q}{\leftarrow} E \stackrel{i}{\leftarrow} F$ as described in §4.2; here $X$ and $Y$ are supposed to be path connected and $Y$ is a CW complex. We then suppose given a commutative diagram of cdga morphisms

$$\begin{array}{ccccc} (\wedge V, d) & \longrightarrow & (\wedge V \otimes \wedge Z, d) & \longrightarrow & (\wedge Z, \overline{d}) \\ {\scriptstyle \varphi_Y}\downarrow & & {\scriptstyle m}\downarrow & & {\scriptstyle m_F}\downarrow \\ A_{PL}(Y) & \longrightarrow & A_{PL}(E) & \longrightarrow & A_{PL}(F) \end{array}$$

in which the upper row is the $\Lambda$-extension above. Note, however that *no assumption is made about the morphisms* $\varphi_Y$, $m$ and $m_F$.

In the case of the path space fibration $Y \leftarrow PY \leftarrow \Omega Y$, we may choose the $\Lambda$-extension to be the acyclic closure $(\wedge V \otimes \wedge U, d)$, thereby defining a morphism $\varphi_\Omega : (\wedge U, 0) \to A_{PL}(\Omega Y)$.

Then, on the one hand, as described in §4.2, the loop space $\Omega Y$ acts on $F$ via a map $\mu_X : F \times \Omega Y \to F$; on the other hand, $\Delta_{\wedge Z} : (\wedge Z, \overline{d}) \to (\wedge Z, \overline{d}) \otimes \wedge U$. Here the algebra translates the topology (Theorem 6.4, p. 188): the diagram

$$\begin{array}{ccc} (\wedge Z, \overline{d}) & \xrightarrow{\Delta_{\wedge Z}} & (\wedge Z, \overline{d}) \otimes (\wedge U, 0) \\ \downarrow{\scriptstyle m_F} & & \downarrow{\scriptstyle m_F \otimes \varphi_\Omega} \\ A_{PL}(F) & \xrightarrow{A_{PL}(\mu_X)} & A_{PL}(F \times \Omega Y) \end{array}$$

is homotopy commutative.

Evaluation with $H_*(\Omega Y)$ determines the holonomy representation

$$H(F) \otimes H_*(\Omega Y) \to H(F)\,, \qquad \gamma \otimes a \mapsto a \cdot \gamma$$

given by

$$a \cdot \gamma = (-1)^{\deg a \cdot \deg \gamma} \langle H(\mu_X)\gamma, a \rangle\,.$$

Moreover, the morphism $\varphi_\Omega$ induces by adjunction a map $|\varphi_\Omega| : |\wedge U| \leftarrow |Sing\, \Omega Y|$, and a homomorphism

$$H_*(|\varphi_\Omega|) : H_*(\Omega Y) \to H_*(|\wedge U|)$$

of graded Hopf algebras. When $\Bbbk = \mathbb{Q}$, we have (Theorem 6.5, p. 193)

$$H(m_F)(H_*(|\varphi_\Omega|)a \cdot \beta) = a \cdot H(m_F)\beta\,, \qquad a \in H_*(\Omega Y), \beta \in H(\wedge Z, \overline{d})\,,$$

thus relating the holonomy representation of $H(\Omega Y)$ to the holonomy representation of $H_*(|\wedge U|)$ and hence to that of $\mathbb{Q}[G_L] \otimes UL_{\geq 1}$.

## Chapter 7: Sullivan spaces

Suppose $(X,*)$ is a path connected, based topological space with universal covering space $\widetilde{X}$. Then (§1.8) a morphism $\varphi : (\wedge V, d) \to A_{PL}(X)$ from a minimal Sullivan algebra determines natural linear maps

$$\pi_k(\varphi) : \pi_k(X) \otimes_{\mathbb{Z}} \Bbbk \to \pi_k(|\wedge V, d|)\,, \qquad k \geq 2\,.$$

Suppose $X$ is simply connected, the graded vector space $H(X)$ has finite type, and $H(\varphi)$ is an isomorphism. Then, (Theorem 1.6, p. 43) the linear maps $\pi_k(\varphi)$ are isomorphisms.

This may not be true if $X$ is not simply connected, which motivates the definition: $X$ is a *Sullivan space* if

(i) $\dim H^1(X) < \infty$ and $H(\widetilde{X})$ has finite type,

and

(ii) the linear maps $\pi_k(m_X)$, $k \geq 2$, induced by a minimal Sullivan model $m_X : (\wedge V, d) \to A_{PL}(X)$, are isomorphisms.

Now for any path connected, based CW complex $(X,*)$ there is (§4.2) a classifying space fibration

$$(B,*) \xleftarrow{\;q\;} (X \times_B MB, *) \xleftarrow{\;j\;} (F,*)\,,$$

together with a based homotopy equivalence $\ell : (X,*) \xrightarrow{\simeq} (X \times_B MB, *)$, in which $(B,*)$ is a classifying space for $\pi_1(X)$, and $\pi_1(q)$ and $\pi_k(j)$, $k \geq 2$, are isomorphisms. Suppose $(\wedge V, d)$ is a minimal Sullivan model for $X$. Then the Example in §4.5 (p. 141) provides a commutative diagram

$$\begin{array}{ccccc}
(\wedge V^1, d) & \longrightarrow & (\wedge V^1 \otimes \wedge V^{\geq 2}, d) & \longrightarrow & (\wedge V^{\geq 2}, \overline{d}) \\
\Big\downarrow m & & \Big\downarrow m_X\,\simeq & & \Big\downarrow \overline{m} \\
A_{PL}(B) & \xrightarrow[A_{PL}(q)]{} & A_{PL}(X \times_B MB) & \xrightarrow[A_{PL}(j)]{} & A_{PL}(F)
\end{array}$$

in which $m_X$ is the minimal Sullivan model for $X \times_B MB$.

When $X$ is a Sullivan space then both $m$ and $\overline{m}$ are quasi-isomorphisms. In fact (Theorem 7.2, p. 197) $X$ is a Sullivan space if and only if the following conditions are satisfied:

(i) $\dim H^1(X)$ and $\dim H^i(\widetilde{X})$, $i \geq 2$ are finite.
(ii) $\overline{m}$ is the minimal Sullivan model for $\widetilde{X}$.

The morphism $\overline{m}: (\wedge V^{\geq 2}, \overline{d}) \to A_{PL}(F)$ belongs to the rational homotopy theory of simply connected spaces, which is laid out in detail in [18]. By contrast, $m: (\wedge V^1, d) \to A_{PL}(B)$ belongs to the rational homotopy theory of classifying spaces (whose universal covering spaces are contractible!), and this is the principal subject of the rest of Chapter 7.

We denote by $BG$ the classical functorial construction of a classifying space for a group $G$, and recall that $G^n = [G, G^{n-1}]$ denotes the lower central series for $G$. The natural isomorphism $G/[G,G] \xrightarrow{\cong} H_1(BG;\mathbb{Z})$ tensors to yield an isomorphism $G/[G,G] \otimes_{\mathbb{Z}} \Bbbk \cong H_1(BG;\Bbbk)$, and so $H^1(BG;\Bbbk)$ is finite dimensional if and only if $G/[G,G] \otimes_{\mathbb{Z}} \Bbbk$ is finite dimensional. As noted as the start of this Introduction, Sullivan's theory requires finiteness hypotheses; thus throughout this Chapter attention is restricted to groups $G$ satisfying

$$\dim G/[G,G] \otimes_{\mathbb{Z}} \Bbbk < \infty\,.$$

In this case (Theorem 7.3, p. 208) there is a commutative diagram

$$\begin{array}{ccccccc}
\cdots \longrightarrow & (\wedge W_{n-1}, d) & \xrightarrow{\lambda_{n-1}} & (\wedge W_n, d) & \longrightarrow \cdots \\
 & \downarrow \simeq & & \downarrow \simeq & \\
\cdots \longrightarrow & A_{PL}(B\,G/G^{n+1}) & \longrightarrow & A_{PL}(B\,G/G^{n+2}) & \longrightarrow \cdots
\end{array}$$

in which the vertical arrows are minimal Sullivan models and each $W_n$ is a finite dimensional vector space concentrated in degree 1. Moreover each $\lambda_n$ is an inclusion and $d: W_n \to \wedge^2 W_{n-1}$. Thus if $(\wedge W, d) = \varinjlim_n (\wedge W_n, d)$ then the diagram above induces a quasi-isomorphism

$$(\wedge W, d) \xrightarrow{\simeq} \varinjlim_n A_{PL}(B\,G/G^n)\,.$$

Composed with the natural morphism $\varinjlim_n A_{PL}(B\,G/G^n) \to A_{PL}(BG)$ this provides (Theorem 7.4) a Sullivan 1-model

$$m: (\wedge W, d) \to A_{PL}(BG)\,.$$

When $\dim H^1(X) < \infty$ and $B = BG$ the diagram at the start of this Introduction to Chapter 7 exhibits the minimal Sullivan model of $X$ in the form $(\wedge V, d) = (\wedge W \otimes \wedge V^{\geq 2}, d)$. Moreover if $X$ is a Sullivan space then $m$ is a quasi-isomorphism and $\varinjlim_n A_{PL}(B\,G/G^n) \xrightarrow{\cong} A_{PL}(BG)$.

Moreover, if $G$ is a group for which $\dim G/[G,G] \otimes_{\mathbb{Z}} \Bbbk < \infty$, then the construction of $(\wedge W, d)$ exhibits $W$ as a filtered vector space $W_0 \subset W_1 \subset$

$\cdots$, and (Proposition 7.5, p. 211) this filtration coincides with the filtration of $W$ defined via the differential in §2.1. In this case,

$$(\wedge W, d) = \varinjlim_n C^*(L_0/L_0^n)\,,$$

where $L_0 = W^\#$ is the homotopy Lie algebra of $(\wedge W, d)$. The Lie algebra $L_0$ is also related to the Lie algebra $\oplus_{k\geq 1} G^k/G^{k+1}$ ([45]), with Lie bracket induced by the commutator map $[\,,\,] : G^k \times G^\ell \to G^{k+\ell}$. Applying $-\otimes_{\mathbb{Z}} k\!\!k$ constructs a Lie algebra $L_G$ and (Theorem 7.7, p. 217),

$$L_G \cong \oplus_{k\geq 1} L_0^k/L_0^{k+1}$$

with Lie bracket on the right induced by the Lie bracket $[\,,\,] : L^k \otimes L^\ell \to L^{k+\ell}$.

On the other hand, the Sullivan 1-model, $m : (\wedge W, d) \to A_{PL}(BG)$, determines a homomorphism,

$$\pi_1(m) : G \to G_L = \pi_1(\wedge W, d)\,,$$

which (Theorem 7.5, p. 214) induces isomorphisms $G^n/G^{n+1} \otimes_{\mathbb{Z}} k\!\!k \xrightarrow{\cong} G_L^n/G_L^{n+1}$. Thus this exhibits $G_L$ as a Malcev-$k\!\!k$-completion of $G$. In particular (Theorem 7.6, p. 216) when $k\!\!k = \mathbb{Q}$ the correspondence $(\wedge W, d) \to \pi_1(\wedge W, d)$ is a bijection from the isomorphism classes of Sullivan 1-algebras satisfying $\dim H^1 < \infty$ to the isomorphism classes of Malcev-complete groups $G$ satisfying $\dim G/[G,G] \otimes \mathbb{Q} < \infty$.

In Chapter 1 we showed that for any minimal Sullivan algebra $(\wedge V, d)$ with $V^1 = 0$ that the graded vector space $V$ has finite type if and only if $m_{|\wedge V,d|} : (\wedge V, d) \to A_{PL}(|\wedge V, d|)$ was a quasi-isomorphism. This Theorem generalizes to the case $V^1 \neq 0$ (Theorem 7.8, p. 218) : $|\wedge V, d|$ is a Sullivan space if and only if $H^1(\wedge V, d)$ and each $V^i$ are finite dimensional, and if $m_{|\wedge V,d|}$ is a quasi-isomorphism. Finally, if $K \to G \to G/K$ is a short exact sequence of groups in which $BK$ and $B(G/K)$ are Sullivan spaces, and if $G/K$ acts nilpotently in each $H^k(BK)$, then (Theorem 7.9) $BG$ is a Sullivan space; moreover, if $BG_1$ and $BG_2$ are Sullivan spaces, then (Theorem 7.10, p. 223) so is the classifying space of the coproduct $G_1 \# G_2$.

## Chapter 8: Examples

This Chapter uses Theorems 7.9 and 7.10 to identify a number of spaces as Sullivan spaces. In particular, these arise classically as follows:

(i) *Rationally nilpotent groups* $G$, defined by the condition $H_*(BG^{n+2}, \mathbb{Q}) = \mathbb{Q}$, for some $n$. Then $BG$ is a Sullivan space if $\dim G/[G,G] \otimes_{\mathbb{Z}} \mathbb{Q} < \infty$.

(ii) *Rationally nilpotent spaces* $(Y, *)$ defined by the conditions that $\pi_1(Y, *)$ is rationally nilpotent and acts nilpotently in each $\pi_n(\widetilde{Y}) \otimes \mathbb{Q}$. These are Sullivan spaces if, in addition, $H(Y)$ has finite type.

The remaining examples go beyond rational nilpotence to cases where the Lie algebra, $L_0$, may be infinite dimensional. Since by Theorem 7.9, $B(\mathbb{Z}\# \cdots \#\mathbb{Z})$ is a Sullivan space, a comparison with a wedge of 3-spheres computes the minimal Sullivan model of $S^1 \vee \cdots \vee S^1 = B(\mathbb{Z}\# \cdots \#\mathbb{Z})$. In particular, if $\mathbb{L}$ is the free Lie algebra on $r$ generators then the homotopy Lie algebra of $BG$ is the inverse limit $E = \varprojlim_n \mathbb{L}/\mathbb{L}^n$.

Next, (Theorem 7.9) the classifying space of the semi-direct product $K \ltimes G$ is a Sullivan space if $BK$ and $BG$ are Sullivan spaces and $G$ acts nilpotently in $K/[K,K] \otimes_{\mathbb{Z}} \mathbb{Q}$.

The original example of a Sullivan model of non-nilpotent space is that of an orientable Riemann surface $S_g$, described by Sullivan ([62]). The proof relies on [13], in which the authors use Hodge theory and Galois descent to prove that $S_g$ is formal. This result is improved here, where it is shown directly from a cochain model that $S_g$ is intrinsically formal. This then gives the minimal Sullivan model (Theorem 8.1, p. 235) from which it follows that $S_g$ is a Sullivan space.

The remaining three examples in this Chapter with Sullivan classifying spaces are: the pure braid groups, the Heisenberg group, and the fundamental groups of those compact Seifert manifolds whose quotient by the action of $S^1$ is an orientable closed Riemann surface.

Finally although $B\pi_1(\mathbb{R}P^{n+1}\#\mathbb{R}P^{n+1}) = B(\mathbb{Z}_2\#\mathbb{Z}_2)$ is a Sullivan space, a simple model calculation shows that the connected sum $\mathbb{R}P^{n+1}\#\mathbb{R}P^{n+1}$ is not a Sullivan space.

## Chapter 9: Lusternik-Schnirelmann category

A topological space $X$ has *LS category* $\leq m$ ($\operatorname{cat} X \leq m$) if it is covered by $m+1$ open sets each contractible in $X$. If $X$ is a normal space, then (Theorem 9.1, p. 245) $\operatorname{cat} X \leq m$ if and only if $X$ is the retract of an $m$-cone. Analogously, if $(\wedge V, d)$ is a minimal Sullivan algebra, then the *LS category of* $(\wedge V, d)$ is the least integer $m$ (or $\infty$) ($\operatorname{cat}(\wedge V, d) = m$) such that the inclusion $(\wedge V, d) \to (\wedge V \otimes \wedge Z, d)$ admits a retraction, where $(\wedge V \otimes$

$\wedge Z, d)$ is the Sullivan model of the surjection $(\wedge V, d) \to (\wedge V / \wedge^{>m} V, d)$. In particular, if $(\wedge V, d)$ is the minimal Sullivan model of a path connected normal space $X$, then (Theorem 9.2, p. 247)

$$\mathrm{cat}\,(\wedge V, d) \leq \mathrm{cat}\, X\,.$$

The hypothesis "finite category" has important implications for the growth and the structure of the homotopy Lie algebra of a minimal Sullivan algebra, and these are the subject of Chapters 10–15. This Chapter 9 extends the results of [18] to the non-simply connected case, beginning with the key mapping Theorem (Theorem 9.3, p. 248). It asserts that if $\varphi : (\wedge V, d) \to (\wedge W, d)$ is a surjection of minimal Sullivan algebras then $\mathrm{cat}(\wedge V, d) \geq \mathrm{cat}\,(\wedge W, d)$. Thus since $\dim H(\wedge V, d) < \infty$ implies that $\mathrm{cat}\,(\wedge V, d) < \infty$, the quotients $(\wedge W, d)$ will also have finite LS category, although they will often have infinite dimensional cohomology.

While finite LS category inherits to quotient Sullivan algebras it does not, usually, inherit to sub Sullivan algebras, as seen immediately from the example $\wedge(a, x), dx = a^2$. There is, however, one important example where this does hold: if $(\wedge V, d)$ is a minimal Sullivan algebra and $\dim V^i < \infty$, $i \geq 2$, then (Proposition 9.6, p. 259) $\mathrm{cat}(\wedge V, d) < \infty$ implies that for some $N$, $H^N(\wedge V^1, d) = 0$, and so $\mathrm{cat}\,(\wedge V^1, d) < \infty$.

There are analogous invariants for ($\mathbb{Z}$-graded) modules $(M, d)$ over a minimal Sullivan algebra $(\wedge V, d)$. Each module admits a $(\wedge V, d)$-*semifree resolution*, namely a quasi-isomorphism $(P, d) \xrightarrow{\cong} (M, d)$ from a semifree $(\wedge V, d)$-module (§1.2). Moreover, the quotient $\rho : (P, d) \to (P/\wedge^{k+1} V \cdot P, d)$ factors as $(P, d) \xrightarrow{\zeta} (P(k), d) \xrightarrow{\cong} (P/\wedge^{k+1} V \cdot P, d)$ in which $(P(k), d)$ is also $(\wedge V, d)$-semifree. With this notation $\mathrm{mcat}\,(M, d)$ is the least integer $k$ (or $\infty$) such that $\zeta$ admits a retraction. The Toomer invariant $e(M, d)$ is then the least integer $k$ such that $H(\zeta)$ or, equivalently $H(\rho)$, is injective.

With these definitions,

$$e(M, d) \leq \mathrm{mcat}(M, d) \leq \mathrm{mcat}(\wedge V, d) \leq \mathrm{cat}(\wedge V, d)\,.$$

The last inequality is improved in the fundamental result of Hess (Theorem 9.4, p. 250): if $(\wedge V, d)$ is a minimal Sullivan algebra then

$$\mathrm{mcat}(\wedge V, d) = \mathrm{cat}(\wedge V, d)\,.$$

If $X$ is a topological space with minimal Sullivan model $(\wedge V, d)$ then the invariant $e(X) = e(\wedge V, d)$ can be defined directly in terms of $X$ ([63]). Originally it was thought that it might be true when $X$ is simply connected that $e(X) = \mathrm{cat}\, X_{\mathbb{Q}}$, where $X_{\mathbb{Q}}$ was the rationalization of $X$. This would imply $e(\wedge V,d) = \mathrm{cat}\,(\wedge V,d)$, but $e(\wedge V,d)$ is in general strictly less than $\mathrm{cat}(\wedge V,d)$. However, if $H(\wedge V,d)$ has finite type then (Theorem 9.5, p. 260) the dual module $(\wedge V,d)^{\#}$ satisfies

$$e\left((\wedge V,d)^{\#}\right) = \mathrm{cat}(\wedge V,d)\,.$$

In particular, if $H(\wedge V,d)$ satisfies Poincaré duality then $(\wedge V,d)^{\#}$ and $(\wedge V,d)$ have isomorphic semifree resolutions. Thus in this case

$$e(\wedge V,d) = \mathrm{cat}(\wedge V,d)\,,$$

and so the original guess is correct in this case.

Finally, suppose a minimal Sullivan algebra decomposes as a Sullivan extension $(\wedge V\otimes\wedge Z,d)$ and that elements $y_1,\cdots,y_r$ of even degree in the homotopy Lie algebra $L$ of $(\wedge V\otimes\wedge Z,d)$ map to linearly independent elements in the homotopy Lie algebra of $(\wedge V,d)$. If $ad_L(y_1),\cdots,ad_L(y_r)$ are nilpotent then a theorem of Jessup (Theorem 9.6, p. 263) asserts that

$$\mathrm{cat}(\wedge V\otimes\wedge Z,d)\geq \mathrm{cat}(\wedge Z,\overline{d})+r\,.$$

A variant of this result (Theorem 9.4, p. 250) asserts that if $L$ is the homotopy Lie algebra of a minimal Sullivan algebra with finite category, and if $\dim L_0=\infty$, then for some $x,y\in L_0$,

$$(ad_L x)^k y\neq 0\,,\qquad k\geq 0\,.$$

## Chapter 10: Depth of a minimal Sullivan algebra and of a Sullivan Lie algebra

As defined in Chapter 2, a quadratic Sullivan algebra is a Sullivan algebra of the form $(\wedge V,d_1)$ in which $d_1: V\to\wedge^2 V$. These arise in two ways: first if $(\wedge V,d)$ is any Sullivan algebra then the associated quadratic Sullivan algebra $(\wedge V,d_1)$ is defined by $d-d_1: V\to\wedge^{\geq 3}V$; second if $L$ is a Sullivan Lie algebra then $\varinjlim_n C^*(L/L^n)$ is the quadratic Sullivan algebra associated to $L$.

Let $(\wedge V\otimes\wedge U,d_1)$ be the acyclic closure of a quadratic Sullivan algebra $(\wedge V,d_1)$. Then the *depth* of $(\wedge V,d_1)$ is the least $p$ (or $\infty$) such that $H^{p,*}(\mathrm{Hom}_{\wedge V}(\wedge V\otimes\wedge U,\wedge V))\neq 0$ where the grading $p$ corresponds to $\mathrm{Hom}(\wedge U,\wedge^p V)$ under the standard isomorphism. If $(\wedge V,d_1)$ is associated with a Sullivan algebra $(\wedge V,d)$ then (Theorem 10.1, p. 273)

$$\mathrm{depth}(\wedge V,d_1)\leq \mathrm{cat}(\wedge V,d)\,.$$

On the other hand, if $L = L_{\geq 0}$ is a graded Lie algebra then, as in Chapter 2, $\widehat{UL} = \varprojlim_n UL/I_L^n$, where $I_L$ is the ideal generated by $L$. Generalizing the definition in [18] when $L = L_{\geq 1}$ has finite type, we set

$$\operatorname{depth} L = \text{least } p \text{ (or } \infty) \text{ such that } \operatorname{Ext}^p_{UL}(\Bbbk, \widehat{UL}) \neq 0\,.$$

Now suppose $L = L_{\geq 0}$ is a Sullivan Lie algebra and that $(\wedge V, d_1)$ is the associated quadratic Sullivan algebra. Then (Proposition 10.3, p. 286) the *depth* of $L$ is the least $p$ (or $\infty$) such that $\operatorname{Tor}^{UL}_p(\Bbbk, \wedge U) \neq 0$, where $L$ acts in $\wedge U$ via the holonomy representation. In this case the pairing $V \times L \to \Bbbk$ induces a morphism $\mu : (\wedge V, d_1) \to C^*(L)$, and if $\mu$ is a quasi-isomorphism then (Theorem 10.2, p. 278)

$$\operatorname{depth} L = \operatorname{depth}(\wedge V, d_1)\,.$$

The main objective in Chapter 10 is the study of Sullivan Lie algebras, $L$, of finite depth, with the aid of Hochschild-Serre spectral sequences recalled in §10.1. In particular, suppose $(\wedge V \otimes \wedge U, d_1)$ is the acyclic closure of the associated quadratic Sullivan algebra $(\wedge V, d_1)$ for $L$. Then for a sub Lie algebra $E \subset L$,

$$\operatorname{depth}_L E := \text{least } p \text{ (or } \infty) \text{ such that } \operatorname{Tor}^{UE}_p(\Bbbk, \wedge U) \neq 0\,.$$

On the other hand, $E$ determines the subspace $K_E \subset V$ defined by

$$K_E = \{\, v \in V \mid \langle v, E\rangle = 0 \,\}\,,$$

and the ideal $K_E \wedge \wedge V$ is preserved by $d_1$. Division by this ideal gives a quadratic Sullivan algebra $(\wedge Z, \overline{d}_1)$ and, if $E$ is itself a Sullivan Lie algebra then there is a natural morphism from $(\wedge Z, \overline{d}_1)$ to the associated quadratic Sullivan algebra $(\wedge W, d_1)$ for $E$. Furthermore, if this morphism is a quasi-isomorphism then (Theorem 10.3, p. 288)

$$\operatorname{depth}_L E = \operatorname{depth} E\,.$$

If $E$ is an ideal then $\operatorname{depth}_L E \leq \operatorname{depth} E$, and it follows that the condition of finite depth inherits to ideals satisfying the hypotheses above.

Again, suppose $E \subset L$ is a sub Lie algebra. The Sullivan closure of $E$ is the Lie sub algebra $\overline{E} \subset L$ defined by

$$\overline{E} = \{ x \in L \mid \langle K_E, x\rangle = 0 \,\}\,,$$

and $E$ is *Sullivan closed* if $E = \overline{E}$. Sullivan Lie sub algebras that are Sullivan closed are the natural Lie sub algebras of a Sullivan Lie algebras. For example, if $L = I \oplus J$ is the sum of two ideals, then each is a Sullivan closed Sullivan Lie algebra, and $\operatorname{depth} L = \operatorname{depth} I +$ depth $J$. On the other hand if an ideal $I \subset L$ is a Sullivan closed Sullivan Lie algebra satisfying $I \supset L_0^n + L_{\geq k}$ for some $n$ and $k$, and if $L$ acts locally nilpotently in $\mathrm{Tor}_*^{UI}(\Bbbk, \Bbbk)$ while $L_0$ acts nilpotently in $I_0/[I_0, I_0]$, then (Theorem 10.5, p. 294)

$$\mathrm{depth}\, L = \dim\,(L/I)_{even} + \mathrm{depth}\, I \quad .$$

Moreover, if $\mathrm{depth}\, L < \infty$ then the subspace $N \subset L_{even}$ spanned by the elements $x$ for which $ad\, x$ is locally nilpotent satisfies (Corollary 10.10 to Theorem 10.4, p. 293)

$$\dim N \leq \mathrm{depth}\, L\, .$$

Finally, as in [18] the *radical*, $rad\, L$, of a Sullivan Lie algebra, $L$, is the sum of its solvable ideals, and if $\mathrm{depth}\, L < \infty$ then (Theorem 10.6, p. 295) $\dim rad\, L < \infty$ and

$$\dim(\mathrm{rad}\, L)_{even} = \mathrm{depth\ rad}\, L \leq \mathrm{depth}\, L\, .$$

## Chapter 11: Depth of a connected graded Lie algebra of finite type

This Chapter contains the results for the depth of a connected graded Lie algebra, $L$, of finite type required for the growth and structure theorems in Chapters 12–14. For instance, (Proposition 11.6, p. 304) if $L$ is abelian, and for some module $M$, $\mathrm{Ext}_{UL}^*(M, UL) \neq 0$ then $M$ contains a submodule of the form $U(L_{\geq n})$ for some $n$.

Generalizing the concept of trivial $L$-modules, an $L$-module $M$ is called *weakly locally finite* if it is the increasing union of finite dimensional subspaces $M(1) \subset M(2) \subset \cdots$ such that $M(k)$ is preserved by $L_{\leq k}$. Then the *weak depth* of $L$ is the least $m$ (or $\infty$) such that $\mathrm{Ext}_{UL}^m(M, UL) \neq 0$ for some weakly locally finite module $M$. By definition, weak $\mathrm{depth}\, L \leq \mathrm{depth}\, L$.

Now suppose $E \subset L$ is a sub Lie algebra with centralizer $Z$. If for some weakly locally finite $L$-module, $M$, the restriction map

$$\mathrm{Ext}_{UL}(M, UL) \to \mathrm{Ext}_{UE}(M, UL)$$

is non-zero, then (Proposition 11.7, p. 306) $Z$ is finite dimensional. It follows that if $L$ has finite depth then there are elements $x_1, \cdots, x_r$ of odd degrees such that for all sufficiently large odd integers $n$, the map

$$\sum_{i=1}^{r} [x_i, -] : L_n \to \oplus_{i=1}^r L_{n + \deg x_i}$$

is injective.

**Chapter 12: Trichotomy**

For a path connected topological space $(X,*)$ there are exactly three, mutually exclusive, possibilities (the trichotomy of the title) for the integers $\mathrm{rk}\,\pi_k(X) = \dim \pi_k(X)\otimes \mathbb{Q}$, $k \geq 2$.

$$(i)\quad \sum_{k\geq 2} rk\,\pi_k(X) < \infty\,.$$

$$(ii)\quad \sum_{k\geq 2} rk\,\pi_k(X) = \infty \ \text{ but } rk\,\pi_k(X) < \infty \quad \text{for } k\geq 2\,.$$

$$(iii)\ \text{For some } k\geq 2\,, rk\,\pi_k(X) = \infty\,.$$

When $X$ is a CW complex with finite LS category then the following results hold for the sequence $\mathrm{rk}(\pi_k(X))$:

*Case (i).*

$$\sum_{k\geq 2} \mathrm{rk}\,\pi_k(X) < \infty\,.$$

In this case, $X$ is *rationally elliptic* and for some $N$ $H^{>N}(\widetilde{X};\mathbb{Q}) = 0$, $\widetilde{X}$ denoting the universal cover. Moreover, (Theorem 12.1 quoting [18]), $\mathrm{rk}\,\pi_k(X) = 0$, $k\geq 2N$ and $\sum_k \mathrm{rk}\,\pi_k(X) \leq 2\,\mathrm{cat}\,X$.

*Case (ii).*

$$\sum_{k\geq 2} \mathrm{rk}\,\pi_k(X) = \infty\,, \quad \text{but rk } \pi_k(X) < \infty\,, \quad k\geq 2\,.$$

In this case $X$ is *rationally hyperbolic*, the name chosen because the $\mathrm{rk}\,\pi_k(X)$ grow exponentially. More precisely, set $\alpha_k(X) = \frac{\log \mathrm{rk}\,(\pi_k(X))}{k}$. Then $\alpha(X) := \limsup_k \alpha_k(X) > 0$, and, if $\alpha(X) < \infty$, then (Theorem 12.2) for some $R$

$$\lim_{n\to\infty} \max_{n\leq k\leq n+R} \alpha_k(X) = \alpha(X)\,.$$

Moreover, in the special case that $\dim X = N$ then (Theorem 12.5) $\alpha(X) < \infty$ and there are constants $\beta, \gamma$, $K > 0$ such that

$$\alpha(X) - \gamma\frac{\log n}{n} \leq \max_{n+2\leq k\leq N} \alpha_k(X) \leq \alpha(X) + \frac{\beta}{n}\,, \qquad n\geq K\,.$$

These results follow from the corresponding results for minimal Sullivan algebras, which are established in Chapter 13.

*Case (iii).* Some $\mathrm{rk}\,\pi_k(X) = \infty$.

In this case $X$ is *rationally infinite* and (Theorem 12.7) for some $K, R > 0$, and all $n \geq K$,

$$\operatorname{rk} \pi_k(X) = \infty\,, \qquad \text{some } k \in [n, n+R]\,.$$

The analogous results for the minimal Sullivan model of $X$ also holds (Theorem 12.6) but does not imply or follow from the topological result.

Finally, if $\dim X = N$, then (Theorem 12.8) the result for $\operatorname{rk} \pi_k(X)$ holds with $R = N$.

## Chapter 13: Exponential growth

Let $L$ be the homotopy Lie algebra of a minimal Sullivan algebra $(\wedge V, d)$ of finite type and satisfying $V = V^{\geq 2}$ and $\dim V = \infty$. Set

$$\alpha_H = \limsup_k \frac{\log \dim H^k(\wedge V, d)}{k} \quad \text{and } \alpha = \limsup_k \frac{\log \dim V^k}{k}\,;$$

since $\dim L_k = \dim V^{k+1}$ it follows that $\alpha = \limsup_k \frac{\log \dim L_k}{k}$. With this notation, (Proposition 13.2, p. 332)

$$\alpha_H \leq \alpha \qquad \text{and } \alpha_H < \infty \Leftrightarrow \alpha < \infty\,.$$

There are three sets of hypotheses, each stronger than the next, which then provide increasing strong assertions about exponential growth for the integers $\dim V^k$. These are

$$H1: \qquad \operatorname{cat}(\wedge V, d) < \infty \ \text{ and } \alpha_H < \infty\,.$$

$$H2: \qquad \operatorname{cat}(\wedge V, d) < \infty \ \text{ and } \alpha_H < \alpha\,.$$

$$H3: \qquad H^{>N}(\wedge V, d) = 0\,.$$

The corresponding conclusions (Theorems 12.9, 12.10 and 12.11) are then

$C1$: For some $R$,

$$\lim_{n\to\infty} \left( \max_{n \leq k \leq n+R}, \frac{\log \dim L_k}{k} \right) = \alpha\,.$$

$C2$: For some $R, K, \beta, \gamma > 0$,

$$\alpha - \gamma \frac{\log n}{n} \leq \max_{n \leq k \leq n+R} \left( \frac{\log \dim L_k}{k} \right) \leq \alpha + \frac{\beta}{n}\,, \quad n \geq K\,.$$

$C3$: For some $\gamma, \beta, K > 0$

$$\alpha - \gamma \frac{\log n}{n} \leq \max_{n+2 \leq k \leq n+N} \left( \frac{\log \dim L_k}{k} \right) \leq \alpha + \frac{\beta}{n}\,, \quad n \geq K\,.$$

The proofs all have the same basic structure, and so are presented as a single proof, except that the technique for estimating the error varies depending on the hypotheses. A detailed outline of the strategy is provided in §13.1.

## Chapter 14: Structure of a graded Lie algebra of finite depth

Let $L = L_{\geq 1}$ be a graded Lie algebra of finite type and finite depth $m$. Then the ideals $I \subset L$ have the same properties, and the set of ideals has a strong structure, analogous to the structure of the ideals in a finite dimensional Lie algebra concentrated in degree zero.

In particular, set

$$\alpha_I = \limsup_k \frac{\log \dim I_k}{k}$$

and let

$$S = \{\, \alpha_I \mid I \text{ an infinite dimensional ideal in } L\,\}\,.$$

Then, (Theorem 14.2, p. 369) $S$ is a finite set with cardinality $\leq m$.

The size of infinite dimensional graded vector spaces $V = V_{\geq 1}$ and $W = W_{\geq 1}$ of finite type can be compared by setting $V \leqslant W$ if for some fixed positive numbers $\lambda, q$ and $N$

$$\dim V_k \leq \lambda \sum_{i=k}^{k+q} \dim W_i\,, \qquad k \geq N\,.$$

With this, an equivalence relation, $\sim_L$, on the set of ideals $I \subset L$ is defined by

$$I \sim_L J \Leftrightarrow I \cap K \leqslant J \cap K \;\text{ and }\; J \cap K \leqslant I \cap K\,,$$

for all ideals $K \subset L$. The set of equivalence classes is a distributive lattice under $\leqslant$, and (Theorem 14.1, p. 368) there are at most $2^m$ equivalence classes with equality holding if and only if $L \sim_L L(1) \oplus \cdots \oplus L(m)$ where each $I(j)$ is an ideal of depth 1.

Finally, if $I \subset L$ is an ideal, then (Theorem 14.4, p. 377) $L$ contains a sub Lie algebra $E$, called a *weak complement* of $I$, and satisfying

(i) $E \cap I = 0$,
(ii) $L/E \leqslant I$, and
(iii) $\operatorname{depth} E \leq \operatorname{depth} F$ for all subalgebras $F$ of $E$ such that $E/F \leqslant I$.

In this case

$$\operatorname{depth} E + \operatorname{depth} I = \operatorname{depth}(E \oplus I) \leq \operatorname{depth} L\,.$$

## Chapter 15: Weight decompositions of a Sullivan Lie algebra

As defined in §7.4, a *weighted graded Lie algebra* is a graded Lie algebra $L = L_{\geq 0}$ satisfying $L = \oplus_{p\geq 1} L(p)$, where the subspaces $L(p)$ satisfy

$$[L(p), L(q)] \subset L(p+q)\,, \qquad p, q \geq 1\,.$$

In particular, each $L^n \subset \oplus_{p\geq n} L(p)$ and so if $L_0/[L_0, L_0]$ and each $L_k$, $k \geq 1$ are finite dimensional then $L$ is a Sullivan Lie algebra.

If $L$ is any graded Lie algebra and $G$ is any group then, as defined in §7.4, the Lie algebras $\mathcal{L}(L) = \oplus_n L^n/L^{n+1}$ and $L_G = \oplus_n G^n/G^{n+1} \otimes \mathbb{Q}$ are weighted Lie algebras with brackets induced by respectively the commutators $[x, y]$ and $[a, b] = aba^{-1}b^{-1}$.

Now suppose a quadratic Sullivan algebra $(\wedge V, d_1)$ is both the associated quadratic Sullivan algebra for a minimal Sullivan algebra $(\wedge V, d)$ and the associated quadratic Sullivan algebra for a Sullivan Lie algebra $L$. If $H(\wedge V, d)$ has finite type and $\mathrm{cat}\,(\wedge V, d) < \infty$, and if $L_0$ has a weight decomposition, then (Theorem 15.1)

$$\mathrm{depth}\, L = \mathrm{depth}\,(\wedge V, d) \leq \mathrm{cat}\,(\wedge V, d)\,.$$

In the particular case of an infinite dimensional Sullivan Lie algebra, $L$, concentrated in even degrees, if the associated weighted Lie algebra $\mathcal{L}$ satisfies $\dim H(C_*(\mathcal{L})) < \infty$, then (Theorem 15.2):

$$\alpha_{\mathcal{L}} = \limsup_{n\geq 1} \frac{\log \dim L^n/L^{n+1}}{n}$$

satisfies $0 < \alpha_{\mathcal{L}} < \infty$, and there are constants $b, c$ and $K$ such that

$$\frac{1}{n^c} e^{n\alpha_{\mathcal{L}}} \leq \dim L^n/L^{n+1} \leq e^b e^{n\alpha_{\mathcal{L}}}\,, \qquad n \geq K\,.$$

Finally, if $(\wedge V, d)$ is a 1-formal Sullivan algebra as defined in §2.10, and if $\dim H^1(\wedge V, d) < \infty$, then $V^1 = \oplus_{n\geq 1} V^1(n)$ and $\mathcal{L} = \oplus_{n\geq 1}\mathcal{L}(n)$ with $\mathcal{L}(n) = (V^1(n))^{\#}$ is a weighted Lie algebra. Here $\mathcal{L}$ is a sub Lie algebra of the fundamental Lie algebra $L_0$ of $(\wedge V, d)$ and $\mathcal{L}(n) = L_0^n/L_0^{n+1}$. In this case (Theorem 15.3) $\mathrm{depth}\, \mathcal{L} = \mathrm{depth}(\wedge V^1, d) \leq \mathrm{cat}(\wedge V^1, d) < \infty$, and the sequence $\dim L_0^n/L_0^{n+1}$ has the same growth properties as in Theorem 15.2.

Moreover, if $\mathbb{L}$ is the free Lie algebra on $\mathcal{L}(1)$ then the inclusion $\mathcal{L}(1) \to \mathcal{L}$ extends to a surjection $\rho : \mathbb{L} \to \mathcal{L}$ of weighted Lie algebras. If $K$ is the ideal generated by $\ker \rho(2)$ then (Theorem 15.4) $\rho$ induces an isomorphism from the Chen Lie algebra $\mathbb{L}/K$ onto $\mathcal{L}$.

## Chapter 1

# Basic definitions and constructions

In this Chapter we review and extend the standard material in [18] from graded algebra, simplicial sets, and topology required for rational homotopy theory. Throughout this entire text all vector spaces and algebras will be defined over a field $\Bbbk$ of characteristic zero.

### 1.1 Graded algebra

A *graded vector space* is a family of vector spaces $V = \{V^i\}_{i\in\mathbb{Z}}$ or a family of vector spaces $X = \{X_i\}_{i\in\mathbb{Z}}$. There is a canonical isomorphism between these categories sending $V \to X$ where

$$X_i = V^{-i}, \qquad i \in \mathbb{Z}.$$

In our review of definitions below we shall largely focus on the "superscript category" leaving it to the reader to translate to the "subscript category". There are constructions which use graded vector spaces from both categories, and for these one must convert the objects in one of the two categories to the other; constructions are always "compatible with conversion".

One could, of course, simply convert the "subscript category" to the "superscript category" and work entirely in the latter, but there are historical reasons for not doing so; the singular chain complex of a space $S$ is always written $\{C_i(S)\}_{i\geq 0}$ while the cochain complex is always written $\{C^i(S)\}_{i\geq 0}$ and this tradition is so deeply embedded in the subject that we have preferred to continue to use both superscripts and subscripts.

Suppose then that $V = \{V^i\}_{i\in\mathbb{Z}}$ is a graded vector space. We write $V^{>n} = \{V^i\}_{i>n}$, $V^+ = V^{>0}$ and define $V^{\geq n}$, $V^{\leq n}$, $V^{<n}$ analogously. An *element* $v \in V$ is an element in some $V^i$: its *degree* is $i$ and we write

$\deg v = i$. We say $V$ is *concentrated* in a subset of degrees if $V^i = 0$ whenever $i$ is not in that subset. We regard our ground field $\Bbbk$ as a graded vector space concentrated in degree 0. A *basis* for $V$ is the union of bases of all the $V^i$. We say that $V$ has *finite type* if each $V^i$ is finite dimensional. In that case, if $V$ is concentrated in degrees $\geq 0$ the *Hilbert series* for $V$ is the formal power series

$$V(z) = \sum_{i=0}^{\infty} (\dim V^i)\, z^i\,.$$

Similarly, if $V = \{V_i\}_{i\geq 0}$ has finite type its Hilbert series is the formal power series

$$V(z) = \sum_{i=0}^{\infty} (\dim V_i)\, z^i\,.$$

Finally we set $\dim V = \sum_i \dim V^i$.

The standard functors from linear algebra (sum, product, quotient space, Hom, tensor product, $\cdots$) extend in the obvious way to graded vector spaces: in particular

$$\left(\prod_\alpha V_\alpha\right)^i = \prod_\alpha V_\alpha^i$$

$$\mathrm{Hom}(U,V)^i = \prod_{k-j=i} \mathrm{Hom}(U^j, V^k)$$

$$(U\otimes V)^i = \bigoplus_{j+k=i} U^j \otimes V^k\,.$$

For simplicity we denote $\mathrm{Hom}(U,\Bbbk)$ simply by $U^\#$. The *suspension*, $sV$, of a graded vector space $V$ is a family of linear isomorphisms $s : V^i \xrightarrow{\cong} (sV)^{i-1}$, written $v \mapsto sv$.

If $U$ and $V$ are graded vector spaces a bilinear map $\langle\ ,\ \rangle : U\times V \to \Bbbk$ is called a *pairing*. A paring determines the obvious linear maps $U \to V^\#$ and $V \to U^\#$.

If $U$ and $V$ are graded vector spaces, the *interchange isomorphism* is the linear isomorphism $U\otimes V \xrightarrow{\cong} V\otimes U$ given by

$$u\otimes v \mapsto (-1)^{\deg u \deg v} v\otimes u\,.$$

If $\leftarrow U(n) \leftarrow U(n+1) \leftarrow \cdots$ and $\leftarrow V(n) \leftarrow V(n+1) \leftarrow \cdots$ are inverse systems of graded vector spaces we write

$$\varprojlim_n (U(n)\otimes V(n)) = \left(\varprojlim_n U(n)\right) \widehat{\otimes} \left(\varprojlim_n V(n)\right).$$

A *graded algebra* is a graded vector space $A$ together with an associative multiplication of degree zero

$$A \otimes A \to A\,, \qquad a \otimes b \mapsto ab\,,$$

for which there is an identity $1 \in A^0$. Morphisms of graded algebras are required to be of degree zero and an *augmentation* is a morphism $A \to \Bbbk$. If $V$ is a graded vector space then composition makes $\mathrm{Hom}(V,V)$ into a graded algebra, which we denote by $\mathrm{End}(V)$. A *representation* of a graded algebra $A$ in a graded vector space $V$ is a morphism $A \to \mathrm{End}(V)$. Thus if $A$ is a graded algebra $\mathrm{End}(A)$ denotes the graded algebra of *all linear maps* $A \to A$.

By contrast, a *derivation* in $A$ is a linear map $\theta \in \mathrm{End}(A)$ satisfying for $a, b \in A$:

$$\theta(ab) = (\theta a)b + (-1)^{\deg\theta\deg a}\, a\theta(b)\,.$$

Thus the derivations form a (graded) subspace $\mathrm{Der}\,A \subset \mathrm{End}(A)$.

The tensor product $A \otimes B$ of graded algebras is a graded algebra with multiplication

$$(a_1 \otimes b_1)(a_2 \otimes b_2) = (-1)^{\deg b_1 \deg a_2} a_1 a_2 \otimes b_1 b_2\,.$$

A simple computation shows that the interchange isomorphism $A \otimes B \xrightarrow{\cong} B \otimes A$ is an isomorphism of graded algebras.

A graded algebra $A$ is *commutative* (a *cga* for short) if for all $a, b \in A$,

$$ab = (-1)^{\deg a \cdot \deg b}\, ba\,,$$

and the tensor product of cga's is again a cga. If $A \to B$ and $A \to C$ are morphisms of cga's then $B \otimes_A C$ is again a cga and the interchange isomorphism factors to give a cga isomorphism $B \otimes_A C \xrightarrow{\cong} C \otimes_A B$ restricting to the identity in $A$. We also call this isomorphism *the interchange isomorphism*.

Given a graded vector space $V$ the tensor algebra

$$TV = \oplus_{k=0}^{\infty} \otimes^k V$$

is a graded algebra with multiplication defined as in the ungraded category, and it has the standard universal property. The elements $v_1 \otimes v_2 - (-1)^{\deg v_1 \cdot \deg v_1} v_2 \otimes v_1$ with $v_1, v_2 \in V$ generate an ideal $I$ and the quotient

$$\wedge V = \frac{TV}{I}$$

is the free graded commutative algebra on $V$. Multiplication in $\wedge V$ is denoted by $\Phi \wedge \Psi$, $\Phi, \Psi \in \wedge V$. The algebra $\wedge V$ has the direct sum decomposition

$$\wedge V = \oplus_{k=0}^{\infty} \wedge^k V$$

in which $\wedge^k V$ is the linear span of elements of the form $v_1 \wedge \cdots \wedge v_k$, $v_i \in V$.

If $W$ is a second graded vector space then the inclusions $V, W \to V \oplus W$ extend uniquely to an isomorphism of graded algebras

$$\wedge V \otimes \wedge W \xrightarrow{\cong} \wedge(V \oplus W).$$

A *graded Lie algebra* is a graded vector space, $L$, together with a Lie bracket

$$[\ ,\ ] : L \otimes L \to L$$

satisfying the graded symmetry and graded Jacobi identities:

$$[x,y] = -(-1)^{\deg x \deg y}[y,x], \text{ and}$$

$$[x,[y,z]] = [[x,y],z] + (-1)^{\deg x \deg y}[y,[x,z]].$$

A graded subspace $E \subset L$ is a *subalgebra* (resp. an *ideal*) if $[x,y] \in E$ whenever $x, y \in E$ (resp. whenever $x \in L$ and $y \in E$). $E$ is *abelian* if $[x,y] = 0$ for all $x, y \in E$, and the *center* of $L$ is the ideal of elements $x \in L$ satisfying $[x,y] = 0$, $y \in L$.

Any graded algebra $A$ (and in particular $\mathrm{End}(V)$) is also a graded Lie algebra: for $\alpha, \beta \in A$,

$$[\alpha, \beta] = \alpha \circ \beta - (-1)^{\deg \alpha \deg \beta} \beta \circ \alpha.$$

Morphisms of graded Lie algebras are required to be of degree 0, but derivations can be of any degree: a *derivation* in $L$ is a linear map $\theta \in \mathrm{End}(L)$ such that for $x, y \in L$,

$$\theta[x,y] = [\theta x, y] + (-1)^{\deg x \deg \theta}[x, \theta y].$$

Now suppose $A$ is a graded algebra and $E$ is a graded Lie algebra. Then

$$\mathrm{Der}\, A \subset \mathrm{End}(A) \qquad \text{and} \qquad \mathrm{Der}\, E \subset \mathrm{End}(E)$$

are sub Lie algebras and a representation of $L$ in the *graded algebra* $A$ (or in the *graded Lie algebra* $E$) is respectively a homomorphism

$$L \to \mathrm{Der}\, A \qquad \text{or} \quad L \to \mathrm{Der}\, E.$$

Associated with a graded Lie algebra $L$ is its *universal enveloping algebra*

$$UL = TL/J$$

where $J$ is the ideal generated by the elements

$$[x, y] - x \otimes y + (-1)^{\deg x \deg y} y \otimes x$$

with $x, y \in L$. The natural map $L \to UL$ is an injection. If $A$ is a graded algebra and $\varphi : L \to A$ is a morphism of graded Lie algebras, then $\varphi$ extends uniquely of a morphism of algebras $UL \to A$.

A *representation of $L$ in a graded vector space $V$* is a linear map of degree 0,

$$\varphi : L \to \mathrm{End}(V)$$

such that

$$\varphi[x, y] = \varphi x \circ \varphi y - \varphi y \circ \varphi x\,, \qquad x, y \in L\,.$$

Thus a representation of $L$ in $V$ is just a homomorphism of graded Lie algebras. A representation of $L$ in $V$ extends uniquely to a representation

$$UL \to \mathrm{End}(V)$$

of $UL$ in $V$.

If $V$ is a graded vector space then, as with any graded algebra, $TV$ inherits a graded Lie algebra structure, and the sub Lie algebra generated by $V$ is the *free graded Lie algebra* $\mathbb{L}_V$ on $V$. It has the standard universal property: a linear map $\alpha : V \to L$ into a graded Lie algebra $L$ extends uniquely to a morphism $\mathbb{L}_V \to L$. It follows that the inclusion $\mathbb{L}_V \to TV$ extends to an isomorphism

$$U\mathbb{L}_V \xrightarrow{\cong} TV$$

of graded algebras.

A *graded Hopf algebra* is a graded algebra $H$ equipped with an augmentation $\varepsilon : H \to \Bbbk$ and a morphism of augmented graded algebras

$$\Delta : H \to H \otimes H$$

that is coassociative (i.e. $(\Delta \otimes id) \circ \Delta = (id \otimes \Delta) \circ \Delta$) and satisfies $(\varepsilon \otimes \varepsilon) \circ \Delta = \varepsilon$. The Hopf algebra is *cocommutative* if $\tau \circ \Delta = \Delta$, where $\tau(x \otimes y) = (-1)^{\deg x \deg y} y \otimes x$. Elements $x$ in a graded Hopf algebra are called *group-like* if $\Delta x = x \otimes x$ and *primitive* if $\Delta x = x \otimes 1 + 1 \otimes x$.

**Definition.** Graded algebras, Lie algebras and Hopf algebras have *finite type* if they have finite type as graded vector spaces.

Note, for example that if $L$ is a graded Lie algebra then $UL$ is the graded Hopf algebra defined by $\Delta x = x \otimes 1 + 1 \otimes x$ for $x \in L$. As a second example, if $G$ is a discrete group then the group ring $\Bbbk[G]$, which has the elements of $G$ as a basis, is a Hopf algebra concentrated in degree zero, with multiplication given by the group multiplication, and diagonal and multiplication given by $\Delta : g \mapsto g \otimes g$ and $\varepsilon(g) = 1$, $g \in G$.

Finally, a *representation* of a group $G$ in a graded vector space $V$ is a morphism of groups $G \to \mathrm{Aut}(V)$ from $G$ to the group of automorphisms (of degree zero) of $V$. It extends uniquely to a representation of $\Bbbk[G]$ in $V$.

**Remark:** The classical categories of (ungraded) vector spaces, algebras, Lie algebras and Hopf algebras coincide with the categories of graded objects concentrated in degree zero - one only suppresses the degree.

## 1.2 Differential graded algebra

A *differential* in a graded vector space $V$ is a linear map $d \in \mathrm{End}^1(V)$ satisfying $d^2 = 0$. The *homology*, $H(V,d)$ is the graded vector space $\ker d/\mathrm{Im}\, d$. Given differentials $d_V$ and $d_W$ in graded vector spaces $V$ and $W$ then $\mathrm{Hom}(V,W)$ is equipped with the differential $d(\varphi) = d_W \circ \varphi - (-1)^{\deg \varphi} \varphi \circ d_V$. If $d(\varphi) = 0$ then $\varphi$ is a *chain map* and induces a linear map $H(\varphi) : H(V,d_V) \to H(W,d_W)$. Two chain maps $\varphi, \psi$ are *homotopic* if $\varphi - \psi = d(h)$ for some linear map $h : V \to W$, and we write $\varphi \sim \psi$. In this case $H(\psi) = H(\varphi)$. A chain map, $\varphi$, is a *quasi-isomorphism* if $\deg \varphi = 0$ and $H(\varphi)$ is an isomorphism.

A *differential graded algebra* (dga for short) is a pair $(A,d)$ in which $A$ is a graded algebra and $d \in \mathrm{Der}^1 A$ is a differential. If $A$ is commutative then $(A,d)$ is a cdga. If $(A,d)$ is a dga then $H(A,d)$ inherits a graded algebra structure from the multiplication in $A$.

If $(A,d)$ is a dga then an *$A$-module* is a pair $(M,d)$ in which $M$ is a graded vector space with a differential, equipped with an $A$-module structure $A \otimes M \to M$ of degree zero, and such that

$$d(am) = (da)m + (-1)^{\deg a} a \cdot dm\,.$$

A morphism $\varphi : (M,d) \to (N,d)$ of $(A,d)$-modules is a linear map of degree zero, commuting with the differentials and with the action of $A$. A second morphism $\psi$ is *homotopic* to $\varphi$ if there is a linear map $h : M \to N$ of degree $-1$ such that $h(am) = (-1)^{\deg a} ah(m)$ and $\varphi - \psi = hd + dh$. In this case we write $\varphi \sim \psi$. When $A = \mathbb{k}$, these definitions coincide with the definitions above for chain maps and chain homotopy.

An $(A, d)$-module, $(P, d)$, of the form $P = A \otimes M$ is *semifree* if $M$ is the increasing union of subspaces $0 = M(0) \subset M(1) \subset \cdots$ such that for each $r$, $A \otimes M(r)$ is preserved by $d$ and the differential in $A \otimes M(r+1)/M(r)$ reduces to $d \otimes id$. For any $(A, d)$-module, $(N, d)$ there is a quasi-isomorphism of $(A, d)$-modules $(P, d) \xrightarrow{\simeq} (N, d)$ in which $(P, d)$ is semifree ([18], Proposition 6.6); such a quasi-isomorphism is a *semifree resolution* of $(N, d)$.

**Proposition 1.1.** *If $\varphi : (A, d) \xrightarrow{\simeq} (B, d)$ is a dga quasi-isomorphism and if $(A \otimes M, d)$ is a semifree $(A, d)$-module then*

$$\varphi \otimes_A M : (A \otimes M, d) \to (B \otimes_A (A \otimes M), d) := (B \otimes M, d)$$

*is a quasi-isomorphism.*

*proof:* This follows directly from Proposition 6.7 in [18].

□

We shall be particularly interested in cochain algebras: a *cochain algebra* is a dga $(A, d)$ with $A = A^{\geq 0}$. An *augmentation* of a cochain algebra is a dga morphism

$$\varepsilon_A : (A, d) \to (\mathbb{k}, 0)\,.$$

A *based morphism* of augmented dga's is a morphism

$$\varphi : (A, d_A, \varepsilon_A) \to (B, d_B, \varepsilon_B)$$

such that $\varphi d_A = d_B \varphi$ and $\varepsilon_A = \varepsilon_B \varphi$. Thus a dga morphism, $\varphi$, is based if and only if $\varphi(\ker \varepsilon_A) \subset \ker \varepsilon_B$.

Spectral sequences do play a role from time to time, and the reader is referred to [18], §18 for a summary of the notation and conventions we will use.

Finally, a differential graded Lie algebra (dgl for short) is a graded Lie algebra $L$ equipped with a differential $d \in \mathrm{Der}_{-1}(L)$. The homology, $H(L, d)$ inherits the structure of a graded Lie algebra from the Lie bracket in $L$.

## 1.3 Simplicial sets

A *simplicial object* in a category $\mathcal{C}$ is a sequence $(C_n)_{n\geq 0}$ of objects in $\mathcal{C}$ together with morphisms (respectively the face and degeneracy morphisms)

$$\partial_i : C_{n+1} \to C_n \qquad \text{and} \quad s_j : C_n \to C_{n+1}$$

with $0 \leq i \leq n+1$ and $0 \leq j \leq n$, and which satisfy the identities $\partial_i\partial_j = \partial_{j-1}\partial_i$ if $i < j$, $s_i s_j = s_{j+1}s_i$ if $i \leq j$ and

$$\partial_i s_j = \begin{cases} s_{j-1}\partial_i & \text{if } i < j \\ id & \text{if } i = j, j+1 \\ s_j\partial_{i-1} & \text{if } i > j+1. \end{cases}$$

The simplicial objects in a category $\mathcal{C}$ form a category whose morphisms are the sequence of morphisms $C_n \to C'_n$ in $\mathcal{C}$ commuting with the face and degeneracy morphisms. If $\mathcal{C}$ has products so does the corresponding simplicial category:

$$K \times L = \{K_n \times L_n\}_{n\geq 0}\,, \qquad \partial_i = \partial_i \times \partial_i\,, \quad s_j = s_j \times s_j\,.$$

We will primarily be concerned with the category of simplical sets in which case morphisms are called *simplicial maps.* If $K$ is a simplicial set then an element $\sigma \in K_n$ is an *$n$-simplex* of $K$; if $n = 0$ then $\sigma$ is a *vertex* of $K$. A pair $(K, \sigma)$ with $\sigma$ a vertex of $K$ is called a *pointed* simplicial set. A simplex is *degenerate* if it is in the image of some $s_j$, otherwise it is *non-degenerate.* If $E_n \subset K_n$ is a sequence of inclusions in which each $\partial_i$ restricts to maps $E_n \to E_{n-1}$, then by adding simplices to $E_n$ of the form $s_{j_1}\circ\cdots\circ s_{j_p}(E_{n-p})$ we obtain a simplicial subset of $K$ whose non-degenerate simplices are all in the sets $E_n$. This is the simplicial subset *generated* by $E = \{E_n\}$. We identify a vertex $v \in K_0$ with the simplicial set it generates, and denote this by $\{v\}$.

Now suppose $L \subset K$ is an inclusion of simplicial sets. Then the simplicial set $K/L$ together with a simplicial map $\rho : K \to K/L$ are defined as follows: let $K_n - L_n$ be the complement of $L_n$ in $K_n$; then

$$(K/L)_n = (K_n - L_n) \cup *_n\,, \qquad \rho_n\sigma = \begin{cases} \sigma & \text{if } \sigma \notin K_n \\ *_n & \text{if } \sigma \in L_n \end{cases}$$

and the set of maps $\partial_i$, $s_j$ in $K/L$ are the unique maps such that $\partial_i\rho = \rho\partial_i$ and $s_j\rho = \rho s_j$.

**Example: The simplicial set $\Delta[n]$.**

Among simplicial sets an important role is played by the simplicial sets $\Delta[n]$ defined as follows. Let $e_0, e_1, \cdots$ be the standard basis of $\mathbb{R}^\infty$ and let $\Delta^n$ denote the Euclidean simplex with vertices $e_0, \cdots, e_n$. In particular we identify $(\Delta^1, e_0, e_1)$ with $(I, 0, 1)$ where $I$ is the unit interval. Then $\Delta[n]_k$ consists of the linear maps

$$\langle e_{\sigma(0)}, \cdots, e_{\sigma(k)} \rangle : \Delta^k \to \Delta^n$$

which map $e_i \mapsto e_{\sigma(i)}$ where $\sigma$ is any order preserving map:

$$0 \leq \sigma(0) \leq \cdots \leq \sigma(k) \leq n\,.$$

The face and degeneracy maps in $\Delta[n]$ are given by

$$\partial_i \langle e_{\sigma(0)}, \cdots, e_{\sigma(k)} \rangle = \langle e_{\sigma(0)}, \cdots, \widehat{e_{\sigma(i)}}, \cdots, e_{\sigma(k)} \rangle \tag{1.1}$$

and

$$s_j \langle e_{\sigma(0)}, \cdots, e_{\sigma(k)} \rangle = \langle e_{\sigma(0)}, \cdots, e_{\sigma(j)}, e_{\sigma(j)}, \cdots, e_{\sigma(k)} \rangle\,.$$

Here "$\widehat{\ }$" means delete.

Almost by definition, the non-degenerate simplices of $\Delta[n]$ are the simplices

$$\langle e_{\sigma(0)}, \cdots, e_{\sigma(k)} \rangle$$

with $\sigma(0) < \cdots < \sigma(k)$; when $k = n - 1$ these have the form $\langle e_0, \cdots, \widehat{e_i}, \cdots, e_n \rangle$; here $\langle e_0, \cdots, \widehat{e_i}, \cdots, e_n \rangle$ is the $i^{th}$ *face* of $\Delta[n]$.

**Definition.** The *fundamental simplex* of $\Delta[n]$ is the identity map

$$c_n = \langle e_0, \cdots, e_n \rangle : \Delta^n \to \Delta^n.$$

In particular, when $n = 0$, for each $k$, $\Delta[0]_k$ consists of a single simplex written as $*_k$; these are all degenerate except for $*_0$, which we denote by $*$. When $n = 1$, $\Delta[1]$ has three non-degenerate simplexes, $c_1 \in \Delta[1]_1$ and $\partial_1 c_1$ and $\partial_0 c_1$ in $\Delta[1]_0$.

The *boundary*, $\partial\Delta[n]$, is the simplicial set generated by the $\{\Delta[n]_k\}_{k<n}$. The quotient, $\Delta[n]/\partial\Delta[n]$ has a single vertex; this and $c_n$ are the only non-degenerate simplices in the quotient.

Finally, an easy check shows that an $n$-simplex $\sigma$ in a simplicial set $K$ determines a unique simplicial map

$$\sigma_* : \Delta[n] \to K$$

such that $\sigma_*(c_n) = \sigma$.

**Definition.** A simplicial set, $K$, is a *Kan simplicial set* if for $n \geq 1$, whenever $L \subset \partial\Delta[n]$ is the subsimplicial set generated by all simplices in $\Delta[n]_k$, $k < n-1$, and by all but one face of $\Delta[n]$ then any simplicial map $L \to K$ extends to a simplicial map from $\Delta[n]$. A Kan simplicial set equipped with a base vertex will be called a pointed Kan simplicial set.

**Definition.** Two morphisms, $\varphi_0, \varphi_1 : L \to K$ between simplicial sets are *homotopic* if there is a simplicial map (called the homotopy)

$$\Phi : \Delta[1] \times L \to K$$

satisfying $\Phi_{|\{\partial_1 c_1\}\times L} = \varphi_0$ and $\Phi_{|\{\partial_0 c_1\}\times L} = \varphi_1$.

If $L$ and $K$ have fixed base points $u_0 \in L_0$ and $v_0 \in K_0$ and if $\varphi_0$ and $\varphi_1$ are base point preserving then $\varphi_0$ and $\varphi_1$ are *based homotopic* if there is a homotopy $\Phi : (\Delta[1] \times L, \Delta[1] \times u_0) \to (K, v_0)$ from $\varphi_0$ to $\varphi_1$.

**Proposition 1.2.** *If $K$ is a Kan simplicial set then homotopy (respectively based homotopy) define equivalence relations in the set of simplicial morphisms (respectively base point preserving simplicial morphisms) from $L$ to $K$.*

*proof:* [47], Corollary 6.11.

□

If $(L, u_0)$ is a simplicial set and $(K, v_0)$ is a Kan simplicial set, and if $(X, x_0)$ and $(Y, y_0)$ are pointed topological spaces we denote by

$$[(L, u_0), (K, v_0)] \quad \text{and} \quad [(X, x_0), (Y, y_0)]$$

the set of based homotopy classes respectively of simplicial maps $(L, u_0) \to (K, v_0)$ and of continuous maps $(X, x_0) \to (Y, y_0)$.

Our next step is to connect simplicial sets with topological spaces via adjoint functors, the *singular simplex functor*

$$Sing : \text{topological spaces} \rightsquigarrow \text{simplicial sets}$$

and the *Milnor realization functor*

$$|\ | : \text{simplicial sets} \rightsquigarrow \text{topological spaces}.$$

The functor *Sing* is defined as follows: if $X$ is a topological space and $\varphi$ is a continuous map then

$$Sing_n(X) = \text{ set of singular simplices } \sigma : \Delta^n \to X\,,$$

$$\partial_i\sigma = \sigma \circ \langle e_0, \cdots, \widehat{e_i}, \cdots, e_n\rangle\,, \qquad s_j\sigma = \sigma \circ \langle e_0, \cdots, e_j, e_j, \cdots, e_n\rangle\,,$$

and

$$Sing_n(\varphi)(\sigma) = \varphi \circ \sigma\,.$$

If $X$ is pointed by $x_0 \in X$ then $Sing X$ is pointed by the vertex $(x_0) : e_0 \to x_0$ which we write simply as $x_0$.

In particular $\Delta[n]$ is a subsimplicial set of $Sing(\Delta^n)$. In the case $n = 1$, $c_1$ is identified with the identity map of $I$ sending $\partial_1 c_1$ and $\partial_0 c_1$ respectively to $\{0\}$ and $\{1\}$.

The *Milnor realization functor*, $|\ |$, is constructed as follows. If $K$ is a simplicial set, give each $K_n$ the discrete topology and set

$$|K| = \bigsqcup_n (K_n \times \Delta^n)\ /\ \sim$$

where $\sim$ is the equivalence relation generated by the relations

$$\partial_i\sigma \times x \sim \sigma \times \mu_i x \qquad \text{and} \quad s_j\sigma \times y \sim \sigma \times \pi_j y\,;$$

here $\mu_i = \langle e_0, \cdots, e_{i-1}, e_{i+1}, \cdots, e_n\rangle : \Delta^{n-1} \to \Delta^n$ and $\pi_j = \langle e_0, \cdots, e_j, e_j, \cdots, e_n\rangle : \Delta^{n+1} \to \Delta^n$. We denote by $\langle \sigma, x\rangle$ the element in $|K|$ represented by $(\sigma, x) \in K_n \times \Delta^n$. If $\varphi : L \to K$ is a simplicial map then $|\varphi| : |L| \to |K|$ is the continuous map induced from the maps $\varphi_n \times id : L_n \times \Delta^n \to K_n \times \Delta^n$. In particular, if $L$ is pointed by a vertex $u_0$ then $|L|$ is pointed by $\langle u_0, c_0\rangle$, which we write simply as $u_0$.

**Proposition 1.3.** *(i) Let $K$ be a simplicial set. Then $|K|$ is a CW complex with $n$-cells indexed by the non-degenerate $n$-simplices of $K$.*

*(ii) If $L$ is a second simplicial set the maps $|L \times K| \to |L|$, $|L \times K| \to |K|$ corresponding to the simplicial projections define a homeomorphism $|L \times K| \xrightarrow{\cong} |L| \times |K|$.*

*proof:* (i) This is [18] (Proposition 17.2). (ii) This is [18] (Proposition 17.4(i)).

□

**Corollary 1.1.** *The map $(c_n, x) \mapsto x$ defines a homeomorphism*

$$|\Delta[n]| \xrightarrow{\cong} \Delta^n\,.$$

**Lemma 1.1.** *If $X$ is a topological space then $Sing X$ is a Kan simplicial set.*

*proof:* ([47], Lemma 1.5). This follows because the union of all but one face of the Euclidean simplex $\Delta^n$ is a retract of $\Delta^n$.

□

Suppose next that $(L, u_0)$ is a pointed simplicial set and that $(X, x_0)$ is a pointed topological space. Then both

$$Top\,(|L|, X) \quad \text{and} \quad Simpl\,(L, Sing\,X)$$

coincide with the set $S$ of continuous maps

$$\bigsqcup_n (L_n \times \Delta^n) \to X$$

compatible with the relations described above. On the one hand this follows for $Top\,(|L|, X)$ by the definition of $|L|$. On the other hand, the maps $\mu_i$ and $\pi_j$ make $\{\,Top\,(\Delta^n, X)\}$ into a simplicial set which by definition is exactly $Sing\,X$. This identification establishes $|\quad|$ and $Sing$ as adjoint functors. It restricts to an identification

$$Top\;((|L|, u_0), (X, x_0)) \cong Simpl\;((L, u_0), (Sing\,X, x_0))\;, \qquad (1.2)$$

giving an adjoint relation between based continuous maps and based simplicial maps.

In particular we denote by

$$\lambda_L : L \to Sing\,|L| \quad \text{and} \quad \lambda_X : |Sing\,X| \to X \qquad (1.3)$$

the natural transformations respectively adjoint to $id_{|L|}$ and $id_{Sing\,X}$.

To convert the relation of homotopy in the simplicial and topological categories we recall that $\Delta[1]$ has a single non-degenerate 1-simplex, $c_1$, and two vertices, $\partial_1 c_1$ and $\partial_0 c_1$. We denote by $e_0, e_1 : \Delta[0] \to \Delta[1]$ the simplicial inclusions corresponding to $\partial_1 c_1$ and $\partial_0 c_1$. As usual, $I$ denotes the unit interval $[0, 1]$ and we identify $\Delta^1 = I$ via the linear isomorphism $te_1 + (1-t)e_0 \mapsto t$. This identifies

$$|\Delta[1], e_0, e_1| = (I, 0, 1)\,.$$

In particular it identifies $\lambda_{\Delta[1]}$ as a simplicial map

$$\lambda_{\Delta[1]} : (\Delta[1], e_0, e_1) \to (\,Sing\,I, 0, 1)\,.$$

**Proposition 1.4.** *(i) Both* $|\quad|$ *and Sing preserve the relations of homotopy and based homotopy.*

*(ii) Two continuous maps* $(|L|, *) \to (X, *)$ *are (based) homotopic if and only if their adjoints are.*

*proof:* (i) We prove only the assertions for based homotopy. The proofs in the non-based case are essentially the same, but simpler.

Suppose $\varphi_0, \varphi_1 : (L, *) \to (K, *)$ are based homotopic via a homotopy $\Phi : (\Delta[1] \times L, \Delta[1] \times *) \to (K, *)$. Then,

$$|\Phi| : (I \times |L|, I \times *) \to (|K|, *)$$

is a based homotopy from $|\varphi_0|$ to $|\varphi_1|$.

Conversely, suppose $\psi_0, \psi_1 : (X, *) \to (Y, *)$ are based homotopic continuous maps via a homotopy $\Psi : (I \times X, I \times *) \to (Y, *)$. Almost by definition, *Sing* preserves products and $Sing\,(pt) = \Delta[0]$. Thus

$$Sing\,\Psi : (Sing\,I \times Sing\,X, Sing\,I \times *) \to (Sing\,Y, *)\,.$$

Since $\lambda_{\Delta[1]} : \Delta[1] \to Sing\,I$ and so $Sing\,\Psi$ restricts to a based homotopy from $Sing\,(\psi_0)$ to $Sing\,(\psi_1)$.

(ii) follows in the same way because $I \times |L| = |\Delta[1]| \times |L| = |\Delta[1] \times L|$.

□

**Corollary 1.2.** *If $(L, *)$ is a pointed simplicial set and if $(X, *)$ is a pointed topological space then* (1.2) *and* (1.3) *induce bijections*

$$[|L|, X] \cong [L, Sing\,X]\,, \quad \text{and} \quad [(|L|, *), (X, *)] \cong [(L, *), (Sing\,X, *)]\,.$$

□

**Proposition 1.5.** *Suppose $(L, *)$ is a pointed simplicial set, $(K, *)$ is a pointed Kan simplicial set, and $X$ is a topological space.*

*(i) The inclusion $\lambda_K : K \to Sing|K|$ has a based homotopy inverse $r_K : Sing|K| \to K$.*

*(ii) The correspondence $\varphi \mapsto |\varphi|$ induces natural bijections*

$$[L, K] \xrightarrow{\cong} [|L|, |K|]\,, \quad \text{and} \quad [(L, *), (K, *)] \xrightarrow{\cong} [(|L|, *), (|K|, *)]\,.$$

*(iii) $\lambda_X : |Sing\,X| \to X$ is a weak homotopy equivalence. If $X$ is a CW complex then $\lambda_X$ is a homotopy equivalence.*

*proof:* (i) This follows from [47], Theorem 16.6(i).

(ii) Corollary 1.2 yields bijections

$$[L, Sing|K|] \xrightarrow{\cong} [|L|, |K|]\,, \text{ and } [(L, *), (Sing|K|, *)] \xrightarrow{\cong} [(|L|, *), (|K|, *)]\,.$$

Precompose with $\lambda_K$ to obtain the desired bijections.

(iii) This is [47], Theorem 16.6(ii).

□

## 1.4 Polynomial differential forms

A second important example of a simplicial object is the *simplicial commutative cochain algebra*, $A_{PL}$, defined as follows:

$$(A_{PL})_n = \frac{\wedge(t_0,\cdots,t_n,dt_0,\cdots,dt_n)}{\sum t_i - 1, \sum dt_i}\,.$$

The morphisms $\partial_i$ and $s_j$ are defined by their restrictions to the basis elements $t_k$.

$$\partial_i t_k = \begin{cases} t_k & \text{if } k<i \\ 0 & \text{if } k=i \\ t_{k-1} & \text{if } k>i \end{cases} \qquad \text{and} \quad s_j t_k = \begin{cases} t_k & \text{if } k<j \\ t_k + t_{k+1} & \text{if } k=j \\ t_{k+1} & \text{if } k>j \end{cases}\,.$$

Note that when $\mathbb{k} \subset \mathbb{R}$ then $(A_{PL})_n$ is the $\mathbb{k}$-subalgebra of the standard cochain algebra of differential forms in $\Delta^n$ generated by the coordinate functions $t_0,\cdots,t_n$. Thus $(A_{PL})_n$ is called the cochain algebra of *polynomial differential forms* in $\Delta^n$.

**Definition.** If $K$ is a simplicial set then the *cochain algebra of polynomial forms on* $K$, $A_{PL}(K)$, is defined by

$$A^k_{PL}(K) = Simpl\,(K, A^k_{PL}) = \{\Phi_\sigma \in (A^k_{PL})_n \,|\, \sigma \in K_n, n \geq 0\}$$

in which the $\Phi_\sigma$ are required to satisfy

$$\partial_i \Phi_\sigma = \Phi_{\partial_i \sigma} \qquad \text{and} \quad s_j\Phi_\sigma = \Phi_{s_j\sigma}\,.$$

The differential is defined in the obvious way: if $\Phi = \{\Phi_\sigma\} \in A^k_{PL}(K)$ then $(d\Phi)_\sigma = d(\Phi_\sigma)$.

The assignment $K \rightsquigarrow A_{PL}(K)$ is a contravariant functor from simplicial sets to commutative cochain algebras: if $\Phi = \{\Phi_\sigma\} \in A_{PL}(K)$ and if $\alpha : L \to K$ is a map of simplicial sets then

$$\left(A_{PL}(\alpha)\Phi\right)_\tau = \Phi_{\alpha(\tau)}\,, \qquad \tau \in L\,.$$

This functor, precomposed with the functor *Sing* yields the contravariant functor

$$X \rightsquigarrow A_{PL}(X)\,, \qquad \varphi \rightsquigarrow A_{PL}(\varphi)$$

from topological spaces to commutative cochain algebras.

**Definition.** If $X$ is a topological space then $A_{PL}(X)$ is the *algebra of polynomial differential forms* on $X$.

Now recall from §1.3 that $c_n$ denotes the fundamental simplex of $\Delta[n]$.

**Proposition 1.6.** *(i) An isomorphism* $A_{PL}(\Delta[n]) \to (A_{PL})_n$ *is given by* $\Phi \mapsto \Phi_{c_n}$. *In particular,* $(A_{PL})(\Delta[0]) = (A_{PL})_0 = 𝕜$.

*(ii)* $H((A_{PL})_n, d) = 𝕜$.

*(iii) If* $\alpha : L \to K$ *is an inclusion of simplicial sets then* $A_{PL}(\alpha)$ *is surjective.*

*proof:* (i) is obvious; (ii) is just [18], Lemma 10.7(ii). (iii) follows from ([18], Lemma 10.7(iii)) applied to [18], Proposition 10.4(ii).

□

Given a zero simplex $v_0 \in K_0$ and a point $x_0 \in X$, then $x_0 \in Sing_0 X$ and so by Proposition 1.6(i) the inclusions $v_0 \hookrightarrow K_0$ and $x_0 \hookrightarrow X$ define augmentations

$$\varepsilon_K : A_{PL}(K) \to 𝕜 \qquad \text{and} \qquad \varepsilon_X : A_{PL}(X) \to 𝕜 \,.$$

Now suppose $L \subset K$ is an inclusion of simplicial sets. We set

$$A_{PL}(K, L) = \{\Phi \in A_{PL}(K) \mid \Phi_{|_L} = 0\}\,.$$

Thus we have the short exact sequence of differential graded objects

$$0 \to A_{PL}(K, L) \to A_{PL}(K) \to A_{PL}(L) \to 0\,,$$

in which $A_{PL}(K, L)$ is an ideal in $A_{PL}(K)$. Thus a commutative graded algebra $A_{PL}(K, L)_*$ is defined by

$$A_{PL}(K, L)_* = A_{PL}(K, L) \oplus 𝕜 \cdot 1\,,$$

and the inclusion $A_{PL}(K, L)_* \hookrightarrow A_{PL}(K)$ may be identified with the morphism

$$A_{PL}(\rho) : A_{PL}(K/L) \to A_{PL}(K)\,.$$

In particular, if $Y \subset X$ is the inclusion of a subspace in a topological space $X$ we write

$$A_{PL}(X, Y) = A_{PL}(Sing\, X, Sing\, Y)$$

and $A_{PL}(X, Y)_* = A_{PL}(Sing\, X, Sing\, Y)_*$.

Finally, if $K$ is a simplicial set we denote by $S_n(K)$ the $𝕜$-vector space with basis $K_n$ and by $D_n(K)$ the subspace with basis the degenerate simplices in $K_n$. The operator $\partial = \sum_{i=0}^n (-1)^i \partial_i$ maps $D_n(K)$ to $D_{n-1}(K)$ and therefore defines a differential, $\partial$, in the graded vector space $C_*(K)$ with

$C_n(K) = S_n(K)/D_n(K)$. The dual, $(C^*(K), d) = \mathrm{Hom}((C_*(K), \partial), \Bbbk)$ is the simplicial cochain complex for $K$. When $K = Sing\, X$ then these are denoted by $C_*(X)$ and $C^*(X)$; they are respectively the singular chain complex and the singular cochain algebra on $X$. (For the multiplication in $C^*(K)$ see [18], §10). Thus $H(C^*(X))$ is the *singular cohomology* of $X$ with coefficients in $\Bbbk$. We write $H(C^*(X)) = H^*(X; \Bbbk)$.

Now recall from Proposition 1.6 (p. 15) that when $\Bbbk = \mathbb{Q}$, $A_{PL}(\Delta[n]) = (A_{PL})_n$ is a sub cochain algebra of the ordinary differential forms on $\Delta^n$. In particular, the classical integral $\oint_n : A^n_{PL}(\Delta[n]) \to \mathbb{R}$ is defined, and as shown in §10(e) of [18], it takes values in $\mathbb{Q}$. Thus for any simplicial set, $K$, and any topological space $X$, we can tensor with $\Bbbk$ to obtain the linear maps,

$$\oint_K : A_{PL}(K) \to C^*(K) \quad \text{and} \quad \oint_X : A_{PL}(X) \to C^*(X),$$

defined by

$$(\oint_K \Phi)(\sigma) = \oint_n \Phi_\sigma \quad \text{and} \quad \oint_X = \oint_{Sing\, X}.$$

(Note that by definition $C^*(X) = C^*(Sing\, X)$.)

**Proposition 1.7.** ([18], Theorem 10.9). *If $K$ is a simplicial set and $X$ is a topological space, then $\oint_K$ and $\oint_X$ are natural quasi-isomorphisms of cochain complexes inducing natural isomorphisms of graded algebras*

$$H(A_{PL}(K)) \xrightarrow{\cong} H^*(C^*(K)) \quad \text{and} \quad H^*(A_{PL}(X)) \xrightarrow{\cong} H^*(X : \Bbbk).$$

**Corollary 1.3.** *The morphisms $C^*(\lambda_K) : C^*(|K|) \to C^*(K)$ and $A_{PL}(\lambda_K) : A_{PL}(|K|) \to A_{PL}(K)$ are quasi-isomorphisms.*

*proof:* The first assertion follows because, as observed in [18], p. 241, $C_*(\lambda_K)$ is a quasi-isomorphism. The second assertion follows from this and the Proposition.

□

Note that for any simplicial set $K$ we have a commutative square of natural isomorphisms

$$\begin{array}{ccc} H(|K|) & \xrightarrow{\cong} & H(C^*(K)) \\ \cong \uparrow & & \uparrow \cong \\ H(A_{PL}(|K|)) & \xrightarrow{\cong} & H(A_{PL}(K)) \end{array}$$

**Convention:** *We shall always identify these four graded algebras, and denote all four simply by* $H|K|$. *In particular, for a topological space* $X$ *and with* $K = Sing\, X$ *we identify* $H^*(X; \Bbbk) = H(A_{PL}(X))$ *and denote this by* $H(X)$.

**Example.** $\Delta[n]$ and $\Delta[n]/\partial\Delta[n]$.

It follows by an easy verification that evaluation at $c_n$ defines an isomorphism

$$A_{PL}(\Delta[n]) \xrightarrow{\cong} (A_{PL})_n\,.$$

Note as well that the inclusions $t_i \mapsto (A_{PL})_n$ extend to an isomorphism

$$\wedge(t_1, \cdots, t_n, dt_1, \cdots, dt_n) \xrightarrow{\cong} (A_{PL})_n\,.$$

On the other hand, for $n \geq 1$, $\Delta[n]/\partial\Delta[n]$ is the simplicial set with two non-degenerate vertices : the single 0-simplex, denoted $\overline{c_0}$, and the image $\overline{c_n}$ of the fundamental simplex $c_n$ of $\Delta[n]$. In particular

$$(\,|\Delta[n]/\partial\Delta[n]|, \overline{c_0}\,) = (S^n, *)\,,$$

and

$$H_*(S^n; \mathbb{Z}) = H_*(\Delta[n]/\partial\Delta[n]; \mathbb{Z}) = C_*(\Delta[n]/\partial\Delta[n]; \mathbb{Z}) = \mathbb{Z}\,\overline{c_n} \oplus \mathbb{Z}\,\overline{c_0}\,.$$

We call $\overline{c_n}$ the *fundamental cycle* of $\Delta[n]/\partial\Delta[n]$ and the *fundamental homology class* of $S^n$.

Next note that the isomorphism $H^*(-, \Bbbk) = \mathrm{Hom}(H_*(-, ; \mathbb{Z}), \Bbbk)$ determines a bilinear map $H^n(A_{PL}(\Delta[n]/\partial\Delta[n]); \Bbbk) \times H_n(\Delta[n]/\partial\Delta[n]; \mathbb{Z}) \to \Bbbk$. Moreover, by definition,

$$A_{PL}(\Delta[n]/\partial\Delta[n]) = \Bbbk \oplus \{\Phi \in (A_{PL})_n \,|\, \partial_i\Phi = 0\,,\ \ 0 \leq i \leq n\}\ .$$

We show now that the class $[dt_1 \wedge \cdots \wedge dt_n] \in H^n(A_{PL}(\Delta^n/\partial\Delta[n]))$ satisfies

$$\langle \overline{c_n}, [dt_1 \wedge \cdots \wedge dt_n] \rangle = 1 = \langle (-1)^n [dt_1 \wedge \cdots \wedge dt_n], \overline{c_n} \rangle \qquad (1.4)$$

and therefore we call $(-1)^n[dt_1 \wedge \cdots \wedge dt_n]$ the *fundamental cohomology class* of $\Delta[n]/\partial\Delta[n]$ and of $S^n$.

For the proof of the equality above, consider the short exact sequence

$$0 \to C_*(e_1) \to C_*(\Delta[1], e_0) \to C_*(\Delta[1], \partial\Delta[1]) \to 0\,.$$

The connecting homomorphism satisfies $\delta[\overline{c_1}] = [e_1]$. Dually the exact sequence

$$0 \leftarrow A_{PL}(e_1) \leftarrow A_{PL}(\Delta[1], e_0) \leftarrow A_{PL}(\Delta[1], \partial\Delta[1]) \leftarrow 0$$

has a connecting homomorphism $\delta^* : [e_1]^* \mapsto [dt_1]$ and so

$$\langle [dt_1], \overline{c_1}\rangle = \langle \delta^*[e_1]^*, \overline{c_1}\rangle = -\langle [e_1^*], \delta[\overline{c_1}]\rangle = -1\,.$$

In particular, when $n = 1$ we denote $t_1$ simply by $t$, thereby identifying $A_{PL}(\Delta[1]) = \wedge(t, dt)$ and $dt$ as a cocycle in $A_{PL}(\Delta[1], \partial\Delta[1])$ satisfying

$$\langle -dt, \overline{c_1}\rangle = 1\,.$$

The rest of the proof follows in the same way by induction from the short exact sequences

$$0 \to C_*(\Delta[n-1], \partial\Delta[n-1]) \xrightarrow{C_*(\lambda_0)} C_*(\Delta[n], J) \to C_*(\Delta[n], \partial\Delta[n]) \to 0$$

in which $J \subset \Delta[n]$ is the subsimplicial set generated by all the faces containing $e_0$.

□

## 1.5 Sullivan algebras

**Definition.** A cochain algebra of the form $(\wedge V, d)$ in which $V = V^{\geq 0}$ satisfies the *Sullivan condition* if $V$ is the union of an increasing family of subspaces

$$V(0) \subset V(1) \subset \cdots \subset V(r) \subset \cdots$$

such that $d = 0$ in $V(0)$ and for $p \geq 1$, $d : V^p(r) \to \wedge(V^{<p} + V(r-1))$.

**Definition.** (i) A *Sullivan algebra* is a cochain algebra of the form $(\wedge V, d)$ with $V = V^{\geq 1}$ and which satisfies the Sullivan condition. It is *minimal* if $d : V \to \wedge^{\geq 2} V$ or, equivalently, $d : V^p(r) \to \wedge(V^{<p} \oplus V^p(r-1))$.

(ii) A $\Lambda$-*algebra* is a cochain algebra of the form $(\wedge Z, d)$ in which $Z = Z^{\geq 0}$ is the union of an increasing family of subspaces $Z(0) \subset \cdots \subset Z(r) \subset \cdots$ such that $d = 0$ in $Z(0)$ and for $p \geq 0$, $r \geq 1$, $d : Z^p(r) \to \wedge(Z^{<p} \oplus Z^p(r-1))$.

In this text we frequently need to impose certain finiteness conditions on a minimal Sullivan algebra $(\wedge V, d)$.

**Definition.** A minimal Sullivan algebra $(\wedge V, d)$ is *1-finite* if $\dim H^1(\wedge V, d) < \infty$. It has *weak finite type* if

$$\dim H^1(\wedge V, d) < \infty \quad \text{ and } \dim V^i < \infty\,, \quad i \geq 2\,. \tag{1.5}$$

**Remarks.** (i) Any cochain algebra $(\wedge V, d)$ with $d : V \to \wedge^{\geq 2} V$ and $V = V^{\geq 2}$ is a minimal Sullivan algebra.

(ii) Minimal Sullivan algebras are precisely the $\Lambda$-algebras $(\wedge Z, d)$ for which $Z^0 = 0$.

(iii) A minimal Sullivan algebra $(\wedge V, d)$ has $(\wedge V^1, d)$ as a sub Sullivan algebra, and $H^1(\wedge V^1, d) \xrightarrow{\cong} H^1(\wedge V, d)$. Thus $(\wedge V, d)$ is 1-finite if and only if $(\wedge V^1, d)$ is 1-finite.

The power of the Sullivan condition comes in part from the Lifting Lemma below. Suppose

$$\begin{array}{ccc} & & (A, d) \\ & & \simeq \big\downarrow \eta \\ (\wedge Z, d) & \xrightarrow[\psi]{} & (C, d) \end{array}$$

is a diagram of commutative cochain algebra morphisms in which $(\wedge Z, d)$ satisfies the Sullivan condition and $\eta$ is a surjective quasi-isomorphism.

**Lemma 1.2.** (Lifting Lemma). *With the hypotheses and notation above there is a morphism $\varphi : (\wedge Z, d) \to (A, d)$ such that $\eta \circ \varphi = \psi$.*

*proof:* The same proof as given for Sullivan algebras in ([18], Lemma 12.4) applies here.

□

**Lemma 1.3.** *A Sullivan algebra $(\wedge V, d)$ admits an isomorphism*

$$(\wedge W, d) \otimes \wedge(U \oplus dU) \xrightarrow{\cong} (\wedge V, d)$$

*in which $(\wedge W, d)$ is a minimal Sullivan algebra and $d : U \xrightarrow{\cong} dU$.*

*proof:* This follows from ([18], Theorem 14.9), with $B = lk$.

□

**Proposition 1.8.** *Every commutative cochain algebra $(A, d)$ satisfying $H^0(A) = lk$ admits a quasi-isomorphism of cochain algebras,*

$$\varphi : (\wedge W, d) \to (A, d),$$

*from a minimal Sullivan algebra.*

*proof:* Proposition 12.1 in [18] provides a quasi-isomorphism $(\wedge V, d) \xrightarrow{\simeq} (A, d)$ from a Sullivan algebra $(\wedge V, d)$. Precompose this with the inclusion $(\wedge W, d) \xrightarrow{\simeq} (\wedge V, d)$ provided by Lemma 1.3. □

**Corollary 1.4.** *If $X$ is a path connected topological space then there is a quasi-isomorphism*

$$(\wedge W, d) \xrightarrow{\simeq} A_{PL}(X)$$

*from a minimal Sullivan algebra. In particular, $(\wedge W, d)$ is 1-finite if and only if $\dim H^1(X) < \infty$.*

Next, consider the commutative cochain algebra $(A_{PL})_1$ identified as $\wedge(t, dt)$ with $\deg t = 0$ in the Example of §1.4. (Since it is not generated in degrees $\geq 1$ this is not a Sullivan algebra.) This cochain algebra has two augmentations $\varepsilon_0$, $\varepsilon_1 : \wedge(t, dt) \to \Bbbk$ corresponding to the end points of an interval, and defined by

$$\varepsilon_0 : t \mapsto 0 \qquad \text{and} \quad \varepsilon_1 : t \mapsto 1\,.$$

**Definition.** (i) Two morphisms, $\varphi_0$, $\varphi_1 : (\wedge Z, d) \to (A, d)$ of commutative cochain algebras in which $(\wedge Z, d)$ satisfies the Sullivan condition are *homotopic* if there is a morphism

$$\Phi : (\wedge Z, d) \to \wedge(t, dt) \otimes (A, d)$$

such that $(\varepsilon_i \otimes id) \circ \Phi = \varphi_i$, $i = 0, 1$. The morphism $\Phi$ is a *homotopy* from $\varphi_0$ to $\varphi_1$ and we write $\varphi_0 \sim \varphi_1$.

(ii) Suppose given augmentations $\varepsilon_A$ in $(A, d)$ and $\varepsilon$ in $(\wedge Z, d)$. If $\varphi_0$ and $\varphi_1$ preserve the augmentations, then $\varphi_0$ is *based homotopic* to $\varphi_1$ with respect to $\varepsilon$ and $\varepsilon_A$ if $\Phi$ can be chosen so that

$$\Phi(\ker \varepsilon) \subset \wedge(t, dt) \otimes \ker \varepsilon_A\,.$$

In this case we write $\varphi_0 \sim_* \varphi_1$ and call $\Phi$ a *based homotopy*.

**Proposition 1.9.** *Suppose $(A, d)$ and $(\wedge Z, d)$ are commutative cochain algebras and $(\wedge Z, d)$ satisfies the Sullivan condition.*

*(i) Homotopy is an equivalence relation in the set of morphisms $(\wedge Z, d) \to (A, d)$.*

*(ii) Based homotopy is an equivalence relation in the set of augmentation preserving morphisms $(\wedge Z, d) \to (A, d)$.*

*(iii) If $\varphi_0 \sim_* \varphi_1$, then $\varphi_0 \sim \varphi_1$.*

*(iv) If $\varphi_0 \sim \varphi_1$, then $H(\varphi_0) = H(\varphi_1)$.*

*proof:* (i) follows from Lemma 1.2 by the identical argument used to prove Proposition 12.7 in [18]. That proof also proves (ii) except that $\wedge(t, dt)$ is written there on the right and the functor $A \otimes -$ must be replaced by $(\ker \varepsilon_A \otimes -) \oplus \Bbbk$.

(iii) is immediate, and (iv) is proved in the same way as Proposition 12.8(i) in [18]. □

If $\varphi : (\wedge Z, d) \to (A, d)$ is a morphism of commutative cochain algebras in which $(\wedge Z, d)$ satisfies the Sullivan condition we denote the homotopy class of $\varphi$ by $[\varphi]$. If augmentations $\varepsilon$ in $\wedge Z$ and $\varepsilon_A$ in $A$ have been fixed the corresponding based homotopy class is denoted by $[\varphi]_*$.

The set of homotopy classes of morphisms is denoted by $[(\wedge Z, d), (A, d)]$ and the set of based homotopy classes is denoted by $[(\wedge Z, d), (A, d)]_*$.

Now suppose $\eta : (A, d) \to (C, d)$ is a morphism of commutative cochain algebras. Then composition with $\eta$ trivially preserves homotopy classes of morphisms; i.e. it defines a set map

$$\eta_\# : [(\wedge Z, d), (A, d)] \to [(\wedge Z, d), (C, d)], \qquad \eta_\#[\varphi] = [\eta \circ \varphi].$$

Moreover if $\eta$ is based with respect to augmentations $\varepsilon_A$ and $\varepsilon_C$ then with respect to an augmentation $\varepsilon$ in $\wedge Z$, it defines a set map

$$\eta_* : [(\wedge Z, d), (A, d)]_* \to [(\wedge Z, d), (C, d)]_*, \qquad \eta_*[\varphi]_* = [\eta \circ \varphi]_*.$$

The Lifting Lemma now has the following critical generalization: Let

$$\begin{array}{ccc} & & (A, d) \\ & & \Big\downarrow{\scriptstyle \eta}\ \simeq \\ (\wedge Z, d) & \xrightarrow[\psi]{} & (C, d) \end{array}$$

be a diagram of commutative cochain algebra morphisms in which $(\wedge Z, d)$ satisfies the Sullivan condition and $\eta$ is a quasi-isomorphism, but in which *$\eta$ may not be surjective.*

**Proposition 1.10.** *With the hypotheses and notation above,*

$$\eta_\# : [(\wedge Z, d), (A, d)] \to [(\wedge Z, d), (C, d)]$$

*is a bijection. If, moreover, $\eta$ and $\psi$ are based with respect to augmentations $\varepsilon_A, \varepsilon_C$, and $\varepsilon$ then*

$$\eta_* : [(\wedge Z, d), (A, d)]_* \to [(\wedge Z, d), (C, d)]_*$$

*is also a bijection.*

*proof:* The first assertion follows in the same way as Proposition 12.9 in [18]. The proof of the second assertion is identical except that the functors $-\otimes A$ and $-\otimes C$ need to be replaced by $(-\otimes \ker \varepsilon_A)\oplus \Bbbk$ and $(-\otimes \ker\varepsilon_C)\oplus \Bbbk$. □

Now suppose $(\wedge V, d)$ is a Sullivan algebra. In addition to the homology algebra, $H(\wedge V, d)$, another fundamental invariant is the *space of indecomposables*

$$Q(\wedge V) = \frac{(\wedge V)^+}{(\wedge V)^+ \wedge (\wedge V)^+}\,.$$

Evidently the differential factors to give a differential $Q(d)$ in $Q(\wedge V, d)$ and a morphism $\varphi : (\wedge W, d) \to (\wedge V, d)$ between Sullivan algebras induces a morphism $Q(\varphi) : (Q(\wedge W), Q(d)) \to (Q(\wedge V), Q(d))$. In summary, $Q$ is a functor from Sullivan algebras to differential graded vector spaces.

The inclusion $V \to \wedge V$ induces an isomorphism $V \xrightarrow{\cong} Q(\wedge V)$ and we shall frequently identify these two spaces. Note that this identifies $Q(d)$ with the "linear part" of $d$.

**Lemma 1.4.** *(i) A Sullivan algebra $(\wedge V, d)$ is minimal if and only if $Q(d) = 0$.*

*(ii) If $\varphi_0 \sim_* \varphi_1 : (\wedge W, d) \to (\wedge V, d)$ are based homotopic morphisms between Sullivan algebras then $H(Q(\varphi_0)) = H(Q(\varphi_1))$. In particular if the Sullivan algebras are minimal then $Q(\varphi_0) = Q(\varphi_1)$.*

*proof:* (i) is immediate from the definition. To prove (ii) let $\Phi : (\wedge V, d) \to [\wedge(t, dt)\otimes \wedge^+ V\,]\oplus \Bbbk$ be a based homotopy from $\varphi_0$ to $\varphi_1$. Then $\Phi$ factors to give a linear map

$$(Q(\wedge V), Q(d)) \xrightarrow{\Phi} \wedge(t, dt)\otimes (\,Q(\wedge V), Q(d)\,)$$

and it follows that

$$H(Q(\varphi_1)) = H(\varepsilon_1\otimes id)\circ H(\Phi) = H(\varepsilon_0\otimes id)\circ H(\Phi) = H(Q(\varphi_0))\,.$$

□

**Example.** Consider the morphism

$$\Phi : (\wedge(x, y, z), d) \to \wedge(t, dt)\otimes(\wedge u, 0)$$

in which $x, y, z$ and $u$ have degree 1, $dz = xy$, $dx = dy = 0$, $\Phi(x) = u$, $\Phi(y) = -dt$ and $\Phi(z) = tu$. Then $\Phi$ is a homotopy

$$\varphi_0 = (\varepsilon_0\otimes id)\Phi \sim \varphi_1 = (\varepsilon_1\otimes id)\Phi : (\wedge(x, y, z), d) \to (\wedge(u, -t), 0)\,.$$

On the other hand, $Q(\varphi_0)z = 0$ and $Q(\varphi_1)z = u$. Thus $Q(\varphi_1) \neq Q(\varphi_0)$ and so $\varphi_0$ and $\varphi_1$ are not based homotopic. □

**Proposition 1.11.** *Suppose*

$$m : (\wedge V, d) \xrightarrow{\simeq} (A, d) \quad \text{and} \quad n : (\wedge W, d) \xrightarrow{\simeq} (A, d)$$

*are quasi-isomorphisms where* $(\wedge V, d)$ *and* $(\wedge W, d)$ *are minimal Sullivan algebras. Then there is a unique homotopy class of isomorphisms*

$$\varphi : (\wedge V, d) \xrightarrow{\cong} (\wedge W, d)$$

*such that* $n \circ \varphi \sim m$. *If* $(A, d_A)$ *is augmented then there is a unique based homotopy class of isomorphisms,* $\varphi$, *such that* $n \circ \varphi \sim_* m$.

*proof:* We consider only the based case. The existence of a unique based homotopy class of morphisms $\varphi$ such that $n \circ \varphi \sim_* m$ follows from Proposition 1.10, since $n$ is a quasi-isomorphism. Since $m$ is also a quasi-isomorphism we also obtain a based homotopy class of morphisms $\psi : (\wedge W, d) \to (\wedge V, d)$ such that $m \circ \psi \sim_* n$.

Since composition respects based homotopy classes, it follows that

$$n \circ \psi \circ \varphi \sim_* m \quad \text{and} \quad n \circ \varphi \circ \psi \sim_* n\,.$$

A final application of Proposition 1.10 yields $\psi\varphi \sim_* id$ and $\varphi\psi \sim_* id$. Thus by Lemma 1.4(ii) $Q(\varphi)$ and $Q(\psi)$ are inverse isomorphisms and it follows that $\varphi$ and $\psi$ are isomorphisms.

□

**Corollary 1.5.** *A quasi-isomorphism between minimal Sullivan algebras is an isomorphism.*

**Definition.** Quasi-isomorphisms $(\wedge V, d) \xrightarrow{\simeq} (A, d_A)$, $(\wedge V, d) \xrightarrow{\simeq} A_{PL}(K)$ and $(\wedge W, d) \xrightarrow{\simeq} A_{PL}(X)$ from (minimal) Sullivan algebras are *(minimal) Sullivan models* for the cdga $(A, d_A)$, the pointed simplicial set $K$ and the pointed topological space $X$.

Finally, suppose $\varphi : (A, d) \to (C, d)$ is a morphism of augmented commutative cochain algebras satisfying $H^0(A, d) = \Bbbk = H^0(C, d)$. If $m : (\wedge V, d) \xrightarrow{\simeq} (A, d)$ and $n : (\wedge W, d) \xrightarrow{\simeq} (C, d)$ are minimal Sullivan models then Proposition 1.10 yields a unique based homotopy class of morphisms

$\psi : (\wedge V, d) \to (\wedge W, d)$ making the diagram

$$\begin{array}{ccc} (\wedge V, d) & \xrightarrow{\psi} & (\wedge W, d) \\ \simeq \big\downarrow m & & \simeq \big\downarrow n \\ (A, d) & \xrightarrow[\varphi]{} & (C, d) \end{array}$$

based homotopy commute.

**Definition.** A morphism in the based homotopy class of $\psi$ is a *Sullivan representative* for the based morphism $\varphi$.

**Remark.** In view of Proposition 1.9 and Lemma 1.4 both $H(\psi)$ and $Q(\psi)$ are independent of the choice of $\psi$.

Finally let $(K, v_0)$ be a connected pointed simplicial set and let $(X, x_0)$ be a pointed path connected topological space. Then the base points define augmentations $\varepsilon_X$ and $\varepsilon_K$ in $A_{PL}(X)$ and in $A_{PL}(K)$ and $H^0(A_{PL}(K)) = \Bbbk = H^0(A_{PL}(X))$.

Propositions 1.8 and 1.11 assert the existence of such minimal Sullivan models and the uniqueness up to isomorphism. Moreover a base point preserving simplicial map $\alpha : (L, u_0) \to (K, v_0)$ and a base point preserving continuous map $f : X \to Y$ define based morphisms

$$A_{PL}(\alpha) : A_{PL}(L) \leftarrow A_{PL}(K) \qquad \text{and} \qquad A_{PL}(f) : A_{PL}(X) \leftarrow A_{PL}(Y)\,.$$

By Proposition 1.10 these lift to unique based homotopy classes of morphisms between the respective minimal Sullivan models.

**Definition.** These morphisms will be called *Sullivan representatives* respectively for the simplicial map $\alpha$ and for the continuous map $f$.

## 1.6 The simplicial and spatial realizations of a $\Lambda$-algebra

A contravariant functor

$$\langle\ ,\ \rangle : \Lambda\text{-algebras} \rightsquigarrow \text{Simplicial sets}$$

is defined as follows:

- $\langle \wedge V, d\rangle_n =$ set of morphisms $(\wedge V, d) \to (A_{PL})_n$ ;
- The face and degeneracy maps are given by

$$\partial_i(\sigma) = \partial_i \circ \sigma \qquad \text{and} \qquad s_j(\sigma) = s_j \circ \sigma\,, \qquad \sigma \in \langle \wedge V, d\rangle_n\,.$$

Further, if $\varphi : (\wedge V, d) \to (\wedge W, d)$ is a morphism of Sullivan algebras then

- $\langle\varphi\rangle : \langle\wedge V, d\rangle \leftarrow \langle\wedge W, d\rangle$ is given by $\langle\varphi\rangle(\sigma) = \sigma \circ \varphi$, for $\sigma \in \langle\wedge W, d\rangle$.

**Definition.** If $(\wedge V, d)$ is a $\Lambda$-algebra then $\langle\wedge V, d\rangle$ is its *simplicial realization* and the CW complex $|\wedge V, d| := |\langle\wedge V, d\rangle|$ is its *spatial realization.* If $\varphi : (\wedge V, d) \to (\wedge W, d)$ is a morphism of $\Lambda$-algebras then $|\langle\varphi\rangle|$ is denoted by $|\varphi|$.

Suppose $L$ is a simplicial set and $(\wedge V, d)$ is a $\Lambda$-algebra. Then both

$$\mathit{Cdga}\,((\wedge V, d), A_{PL}(L)) \qquad \text{and} \qquad \mathit{Simpl}\,(L, \langle\wedge V, d\rangle)$$

can be identified with the set of maps $\wedge V \times L \to A_{PL}$ which for fixed $\sigma \in L_n$ are cdga morphisms $(\wedge V, d) \to (A_{PL})_n$ and for fixed $\Omega \in \wedge V$ are maps $\Omega \times L \to A_{PL}$ compatible with the face and degeneracy morphisms. This, then, defines a bijection, natural in $(\wedge V, d)$ and in $L$,

$$\langle\ \rangle : \mathrm{Cdga}\,((\wedge V, d), A_{PL}(L)) \xrightarrow{\cong} \mathrm{Simpl}\,(L, \langle\wedge V, d\rangle)\,. \tag{1.6}$$

In particular if $X$ is a topological space, (1.6) reduces to

$$\langle\ \rangle : \mathrm{Cdga}\,((\wedge V, d), A_{PL}(X)) \xrightarrow{\cong} \mathrm{Simpl}\,(\mathit{Sing}\,X, \langle\wedge V, d\rangle)\,. \tag{1.7}$$

**Definition**. If $\varphi : (\wedge V, d) \to A_{PL}(L)$ is a cdga morphism from a $\Lambda$-algebra then $\langle\varphi\rangle$ and $\varphi$ are said to be *adjoint* to each other, and the spatial realization $|\langle\varphi\rangle|$ is denoted simply by

$$|\varphi| : |L| \to |\wedge V, d|\,.$$

**Note:** We use the notation $\langle\varphi\rangle$ and $|\varphi|$ in two different contexts, once when $\varphi$ is a morphism of $\Lambda$-algebras and once when it maps a $\Lambda$-algebra to the polynomial differential forms on a simplicial set. In particular, given morphisms

$$\langle\wedge V, d\rangle \xrightarrow{\psi} \langle\wedge W, d\rangle \xrightarrow{\varphi} A_{PL}(L)$$

we have that

$$|\varphi \circ \psi| = |\psi| \circ |\varphi| : |L| \to |\wedge V, d|\,.$$

Recall from Proposition 1.5 that if $(X, *)$ is a CW complex then $\lambda_X : (|\,\mathit{Sing}\,X|, *) \to (X, *)$ is a homotopy equivalence. For each such $X$ we fix a homotopy inverse $\nu_X : (X, *) \to |\,\mathit{Sing}\,X, *|$ and introduce the following

**Notation:** If $\varphi : (\wedge V, d) \to A_{PL}(X)$ is a morphism from a $\Lambda$-algebra to the polynomial forms on a CW complex $(X, *)$ then we write

$$||\varphi|| = |\varphi| \circ \nu_X : (X, *) \to |\wedge V, d|\,. \tag{1.8}$$

Next, note that morphisms from Sullivan algebras $\varphi : (\wedge V, d) \to (A_{PL})_n$ and $\psi : (\wedge W, d) \to (A_{PL})_n$ determine the morphism

$$\varphi \wedge \psi : (\wedge V, d) \otimes (\wedge W, d) \to (A_{PL})_n$$

given by $(\varphi \wedge \psi)(\alpha \otimes \beta) = \varphi(\alpha) \wedge \psi(\beta)$.

**Proposition 1.12.** *Suppose $(\wedge V, d)$ and $(\wedge W, d)$ are $\Lambda$-algebras. Then*

*(i) If $V^0 = 0$ then $\langle \wedge V, d\rangle$ has a single vertex, $x$, defined by the unique augmentation $(\wedge V, d) \to \Bbbk$.*
*(ii) The correspondence $(\varphi, \psi) \mapsto \varphi \wedge \psi$ defines a natural simplicial isomorphism*

$$\langle \wedge V, d\rangle \times \langle \wedge W, d\rangle \xrightarrow{\cong} \langle\, (\wedge V, d) \otimes (\wedge W, d)\,\rangle\,.$$

*(iii) $\langle \wedge V, d\rangle$ is a Kan simplicial set.*

*proof:* (i) and (ii) are immediate. To prove (iii), let $L \subset \partial\Delta[n]$, some $n \geq 1$, be the simplicial set obtained by omitting a single face of $\Delta[n]$. Then, using (1.6), identify a simplicial map $\varphi : L \to \langle \wedge V, d\rangle$ with a cdga morphism $(\wedge V, d) \to A_{PL}(L)$.

By Proposition 1.6(iii), $A_{PL}(\Delta[n]) \to A_{PL}(L)$ is surjective. Because $|L|$ is contractible to the vertex in $\Delta[n]$ opposite to the missing face, it follows that $H^*(\Delta[n]) = \Bbbk = H^*(L)$. Thus by Proposition 1.7, $A_{PL}(\Delta[n]) \to A_{PL}(L)$ is also a quasi-isomorphism.

Now it follows from the Lifting Lemma 1.2 (p. 19) that $\varphi$ lifts to a morphism $(\wedge V, d) \to A_{PL}(\Delta[n])$. This is identified by (1.6) with a simplicial map $\Delta[n] \to \langle \wedge V, d\rangle$, and this is the desired extension of $\varphi$.

□

Suppose $(\wedge V, d)$ is a $\Lambda$-algebra. Adjoint to the identity map of $\langle \wedge V, d\rangle$ is the natural morphism

$$m_{\langle \wedge V, d\rangle} : (\wedge V, d) \to A_{PL}(\langle \wedge V, d\rangle)\,. \tag{1.9}$$

If $L$ is a simplicial set and $\varphi : (\wedge V, d) \to A_{PL}(L)$ is a morphism, then adjoint to $\varphi$ is a simplical map $\langle \varphi\rangle : L \to \langle \wedge V, d\rangle$, and it follows from the definitions that

$$\varphi = A_{PL}(\langle \varphi\rangle) \circ m_{\langle \wedge V, d\rangle}\,. \tag{1.10}$$

In general $m_{\langle \wedge V, d\rangle}$ may not be a quasi-isomorphism. However, we have

**Proposition 1.13.** *Suppose $\varphi : (\wedge V, d) \xrightarrow{\simeq} A_{PL}(L)$ is a Sullivan model of a simplicial set $L$. Then $m_{\langle \wedge V, d\rangle}$ has a homotopy left inverse.*

*proof:* Let $\psi : (\wedge W, d) \to A_{PL}(\langle \wedge V, d\rangle)$ be a Sullivan model. By Proposition 1.10 (p. 21) there are morphisms

$$\chi_1 : (\wedge V, d) \to (\wedge W, d) \quad \text{and} \quad \chi_2 : (\wedge W, d) \to (\wedge V, d)$$

such that $\psi \circ \chi_1 \sim m_{\langle \wedge V, d\rangle}$ and $\varphi \circ \chi_2 \sim A_{PL}(\langle \varphi \rangle) \circ \psi$. Then $\varphi \circ \chi_2 \circ \chi_1 \sim A_{PL}(\langle \varphi \rangle) \circ m_{\langle \wedge V, d\rangle} = \varphi$, and so $\chi_2 \circ \chi_1 \sim id_{\wedge V}$.

□

**Corollary 1.6.** *If $\varphi : (\wedge V, d) \xrightarrow{\simeq} A_{PL}(X)$ is a Sullivan model of a topological space $X$ then $m_{\langle \wedge V, d\rangle}$ has a left homotopy inverse. In particular $H(X)$ is a natural retract of $H^*(|\wedge V, d|)$.*

*proof:* By definition $A_{PL}(X) = A_{PL}(Sing\, X)$. □

**Proposition 1.14.** *Let $\varphi : (\wedge V, d) \to (\wedge W, d)$ be a Sullivan representative for a continuous map $f : (X, *) \to (Y, *)$. Then $H(f)$ is an isomorphism if and only if $\langle \varphi \rangle : \langle \wedge W, d\rangle \to \langle \wedge V, d\rangle$ is a homotopy equivalence.*

*proof:* If $H(f)$ is an isomorphism, then $X$ and $Y$ have the same minimal Sullivan model and we can choose $\varphi = id_{(\wedge V, d)}$.

Conversely from the naturality property of Corollary 1.6, we have the commutative diagram

$$\begin{array}{ccc}
H(X) & \underset{r_X}{\overset{i_X}{\rightleftarrows}} & H(\langle \wedge W, d\rangle) \\
{\scriptstyle H(f)}\uparrow & & \uparrow{\scriptstyle H(\langle \varphi \rangle)} \\
H(Y) & \underset{r_Y}{\overset{i_Y}{\rightleftarrows}} & H(\langle \wedge V, d\rangle)
\end{array}$$

with $r_X \circ i_X = id_{H(X)}$ and $r_Y \circ i_Y = id_{H(Y)}$. Since $H(\langle \varphi \rangle)$ is an isomorphism, $r_Y \circ H(\langle \varphi \rangle)^{-1} \circ i_X$ is inverse to $H(f)$.

□

If $(\wedge V, d)$ is a $\Lambda$-algebra then, since $\langle \wedge V, d\rangle$ is a Kan simplicial set, it follows from Proposition 1.5(i) (p. 13) that the natural map

$$\lambda_{\langle \wedge V, d\rangle} : \langle \wedge V, d\rangle \to Sing\, |\wedge V, d|$$

has a based homotopy inverse. Since $A_{PL}|\wedge V,d| = A_{PL}(Sing\,|\wedge V,d|)$ this implies that the corresponding surjective morphism

$$A_{PL}(\lambda_{\langle\wedge V,d\rangle}) : A_{PL}\langle\wedge V,d\rangle \leftarrow A_{PL}|\wedge V,d|$$

is a quasi-isomorphism. In particular, (Proposition 1.10, p. 21) $m_{\langle\wedge V,d\rangle}$ lifts through the surjective quasi-isomorphism $A_{PL}(\lambda_{\langle\wedge V,d\rangle})$ to a unique based homotopy class of morphisms

$$m_{|\wedge V,d|} : (\wedge V,d) \to A_{PL}|\wedge V,d|\,. \tag{1.11}$$

Moreover, if $\psi : (\wedge V,d) \to (\wedge W,d)$ is a morphism of augmented Sullivan algebras, then the diagram

$$\begin{array}{ccccc}
(\wedge V,d) & \xrightarrow{m_{\langle\wedge V,d\rangle}} & A_{PL}\langle\wedge V,d\rangle & \xleftarrow{\simeq} & A_{PL}|\wedge V,d| \\
\Big\downarrow{\scriptstyle\psi} & & \Big\downarrow{\scriptstyle A_{PL}(\langle\psi\rangle)} & & \Big\downarrow{\scriptstyle A_{PL}(|\psi|)} \\
(\wedge W,d) & \xrightarrow{m_{\langle\wedge W,d\rangle}} & A_{PL}\langle\wedge W,d\rangle & \xleftarrow{\simeq} & A_{PL}|\wedge W,d|
\end{array} \tag{1.12}$$

commutes. Thus by Proposition 1.10 (p. 21)

$$A_{PL}(|\psi|)\circ m_{|\wedge V,d|} \sim_* m_{|\wedge W,d|}\circ\psi\,.$$

**Caution.** While $m_{\langle\wedge V,d\rangle}$ is a quasi-isomorphism if and only if $m_{|\wedge V,d|}$ is, this will in general *not* be the case. (If $X$ is simply connected and $H(X)$ has finite type then Theorem 5.4, p. 166, asserts that $m_{|\wedge V,d|}$ is a quasi-isomorphism.) However, if $m_{\langle\wedge V,d\rangle}$ and $m_{\langle\wedge W,d\rangle}$ are quasi-isomorphisms then the diagram above exhibits $\psi$ as a Sullivan representative for $|\psi|$.

Next, suppose $X$ is a path connected based topological space. Since $Sing\,X$ is a Kan simplicial set (Lemma 1.1, p. 11) it follows from Proposition 1.5 (p. 13) that $\lambda_{Sing\,X} : Sing\,X \to Sing\,|Sing\,X|$ is a based homotopy equivalence. Thus if $\varphi : (\wedge V,d) \to A_{PL}(X)$ is an augmentation preserving morphism from an augmented Sullivan algebra then we have the commutative diagram

$$\begin{array}{ccccc}
(\wedge V,d) & \xrightarrow{m_{\langle\wedge V,d\rangle}} & A_{PL}(\langle\wedge V,d\rangle) & \xleftarrow[A_{PL}(\lambda_{\langle\wedge V\rangle})]{\simeq} & A_{PL}(|\wedge V,d|) \\
\Big\downarrow{\scriptstyle\varphi} & & \Big\downarrow{\scriptstyle A_{PL}(\langle\varphi\rangle)} & & \Big\downarrow{\scriptstyle A_{PL}(|\varphi|)} \\
A_{PL}(X) & \xrightarrow{=} & A_{PL}(Sing\,X) & \xleftarrow[A_{PL}(\lambda_{Sing\,X})]{\simeq} & A_{PL}(|Sing\,X|)\,.
\end{array} \tag{1.13}$$

Now recall from Proposition 1.5 (p. 13) that $\lambda_X : |Sing\,X| \to X$ is a weak homotopy equivalence.

**Proposition 1.15.** *With the hypotheses and notation above,*

$$A_{PL}(|\varphi|)\circ m_{|\wedge V,d|} \sim_* A_{PL}(\lambda_X)\circ\varphi : (\wedge V,d)\to A_{PL}|Sing\,X|\,.$$

*In particular, if $X$ is a CW complex,*

$$A_{PL}(||\varphi||)\circ m_{|\wedge V,d|} \sim_* \varphi\,,$$

*where $||\varphi||$ is defined in (1.8).*

*proof:* This follows from (1.13) and the fact that $Sing(\lambda_X)\circ\lambda_{Sing\,X} = id_{Sing\,X}$.

□

**Definition.** (i) A continuous map, $f$, between path connected topological spaces is a *rational homology equivalence* if $H^*(f;\mathbb{Q})$ is an isomorphism. (By Corollary 1.5, p. 23, this is equivalent to requiring that a Sullivan representative of $A_{PL}(f)$ between minimal Sullivan models be an isomorphism.)

(ii) Two path connected topological spaces have the *same rational homology type* if they are connected by a finite chain of rational homology equivalences.

(iii) Two path connected topological spaces are *rationally homologically equivalent* if their minimal Sullivan models are isomorphic.

**Remark:** If a minimal Sullivan model $\varphi : (\wedge V,d)\to A_{PL}(X)$ satisfies $m_{|\wedge V,d|}$ is a quasi-isomorphism, then it follows from Proposition 1.15 that $X$ is connected to $|\wedge V,d|$ by a chain a rational homology equivalences. Thus two such spaces are rationally homologically equivalent if and only if they are connected by a chain of rational homology equivalences.

However, this may not always be the case for non-simply connected spaces. Thus while "same rational homology type" always implies "rational homology equivalence" the cautionary note above suggests that the reverse implication may not be true for non-simply connected spaces.

Finally, recall from Proposition 1.7 that $\oint_K$ and $\oint_X$ induce natural isomorphisms $H(A_{PL}(K))\xrightarrow{\cong} H^*(K)$ and $H(A_{PL}(X))\xrightarrow{\cong} H^*(X)$ where $K$ and $X$ are respectively a simplicial set and a topological space. In particular, we obtain pairings

$$\langle\,,\,\rangle : H(A_{PL}(K))\times H_*(K)\to I\!k \qquad\text{and } \langle\,,\,\rangle : H(A_{PL}(X))\times H_*(X)\to I\!k\,,$$

where $H_*(K)$ and $H_*(X)$ denote simplicial homology and singular homology respectively.

**Proposition 1.16.** *Suppose* $\varphi : (\wedge V, d) \to A_{PL}(L)$ *is a morphism from a* $\Lambda$*-algebra to the forms on a simplicial set* $L$*. Then*

$$\langle H(m_{\langle \wedge V,d\rangle})\alpha, H_*(\langle \varphi \rangle)\beta\rangle = \langle H(\varphi)\alpha, \beta\rangle\,, \qquad \alpha \in H(\wedge V, d), \beta \in H_*(L)\,.$$

*proof:* For simplicity, denote $\langle \wedge V, d\rangle$ simply by $K$. Then the Proposition follows via passage to homology from the identity

$$\langle \oint_K (m_{\langle \wedge V,d\rangle}\Phi), \langle\varphi\rangle\sigma\rangle = \langle \oint_L (\varphi\Phi), \sigma\rangle\,, \qquad \Phi \in (\wedge V)^n, \sigma \in L_n, n \geq 0\,.$$

But by definition,

$$\begin{aligned}\langle \oint_K (m_{\langle \wedge V,d\rangle}\Phi), \langle\varphi\rangle\sigma\rangle &= \oint_n ((m_{\langle \wedge V,d\rangle}\Phi)_{\langle\varphi\rangle\sigma}) \\ &= \oint_n (\varphi\Phi)_\sigma = \langle \oint_L (\varphi\Phi), \sigma\rangle\,.\end{aligned}$$

□

**Corollary 1.7.** *Suppose* $\varphi : (\wedge V, d) \to A_{PL}(X)$ *is a morphism to the forms on a topological space* $X$*. Then*

$$\langle H(m_{|\wedge V,d|})\alpha, H_*(|\varphi|)\beta\rangle = \langle H(\varphi)\alpha, \beta\rangle\,, \qquad \alpha \in H(\wedge V, d), \beta \in H_*(X)\,.$$

*proof:* This follows at once from the Proposition together with (1.13).

□

## 1.7 Homotopy and based homotopy

To describe how homotopy in each of the categories of topological spaces or simplicial sets is related to that of Sullivan algebras, we first recall the respective notion of "interval" in each category, and how these are related. Here we recall and add to terminology to be used throughout the rest of the book:

- $I = [0,1]$ is the *classical unit interval* with initial end point 0 and terminal end point 1.
- $\Delta[1]$ is the *simplicial interval* with initial vertex $e_0 = \partial_1 c_1$ and terminal vertex $e_1 = \partial_0 c_1$.
- $\wedge(t, dt)$ plays the role of interval in the cdga category (here $\deg t = 0$) with initial and terminal augmentations

$$\varepsilon_0 : t \mapsto 0 \quad \text{and} \quad \varepsilon_1 : t \mapsto 1\,.$$

- As observed in §1.4 (p. 18), we identify

$$|\,\Delta[1], e_0, e_1\,| \xrightarrow{\cong} (I, 0, 1)$$

via $|\Delta[1]| = \Delta^1$.
- As noted in the Example of §1.4, an isomorphism

$$(\,\wedge(t, dt), \varepsilon_0, \varepsilon_1\,) \xrightarrow{\cong} A_{PL}(\Delta[1], e_0, e_1)$$

is defined by $t \mapsto t_1$.

Let $L$ be a simplicial set and denote by $\varepsilon_L$ the augmentation in $A_{PL}(L)$ determined by a base vertex, $u_0$. Let $\pi_L : \Delta[1]\times L \to L$ and $\pi_\Delta : \Delta[1]\times L \to \Delta[1]$ be the projections. Then, since $\pi_L$ is a homotopy equivalence, it follows from Proposition 1.7 that a quasi-isomorphism

$$\gamma : \wedge(t, dt) \otimes A_{PL}(L) \xrightarrow{\simeq} A_{PL}(\Delta[1] \times L) \tag{1.14}$$

is given by $\beta \otimes \alpha \mapsto A_{PL}(\pi_\Delta)\beta \wedge A_{PL}(\pi_L)\alpha$. This restricts to a based quasi-isomorphism

$$\gamma_* : (\wedge(t, dt) \otimes \ker \varepsilon_L) \oplus k\!\!k \xrightarrow{\simeq} A_{PL}(\Delta[1] \times L, \Delta[1] \times u_0) \oplus k\!\!k\,. \tag{1.15}$$

**Proposition 1.17.** *(i) Suppose $\varphi : L \to K$ is a simplicial map of connected simplicial sets and that $K$ is a Kan simplicial set. Then the homotopy class of a Sullivan representative of $\varphi$ depends only on the homotopy class of $\varphi$.*

*(ii) If $\varphi$ is base point preserving then the based homotopy class of a Sullivan representative depends only on the based homotopy class of $\varphi$.*

*proof:* As always, we provide only the proof in the based case, noting that the proof in the non-based case is essentially the same. Let

$$\Phi : (\Delta[1] \times L, \Delta[1] \times u_0) \to (K, v_0)$$

be a based simplicial homotopy from $\varphi_0$ to $\varphi_1$, and let $m : (\wedge V, d) \to A_{PL}(K)$ and $n : (\wedge W, d) \to A_{PL}(L)$ be minimal Sullivan models. These morphisms fit in the diagram

$$\begin{array}{ccc} & & (\wedge(t, dt) \otimes (\wedge^+ W, d)) \oplus k\!\!k \\ & & \simeq \big\downarrow \gamma_*\circ(id\otimes n) \\ (\wedge V, d) & \xrightarrow{A_{PL}(\Phi)\circ m} & A_{PL}(\Delta[1] \times L, \Delta[1] \times u_0) \oplus k\!\!k \end{array}$$

in which the vertical arrow is a quasi-isomorphism. Thus Proposition 1.10 yields a morphism

$$\lambda : (\wedge V, d) \to \big(\wedge(t, dt) \otimes (\wedge^+ W, d)\big) \oplus \Bbbk$$

which makes the diagram homotopy commute.

Then for $i = 0, 1$ we have

$$n \circ (\varepsilon_i \otimes id) \circ \lambda = A_{PL}(j_i) \circ \gamma \circ (id \otimes n) \circ \lambda \sim_* A_{PL}(j_i) \circ A_{PL}(\Phi) \circ m$$

and so

$$n \circ (\varepsilon_i \otimes id) \circ \lambda \sim_* A_{PL}(\varphi_i) \circ m\,.$$

This identifies $(\varepsilon_i \otimes id) \circ \lambda$ as a Sullivan representative of $\varphi_i$.

On the other hand, a quasi-isomorphism

$$\tau : (\wedge W, d) \xrightarrow{\simeq} \big(\wedge(t, dt) \otimes (\wedge^+ W, d)\big) \oplus \Bbbk$$

is given by $\tau(1) = 1$ and $\tau(w) = 1 \otimes w$. Thus it follows from Proposition 1.10 that there is a morphism $\mu : (\wedge V, d) \to (\wedge W, d)$ such that $\tau \circ \mu \sim_* \lambda$. In particular for $i = 0, 1$ we have

$$(\varepsilon_i \otimes id) \circ \lambda \sim_* (\varepsilon_i \otimes id) \circ \tau \circ \mu = \mu\,.$$

This shows that $(\varepsilon_0 \otimes id) \circ \lambda \sim_* (\varepsilon_1 \otimes id) \circ \lambda$ as desired.

□

**Corollary 1.8.** *Suppose $f : (X, *) \to (Y, *)$ is a (base point preserving) continuous map between path connected topological spaces. Then the (based) homotopy class of a Sullivan representative of $f$ depends only on the (based) homotopy class of $f$.*

*proof:* By definition a Sullivan representative of $f$ is just a Sullivan representative of $Sing\, f$. But if $f \sim_* g$ then according to Proposition 1.4(i), $Sing\, f \sim_* Sing\, g$. Thus the Corollary follows from Proposition 1.17.

□

**Proposition 1.18.** *Suppose $m : (\wedge V, d) \to A_{PL}(X)$ and $n : (\wedge W, d) \to A_{PL}(Y)$ are minimal Sullivan models of pointed CW complexes. If $\psi : (\wedge W, d) \to (\wedge V, d)$ is a Sullivan representative of a continuous based map $f : (X, *) \to (Y, *)$ then*

$$||\psi|| \circ |m| \sim_* |n| \circ f : (X, *) \to (|\wedge W, d|, *)\,.$$

*proof:* This follows from formula (1.12) together with the relations used in the proof of Proposition 1.15.

□

**Theorem 1.1.** *Suppose $(\wedge V, d)$, $(L, *)$ and $(X, *)$ are respectively a Sullivan algebra, a pointed simplicial set and a pointed topological space. Then,*

*(i) Two morphisms $(\wedge V, d) \to A_{PL}(L)$ are (based) homotopic if and only if their adjoints are.*

*(ii) The correspondence $\varphi \mapsto |\varphi|$ induces natural bijections,*

$$[(\wedge V, d), A_{PL}(L)]_* \xrightarrow{\cong} [(|L|, *), (|\wedge V, d|, *)] \cong [(L, *), (\langle \wedge V, d \rangle, *)]$$

*and*

$$[(\wedge V, d), A_{PL}(X)]_* \xrightarrow{\cong} [(|\,Sing\, X\,|, *), (|\wedge V, d|, *)]$$

*where the augmentations of $A_{PL}(L)$ and $A_{PL}(X)$ are determined by their base points.*

*(iii) If $X$ is a CW complex the correspondence $\varphi \mapsto ||\varphi||$ induces natural bijections*

$$[(\wedge V, d), A_{PL}(X)]_* \xrightarrow{\cong} [(X, *), (|\wedge V, d|, *)]\,.$$

*proof:* (i) Suppose that $\langle \varphi_0 \rangle \sim_* \langle \varphi_1 \rangle : (L, *) \to \langle \wedge V, d \rangle$ and let

$$\Phi : (\Delta[1] \times L, \Delta[1] \times *) \to \langle \wedge V, d \rangle$$

be a based homotopy. Then $\Phi = \langle \Psi \rangle$, with

$$\Psi : (\wedge V, d) \to A_{PL}\,(\Delta[1] \times L, \Delta[1] \times *) \oplus \Bbbk\,.$$

Use the quasi-isomorphism $\gamma_*$ (cf (1.15) above),

$$\gamma_* : (\wedge(t, dt) \otimes \ker \varepsilon_L) \oplus \Bbbk \xrightarrow{\simeq} A_{PL}\,(\Delta[1] \times L, \Delta[1] \times *) \oplus \Bbbk$$

to obtain via Proposition 1.10 a morphism

$$\lambda : (\wedge V, d) \to (\wedge(t, dt) \otimes \ker \varepsilon_L) \oplus \Bbbk$$

such that $\gamma_* \circ \lambda \sim_* \Psi$. A straightforward check shows that $\lambda$ is a based homotopy from $\varphi_0$ to $\varphi_1$.

Conversely, suppose $\varphi_0 \sim_* \varphi_1 : (\wedge V, d) \to A_{PL}(L, *) \oplus \Bbbk = \ker \varepsilon_L \oplus \Bbbk$. Then there is a based homotopy from $\varphi_0$ to $\varphi_1$,

$$\Psi : (\wedge V, d) \to (\wedge(t, dt) \otimes \ker \varepsilon_L) \oplus \Bbbk\,.$$

Thus

$$\gamma_* \circ \Psi : (\wedge V, d) \to A_{PL}(\Delta[1] \times L, \Delta[1] \times *) \oplus \Bbbk \subset A_{PL}(\Delta[1] \times L)\,.$$

A straightforward check shows that $\langle \gamma_* \circ \Psi \rangle : (\Delta[1] \times L, \Delta[1] \times *) \to (\langle \wedge V, d \rangle, *)$ is a based homotopy from $\langle \varphi_0 \rangle$ to $\langle \varphi_1 \rangle$.

(ii) Assertion (i) shows that $\varphi \mapsto \langle \varphi \rangle$ induces a bijection

$$[(\wedge V, d), A_{PL}(L)]_* \xrightarrow{\cong} [(L, u_0), (\langle \wedge V, d \rangle, *)]\,.$$

On the other hand (Proposition 1.12), since $\langle \wedge V, d \rangle$ is a Kan simplicial set, Proposition 1.5(iii) asserts that the correspondence $\psi \mapsto |\psi|$ induces a bijection

$$[(L, *), (\langle \wedge V, d \rangle, *)] \xrightarrow{\cong} [(|L|, *), (| \wedge V, d|, *)]\,.$$

The second bijection is the special case of the first with $L = Sing\, X$.

(iii) This is immediate from (ii).

□

**Theorem 1.2.** *Let $f, g : (K, *) \to (L, *)$ be base point preserving, based homotopic, simplicial maps between pointed simplicial Kan complexes, then their Sullivan representatives $\lambda_f, \lambda_g : (\wedge V, d) \to (\wedge W, d)$ are based homotopic.*

*proof:* Consider the diagram

$$\begin{array}{ccc} A_{PL}(L) & \overset{A_{PL}(f)}{\underset{A_{PL}(g)}{\rightrightarrows}} & A_{PL}(K) \\ m \uparrow & & \uparrow n \\ (\wedge V, d) & \overset{\lambda_f}{\underset{\lambda_g}{\rightrightarrows}} & (\wedge W, d) \end{array}$$

where $m$ and $n$ are quasi-isomorphisms. We denote by $\theta : (L, *) \to (\langle \wedge V, d \rangle), *)$ the map corresponding by adjunction to $m$. The classes $[n\lambda_f], [n\lambda_g] \in [((\wedge V, d), *), (A_{PL}(K), *)]_*$ correspond by adjunction to the classes $[\theta \circ f]$ and $[\theta \circ g] \in [(K, *), (\langle \wedge V, d \rangle, *)]_*$ (Theorem 1.1). Since $f \sim_* g$, $\theta \circ f \sim_* \theta \circ g$ and $n\lambda_f \sim_* n\lambda_g$. Since $\psi$ is a quasi-isomorphism, it follows from Proposition 1.10 that $\lambda_f \sim_* \lambda_g$. □

**Corollary 1.9.** *Let $f, g : X \to Y$ be base point preserving continuous maps that are based homotopic, then their representatives are also based homotopic.*

## 1.8 The homotopy groups of a minimal Sullivan algebra

Recall from §1.5 that $Q(\wedge V) = \wedge^+ V / \wedge^+ V \wedge \wedge^+ V$ denotes the space of indecomposables of a minimal Sullivan algebra $(\wedge V, d)$. In §2.1 we shall define a graded Lie algebra $L_{(\wedge V,d)}$ with $\left(L_{(\wedge V,d)}\right)_k \cong (Q^{k+1}(\wedge V, d))^{\#}$, to be called the *homotopy Lie algebra* of $(\wedge V, d)$, and the component $L_0 = \left(L_{(\wedge V,d)}\right)_0$ to be called the *fundamental Lie algebra* of $(\wedge V, d)$.

**Definition.** If $(\wedge V, d)$ is a minimal Sullivan algebra we define

$$\pi_n(\wedge V, d) = Q^n(\wedge V, d)^{\#}\,, \qquad n \geq 2\,.$$

These groups are the *higher homotopy groups* of $(\wedge V, d)$.

To extend the definition of homotopy groups to the case $n = 1$ we need to restrict to 1-finite minimal Sullivan algebras $(\wedge V, d)$; i.e. to those satisfying $\dim H^1(\wedge V, d) < \infty$. In §2.7 we shall construct a group $G_L$ associated with the fundamental Lie algebra $L_0$ of a 1-finite minimal Sullivan model $(\wedge V, d)$, together with inverse bijections

$$L_0 \underset{\log_{L_0}}{\overset{\exp_{L_0}}{\rightleftarrows}} G_L\,.$$

$G_L$ will be called the *fundamental group* of $(\wedge V, d)$, and we set

$$\pi_1(\wedge V, d) = G_L\,.$$

Thus when $n = 1$, $\exp_{L_0}$ is a bijection from $Q^1(\wedge V)^{\#}$ to $G_L$.

If $\psi : (\wedge V, d) \to (\wedge W, d)$ is a morphism of minimal Sullivan algebras we write

$$\pi_n(\psi) = (Q^n(\psi))^{\#}\,, \qquad n \geq 2\,,$$

and define $\pi_1(\psi) : G_{L_{\wedge W}} \to G_{L_{\wedge V}}$ to be the map corresponding to $(Q^1(\psi))^{\#}$ when the Sullivan algebras are 1-finite.

We shall also denote the pairing between $Q(\wedge V)$ and $(Q(\wedge V))^{\#}$ by

$$\langle v, f \rangle\,, \qquad v \in V^n, f \in (Q^n(\wedge V))^{\#}\,.$$

Now consider a morphism

$$\varphi : (\wedge V, d) \to A_{PL}(X)\,,$$

from a minimal Sullivan algebra to the forms on a based topological space $(X, *)$. We shall construct set maps

$$\pi_n(\varphi) : \pi_n(X) \to \pi_n(\wedge V, d)\,, \qquad n \geq 1\,.$$

Indeed, if $\sigma : (S^n, *) \to (X, *)$ represents $\alpha \in \pi_n(X, *)$ and if $m(n) : (\wedge W, d) \to A_{PL}(S^n)$ is a minimal Sullivan model, then $A_{PL}(\sigma) \circ \varphi$ is based homotopic to a morphism

$$\chi : (\wedge V, d) \to (\wedge W, d)$$

whose based homotopy class depends only on $\alpha$ and the based homotopy class of $m(n)$. Moreover, $(\wedge W, d)$ has the form

$$(\wedge W, d) = \begin{cases} (\wedge w, 0)\,, & \text{if } n \text{ is odd, and} \\ (\wedge(w, u), du = w^2)\,, & \text{if } n \text{ is even}\,. \end{cases}$$

Further, we fix the based homotopy class of $m(n)$ simply by requiring that $z = m(n)(w)$ represent the fundamental cohomology class of $H^n(S^n)$ (cf. §1.4, p. 20).

Thus with these conventions, it follows from Lemma 1.4(ii) (p. 22) that a linear map $f \in (Q^n(\wedge V))^\#$ depending only on $\alpha$ is defined by

$$\chi(v) = \langle v, f\rangle\, z\,, \qquad v \in V^n\,.$$

**Definition.** The set maps

$$\pi_n(\varphi) : \pi_n(X) \to \pi_n(\wedge V, d)$$

determined by $\varphi$ are defined by

$$\pi_n(\varphi)(\alpha) = \begin{cases} f\,, & n \geq 2 \\ \exp_{L_0} f\,, & n = 1\,, \end{cases} \tag{1.16}$$

where $\pi_1(\varphi)$ is only defined if $(\wedge V, d)$ is 1-finite.

Recall that if $\overline{c_n}$ is the fundamental homology class of $S^n$ (§1.4) then $\langle \overline{c_n}, z\rangle = 1$. It follows that

$$\langle \overline{c_n}, \varphi(v)\rangle = \langle v, \pi_n(\varphi)\alpha\rangle\,, \qquad v \in V^n, n \geq 2\,. \tag{1.17}$$

It is immediate from the construction that the maps $\pi_n(\varphi)$ are natural with respect to morphisms of minimal Sullivan algebras and with respect to based maps between topological spaces.

Now in the case of $m_{|\wedge V,d|}$ the correspondences $\alpha \mapsto f$ constructed above are set maps

$$\iota_n : \pi_n(|\wedge V, d|, *) \to (Q^n(\wedge V))^\#\,,$$

and

$$\pi_n(m_{|\wedge V,d|}) = \begin{cases} \iota_n\,, & n \geq 2 \\ \exp_{L_0} \circ \iota_1\,, & n = 1\,, \end{cases} \tag{1.18}$$

the case $n = 1$ requiring that $(\wedge V, d)$ be 1-finite.

If $\varphi : (\wedge V, d) \to A_{PL}(X)$ is a morphism from a minimal Sullivan algebra, then $\varphi$ decomposes in the form (see (1.10), p. 26)

$$\varphi = A_{PL}(\langle\varphi\rangle) \circ m_{\langle\wedge V,d\rangle} .$$

Thus by Proposition 1.15 (p. 29) we have

$$A_{PL}(\lambda_X) \circ \varphi \sim_* A_{PL}(|\varphi|) \circ m_{|\wedge V,d|} : (\wedge V, d) \to A_{PL}(|\, Sing\, X|) ,$$

where $\lambda_X : |Sing\, X| \to X$ is defined in (1.3), p. 12. We deduce that

$$\pi_n(\varphi) = \pi_n(m_{|\wedge V,d|}) \circ \pi_n(|\varphi|) : \pi_n(X) \to \pi_n(\wedge V, d) ,$$

where we have identified $\pi_n(X)$ with $\pi_n(|Sing\, X|)$ via the weak homotopy equivalence $\lambda_X$.

The principal Theorems in this section are Theorems 1.3 and 1.4 below. However, we shall only prove Theorem 1.3 and Theorem 1.4(i) here; the proof of Theorem 1.4(ii) is deferred to Theorem 2.4 (p. 75).

**Theorem 1.3.** *If $(\wedge V, d)$ is a minimal Sullivan algebra then the set maps $\iota_n, n \geq 1$, are natural bijections.*

**Theorem 1.4.** *(i) When $n \geq 2$, the bijections $\iota_n$ are isomorphisms of abelian groups.*

*(ii) If $H^1(\wedge V, d)$ is finite dimensional then $\exp_L \circ \iota_1$ is an isomorphism of groups.*

**Corollary 1.10.** *If $\varphi : (\wedge V, d) \to A_{PL}(X)$ is a morphism from a minimal Sullivan algebra to the forms on a topological space then $\pi_n(\varphi)$, $n \geq 2$ is a linear map. If $dim\, H^1(\wedge V, d) < \infty$ then $\pi_1(\varphi)$ is a morphism of groups.*

**Remarks.**

(1) When $V^1 = 0$ and $V$ has finite type, Theorem 1.4 is just [18], Theorem 17.10(i).

(2) When $n \geq 2$ it is immediate from the definition that if $\psi : (\wedge V, d) \to (\wedge W, d)$ is a morphism of Sullivan algebras then $\pi_n(\psi)$ is a homomorphism. When $n = 1$, this follows from Theorem 2.4 (p. 75).

*proof of Theorem 1.3.* The naturality is immediate from Proposition 1.18 (p. 32). To prove that $\iota_n$ are bijections we shall construct bijections

$$\tau_n : (Q^n(\wedge V))^\# \to \pi_n(|\wedge V, d|, *) ,$$

and show they are inverse to the morphisms $\iota_n$. For this, let $f \in (Q(\wedge V))^{\#}$. Recall from the Example of §1.4 that

$$A_{PL}(\Delta[n]/\partial\Delta[n]) = \mathbb{k} \oplus \{\Phi \in (A_{PL})_n \,|\, \partial_i \Phi = 0\,, \quad 0 \le i \le n\}\,.$$

In particular, $A^k_{PL}(\Delta[n]/\partial\Delta[n]) = 0$, $k > n$, and so $f$ determines the morphism

$$\varphi_f : (\wedge V, d) \to A_{PL}(\Delta[n]/\partial\Delta[n])$$

given by $\varphi_f(v) = (-1)^n \langle v, f\rangle\, dt_1 \wedge \cdots \wedge dt_n$.

Then recall from §1.6 that the adjoint of $\varphi_f$ is a based simplicial map $\langle \varphi_f\rangle : \Delta[n]/\partial\Delta[n] \to \langle \wedge V, d\rangle$. This realizes to a continuous map

$$|\varphi_f| : (S^n, *) \to |(|\wedge V, d|, *)\,,$$

since, as observed in the Example of §1.4, $(S^n, *) = (|\Delta[n]/\partial\Delta[n]|, *)$. Now let $\tau_n : \pi_n(\wedge V, d) \to \pi_n(|\wedge V, d|, *)$ be the map that associates to $f$ the based homotopy class of $|\varphi_f|$.

It follows from Theorem 1.1(iii) that the correspondence $\chi \mapsto |\chi|$ induces a bijection

$$\begin{aligned}[]
[(\wedge V, d), A_{PL}(\Delta[n]/\partial\Delta[n])]_* &\xrightarrow{\cong} [(|\Delta[n]/\partial\Delta[n]|, *), (|\wedge V, d|, *)] \\
&= \pi_n(|\wedge V, d|, *)\,.
\end{aligned}$$

It remains to show that the correspondence $f \mapsto \varphi_f$ induces a bijection $\pi_n(\wedge V, d) \xrightarrow{\cong} [(\wedge V, d), A_{PL}(\Delta[n]/\partial\Delta[n])]_*$.

Recall that the elements 1 and $[-dt_1 \wedge \cdots \wedge dt_n]$ form a basis of $H(\Delta[n]/\partial\Delta[n])$ and so a quasi-isomorphism $\sigma_n : H(\Delta[n]/\partial\Delta[n]) \xrightarrow{\simeq} A_{PL}(\Delta[n]/\partial\Delta[n])$ is given by $[dt_1 \wedge \cdots \wedge dt_n] \to dt_1 \wedge \cdots \wedge dt_n$. Moreover, $\varphi_f$ factors as $\sigma_n \circ \psi_f$ where $\psi_f : (\wedge V, d) \to H(\Delta[n]/\partial\Delta[n])$ is given by $\psi_f(v) = -f(v)[dt_1 \wedge \cdots \wedge dt_n]$. Since $\sigma_n$ is a quasi-isomorphism, by Proposition 1.10 it induces a bijection $[(\wedge V, d), H(\Delta[n]/\partial\Delta[n])]_* \xrightarrow{\cong} [(\wedge V, d), A_{PL}(\Delta[n]/\partial\Delta[n])]_*$. The correspondence $f \mapsto \psi_f$ is then easily seen to be a bijection from $\pi_n(\wedge V, d)$ to $[(\wedge V, d), H(\Delta[n]/\partial\Delta[n])]_*$ via an argument like that for Lemma 1.4(ii).

Finally, it is immediate from the definitions that $\tau_n$ and $\iota_n$ are inverse maps.

□

*proof of Theorem 1.4(i)*: Fix $n \ge 2$ and recall that $\pi_n(m_{|\wedge V, d|}) = \iota_n$. Then there are cdga morphisms

$$(\wedge V, d) \xrightarrow{\rho} \mathbb{k} \otimes_{\wedge V^{<n}} (\wedge V, d) = (\wedge V^{\ge n}, \bar{d}) \xleftarrow{j} (\wedge V^n, 0)\,.$$

Via the identification $V = Q(\wedge V)$ it is visible that $Q^n(\rho)$ and $Q^n(j)$ are linear isomorphisms. Now the naturality of the bijections $\iota_n$ of Theorem 1.3 gives the commutative diagram

$$\begin{array}{ccccc} (Q^n(\wedge V, d))^\# & \xleftarrow{\cong} & (Q^n(\wedge V^{\geq n}, \overline{d}))^\# & \xleftarrow{\cong} & (Q^n(\wedge V^n, 0))^\# \\ \iota_n \uparrow \cong & & \iota_n \uparrow \cong & & \iota_n \uparrow \cong \\ \pi_n(|\wedge V, d|) & \longleftarrow & \pi_n(|\wedge V^{\geq n}, \overline{d}|) & \longleftarrow & \pi_n(|\wedge V^n, 0|) \end{array}$$

in which the upper horizontal arrows are isomorphisms, the vertical maps are set bijections, and the lower horizontal arrows are homomorphisms of abelian groups. It follows that the lower horizontal lines are isomorphisms of abelian groups.

Thus it is sufficient to show that

$$\pi_n(|\wedge V^n, 0|) \to (Q^n(\wedge V^n)^\#, 0)$$

is an abelian group homomorphism.

Consider the morphism $\Delta : (\wedge V^n, 0) \longrightarrow (\wedge V^n, 0) \otimes (\wedge V^n, 0)$, $v \mapsto v \otimes 1 + 1 \otimes v$. Use Propositions 1.3 and 1.12 to identify the corresponding continuous map as a map $|\Delta| : |\wedge V^n, 0| \leftarrow |\wedge V^n, 0| \times |\wedge V^n, 0|$. It follows from the definition of $\Delta$ that the restriction of $|\Delta|$ to $|\wedge V^n, 0| \times *$ and $* \times |\wedge V^n, 0|$ is the identity. Since $n \geq 2$, $\pi_n$ takes values in abelian group and so this implies that

$$\pi_n(|\Delta|) : \pi_n(|\wedge V^n, 0|) \leftarrow \pi_n(|\wedge V^n, 0|) \oplus \pi_n(|\wedge V^n, 0|)$$

is simply addition.

Now the commutative diagram

$$\begin{array}{ccc} \pi_n(|\wedge V^n, 0|) & \longleftarrow & \pi_n(|\wedge V^n, 0|) \oplus \pi_n(|\wedge V^n, 0|) \\ \iota_n \downarrow \cong & & \cong \downarrow \iota_n \oplus \iota_n \\ Q^n(\wedge V)^\# & \longleftarrow & Q^n(\wedge V)^\# \oplus Q^n(\wedge V)^\# \end{array}$$

shows that $\iota_n$ is an isomorphism of abelian groups.

□

**Corollary 1.11.** *The abelian groups $\pi_n(|\wedge V, d|)$, $n \geq 2$, are vector spaces over $\Bbbk$.*

□

Now let $(X, *)$ be a based topological space with minimal Sullivan model

$$m_X : (\wedge W, d) \to A_{PL}(X).$$

**Definition.** The continuous map

$$|m_X| : |Sing\, X| \to |\wedge W, d|$$

is the *rational completion* of $X$.

**Example.** In the literature ([55], [15], [57]), a connected CW complex $X$ is called a *rational* $K(\pi, 1)$ *space* if its minimal Sullivan model $(\wedge V, d)$ is a Sullivan 1-algebra, i.e. $V = V^1$. In view of Theorem 1.3 (p. 37) $X$ is a rational $K(\pi, 1)$ space if and only if the rational completion of $X$ is a $K(\pi, 1)$.

Again, suppose $m_X : (\wedge W, d) \to A_{PL}(X)$ is the minimal Sullivan model of a connected space $(X, *)$. Then a morphism $\varphi : (\wedge V, d) \to A_{PL}(X)$ lifts up to based homotopy through $m_X$ to a morphism $\psi : (\wedge V, d) \to (\wedge W, d)$ such that $m_X \psi \sim_* \varphi$, and so we have the diagram

$$\begin{array}{ccc} (\wedge V, d) & \xrightarrow{\quad\psi\quad} & (\wedge W, d) \\ {\scriptstyle m_{|\wedge V, d|}}\big\downarrow & \searrow^{A_{PL}(\lambda_X)\circ\varphi} & \big\downarrow{\scriptstyle A_{PL}(\lambda_X)\circ m_X} \\ A_{PL}(|\wedge V, d|) & \xrightarrow[A_{PL}(|\varphi|)]{} & A_{PL}(|\, Sing\, X|)\,. \end{array} \tag{1.19}$$

The upper triangle is based homotopy commutative by the definition of $\psi$ and the lower triangle is based homotopy commutative as observed in Proposition 1.15 (p. 29).

For cohomology this yields the commutative diagram

$$\begin{array}{ccc} H(\wedge V, d) & \xrightarrow{\quad H(\psi)\quad} & H(\wedge W, d) \\ {\scriptstyle H(m_{|\wedge V, d|})}\big\downarrow & \searrow^{H(\varphi)} & \cong\big\downarrow{\scriptstyle H(m_X)} \\ H(|\wedge V, d|) & \xrightarrow[H(|\varphi|)]{} & H(X)\,, \end{array} \tag{1.20}$$

in which we have used $H(\lambda_X)$ to identify $H(X)$ with $H(|\, Sing\, X|)$.

On the other hand, since $\lambda_X : (|\, Sing\, X|, *) \to (X, *)$ is a weak homotopy equivalence (Proposition 1.5, p. 13), each $\pi_n(\lambda_X)$ is an isomorphism. We use these isomorphisms to identify $\pi_n(X, *)$ with $\pi_n(|\, Sing\, X|, *)$, so that $\pi_n(|\varphi|)$ becomes a homomorphism $\pi_n(|\wedge V, d|, *) \to \pi_n(X, *)$. Now identify $\pi_n(|\, Sing\, X|, *)$ with $\pi_n(X, *)$ via $\pi_n(\lambda_X)$. Then from the naturality of $\pi_n(\varphi)$ and (1.19) we obtain the commutative diagrams for $n \geq 2$

$$\begin{array}{ccc} \pi_n(\wedge V, d) & \xleftarrow{\pi_n(\psi)} & \pi_n(\wedge W, d) \\ \cong \Big\uparrow \iota_n & \nwarrow^{\pi_n(\varphi)} & \Big\uparrow \pi_n(m_X) \\ \pi_n(|\wedge V, d|, *) & \xleftarrow[\pi_n(|\varphi|)]{} & \pi_n(X, *)\,, \end{array} \tag{1.21}$$

and, if $(\wedge V, d)$ is 1-finite and $\dim H^1(X) < \infty$,

$$\begin{array}{ccc} \pi_1(\wedge V, d) & \xleftarrow{\pi_1(\psi)} & \pi_1(\wedge W, d) \\ \cong \Big\uparrow \exp_{L_0} \circ \iota_1 & \nwarrow_{\pi_1(\varphi)} & \Big\uparrow \pi_1(m_X) \\ \pi_1(|\wedge V, d|, *) & \xleftarrow[\pi_1(|\varphi|)]{} & \pi_1(X, *)\,, \end{array} \tag{1.22}$$

where $L_0$ is the fundamental Lie algebra of $(\wedge V, d)$.

In particular, if $X$ is a CW complex recall that we write $||\varphi|| = \nu_X \circ |\varphi| : |\wedge V, d| \to X$, where $\nu_X$ is the assigned homotopy inverse to $\lambda_X$. Now $\pi_n(||\varphi||)$ depends only on the based homotopy class of $\varphi$. Moreover, in this case (1.19) simplifies to the based homotopy commutative diagram

$$\begin{array}{ccc} (\wedge V, d) & \xrightarrow{\psi} & (\wedge W, d) \\ \Big\downarrow m_{|\wedge V, d|} & \searrow^{\varphi} & \Big\downarrow m_X \\ A_{PL}(|\wedge V, d|) & \xrightarrow{A_{PL}(||\varphi||)} & A_{PL}(X)\,, \end{array}$$

and (1.20), (1.21) and (1.22) reduce to the equations

$$\left.\begin{array}{l} H(m_X) \circ H(\psi) = H(\varphi) = H(||\varphi||) \circ H(m_{|\wedge V, d|})\,, \\ \pi_n(\psi) \circ \pi_n(m_X) = \pi_n(\varphi) = \iota_n \circ \pi_n(||\varphi||)\,, \qquad n \geq 2 \\ \text{and} \\ \pi_1(\psi) \circ \pi_1(m_X) = \pi_1(\varphi) = \exp_L \circ \iota_1 \circ \pi_1(||\varphi||)\,. \end{array}\right\} \tag{1.23}$$

Suppose again that $\varphi : (\wedge V, d) \to A_{PL}(X)$ is a morphism from a minimal Sullivan algebra to the forms on a path connected based topological space. Recall that $\rho : \wedge^+ V \to Q(\wedge V)$ induces a linear map $\xi : H^+(\wedge V, d) \to Q(\wedge V)$. Recall further that the Hurewicz homomorphism $hur : \pi_n(X, *) \to H_n(X; \mathbb{Z})$ is defined by $hur\,(\alpha) = H_n(\sigma)[e_n]$, where $\sigma : (S^n, *) \to (X, *)$ represents $\alpha$ and $e_n$ denotes the fundamental cycle of $\Delta[n]/\partial\Delta[n]$ as defined in the Example of §1.4.

**Proposition 1.19.** *If $\beta \in H^n(\wedge V, d)$ and $\alpha \in \pi_n(X,*)$ then*

$$\langle H^n(\varphi)\beta, hur\,\alpha\rangle = \begin{cases} \langle \xi(\beta), \pi_n(\varphi)\alpha\rangle\,, & n \geq 2 \\ \langle \xi(\beta), \log_{L_0} \pi_1(\varphi)\alpha\rangle\,, & n = 1\,, \end{cases}$$

*again supposing $(\wedge V, d)$ is 1-finite when $n = 1$.*

*proof:* A naturality argument reduces the Proposition to the case $X = S^n$, $\sigma$ is the identity map of $S^n$, and $m_X$ is the minimal model $m(n) : (\wedge W, d) \to A_{PL}(S^n)$ defined as the start of §1.8. Thus $w \in W^n$ is the cycle mapped to a representative of the fundamental cohomology class of $S^n$.

Now let $\psi : (\wedge V, d) \to (\wedge W, d)$ be a lift through $m(n)$ of $\varphi$. Then $f \in (V^n)^{\#}$ is defined (with the notation above) by

$$\psi v = \langle v, f\rangle w\,, \qquad v \in V\,,$$

and, by the definition at the start of §1.8,

$$\pi_n(\varphi)(\alpha) = \begin{cases} f\,, & \text{if } n \geq 2 \\ \exp_{L_0} f\,, & \text{if } n = 1 \end{cases}.$$

Thus if $\beta$ is represented by a cycle $\Phi \in (\wedge V)^n$ then $\xi(\beta) = \rho\Phi$ and so the right-hand sides of the equation of the Proposition reduce to $\langle \rho\Phi, f\rangle$.

On the other hand, since the identity map of $S^n$ represents $\alpha$, $hur\,\alpha$ is the fundamental class $[e_n]$ in $H_n(S^n)$. Thus

$$\begin{aligned}\langle H^n(\varphi)\beta, hur\,\alpha\rangle &= \langle H^n(m(n))H^n(\psi)\beta, [e_n]\rangle \\ &= \langle H^n(m(n))\,(\psi(\rho\Phi)[e_n])\rangle = \langle \rho\Phi, f\rangle\,.\end{aligned}$$

□

**Remark:** In (1.21), $\pi_n(m_X)$ may not be a bijection and in (1.20) $H(m_{|\wedge V,d|})$ may not be an isomorphism. However, we have

**Theorem 1.5.** *Suppose $(\wedge V, d)$ is a minimal Sullivan algebra. If $V^1 = 0$ and $\Bbbk = \mathbb{Q}$, then*

$$m_{|\wedge V,d|} : (\wedge V, d) \to A_{PL}(|\wedge V, d|)$$

*is a quasi-isomorphism if and only if $V$ has finite type.*

*proof:* If $V$ has finite type this is Theorem 17.10 (ii) of [18].

Conversely, suppose $m_{|\wedge V,d|}$ is a quasi-isomorphism and for simplicity write

$$|\wedge V, d| = Y \quad \text{and } m_{|\wedge V,d|} = \varphi\,.$$

Then suppose $\dim V^i < \infty$, $i < k$. In this case the linear map $\xi^k : H^k(\wedge V, d) \to V^k$ satisfies

$$\dim \ker \xi^k < \infty \quad \text{and } \operatorname{codim} \operatorname{Im} \xi^k < \infty\,. \tag{1.24}$$

We say two vector spaces $S$ and $T$ are *commensurable* and write $S \sim T$ if there are finite dimensional vector spaces $E$ and $F$ such that $S \oplus E \cong T \oplus F$. Thus it follows from (1.24) that $H^k(\wedge V, d) \sim V^k$ and also that

$$\dim \ker (\xi^k)^{\#} < \infty \quad \text{and codimension } \operatorname{Im} (\xi^k)^{\#} < \infty\,,$$

so that $\left[H^k(\wedge V, d)\right]^{\#} \sim \left(V^k\right)^{\#}$.

Now by Proposition 1.19 the diagram

$$\begin{array}{ccc} \pi_k(Y) & \xrightarrow[\cong]{\pi_k(\varphi)} & (V^k)^{\#} \\ \big\downarrow{\scriptstyle hur} & & \big\downarrow{\scriptstyle (\xi^k)^{\#}} \\ H_k(Y) & \xrightarrow{H^k(\varphi)^{\#}} & \left[H^k(\wedge V, d)\right]^{\#} \end{array}$$

commutes. It follows that $\operatorname{Im} H^k(\varphi)^{\#}$ has finite codimension in $\left[H^k(\wedge V, d)\right]^{\#}$ and so the double dual of $H^k(\varphi)$ is a linear map,

$$\left[\left[H^k(\wedge V, d)\right]^{\#}\right]^{\#} \to H^k(Y) \cong H^k(\wedge V, d),$$

with finite dimensional kernel. Since $V^k \sim H^k(\wedge V, d)$ it follows that $\left[\left(V^k\right)^{\#}\right]^{\#} \sim S$ for some subspace $S \subset V^k$, and this is impossible unless $\dim V^k < \infty$. Thus the Theorem follows by induction on $k$.

□

**Theorem 1.6.** *If $(X, *)$ is a simply connected space then $H^*(X; \Bbbk)$ has finite type if and only if $\pi_*(X) \otimes \Bbbk$ has finite type. In this case the homomorphisms $\pi_n(|m_X|)$ extend to isomorphisms*

$$\pi_n(X, *) \otimes \Bbbk \xrightarrow{\cong} \pi_n(|\wedge W, d|)\,, \qquad n \geq 2\,.$$

*In particular, the maps*

$$\pi_n(m_X) : \pi_n(X, *) \otimes \Bbbk \to \pi_n(\wedge W, d)\,, \qquad n \geq 2$$

*are isomorphisms.*

*proof:* This follows from Theorem 15.11 of [18], together with Theorem 1.5 and ([35], Theorem 2.16, p. 69).

□

# Chapter 2

# Homotopy Lie algebras and Sullivan Lie algebras

## 2.1 The homotopy Lie algebra of a minimal Sullivan algebra

Fix a minimal Sullivan algebra, $(\wedge V, d)$. Then $d$ is the sum of derivations $d_k$, $k \geq 1$, defined by the condition : $d_k : V \to \wedge^{k+1} V$. In particular $d_1^2 = 0$ and so $(\wedge V, d_1)$ is also a minimal Sullivan algebra: $d_1$ is called the *quadratic part* of $d$, and $(\wedge V, d_1)$ is the *associated quadratic Sullivan algebra.*

Now consider the graded vector space $L = \{ L_k \}_{k \geq 0}$ defined by

$$L_k = (Q^{k+1}(\wedge V))^{\#}, \qquad k \geq 0 .$$

The choice of the generating space $V$ gives an isomorphism $V \cong Q(\wedge V)$ and hence an isomorphism $L_k \cong (V^{k+1})^{\#}$. We fix that choice and use this isomorphism for ease of notation. Then we define the pairing $V \times sL \to \Bbbk$ ($sL$ is the suspension of $L$) by

$$\langle v, sx \rangle = (-1)^{\deg v} x(v)$$

and extend this as in [18], Chapter 21(e) to the pairings

$$\wedge^p V \times (sL \times \cdots \times sL) \to \Bbbk$$

given by

$$\langle v_1 \wedge \cdots \wedge v_p, sx_p, \cdots, sx_1 \rangle = \sum_{\sigma \in \Sigma_p} \varepsilon_\sigma \langle v_{\sigma(1)}, sx_1 \rangle \cdots \langle v_{\sigma(p)}, sx_p \rangle ,$$

where $\Sigma_p$ is the permutation group on $p$ symbols and

$$v_{\sigma(1)} \wedge \cdots \wedge v_{\sigma(p)} = \varepsilon_\sigma v_1 \wedge \cdots \wedge v_p .$$

We shall further abuse notation and denote these pairings by

$$\langle v, x \rangle \qquad \text{and by } \langle v_1 \wedge \cdots \wedge v_p, x_p, \cdots, x_1 \rangle .$$

It then follows as in ([18], Chapter 21(e)) from the equation $d_1^2 = 0$ that a graded Lie algebra $(L, [\,-\,])$ is defined by

$$\langle v, [x, y]\rangle = (-1)^{\deg y + 1}\langle d_1 v, x, y\rangle \,.$$

It is immediate from the definitions that the Lie bracket in $L$ is independent of the choice of the generating space $V$. Thus we may make the

**Definition.** The graded Lie algebra $(L, [\,-\,])$ is the *homotopy Lie algebra* of $(\wedge V, d)$. If $(\wedge V, d)$ is a minimal Sullivan model for a based topological space $(X, *)$ then $(L, [\,-\,])$ is the *homotopy Lie algebra of* $(X, *)$.

Note that $(\wedge V^1, d)$ is a sub minimal Sullivan algebra of $(\wedge V, d)$ and in this subalgebra $d = d_1$ and the corresponding homotopy Lie algebra is just the component $L_0$ of the homotopy Lie algebra of $(\wedge V, d)$.

**Definition.** The homotopy Lie algebra of $(\wedge V^1, d)$ is the *fundamental Lie algebra* of $(\wedge V, d)$; if $(\wedge V, d)$ is a minimal Sullivan model for a space $(X, *)$ then this is the *fundamental Lie algebra of* $(X, *)$.

**Notation:** In any graded Lie algebra $L$ an *iterated bracket* of length $r$ is an element of the form $[x_1, \cdots, x_r] = [x_1, [x_2, \cdots [x_{r-1}, x_r] \cdots]$ with each $x_i \in L$. The linear span of such Lie brackets in which each $x_i$ is in a subspace $S_i$ will be denoted $[S_1, \cdots, S_r]$.

**Definition.** The *lower central series* of a graded Lie algebra $L$ is the sequence of ideals

$$L = L^1 \supset L^2 \supset \cdots \supset L^n \supset \cdots$$

defined by $L^{n+1} = [L, L, \cdots, L] = [L, L^n]$. The inverse limit $\widehat{L} = \varprojlim_n L/L^n$ is the *completion* of $L$.

**Definition.** A graded Lie algebra $L$ is *nilpotent* if some $L^n = 0$ and *pronilpotent* if the natural map $L \to \varprojlim_n L/L^n$ is an isomorphism.

Note that nilpotent Lie algebras are pronilpotent and that for any Lie algebra $L$, $\widehat{L}$ is pronilpotent.

Recall that if $(\wedge V, d)$ is a minimal Sullivan algebra then $V$ is the increasing union of subspaces $V(0) \subset \cdots \subset V(r) \subset \cdots$ such that $d = 0$ in $V(0)$ and $d : V(r+1) \to \wedge V(r)$.

**Lemma 2.1.** *With the notation above, the homotopy Lie algebra $L$ of $(\wedge V, d)$ satisfies*

$$\langle\, V(r), L^{r+2}\,\rangle = 0\,, \qquad r \geq 0\,.$$

*Moreover, the natural map $i : L \to \widehat{L}$ has a left inverse Lie algebra homomorphism $\rho : \widehat{L} \to L$. In particular, $i$ is injective.*

*proof:* The proof of the lemma is by induction on $r$. If $x, y \in L$ and $v \in V(0)$ then $\langle v, [x,y]\rangle = \pm\langle dv, x, y\rangle = 0$. Assume the equation holds for some $r$ and let $x \in L$, $y \in L^{r+2}$. Then for $v \in V(r+1)$ write $dv = \sum v_i \wedge v_i'$ with $v_i, v_i' \in V(r)$, and observe that by our induction hypothesis,

$$\langle v, [x,y]\rangle = \pm\langle dv, x, y\rangle = \sum \pm\langle v_i, x\rangle\langle v_i', y\rangle + \sum \pm\langle v_i, y\rangle\langle v_i', x\rangle = 0\,.$$

This completes the proof of the equation.

Now, denoting the homotopy Lie algebra of $(\wedge V(r), d)$ by $L(r)$ we have that the induced surjection $L \to L(r)$ extends to a Lie algebra homomorphism $\widehat{L} \to L(r)$. These morphisms define a Lie algebra homomorphism

$$\rho : \widehat{L} \to \varprojlim_r L(r) = L$$

which is, almost by definition, a left inverse for $i$.

□

Finally, recall from §2 in [18] that for any graded Lie algebra $L = L_{\geq 0}$ the Cartan-Chevalley-Eilenberg chain coalgebra $C_*(L)$ is defined by

$$C_*(L) = (\wedge sL, \partial_2)$$

with

$$\partial_2(sx_1\wedge\cdots\wedge sx_p) = -\sum_{i<j}(-1)^{n_{ij}+\deg x_i} s[x_i, x_j]\wedge sx_1\wedge\cdots\widehat{sx_i}\cdots\widehat{sx_j}\cdots\wedge sx_p\,,$$

with $x_\ell \in L$ and where

$$sx_1 \wedge \cdots \wedge sx_p = (-1)^{n_{ij}} sx_i \wedge sx_j \wedge sx_1 \wedge \cdots \widehat{sx_i} \cdots \widehat{sx_j} \cdots \wedge sx_p\,.$$

More generally, if $M$ is any graded $L$-module then $C_*(L, M) = C_*(L) \otimes M$ is the complex defined by

$$\begin{aligned}\partial(sx_1\wedge\cdots\wedge sx_p \otimes a) &= \partial_2(sx_1\wedge\cdots\wedge sx_p) \otimes a \\ &\quad + \sum_{i=1}^{p}(-1)^{\varepsilon_i} sx_1\wedge\cdots\wedge\widehat{sx_i}\wedge\cdots\wedge sx_p \otimes x_i \cdot a\end{aligned}$$

where $\varepsilon_i = (\deg sx_i)\deg(sx_{i+1}\wedge\cdots\wedge sx_p) + \sum_{j=1}^{p} \deg sx_j$.

Observe that $C_*(L)$ is a bigraded coalgebra; we write $C_p(L) = \wedge^p sL$ and $C_{p,q}(L) = (\wedge^p sL)_{p+q}$; here $p$ is the *homological degree* and $\partial$ is homogeneous of bidegree $(-1,0)$. Similarly $C_*(L)\otimes M$ is a bigraded complex with $C_p(L)\otimes M$ the component of homological degree $p$.

The bigraded cochain algebra dual to $C_*(L)$ is denoted by $(C^*(L),d)$, with $C^p(L)=C_p(L)^\#$ and

$$\langle d\alpha, a\rangle + (-1)^{\deg\alpha}\langle \alpha, \partial a\rangle = 0\,, \qquad \alpha\in C^p(L) \text{ and } a\in C_{p+1}(L)\,.$$

In particular, suppose $L$ is the homotopy Lie algebra of a quadratic Sullivan algebra $(\wedge V, d_1)$. Then the isomorphism $sL = V^\#$ defines a pairing $V\times sL\to \Bbbk$ and hence a linear map $V\to (sL)^\#$. This extends uniquely to a morphism

$$\mu : \wedge V \to C^*(L)$$

of bigraded algebras with the wedge degree in $\wedge V$ corresponding to the homological degree in $C^*(L)$.

**Lemma 2.2.** *With the hypotheses and notation above,*

$$\mu : (\wedge V, d_1) \to (C^*(L), d)$$

*is a morphism of commutative cochain algebras. Moreover, if $V$ has finite type then $\mu$ is an isomorphism.*

*proof*: A straightforward calculation from the definition shows that $\mu\circ d_1 = d\circ\mu$. If $V$ has finite type then so does $sL$, and then each $(\wedge^p sL)^\# = \wedge^p (sL)^\#$, and it follows that $\mu$ is an isomorphism.

□

## 2.2 The fundamental Lie algebra of a Sullivan 1-algebra

**Definition.** A *Sullivan* 1-*algebra* is a Sullivan algebra $(\wedge W, d)$ in which $W = W^1$.

**Remark.** For Sullivan 1-algebra $(\wedge W, d)$ the homotopy Lie algebra, $L$ is concentrated in degree zero. Thus in this case the homotopy Lie algebra coincides with the fundamental Lie algebra.

The differential in a Sullivan 1-algebra increases the wedge degree by exactly 1:

$$d : W \to \wedge^2 W\,.$$

Further, associated with $(\wedge W, d)$ is the canonical filtration

$$W_0 \subset W_1 \subset \cdots \subset W_n \subset \cdots$$

dcfined by

$$W_0 = W \cap \ker d \quad \text{and} \quad W_{n+1} = W \cap d^{-1}(\wedge^2 W_n)\,.$$

The Sullivan condition implies that $W = \cup_n W_n$. Moreover, it follows from Lemma 2.1 (p. 46) that

$$\langle\, W_n, L^{n+2}\,\rangle = 0\,, \qquad n \geq 0\,. \tag{2.1}$$

Given a Sullivan 1-algebra $(\wedge W, d)$ we may choose subspaces $M_n \subset W_n$ by setting $M_0 = W_0$ and requiring $M_n \oplus W_{n-1} = W_n$, $n \geq 1$. Then $W = \oplus_{n\geq 0} M_n$ and this in turn defines direct sum decompositions of each $\wedge^p W$. If $z \in \wedge^p W$ then $z$ is the sum of its *components* in these subspaces. In particular, for $w \in M_n$, $n \geq 2$, we denote by $\delta w$ the component of $dw$ in $M_0 \otimes M_{n-1}$ and this then defines linear maps

$$\delta : M_n \to M_0 \otimes M_{n-1}\,, \qquad n \geq 2\,.$$

**Lemma 2.3.** *Let $(\wedge W, d)$ be a Sullivan 1-algebra. Then with the notation above*

*(i) $d : M_1 \to \wedge^2 M_0$ is injective.*
*(ii) For $n \geq 2$, $\delta : M_n \to M_0 \otimes M_{n-1}$ is injective, and for $w \in M_n$,*

$$dw = \delta w + \Omega\,, \qquad \text{some } \Omega \in \wedge^2 W_{n-2}\,.$$

*Moreover, $\Omega = \sum \Omega_{ij}$ with $\Omega_{ij} \in M_i \wedge M_j$ and $i + j \leq n - 1$.*

*proof:* (i) is immediate because $\ker d \cap M_1 \subset M_0 \cap M_1 = 0$.

ii) We induct on $n$. When $n = 2$, let $w \in M_2$. Then

$$dw \in \wedge^2 W_1 = \wedge^2 M_1 \oplus (M_0 \otimes M_1) \oplus \wedge^2 M_0\,.$$

If the component of $dw$ in $\wedge^2 M_1$ is non-zero it may be written $\sum w_{2i} \wedge w_{2i+1}$ with the $w_j$ all linearly independent. Then

$$d(\sum w_{2i} \wedge w_{2i+1}) = \sum_i (dw_{2i}) \otimes w_{2i+1} - \sum_i dw_{2i+1} \otimes w_{2i} \in \wedge^2 M_0 \otimes M_1\,.$$

But this is clearly the component of $0 = d^2 w$ in $\wedge^2 M_0 \otimes M_1$ and so it must vanish. Since the $w_j$ are all linearly independent, each $dw_j = 0$, which is impossible since each $w_j \in M_1$. Thus

$$dw = \delta w + \Omega\,, \qquad \text{some } \Omega \in \wedge^2 M_0\,,$$

and $\delta w \neq 0$ since this would imply $w \in M_1$.

As above we first show that the component in $M_i \wedge M_j$ of a cycle $z \in \wedge^2 W_n$ is zero if $i + j > n$. For this choose the maximum $k$ such that $z$ has a non-trivial component in some $M_i \wedge M_j$ with $i + j = k$; then choose the largest $j$ such that the component of $z$ in $M_i \wedge M_j$ with $i + j = k$ is non-zero. If $i < j$ write the component as $\sum w_\alpha \otimes w'_\alpha$ with the $w_\alpha$ and the $w'_\alpha$ respectively linearly independent elements in $M_i$ and $M_j$. Thus the component of $dz$ in $M_0 \otimes M_{i-1} \otimes M_j$ is just $\sum \delta w_\alpha \otimes w'_\alpha$. Since the $w'_\alpha$ are linearly independent it follows that $\delta w_\alpha = 0$, each $\alpha$, and so by induction $w_\alpha \in M_0$. Thus $i = 0$ and $i + j \leq n$. If $i = j$ write this component as $\sum w_\alpha \wedge w'_\alpha$ in which the $w_\alpha$ together with the $w'_\alpha$ are linearly independent elements of $M_j$ and apply the same argument to conclude that $j = 0$.

In particular if $w \in M_{n+1}$, $dw \in (M_0 \otimes M_n) \oplus \wedge^2 W_{n-1}$. Thus if $\delta w = 0$ then $dw \in \wedge^2 W_{n-1}$ and so $w \in M_{n+1} \cap W_n = 0$. This completes the proof. □

**Lemma 2.4.** *A Sullivan* 1*-algebra* $(\wedge W, d)$ *is* 1*-finite if and only if*

$$\dim W_k < \infty\,, \qquad k \geq 0\,.$$

*proof:* By definition, $W_0 \cong H^1(\wedge W, d)$. Thus we have only to show that if $\dim W_0 < \infty$ then $\dim W_k < \infty$ for all $k \geq 1$. Let $W = \oplus_k M_k$ be the direct sum decomposition above. Then it follows from Lemma 2.3 that $d : M_k \to \wedge W_{<k}$, and by definition this map is injective for $k \geq 1$. The lemma follows by induction on $k$.

□

Now recall from §2.1 the lower central filtration $L^{n+1} = [L, L^n]$ of a graded Lie algebra $L$, and let $L$ be the homotopy Lie algebra of a Sullivan 1-algebra $(\wedge W, d)$.

It follows from Lemma 2.1 that $L^{n+1}$ vanishes on $W_{n-1}$ and $L^{n+2}$ vanishes on $W_n$. Thus we obtain natural maps $\lambda_n : L/L^{n+2} \to W_n^\#$, which induce maps

$$\overline{\lambda}_n : L^{n+1}/L^{n+2} \to (W_n/W_{n-1})^\#\,, \qquad n \geq 1\,.$$

**Theorem 2.1.** *If* $(\wedge W, d)$ *is a* 1*-finite Sullivan* 1*-algebra then each* $\lambda_n$ *and each* $\overline{\lambda}_n$ *is an isomorphism. Moreover, in this case the natural map* $i : L \to \widehat{L}$ *is an isomorphism, and so* $L$ *is pronilpotent.*

**Corollary 2.1.** *If $L$ is the homotopy Lie algebra of a minimal Sullivan algebra $(\wedge V, d)$ then $\dim L_0/[L_0, L_0] < \infty$ if and only if $\dim H^1(\wedge V, d) < \infty$ and in this case $\lambda_0$ is an isomorphism*

$$L_0/[L_0, L_0] \xrightarrow{\cong} H^1(\wedge V, d)^{\#} .$$

*proof:* It follows from Lemma 2.1 (p. 46) that the isomorphism $L_0 \stackrel{\cong}{\to} V_1^{\#}$ induces a surjection $L_0/[L_0, L_0] \to H^1(\wedge V, d)^{\#}$. Thus if $\dim L_0/[L_0, L_0] < \infty$, $\dim H^1(\wedge V, d) < \infty$. On the other hand, if $\dim H^1(\wedge V, d) < \infty$, Theorem 2.1 states that $\lambda_0$ is an isomorphism and so dim $L_0/[L_0, L_0] < \infty$.

□

**Corollary 2.2.** *If $L$ is the homotopy Lie algebra of a 1-finite minimal Sullivan algebra $(\wedge V, d)$, then $L_0$ is nilpotent if and only if $\dim V^1 < \infty$.*

*proof of Theorem 2.1:* For the first assertion it is sufficient to prove that the $\overline{\lambda}_n$ are isomorphisms. We again use the direct sum decomposition $W = \oplus_n M_n$ defined earlier, and the linear map $\delta$ of Lemma 2.3. Fix a subspace $S \subset L$ such that restriction to $M_0$ defines an isomorphism

$$\mu_0 : S \stackrel{\cong}{\to} (M_0)^{\#} .$$

Since $(\wedge W, d)$ is 1-finite, $S$ is finite dimensional.

Denote by $E$ the Lie algebra generated by $S$, and by

$$E = E^1 \supset E^2 \supset \cdots$$

the lower central series for $E$. Then define subspaces $S^n \subset E$ by $S^1 = S$ and $S^{n+1} = [S, S^n]$. Since $S$ generates $E$, it follows that

$$E^n = S^n + E^{n+1} , \qquad n \geq 1 .$$

Now identify $\overline{\lambda}_n$ with a linear map $L^{n+1}/L^{n+2} \to (M_n)^{\#}$. To prove $\lambda_n$ is surjective it is sufficient to show that the linear maps,

$$\mu_n : S^{n+1} \to (M_n)^{\#} ,$$

obtained by restricting $\overline{\lambda}_n(S^{n+1})$ to $M_n$, are surjective. We establish this by induction on $n$, noting that for finite dimensional vector spaces the functors $(-)^{\#}$ and $\otimes_{\mathbb{k}}$ commute.

First note that by construction, $\mu_0$ is an isomorphism. For $n = 1$, an injection $\wedge^2 M_0 \to M_0 \otimes M_0$ is given by $v \wedge w \to v \otimes w - w \otimes v$. It dualizes to a surjection

$$(M_0)^{\#} \otimes (M_0)^{\#} \to (\wedge^2 M_0)^{\#} .$$

By the definition of the Lie bracket, the diagram

$$\begin{array}{ccc} S^2 & \xleftarrow{[\,,]} & S \otimes S \\ & & \big\downarrow \cong \\ \Big\downarrow \mu_1 & & (M_0)^\# \otimes (M_0)^\# \\ & & \big\downarrow \\ (M_1)^\# & \xleftarrow{(d)^\#} & (\wedge^2 M_0)^\# \end{array}$$

commutes. Since $d$ is injective, (Lemma 2.3), $(d)^\#$ is surjective and it follows that $\mu_1$ is surjective.

Now suppose by induction that some $\mu_n$ is surjective. As above we have the commutative diagram

$$\begin{array}{ccc} S^{n+2} & \longleftarrow & S \otimes S^{n+1} \\ \Big\downarrow \mu_{n+1} & & \Big\downarrow \mu_0 \otimes \mu_n \\ (M_{n+1})^\# & \xleftarrow{\delta^\#} & (M_0)^\# \otimes (M_n)^\# \end{array}$$

in which $\delta^\#$ is surjective because $\delta$ is injective. It follows by induction that each $\mu_n$ is surjective.

Next, given a sequence of elements $x_k \in L^k$, $k \geq 1$ we may define an element $x \in L$ by

$$\langle w, x \rangle = \sum_k \langle w, x_k \rangle \,,$$

since, as observed in (2.1), $\langle W_n, L^{n+2} \rangle = 0$, $n \geq N$. We denote $x$ by $\sum x_k$. Our next step is to show that every $x \in L$ has the form

$$x = \sum_{k \geq 1} x_k \qquad \text{with } x_k \in S^k \,.$$

In fact, choose $x_1 \in S$ so that $x|_{M_0} = x_1|_{M_0}$. Then suppose $x_i \in S^i, i \leq n$, have been chosen so that $x$ and $\sum_{i \leq n} x_i$ restrict to the same function in $W_{n-1}$. Because $S^{n+1}$ restricts to zero in $W_{n-1}$ and restricts surjectively to $(M_n)^\#$ we may find some $x_{n+1} \in S^{n+1}$ such that $\sum_{i \leq n+1} x_i$ and $x$ restrict to the same function in $W_n$.

This procedure yields an infinite sequence of elements $x_k \in S^k$ such that $\sum x_k$ and $x$ restrict to the same function in each $W_n$, i.e. $x = \sum_k x_k$.

The identical argument shows that if $x \in L$ vanishes in some $W_n$ then $x$ can be written as

$$x = \sum_{k \geq n+2} x_k$$

with $x_k \in S^k$.

To complete the proof of the theorem we need to show that if $x \in L^{n+1}$ vanishes on $W_n$ then $x \in L^{n+2}$, because then $\overline{\lambda}_n$ will be injective. We show this by induction on $n$ beginning with $n = -1$ and $W_{-1} = 0$, where the assertion is a tautology.

Suppose the assertion established for some $n - 1 \geq -1$. Recall that if $x \in L$ vanishes on $W_n$ then we may write

$$x = \sum_{k \geq n+2} x_k \qquad \text{with } x_k \in S^k \,.$$

Fix a basis $y_1, \ldots, y_r$ of $S$ and write

$$x_k = \sum_{i=1}^{r} [y_i, z_{ik}] \quad \text{with } z_{ik} \in S^{k-1} \,.$$

Set $z_i = \sum_k z_{ik}$. Since each $z_{ik} \in S^{k-1}$ vanishes on $W_{k-3}$ and hence on $W_{n-1}$. Thus $z_i$ vanishes on $W_{n-1}$, and so by the induction hypothesis $z_i \in L^{n+1}$. But then

$$x = \sum_{k,i} [y_i, z_{ik}] = \sum_{i=1}^{r} [y_i, z_i] \in L^{n+2} \,.$$

This completes the proof of the first part of the theorem.

Finally, recall from Lemma 2.1 that $i : L \to \widehat{L}$ is injective. On the other hand, suppose $y = (y_n)$ is an element of $\widehat{L}$ with $y_n \in L/L^n$ and $y_{n+1} \to y_n$, $n \geq 1$. Define $x \in L$ by setting $\langle w, x \rangle = \langle w, y_{n+2} \rangle$, $w \in W_n$. Let $x_{n+2} \in L/L^{n+2}$ be the image of $x$. Then $\lambda_n(x_{n+2}) = \lambda_n(y_{n+2})$ and so $x_{n+2} = y_{n+2}$. It follows that $i(x) = y$ and so $i$ is surjective.

□

**Example.** Theorem 2.1 asserts that if a minimal Sullivan algebra $(\wedge V, d)$ satisifes $\dim H^1(\wedge V, d) < \infty$ then $L_0/[L_0, L_0] \xrightarrow{\cong} H^1(\wedge V, d)^{\#}$. By contrast, let $(\wedge V, d) \to (A, 0)$ be the minimal Sullivan model of the cdga $(A, 0)$ where $A$ is the quotient of the exterior algebra $\wedge(x_n, n \geq 1)$, $\deg x_n = 1$, by the ideal generated by the products $x_n x_m$, $n, m \geq 1$, and let $W = V^1$. Denote by $L$ the fundamental Lie algebra of $(\wedge W, d_1)$. We show that the canonical map $L/[L, L] \to (W_0)^{\#}$ is surjective but not injective.

Denote by $v_n, n \geq 1$ a linear basis for $W_0$. Then $W_1$ has a basis given by the $v_i$, together with elements $v_{i,j}$, $1 \leq i < j$, with $d(v_{i,j}) = v_i v_j$. Let $y \in L$ be an element such that $\langle W_0, y \rangle = 0$ and

$$\langle v_{ij}, y \rangle = \begin{cases} 1 \text{ if } (i,j) = (2p+1, 2p+2) \text{ some } p \geq 0 \\ 0 \text{ otherwise.} \end{cases}$$

We show that $y \notin [L, L]$.

Indeed, if $y \in [L, L]$, then $y = \sum_{k=1}^n [y_k, y_k']$. Thus,

$$\langle v_{ij}, y\rangle = \sum_{k=1}^{n} \left( \langle v_i, y_k\rangle\langle v_j, y_k'\rangle - \langle v_i, y_k'\rangle\langle v_j, y_k\rangle \right).$$

Now for each $i \geq 1$, define the following sequences of elements in $\Bbbk$:

$$\overline{y}(i) = (\langle v_{1i}, y\rangle, \langle v_{2i}, y\rangle, \cdots, \langle v_{i-1,i}, y\rangle, 0, \langle -v_{i+1,i}, y\rangle, -\langle v_{i+2,i}, y\rangle, \cdots)$$

$$\overline{y_k} = (\langle v_1, y_k\rangle, \langle v_2, y_k\rangle, \cdots) \qquad \text{and } \overline{y_k'} = (\langle v_1, y_k'\rangle, \langle v_2, y_k'\rangle, \cdots).$$

It follows from the equalities above that for each $i$,

$$\overline{y}(i) = \sum_{k=1}^{n} \left( -\langle v_i, y_k'\rangle\, \overline{y_k} + \langle v_i, y_k\rangle \overline{y_k'} \right).$$

Now observe that $\overline{y}(2p+1) = (0, \cdots, 0, 1, 0, \cdots)$ with only one 1 in position $2p$. This implies that the $\overline{y}(i)$, $i \geq 1$, form a vector space of infinite dimension, which cannot be in the linear span of the sequence $\overline{y_k}$ and $\overline{y_k'}$, $1 \leq k \leq n$.

□

## 2.3 Sullivan Lie algebras

Suppose $(\wedge V, d_1)$ is a minimal Sullivan algebra for which

$$\dim H^1(\wedge V, d_1) < \infty \qquad \text{and} \quad \dim V^i < \infty\,, \quad i \geq 2\,, \tag{1.5}$$

and let $L$ be its homotopy Lie algebra. Then it follows from Theorem 2.1 that $H^1(\wedge V, d_1) \cong (L_0/[L_0, L_0])^\#$ and that $L$ is pronilpotent. In particular, $L$ satisfies

$$\dim L_0/[L_0, L_0] < \infty\,, \quad \dim L_i < \infty\,, \quad i \geq 1 \quad \text{and} \quad \cap_n L^n = 0\,. \tag{2.2}$$

**Definition.** A *Sullivan Lie algebra* is a graded Lie algebra $L = L_{\geq 0}$ which satisfies (2.2).

**Lemma 2.5.** *If $L$ is a Sullivan Lie algebra then the graded vector spaces $L/L^n$ all have finite type.*

*proof:* The Lie bracket $[\,,\,] : L \otimes L \to L$ induces linear surjections

$$L/[L,L] \otimes L^n/L^{n+1} \to L^{n+1}/L^{n+2}\,.$$

It follows by induction that each $L^n/L^{n+1}$ has finite type, and thus a short exact sequence argument completes the proof.

□

As observed above, if $(\wedge V, d)$ is a minimal Sullivan algebra satisfying (1.5) then its homotopy Lie algebra is a pronilpotent Sullivan Lie algebra. In the reverse direction we have

**Proposition 2.1.** *Let $L$ be a Sullivan Lie algebra. Then the cochain algebra $\varinjlim_n C^*(L/L^n)$ is a quadratic Sullivan algebra, $(\wedge V, d_1)$ satisfying (1.5), with $V = \varinjlim_n (L/L^n)^\#$.*

*Moreover, the induced map $L \to V^\#$ extends to an isomorphism from $\widehat{L} = \widehat{L_0} \oplus L_{\geq 1}$ to the homotopy Lie algebra of $(\wedge V, d_1)$.*

*proof:* Since $L/L^n$ is finite dimensional, $C^*(L/L^{n+1})$ is a quadratic Sullivan algebra $(\wedge V(n), d_1)$ and $V(n) = (sL/L^{n+1})^\#$. Moreover the surjection $L/L^{n+2} \to L/L^{n+1}$ yields an isomorphism $(\wedge V(n+1), d_1) = (\wedge V(n) \otimes \wedge M(n+1), d_1)$ with $d_1 : M(n+1) \to \wedge^2 V(n)$. Thus $(\wedge V, d_1) = \varinjlim_n C^*(L/L^n)$ is a quadratic Sullivan algebra.

Now, since $L/L^{n+1}$ is finite dimensional, the isomorphism $V(n) = \left(sL/L^{n+1}\right)^\#$ can be rewritten as $L/L^{n+1} = V(n)^\#$. Thus the homotopy Lie algebra $E$ of $(\wedge V, d_1)$ is given by

$$E = V^\# = \varprojlim L/L^n = \widehat{L}\,.$$

It follows from Lemma 2.1 (p. 46) that the isomorphism $E = V^\#$ induces a surjection $L_0/[L_0, L_0] \to V_0^1 = H^1(\wedge V, d)$. Since $E^i = L^i$, $i \geq 1$, it follows that $(\wedge V, d_1)$ satisfies (1.5).

□

**Definition.** If $L$ is a Sullivan Lie algebra then the corresponding quadratic Sullivan algebra $(\wedge V, d_1)$ with homotopy Lie algebra $\widehat{L}$ is called the *associated quadratic Sullivan algebra* for $L$.

**Remark.** If $L$ is the homotopy Lie algebra of a minimal Sullivan algebra $(\wedge V, d)$ satisfying (1.5) then $L$ is a Sullivan Lie algebra and $(\wedge V, d_1)$ is its associated quadratic Sullivan algebra.

**Example.** If $L$ is a Sullivan Lie algebra, the associated quadratic Sullivan algebra $(\wedge V, d_1)$ satisfies $\dim H^1(\wedge V, d_1) < \infty$. But the vector spaces $H^q(\wedge V, d_1)$ can be infinite dimensional for $q \geq 2$. Consider for instance the homotopy Lie algebra $L$ of $(\wedge V, d_1)$ where $V = V^1$ has a basis $w, v_0, v_1, \cdots$ with $dw = dv_0 = 0$ and $dv_n = wv_{n-1}$ for $n \geq 1$. For $n > 0$ the elements

$$z_n = \sum_{i>j, i+j=2n+1} v_i \wedge v_j$$

are cycles and generate a sub vector space of infinite dimension in $H^2(\wedge V, d_1)$.

□

## 2.4 Primitive Lie algebras and exponential groups

Let $I_L$ be the augmentation ideal in the universal enveloping algebra $UL$ of a graded Lie algebra $L$: $I_L$ is the two sided ideal generated by $L$. This determines the filtration

$$UL = I_L^0 \supset I_L^1 \supset \cdots \supset I_L^n \supset \cdots$$

in which $I_L^1 = I_L$ and $I_L^{n+1} = I_L \cdot I_L^n$. The algebra $\widehat{UL} = \varprojlim_n UL/I_L^n$ is the *completion* of $UL$. It is augmented with augmentation ideal $\widehat{I}_L = \varprojlim_n I_L/I_L^n$.

Fix a Lie algebra, $L = L_{\geq 0}$. The direct sum $L \oplus L$ is a graded Lie algebra in which the sub Lie algebras $L \oplus 0$ and $0 \oplus L$ commute. These two inclusions extend to inclusions $\lambda, \mu : UL \to U(L \oplus L)$ and we identify $UL \otimes UL = U(L \oplus L)$ via the isomorphism $\alpha \otimes \beta \to \lambda(\alpha)\mu(\beta)$. The *diagonal* is the Lie algebra morphism $\Delta : L \to L \oplus L$, $x \mapsto (x, 0) + (0, x)$, and it extends to the algebra morphisms $U\Delta : UL \to UL \otimes UL$ and to (cf. §1.1, p. 6)

$$\widehat{U\Delta} : \widehat{UL} \to \widehat{U(L \oplus L)} = \widehat{UL} \widehat{\otimes} \widehat{UL}\,.$$

A formal series calculation gives

**Lemma 2.6.** ([60], Chapter 4, [58], Appendix A) *For any Lie algebra $L$ (concentrated in degree zero),*

$$\exp : \widehat{I}_L \to 1 + \widehat{I}_L \quad \textit{and} \quad \log : 1 + \widehat{I}_L \to \widehat{I}_L$$

*are inverse bijections, where*

$$\exp x = \sum_{n=0}^{\infty} \frac{1}{n!} x^n \qquad \textit{and} \quad \log(1+x) = \sum_{n \geq 1} \frac{(-1)^{n+1}}{n} x^n\,.$$

**Lemma 2.7.** *Let $\varphi : E \to L$ be a homomorphism of Lie algebras (concentrated in degree zero), and let $I_E \subset UE$ be the augmentation ideal. Then the diagram*

$$\begin{array}{ccc} \widehat{I}_E & \underset{\log}{\overset{\exp}{\rightleftarrows}} & 1+\widehat{I}_E \\ \widehat{U\varphi}\downarrow & & \downarrow\widehat{U\varphi} \\ \widehat{I}_L & \underset{\log}{\overset{\exp}{\rightleftarrows}} & 1+\widehat{I}_L \end{array}$$

*commutes.*

*proof:* Immediate from the fact that $U(\varphi)(x^n) = [U(\varphi)x]^n$.

□

**Lemma 2.8.** *For any Lie algebra $L$ (concentrated in degree zero), if $x, y$ are commuting elements in $\widehat{I}_L$ then*

$$\exp(x+y) = \exp x \cdot \exp y \quad \textit{and} \quad \log[(1+x)(1+y)] = \log(1+x) + \log(1+y)\,.$$

*proof:* The first assertion is immediate from a series calculation. For the second note first that a series calculation shows that $\log(1+x)$ commutes with $\log(1+y)$. Thus

$$\begin{aligned} \exp(\log(1+x)+\log(1+y)) &= \exp\log(1+x)\cdot\exp\log(1+y) \\ &= (1+x)(1+y) \\ &= \exp\log[(1+x)(1+y)]. \end{aligned}$$

Since exp is injective in $\widehat{I}_L$ the second assertion follows.

□

Next, let $L = L_{\geq 0}$ be a graded Lie algebra, and define $P_L \subset \widehat{I_L}$ and $G_L \subset 1 + \widehat{I_{L_0}}$ by

$$P_L = \{\, x \in \widehat{I_L} \mid \widehat{U\Delta}(x) = x\widehat{\otimes}1 + 1\widehat{\otimes}x \,\}, \qquad \text{and}$$

$$G_L = \{\, 1+y \in 1+\widehat{I_{L_0}} \mid \widehat{U\Delta}(1+y) = (1+y)\widehat{\otimes}(1+y) \,\}\,.$$

**Lemma 2.9.** exp *and* log *restrict to inverse bijections*

$$P_{L_0} \to G_L \qquad \textit{and} \qquad G_L \to P_{L_0}\,.$$

*proof:* If $x \in P_{L_0}$ then by naturality of exp and log (Lemma 2.7), $\widehat{U\Delta}(\exp x) = \exp(x\widehat{\otimes}1 + 1\widehat{\otimes}x)$. Since $1\widehat{\otimes}x$ and $x\widehat{\otimes}1$ commute, $\exp(x\widehat{\otimes}1 + 1\widehat{\otimes}x) = [\exp x\widehat{\otimes}1][1\widehat{\otimes}\exp x] = \exp x\widehat{\otimes}\exp x$. Thus $\exp x \in G_L$.

If $1+y \in G_L$ then by Lemma 2.7,

$$\begin{aligned}(\widehat{U\Delta})(\log(1+y)) &= \log \widehat{U\Delta}(1+y) = \log\left((1+y)\widehat{\otimes}(1+y)\right) \\ &= \log\left([(1+y)\widehat{\otimes}1][1\widehat{\otimes}(1+y)]\right) \\ &= \log(1+y)\widehat{\otimes}1 + 1\widehat{\otimes}\log(1+y)\,,\end{aligned}$$

and so $\log(1+y) \in P_{L_0}$.

Now if $x \in P_{L_0}$ then $x = \log\exp x \in \log(G_L)$ and so log maps $G_L$ onto $P_{L_0}$. Since log is injective it maps $G_L$ bijectively to $P_{L_0}$. Similarly, exp maps $P_{L_0}$ bijectively to $G_L$.

□

**Proposition 2.2.** *Let $L = L_{\geq 0}$ be a graded Lie algebra.*

*(i) $P_L$ is a sub Lie algebra of $\widehat{I}_L$, and the map $L \to \widehat{UL}$ factors through a Lie algebra homomorphism $L \to P_L$ of graded Lie algebras.*

*(ii) Multiplication in $\widehat{UL}$ restricts to a product $(a,b) \mapsto a \star b$, in $G_L$ which makes $G_L$ into a group.*

*(iii) If $\varphi : E \to L$ is a homomorphism of graded Lie algebras then $U\varphi$ extends to a morphism $\widehat{U\varphi} : \widehat{UE} \to \widehat{UL}$ which in turn restricts to homomorphisms $\varphi_P : P_E \to P_L$ and $\varphi_G : G_E \to G_L$ commuting with* exp *and* log.

*proof :* (i) If $x, y \in P_L$ then

$$\begin{aligned}\widehat{U\Delta}([x,y]) &= \left[\widehat{U\Delta}(x), \widehat{U\Delta}(y)\right] \\ &= [x\widehat{\otimes}1 + 1\widehat{\otimes}x, y\widehat{\otimes}1 + 1\widehat{\otimes}y] \\ &= [x,y]\widehat{\otimes}1 + 1\widehat{\otimes}[x,y]\,.\end{aligned}$$

since $x\widehat{\otimes}1$ commutes with $1\widehat{\otimes}y$ and $1\widehat{\otimes}x$ commutes with $y\widehat{\otimes}1$.

(ii) If $1+x, 1+y \in G_L$ then

$$\begin{aligned}\widehat{U\Delta}\left((1+x)(1+y)\right) &= \widehat{U\Delta}(1+x)\cdot\widehat{U\Delta}(1+y) \\ &= \left[(1+x)\widehat{\otimes}(1+x)\right]\cdot\left[(1+y)\widehat{\otimes}(1+y)\right] \\ &= (1+x)(1+y)\widehat{\otimes}(1+x)(1+y)\,.\end{aligned}$$

Then multiplication in $\widehat{UL}$ restricts to a product in $G_L$. Moreover, if $1+y \in G_L$ we may write $1+y = \exp x$ for some $x \in P_L$. Then $\exp(-x) = 1+z \in G_L$ and

$$(1+y) \star (1+z) = \exp x \star \exp(-x) = 1\,.$$

Thus $1+z = (1+y)^{-1}$ and $G_L$ is a group.

(iii) This is immediate from the definitions.

□

**Definition.** $P_L$ is the *primitive Lie algebra* and $G_L$ is the *exponential group* associated with $L$. Multiplication in $G_L$ will be sometimes denoted by $(a,b) \mapsto a \star b$.

**Proposition 2.3.** *For any graded Lie algebra $L = L_{\geq 0}$, the inclusion $L \to P_L$ extends to an isomorphism $\widehat{L} \xrightarrow{\cong} P_L$. In particular, the primitive Lie algebra $P_L$ is pronilpotent.*

*Thus if $L$ itself is pronilpotent then the homomorphism $L \to P_L$ is an isomorphism.*

Recall the condition,

$$\dim H^1(\wedge V, d) < \infty \qquad \text{and} \ \dim V^i < \infty\,, \qquad i \geq 2\,, \tag{1.5}$$

on a minimal Sullivan algebra considered on p. 54.

**Corollary 2.3.** *Suppose $L$ is the homotopy Lie algebra of a minimal Sullivan algebra.*

*(i) Then the Lie algebra homomorphism $L \to P_L$ is injective.*
*(ii) If the Sullivan algebra satisfies (1.5) then $L \to P_L$ is an isomorphism.*

*proof:* According to Lemma 2.1 (p. 46) the natural map $i : L \to \widehat{L}$ is injective. It follows from the Proposition that $\widehat{L} \to P_{\widehat{L}}$ is injective. Since $L \to \widehat{L} \to P_{\widehat{L}}$ is also the composite $L \to P_L \to P_{\widehat{L}}$, it follows that $L \to P_L$ is injective.

Finally, if (1.5) is satisfied then Proposition 2.1 states that $L \to \widehat{L}$ is an isomorphism and hence $L \to P_L$ is also an isomorphism.

□

**Important Notation:** If $L = L_{\geq 0}$ is a pronilpotent graded Lie algebra we identify $P_L$ with $L$ via the isomorphism of Proposition 2.3. With this identification the inverse bijections of Lemma 2.9 become inverse bijections $\exp_L$ and $\log_L$

$$\mathbf{L_0} \overset{\exp_\mathbf{L}}{\underset{\log_\mathbf{L}}{\rightleftarrows}} \mathbf{G_L}\,.$$

The proof of the proposition requires a number of preliminary lemmas. As usual we identify any Lie algebra $L$ with its image in $UL$.

**Lemma 2.10.** *For any graded Lie algebra $L = L_{\geq 0}$, $L^n = L \cap I_L^n$.*

*proof :* Clearly $L^n \subset L \cap I_L^n$. To prove the reverse inclusion, write $L^k = S^k \oplus L^{k+1}$, $k < n$. Choose a well-ordered basis $w_\alpha$ of $L$ compatible with the direct decomposition

$$L = \overset{n-1}{\underset{i=1}{\oplus}} S^i \;\oplus\; L^n$$

and set $|w_\alpha| = i$ if $w_\alpha \in S^i$ and $|w_\alpha| = n$ if $w_\alpha \in L^n$. Then a vector space basis of $UL$ is provided by the elements $w_{\alpha_1} \dots w_{\alpha_r}$ with $\alpha_1 \leq \cdots \leq \alpha_r$. Define

$$\pi : UL \to L$$

by setting $\pi(w_{\alpha_1} \dots w_{\alpha_r})$ to be the iterated Lie bracket $[w_{\alpha_1}, \cdots, w_{\alpha_r}]$. We shall show that

$$\pi : I_L^n \to L^n\,. \tag{2.3}$$

Since $\pi x = x$, $x \in L$, we will have for $x \in I_L^n \cap L$ that

$$x = \pi x \in \pi(I_L^n) \subset L^n\,.$$

To establish (2.3) we show that for any monomial $w_{\beta_1} \dots w_{\beta_r}$ (without hypothesis on the ordering) that if $\sum_{i=1}^r |w_{\beta_i}| \geq n$ then $\pi(w_{\beta_1} \dots w_{\beta_r}) \in L^n$. This is true by definition when $r = 1$. Suppose it is true for some $r - 1 \geq 1$.

Then for any $w_{\beta_1} \dots w_{\beta_r}$, each $[w_{\beta_i}, w_{\beta_j}]$ is a linear combination of the $w_\alpha$ with $|w_\alpha| \geq |w_{\beta_i}| + |w_{\beta_j}|$. Thus

$$w_{\beta_1} \dots w_{\beta_r} - w_{\beta_1} \dots w_{\beta_{i+1}} w_{\beta_i} w_{\beta_{i+2}} \dots w_{\beta_r} = w_{\beta_1} \dots w_{\beta_{i-1}} [w_{\beta_i}, w_{\beta_{i+1}}] \dots w_{\beta_r}$$

and so by our induction hypothesis, if $\sum |w_{\beta_i}| \geq n$ then

$$\pi(w_{\beta_1} \dots w_{\beta_r} - w_{\beta_1} \dots w_{\beta_{i+1}} w_{\beta_i} w_{\beta_{i+2}} \dots w_{\beta_r}) \in L^n\,.$$

But by definition, if $\beta_{\sigma(1)} \leq \cdots \leq \beta_{\sigma(r)}$, then

$$\pi(w_{\beta_{\sigma(1)}} \dots w_{\beta_{\sigma(r)}}) = [w_{\beta_{\sigma(1)}}, \cdots, w_{\beta_{\sigma(r)}}]\,.$$

Now if $|w_{\beta_{\sigma(i)}}| = k_i$ then $w_{\beta_{\sigma(i)}} \in L^{k_i}$ and so

$$\pi(w_{\beta_{\sigma(1)}} \dots w_{\beta_{\sigma(r)}}) \in L^{k_1+\cdots+k_r},$$

and by hypothesis $\sum k_i \geq n$.

□

**Lemma 2.11.** *For any graded Lie algebra $L = L_{\geq 0}$ and any $n \geq 1$, suppose subspaces $Y^k \subset I_L^k$, $k < n$ satisfy*

$$Y^k \oplus I_L^{k+1} = I_L^k,$$

*and set $Y^n = I_L^n$. Then*

$$I_{L\oplus L}^n = \left(\oplus_{k=1}^n Y^k \otimes I_L^{n-k}\right) \oplus (1 \otimes I_L^n).$$

*proof:* Let $\rho_k : UL \to UL/I_L^{k+1}$ be the projection. Then we may regard $\rho_k \otimes id$ as a linear map

$$\rho_k \otimes id : I_{L\oplus L}^n \to \oplus_{i=1}^k Y^i \otimes I_L^{n-i},$$

since $I_{L\oplus L}^n = \sum_{i+j=n} I_L^i \otimes I_L^j$. Thus if $\Omega \in (I_L^k \otimes I_L) \cap I_{L\oplus L}^n$ we have

$$\rho_k \otimes id(\Omega) \in Y^k \otimes I_L^{n-k} \subset I_{L\oplus L}^n.$$

In other words,

$$\Omega = \Omega_1 + \Omega_2, \qquad \text{with } \Omega_1 \in Y^k \oplus I_L^{n-k}$$

and $\Omega_2 \in (I^{k+1} \otimes I) \cap I_{L\oplus L}^n$. The Lemma follows.

□

Next let $L = L_{\geq 0}$ be a graded Lie algebra and fix $n \geq 1$. Choose a well-ordered basis of $L$ as in the proof of Lemma 2.10 above with $w_\alpha < w_\beta$ if $|w_\alpha| < |w_\beta|$. Then the elements $w_{\alpha_1} w_{\alpha_2} \cdots w_{\alpha_r}$ with $w_{\alpha_1} \leq w_{\alpha_2} \leq \cdots \leq w_{\alpha_r}$ are a basis of $UL$, with 1 corresponding to the empty word.

**Lemma 2.12.** *The elements $w_{\alpha_1} \cdots w_{\alpha_r}$ with $\sum |w_{\alpha_i}| < n$ represent a basis of $UL/I_L^n$. In particular, if $\widehat{L} = \varprojlim_n L/L^n$ then*

$$UL/I_L^n \xrightarrow{\cong} U\widehat{L}/I_{\widehat{L}}^n.$$

*proof:* The diagonal $\Delta : L \to L \oplus L$ induces maps

$$(U\Delta)_n : UL/I_L^n \to (UL \otimes UL)/I_{L\oplus L}^n.$$

Denote by $\rho_n : UL \to UL/I_L^n$ the canonical projection. Since if $\sum |w_{\alpha_i}| > n$ we have $\rho_n(w_{\alpha_1} \cdots w_{\alpha_r}) = 0$, we see that $\rho_n$ maps the linear span of the $w_{\alpha_1} \cdots w_{\alpha_r}$ with $|w_{\alpha_1}| + \cdots + |w_{\alpha_r}| < n$ onto $UL/I_L^n$. Thus we need only to show that the $\rho_n(w_{\alpha_1} \cdots w_{\alpha_r})$ are linearly independent in $UL/I_L^n$.

First, fix a finite set of basis elements $z_i = w_{\alpha_i}$ from the basis $\{w_\alpha\}$, numbered so that $z_1 < \cdots < z_r$. Then let $f = f(z_1, \cdots, z_r)$ be a linear combination of monomials of the form $z_1^{\ell_1} \cdots z_r^{\ell_r}$ with $\sum \ell_i |z_i| < n$, and in which $z_1$ occurs non-trivially. We need to show that

$$f \notin I_L^n .$$

Set $k = |z_1|$ and let $z_1, \ldots, z_q$ be the basis elements with $|z_i| = k$. Then we may write

$$f = \sum_{i=1}^{n} z_1^i \, p_i(z_2, \ldots, z_r) + p_0(z_2, \ldots, z_r)$$

where the polynomials $p_i$, $1 \le i \le n$, may have constant term. Thus

$$U\Delta f = \sum_{j=1}^{q} z_j \otimes q_j(z_1, \ldots, z_r) + I_L^{k+1} \otimes UL ,$$

with $q_1(z_1, \ldots, z_r) = \sum_{i=1}^n z_1^{i-1} p_i(z_2, \ldots, z_r)$.

On the other hand, by Lemma 2.10, the inclusion $L \to UL$ induces injections

$$S^k \to UL/I_L^{k+1} .$$

It follows that $\rho_n z_1, \ldots, \rho_n z_q$ are linearly independent elements in $UL/I_L^{k+1}$. Thus if $f \in I_L^n$ it follows from Lemma 2.11 that each $q_j \in I_L^{n-k}$. But the formula above shows that $q_1$ is a combination of monomials $z_1^{\zeta_1} \ldots z_r^{\zeta_r}$ with $\sum \zeta_i |z_i| < n - k$. This implies by induction on $n$ that $q_1 = 0$, whereas it is clear from the hypothesis above that $q_1 \neq 0$, and so $f \notin I_L^n$.

□

*proof of Proposition 2.3:* It follows from Lemma 2.10 (p. 60) that the inclusion $L \to UL$ maps $L^n$ into $I_L^n$. Hence it extends to a homomorphism $\widehat{L} \to \widehat{UL}$ whose image, clearly, is contained in $P_L$. Now recall the projections $\rho_n : UL \to UL/I_L^{n+1}$. Suppose $x \in P_L \subset \widehat{UL}$. Then $x$ is a sequence of elements $x_n \in UL/I^n$ such that

$$x_{n+1} \to x_n$$

and

$$U\Delta(x_n) - (x_n \otimes 1 + 1 \otimes x_n) \in I_{L \oplus L}^n .$$

Note that $U\Delta$ may be regarded as a sequence of maps

$$UL/I_L^n \to (UL \otimes UL)\,/I_{L\oplus L}^n\,.$$

By Lemma 2.12 (p. 61), the elements $w_{\alpha_1}\cdots w_{\alpha_r}$ with

$$w_{\alpha_1} \le \cdots \le w_{\alpha_r} \quad \text{and} \quad \sum |w_{\alpha_i}| < n$$

represent a basis of $UL/I_L^n$.

The identical argument as given above shows that if $f(w_{\alpha_1},\ldots,w_{\alpha_r})$ is a polynomial with no linear term then $(\rho_n\otimes\rho_n)\circ(U\Delta)(f)$ has a non-trivial component in $\rho_n(I_L)\otimes\rho_n(I_L)$, whereas clearly $U\Delta(\rho_n y) = \rho_n y\otimes 1 + 1\otimes\rho_n y$ for $y\in L$.

It follows that $x_n = \rho_n y_n$ for some $y_n \in L$. Now we have

$$\rho_{n+1}y_{n+1} = x_{n+1} \to x_n = \rho_n y_n\,.$$

Thus $y_{n+1} - y_n \in I_L^{n+1}$. By Lemma 2.10, $y_{n+1} - y_n \in L^{n+1}$. Thus the sequence $(y_n)$ is an element of $\widehat{L}$. It follows that the map $\widehat{L} \to P_L$ is surjective.

On the other hand, by Lemma 2.10 (p. 60), the maps $L/L^n \to UL/I_L^n$ are injective. It follows that $\widehat{L}\to P_L$ is injective and hence an isomorphism. In particular, $P_L$ is pronilpotent.

Finally, if $L$ is pronilpotent then $L\to\widehat{L}$ is an isomorphism, and so $L\to P_L$ is an isomorphism as well.

□

## 2.5 The lower central series of a group

*In this section all Lie algebras are ungraded or, equivalently, graded and concentrated in degree* 0.

If $G$ is any group we denote by $[a,b]$ the commutator $aba^{-1}b^{-1}$ of the elements $a,b\in G$. The *lower central series* of $G$ is the sequence of normal subgroups

$$G = G^1 \supset G^2 \supset \cdots$$

defined inductively by setting $G^n$, $n\ge 2$, to be the subgroup generated by the commutators $[a,b]$ with $a\in G$ and $b\in G^{n-1}$. In particular, $G^2$ is often denoted by $[G,G]$. It is straightforward to check that $G^n$ is generated by iterated commutators $[g_1,[g_2,[\cdots[g_{n-1},g_n]\cdots]$ of length $n$. More generally, if $x_1,\cdots,x_n$ and $g_1,\cdots,g_n$ are elements respectively in a Lie algebra $L$ and a group $G$ consistent with §2.1, we denote their iterated commutators of length $n$ by

$$[x_1, \cdots, x_n] = [x_1, [x_2, \cdots, [x_{n-1}, x_n] \cdots]]$$

and

$$[g_1, \cdots, g_n] = [g_1, [g_2, \cdots [g_{n-1}, g_n] \cdots]] .$$

**Definition.**

(i) A *convergent* sequence $(x_i)$ of elements in a pronilpotent Lie algebra $L$ is a sequence such that $x_{i+1} - x_i \in L^{i+1}$, $i \geq 1$. Its *limit* is the unique $x \in L$ such that $x - x_i \in L^{i+1}$, $i \geq 1$, and we write $x = \lim_i (x_i)$ or $(x_i) \to x$. If $y_i \in L^i$ is a sequence of elements in $L$ we write

$$\sum_i y_i = \lim_n \left( \sum_{i=1}^{n} y_i \right) .$$

(ii) A subspace $E \subset L$ is *closed* if every convergent sequence of elements $x_i \in E$ converges to a limit in $E$.

(iii) A sequence $(g_i)$ of elements in $G_L$ *converges* to $g \in G_L$ if $\log_L g_i$ converges to $\log_L g$. In this case we write $\lim_i g_i = g$ or $(g_i) \to g$.

The objective of this section is to prove

**Theorem 2.2.** *Let $L = L_0$ be a pronilpotent Lie algebra for which $dim\, L/[L, L] < \infty$. Then for $n \geq 1$,*

$$G_L^n = \exp(L^n) \qquad \text{and} \quad G_L \to \varprojlim_n G_L/G_L^n$$

*is an isomorphism.*

We begin with some preliminary results.

**Lemma 2.13.** *Let $L = L_0$ be a pronilpotent Lie algebra for which $dim\, L/[L, L] < \infty$. Then each $L^n$ is closed in $L$.*

*proof:* Let $x_1, \ldots, x_r \in L$ represent a basis of $L/[L, L]$. Suppose by induction that $L^n$ is closed, and let $y_k \in L^{n+1} \cap L^k = L^k$, $k \geq n+1$. Then we may write

$$y_k = \sum_{i=1}^{r} [x_i, z_{k_i}]$$

with $z_{k_i} \in L^{k-1}$. By our induction hypothesis, $\sum_k z_{k_i} \in L^n$. Thus $\sum y_k = \sum_{i,k} [x_i, z_{k_i}] = \sum_{i=1}^r [x_i, \sum_k z_{k_i}] \in L^{n+1}$, and so $L^{n+1}$ is closed.

□

Next, let $\mathbb{L}(u,v) = \mathbb{L}$ denote the free Lie algebra in $u$ and $v$, and use the inclusion $\mathbb{L} \to P_{\mathbb{L}}$ to identify $u$ and $v$ as elements of $P_{\mathbb{L}}$. Since $P_{\mathbb{L}} \cong \widehat{\mathbb{L}}$ (Proposition 2.3), we have

$$\log(\exp u \star \exp v) = u + v + \xi(u,v)$$

where $\xi(u,v) = \sum_{i\geq 2} \xi_i(u,v)$ and $\xi_i(u,v)$ is a linear combination of iterated Lie brackets of length $i$ in the elements $u$ and $v$. In particular $\xi_2(u,v) = \frac{1}{2}[u,v]$.

**Proposition 2.4.** *Let $L = L_0$ be a pronilpotent Lie algebra. Then for $x, y \in L$,*

$$\log_L(\exp_L x \star \exp_L y) = x + y + \xi(x,y)\,,$$

*and*

$$\log_L\left([\exp_L x, \exp_L y]\right) = [x,y] + \eta(x,y)\,,$$

*where $\xi(x,y) = \sum_{i\geq 2} \xi_i(x,y)$ and $\eta(x,y) = \sum_{i\geq 3} \eta_i(x,y)$, $\xi_i(x,y)$ and $\eta_i(x,y)$ are linear combinations of iterated Lie brackets of length $i$ in $x$ and $y$.*

*proof:* The first formula follows from the formula above applied to the homomorphism $\mathbb{L}(u,v) \to L$ sending $u$ to $x$ and $v$ to $y$. The second is a straightforward calculation from the first. □

**Lemma 2.14.** *Suppose $L = L_0$ is a pronilpotent Lie algebra for which $\dim L/[L,L] < \infty$. Then the natural homomorphisms*

$$\sigma : \exp_L(L^k) \to \varprojlim_{n\geq k} \exp_L(L^k)/\exp_L(L^n)$$

*are isomorphisms.*

*proof:* Fix $k$. If $\sigma(g) = 1$ then $g \in \exp_L(L^n)$ for all $n \geq k$ and so $\log_L(g) \in \cap_{n\geq k} L^n = 0$. Thus $g = 1$ and $\sigma$ is injective.

Now suppose $(g_n)_{n\geq k}$ is a sequence of elements in $\exp_L(L^k)$ such that the sequence $\overline{g_n} \in \exp_L(L^k)/\exp_L(L^n)$ is an element in the inverse limit. Recall from Lemma 2.13 that each $L^n$ is closed. Then for each $n$, $g_{n+1} =$

$g_n a_n$ for some $a_n \in \exp_L(L^n)$. Write $g_n = \exp_L(x_n)$ and $a_n = \exp_L(y_n)$, where $x_n \in L$ and $y_n \in L^n$. Then, as in the proof of Proposition 2.4, $\xi(x_n, y_n) \in L^{n+1}$ and so

$$x_{n+1} = x_n + y_n + \xi(x_n, y_n) = x_n + z_n$$

with $z_n \in L^n$.

Thus $x = x_n + \sum_{j\geq n} z_j$ is an element in $L$ independent of the choice of $n$. Moreover

$$\exp_L(-x_n) \star \exp_L(x) = \exp_L(x - x_n + \xi(-x_n, x)) .$$

Since $x = x_n + \sum_{j\geq n} z_n$, $\xi(-x_n, x)$ is an infinite sum of iterated brackets of $x_n$ and $\sum_{j\geq n} z_j$, each of which involves at least one $\sum_{j\geq n} z_j$. Thus $\xi(-x_n, x) \in L^n$, as does $x - x_n$. It follows that

$$\exp_L x = (\exp_L(x_n)) \star b_n = g_n \star b_n$$

for some $b_n \in \exp_L(L^n)$. Thus $\sigma(x) = (\overline{g_n})$ and $\sigma$ is surjective.

□

**Lemma 2.15.** *Suppose $L = L_0$ is a pronilpotent Lie algebra such that $(a_i) \to a$ and $(b_i) \to b$ are convergent sequences in $G_L$. Then*

$$(a_i \star b_i) \to a \star b \quad \text{and} \quad [a_i, b_i] \to [a, b] .$$

*proof:* Write $x_i = \log_L a_i$, $y_i = \log_L b_i$, $x = \log_L a$ and $y = \log_L b$. Then it follows from Proposition 2.4 that

$$\log_L(a\star b) - \log_L(a_i \star b_i) = x + y + \xi(x,y) - x_i - y_i - \xi(x_i, y_i) \in L^{i+1} , \qquad i \geq 1 .$$

This proves the first assertion and the second follows from the first.

□

*proof of Theorem 2.2:* By Lemma 2.13, each $L^k$ is closed. Thus it follows from Proposition 2.4 that $G^n \subset \exp L^n$ for all $n$. Now we show by induction on $n$ that

$$\exp L^n \subset G^n , \qquad n \geq 1 .$$

This is true by definition for $n = 1$. Suppose this is established for some $n \geq 1$ and fix an element $x \in L^{n+1}$. Then for some finite set of elements $x_1, \cdots, x_r$ in $L$ we may write $x$ as a sum of iterated Lie brackets in the $x_i$ of length $n + 1$.

Let $E(i)$ be the linear span of the iterated Lie brackets of length $i$ in $x_1, \cdots, x_r$. Define a descending sequence of Lie algebras

$$E = E^1 \supset \cdots \supset E^k \supset \cdots$$

by setting

$$E^k = \left\{ \sum_{i\geq k} u_i \,|\, u_i \in E(i) \right\}.$$

In particular, each $E^k$ is an ideal in $E$.

Now set $a_i = \exp_L x_i$, $1 \leq i \leq r$. For each $i$, $1 \leq i \leq r$, we shall define a convergent sequence $y_{i,k}$ of elements in $E^n$ such that with $b_{j,k} = \exp_L y_{j,k}$ we have

$$x - \log_L([a_1, b_{1,k}] \star \cdots \star [a_r, b_{r,k}]) \in E^{n+k+1}. \tag{2.4}$$

For this first note that by definition there are elements $y_{i,1} \in E^n$ such that

$$x = \sum_{j=1}^{r} [x_i, y_{i,1}].$$

It follows from Proposition 2.4 that

$$x - \log_L([a_1, b_{1,1}] \star \cdots \star [a_r, b_{r,1}]) \in E^{n+2}.$$

Now suppose we have constructed $y_{1,k}, \cdots, y_{r,k}$ to satisfy (2.4). Then we may write

$$x - \log_L([a_1, b_{1,k}] \star \cdots \star [a_r, b_{r,k}]) = \sum_{i=1}^{r} [x_i, z_i]$$

with $z_i \in E^{n+k}$. Set $c_i = \exp_L z_i$ and

$$d_i = b_{i,k} \star [a_i, c_i] \star b_{i,k}^{-1}.$$

Then $\log_L d_i - [x_i, z_i] \in E^{n+2+k}$. It follows that

$$x - \log_L([a_1, b_{1,k}] \star \cdots \star [a_r, b_{r,k}] \star d_1 \star \cdots \star d_r) \in E^{n+k+2}.$$

On the other hand, it follows from Proposition 2.4 that

$$\log_L[[a_j, b_{j,k}], d_i] \in E^{n+k+2}.$$

Thus

$$x - \log_L([a_1, b_{1,k}] \star d_1 \star [a_2, d_{2,k}] \star d_2 \star \cdots \star [a_r, b_{r,k}] \star d_r) \in E^{n+k+2}.$$

But for any elements $a, b, c$ in an arbitrary group $G$,

$$[a, bc] = [a, b]b[a, c]b^{-1}.$$

Apply this with $a_i$, $b_{i,k}$ and $c_i$ to conclude that

$$x - \log_L([a_1, b_{1,k} \star c_1] \star \cdots \star [a_r, b_{r,k} \star c_r]) \in E^{n+k+2}.$$

Now set $y_{i,k+1} = \log_L(b_{i,k} \star c_i)$.

This completes the construction of the sequences $y_{i,k}$. By Proposition 2.4, $y_{i,k+1} - y_{i,k} \in E^{n+k}$ and so the sequences are convergent. Let $y_i = \lim_k y_{i,k}$ and $b_i = \exp_L y_i$. Then

$$x = \log_L\left([a_1, b_1] \star \cdots \star [a_r, b_r]\right) .$$

But each $y_{i,k} \in E^n$ and so, since by Lemma 2.13 $L^n$ is closed, it follows that $y_i \in L^n$. By our induction hypothesis, $\exp_L(L^n) \subset G_L^n$ and so the formula above shows that $\exp_L(L^{n+1}) \subset G_L^{n+1}$. This closes the induction, and completes the proof that $G_L^n = \exp_L(L^n)$ for all $n$. □

**Proposition 2.5.** *Let $L = L_0$ be a pronilpotent Lie algebra.*

*(i) If $E \subset L$ is a closed Lie subalgebra then $\exp_L(E)$ is a subgroup of $G_L$.*
*(ii) If $I \subset L$ is a closed ideal, then $\exp_L(I)$ is a normal subgroup of $G_L$.*
*(iii) If $I \subset L$ is a closed ideal and $S \subset L$ is a subspace such that $S \cap I = 0$ and $S \oplus I$ is a closed sub Lie algebra $E \subset L$, then $\exp_E$ restricts to bijections*

$$y + I \mapsto \exp_E y \star G_I\,, \qquad y \in E\,.$$

*(iv) With the hypotheses of (iii) there is a bijection $\sigma : E/I \to G_E/G_I$ and there is an isomorphism $G_{E/I} \xrightarrow{\cong} G_E/G_I$, uniquely determined such that the diagrams*

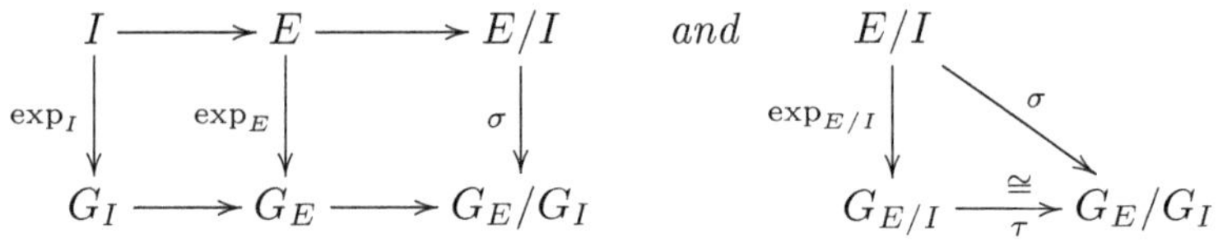


*commute.*

*proof:* (i) If $u, v \in E$ then $u + v + \xi(u, v) \in E$ because $E$ is a closed sub Lie algebra. Thus it follows from Proposition 2.4 that $\exp_L(E)$ is closed under products. It also follows from Proposition 2.4 that $\exp_L(u)^{-1} = \exp_L(-u)$ for all $u \in E$ and so $\exp_L(E)$ is a subgroup.

(ii) It follows from (i) that $\exp_L(I)$ is a subgroup of $G_L$. Furthermore, it follows from Proposition 2.4 that if $x \in L$ and $y \in I$ then for some $z \in I$,

$$\exp_L(x) \star \exp_L(y) = \exp_L(x + z)\,.$$

Thus for some $u \in I$,

$$\begin{aligned}\exp_L(x) \star \exp_L(y) \star \exp_L(x)^{-1} &= \exp_L(x + z) \star \exp_L(-x)\\ &= \exp_L(z + u) \in \exp_L(I)\,.\end{aligned}$$

Thus $\exp_L(I)$ is a normal subgroup.

(iii) Since $E = S \oplus I$ is a closed sub Lie algebra of $L$, $G_E = \exp_L(S \oplus L)$ is a subgroup of $G_L$. Now apply Proposition 2.4 (p. 65) as follows.

Suppose $y \in S$ and $x \in I$. Then $\xi(-y, y+x)$ is an infinite sum of iterated brackets of $x$ and $y$ of the form $[x_{i_1}, \ldots, x_{i_r}]$ with some $x_{i_\lambda} = x$. Since $I$ is a closed ideal it follows that $\xi(-y, y+x) \in I$. Thus

$$\log_E(\exp_E(-y) \star \exp_E(y+x)) = x + \xi(-y, y+x) \in I\,.$$

In particular

$$\exp_E(y+x) = \exp_E(y) \star \exp_E(x + \xi(-y, y+x)) \in \exp_E y \star \exp_E(I)\,.$$

On the other hand, if $z \in I$ then

$$\log_E(\exp_E y \star \exp_E z) = y + z + \xi(y, z) \in y + I\,.$$

Thus $\exp_E(y + I) = \exp_E y \star G_I$.

(iv) It follows immediately from (iii) that the bijections $y + I \to \exp_E y \star G_I$ determine a bijection $\sigma : E/I \to G_E/G_I$ which makes the first diagram commute. Moreover, if $\rho : E \to E/I$ is the projection then $\exp_{E/I} \circ \rho = \sigma \circ \rho$. It follows that

$$\begin{array}{ccc} G_E & & \\ \downarrow & \searrow & \\ G_{E/I} & \xrightarrow[\tau]{} & G_E/G_I \end{array}$$

commutes. The two arrows from $G_E$ are surjective group homomorphisms (respectively by Proposition 2.2(iii), p. 58 and by definition). Thus the bijection $\tau$ is an isomorphism of groups. □

**Corollary 2.4.** *Let $L = L_0$ be a pronilpotent Lie algebra with $\dim L/[L,L] < \infty$, and for $n \leq k$ denote by $\sigma^{n,k} : L^n/L^k \to G_L^n/G_L^k$ the bijections of Proposition 2.5(iii). Then*

*(i) For $m \leq n \leq k \leq \ell$ the diagrams*

$$\begin{array}{ccc} L^n/L^k & \xrightarrow{\sigma^{n,k}} & G_L^n/G_L^k \\ \downarrow & & \downarrow \\ L^m/L^\ell & \xrightarrow{\sigma^{m,\ell}} & G_L^m/G_L^\ell \end{array}$$

*commute.*

*(ii) For each $n \geq 1$, $\sigma^{n,n+1} : L^n/L^{n+1} \to G_L^n/G_L^{n+1}$ is an isomorphism of abelian groups.*

*proof:* (i) By Lemma 2.13, each $L^n$ is closed in $L$. Thus (i) follows from Theorem 2.2.

(ii) This is immediate from Proposition 2.4.

□

For any group $G$ the quotients $G^q/G^{q+1}$ are abelian groups, and it follows from a classical identity of Hall ( [45], Theorem 5.3) that conjugation $[\,,\,] : (a,b) \mapsto aba^{-1}b^{-1}$ induces additive maps

$$[\,,\,] : G^q/G^{q+1} \otimes G^n/G^{n+1} \to G^{q+n}/G^{q+n+1}\,,$$

which make the direct sum, $\oplus_{k\geq 1} G^k/G^{k+1}$, into a Lie algebra.

**Corollary 2.5.** *For $q, n \geq 1$, the diagram*

$$\begin{array}{ccc} L^q/L^{q+1} \otimes L^n/L^{n+1} & \xrightarrow{\sigma^{q,q+1}\otimes\sigma^{n,n+1}} & G_L^q/G_L^{q+1} \otimes G_L^n/G_L^{n+1} \\ \big\downarrow{\scriptstyle [-,-]} & & \big\downarrow{\scriptstyle [-,-]} \\ L^{n+q}/L^{n+q+1} & \xrightarrow[\cong]{\sigma^{n+q,n+q+1}} & G_L^{n+q}/G_L^{n+q+1} \end{array}$$

*commutes. In particular, $[\,,\,]$ makes the direct sum, $\oplus_{k\geq 1} L^k/L^{k+1}$ into a Lie algebra.*

*proof:* This follows directly from Corollary 2.4 and Proposition 2.4.

□

**Example.** Using the bijection $\exp_L : L \to G_L$ we can equip a pronilpotent graded Lie algebra $L$ with the structure of a group which makes $\exp_L$ an isomorphism of groups. Denote this product $a \diamond b$. By naturality, in the case $L = \mathbb{L}(a,b)/\mathbb{L}(a,b)^4$, where $\mathbb{L}(a,b)$ is the free Lie algebra on generators $a, b$, we deduce that there are natural constants $\alpha, \beta$ and $\gamma$ such that in $L$

$$a \diamond b = a + b + \alpha[a,b] + \beta[a,[a,b]] + \gamma[b,[a,b]]\,.$$

Now a straightforward computation, using $\exp_L(a \diamond b) = \exp_L a \star \exp_L b$ shows that $\alpha = \frac{1}{2}$, $\beta = \frac{1}{12}$ and $\gamma = -\frac{1}{12}$. Thus in $L$

$$a \diamond b = a + b + \frac{1}{2}[a,b] + \frac{1}{12}[a,[a,b]] - \frac{1}{12}[b,[a,b]]\,.$$

This formula is a particular case of the Campbell-Baker-Hausdorff formula ([48], [54], [58]) for pronilpotent Lie algebras. The general formula for pronilpotent Lie algebras can be found in the literature, for instance in [60].

Consider now the Sullivan algebra $(\wedge(x,y,z,t),d)$ with all generators in degree 1 and $dx = dy = 0$, $dz = xy$ and $dt = xz$. The associated Lie algebra $L$ is

$$L = \mathbb{L}(a,b)/(\mathbb{L}(a,b)^4, [b,[a,b]])\,,$$

where $\langle x, sa\rangle = \langle y, sb\rangle = 1$ and the other pairings are zero. Then with $G_L$ identified with $L$, the multiplication is given by

$$a \diamond b = a + b + \frac{1}{2}[a,b] + \frac{1}{12}[a,[a,b]]$$

and

$$b \diamond a = a + b - \frac{1}{2}[a,b] + \frac{1}{12}[a,[a,b]]\,.$$

□

## 2.6 The linear isomorphism $(\wedge sV)^\# \cong \widehat{UL}$

Recall that the suspension of a graded vector space $V$ is the graded vector space $sV$ defined by $(sV)^p = V^{p+1}$. Thus if $(\wedge V,d)$ is a minimal Sullivan algebra then as a graded vector space its homotopy Lie algebra, $L$, is given by $L = (sV)^\#$. Recall as well that the homotopy Lie algebra of $(\wedge V,d)\otimes(\wedge V,d)$ is given by $L\oplus L$ and (§2.4) $\widehat{U(L\oplus L)} = \widehat{UL}\widehat{\otimes}\widehat{UL}$.

**Theorem 2.3.** *Let $L$ be the homotopy Lie algebra of a minimal Sullivan algebra $(\wedge V,d)$ satisfying*

$$\dim H^1(\wedge V,d) < \infty \qquad \text{and} \quad \dim V^i < \infty\,, \qquad i \geq 2\,. \tag{1.5}$$

*Then*

*(i) There is a natural isomorphism of graded vector spaces*

$$\gamma_L : (\wedge sV)^\# \xrightarrow{\cong} \widehat{UL}\,.$$

*(ii) The isomorphism*

$$\gamma_{L\oplus L} : (\wedge sV \otimes \wedge sV)^\# \to \widehat{UL}\widehat{\otimes}\widehat{UL}$$

*identifies the dual of the multiplication map $\mu : \wedge sV\otimes\wedge sV \to \wedge sV$ with the diagonal $\Delta : \widehat{UL} \to \widehat{UL}\widehat{\otimes}\widehat{UL}$.*

Before beginning the proof we need two lemmas about completions. For this we first suppose a graded Lie algebra $L = \{L_p\}_{p\geq 0}$ is the inverse limit $L = \varprojlim_n E(n)$ of a sequence of surjections of graded Lie algebras

$$0 \leftarrow E(1) \leftarrow E(2) \leftarrow \cdots \leftarrow E(n) \leftarrow \cdots$$

with the property that given integers $q$ and $p$ there is an integer $k = k(q,p)$ such that for $n \geq k$,

$$\ker (L \to E(n))_p \subset (L^q)_p\,. \tag{2.5}$$

**Lemma 2.16.** *With the hypotheses and notation above, the surjections $UL \to UE(n)$ induce an isomorphism*

$$\widehat{UL} \xrightarrow{\cong} \varprojlim_n \widehat{UE(n)}\,.$$

*proof:* Consider first the surjection $\rho : L \to L/L^q$, some $q \geq 2$, and for simplicity denote $L/L^q$ by $F$. We show first that for $r \leq q$ the induced homomorphism

$$UL/I_L^r \to UF/I_F^r \tag{2.6}$$

is an isomorphism.

In fact, it follows from the Poincaré-Birkoff-Witt theorem ( [18], Theorem 21.1) that if $\sigma : UF \to UL$ is a right inverse for $U\rho$ then multiplication defines an isomorphism

$$\sigma(UF) \otimes UL^q \xrightarrow{\cong} UL\,.$$

We identify $UL$ with $\sigma(UF) \otimes UL^q$ via this isomorphism. This identifies

$$\ker U\rho = \sigma(UF) \otimes I_{L^q}\,, \qquad \text{and } I_L = \sigma(I_F) \oplus (\sigma(UF) \otimes I_{L^q})\,.$$

Next we show that

$$I_L^r = \sigma(I_F^r) \oplus (\sigma(UF) \otimes I_{L^q})\,. \tag{2.7}$$

In fact, since $q \geq r$, $I_{L^q} \subset I_L^r$. It follows that $\sigma(UF) \otimes I_{L^q} \subset I_L^r$. Moreover, since for $x_1, \cdots, x_r \in F$,

$$\sigma(x_1 \cdots x_r) - \sigma(x_1) \cdots \sigma(x_r) \in \ker U\rho = \sigma(UF) \otimes I_{L^q}$$

it follows that $\sigma(I_F^r) \subset I_L^r$. On the other hand, if $y_i \in L$, $1 \leq i \leq r$, write $y_i = \sigma(x_i) + z_i$ with $z_i \in L^q$. Then

$$y_1 \cdots y_r = \sigma(x_1 \cdots x_r) + \Phi$$

with $\Phi \in \sigma(UF) \otimes I_{L^q}$. Thus $I_L^r \subset \sigma(I_F^r) + \sigma(UF) \otimes I_{L^q}$. This establishes (2.7) and (2.6) follows immediately.

Now consider the situation of the Lemma. Given $p$ and $q$ there is an integer $k = k(q,p)$ such that for $n \geq k$,

$$\ker (L \to E(n))_p \subset (L^q)_p \,.$$

This implies that if $n \geq k$ the induced homomorphism

$$L/L^q \to E(n)/E(n)^q$$

is an isomorphism in degrees $\leq p$. As above simplify notation by writing $F = L/L^q$ and $F(n) = E(n)/E(n)^q$. Thus if $n \geq k$, $F_{\leq p} \xrightarrow{\cong} F(n)_{\leq p}$ and we get the commutative diagram for $r \leq q$,

$$\begin{array}{ccc}
(UL/I_L^r)_{\leq p} & \longrightarrow & \left(UE(n)/I^r_{E(n)}\right)_{\leq p} \\
\downarrow \cong & & \downarrow \cong \\
(UF/I_F^r)_{\leq p} & \xrightarrow{\cong} & \left(UF(n)/I^r_{F(n)}\right)_{\leq p} ,
\end{array}$$

in which the vertical arrows are isomorphisms by (2.6).

Since this holds for $n \geq k$ it follows that for every $p$

$$(UL/I_L^r)_{\leq p} \xrightarrow{\cong} \varprojlim_n \left(UE(n)/I^r_{E(n)}\right)_{\leq p} ,$$

and hence that

$$\varprojlim_r UL/I_L^r \xrightarrow{\cong} \varprojlim_r \varprojlim_n UE(n)/I^r_{E(n)} = \varprojlim_n \varprojlim_r UE(n)/I^r_{E(n)} \,.$$

□

On the other hand (Proposition 21.2 in [18]) for any graded Lie algebra $L$ a natural linear isomorphism

$$\gamma : \wedge L \xrightarrow{\cong} UL \tag{2.8}$$

is given by

$$\gamma(x_1 \wedge \cdots \wedge x_k) = \frac{1}{k!} \sum_{\sigma \in \Sigma_k} \varepsilon_\sigma x_{\sigma(1)} \cdot \cdots \cdot x_{\sigma(k)}$$

where $x_{\sigma(1)} \wedge \cdots \wedge x_{\sigma(k)} = \varepsilon_\sigma x_1 \wedge \cdots \wedge x_k$. The naturality of $\gamma$ implies that it is an isomorphism of graded coalgebras.

By definition, $\gamma$ restricts to linear maps $\gamma_r : \wedge^{\geq r} L \to I_L^r$ and so induces a linear map

$$\widehat{\gamma} : \varprojlim_r \wedge L / \wedge^{\geq r} L \to \varprojlim_r UL/I_L^r \,.$$

**Lemma 2.17.** *If $L$ is a nilpotent graded Lie algebra then $\widehat{\gamma}$ is an isomorphism of graded vector spaces.*

*proof:* Write $L = S^1 \oplus \cdots \oplus S^n$ where $L^k = S^k \oplus L^{k+1}$. Then define

$$S(r) = \bigoplus_{\sum ik_i < r} \wedge^{k_1} S^1 \otimes \cdots \otimes \wedge^{k_n} S^n \subset \wedge L .$$

Obviously $S(r) = \oplus_m S(r) \cap \wedge^m L$. An ordered basis of each $S^i$ determines an ordered basis of monomials of $S(r)$, and by Lemma 2.12 (p. 61) multiplication in $UL$ then defines isomorphisms

$$\omega(r) : S(r) \xrightarrow{\cong} UL/I_L^r .$$

Moreover, if $\Phi \in S(r) \cap \wedge^m L$ it is straightforward to check that $\gamma\Phi = \omega(r)\Psi$ where $\Psi \in S(r)$ and $\Phi - \Psi \in \wedge^{<m} L$. Let $\rho_r : UL \to UL/I_L^r$ be the quotient map. Then this implies that

$$\rho_r \circ \gamma : S(r) \xrightarrow{\cong} UL/I_L^r .$$

Next, define $F(r) = \oplus_{j>r} S(j)$. We have shown that $\gamma$ induces an isomorphism

$$\varprojlim_r \wedge L/F(r) \xrightarrow{\cong} \varprojlim_r UL/I_L^r .$$

Since $\wedge^{nr} L \subset F(r) \subset \wedge^{>r} L$, it follows that $\varprojlim_r \wedge L/F(r) = \varprojlim_r \wedge L/\wedge^r L$. This establishes the Lemma.

□

*proof of Theorem 2.3:* (i) We first consider the case that $V$ (and hence $L$) are finite dimensional. In this case $\wedge^{<r} sV$ and $\wedge L/\wedge^{\geq r} L$ are finite dimensional and so the equality $L = V^\#$ extends to an isomorphism $(\wedge^{<r} sV)^\# \cong \wedge L/\wedge^{\geq r} L$. Thus the isomorphism $\widehat{\gamma}$ of Lemma 2.17 becomes an isomorphism

$$\gamma_L : (\wedge sV)^\# = \varprojlim_r (\wedge^{<r} sV)^\# = \varprojlim_r \wedge L/\wedge^{\geq r} L \xrightarrow{\cong} \widehat{UL} .$$

It is of the form $1 + \widetilde{\gamma_L}$ by construction. This proves (i) when $\dim V < \infty$.

We now establish the Theorem for any $(\wedge V, d)$ satisfying (1.5).

Recall from Lemma 2.3 and Theorem 2.1 that $\wedge V^1$ is the increasing union of minimal Sullivan algebras $(\wedge V_n^1, d_1)$ such that $\dim V_n^1 < \infty$ and the homotopy Lie algebra of $(\wedge V_n^1, d_1)$ is $L_0/L_0^n$. Since $V^{\geq 2}$ has finite type, $(\wedge V, d_1)$ is then the increasing union of sub minimal Sullivan algebras such that $\dim V_n < \infty$ and $V_n^1$ is the subspace just identified.

Denote the homotopy Lie algebra of $(\wedge V_n, d_1)$ by $E(n+1)$. Thus $E_0(n+1) = L_0/L_0^n$. Now by definition, $L$ is the inverse limit of the surjections

$L \to E(n+1)$ dual to the inclusions $V^n \to V$. Moreover, since $V^{\geq 2}$ has finite type it follows that given integers $q$ and $p$ there is a $k = k(p,q)$ such that for $n \geq k$,

$$\ker(L \to E(n+1))_p \subset (L^q)_p\,.$$

Let $\gamma_{E(n+1)} : (\wedge sV_n)^\# \overset{\cong}{\to} \widehat{UE}(n+1)$ be the isomorphism constructed above. It follows then from Lemma 2.17 that

$$\varprojlim_n \gamma_{E(n+1)} : (\wedge sV)^\# \overset{\cong}{\to} \varprojlim_n \widehat{UE}(n+1) = \widehat{UL}\,.$$

Set $\gamma_L = \varprojlim_n \gamma_{E(n+1)}$.

(ii) By naturality with respect to the multiplication morphism $\mu : (\wedge V, d) \otimes (\wedge V, d) \to (\wedge V, d)$ we obtain the commutative diagram

$$\begin{array}{ccc} (\wedge sV)^\# & \xrightarrow{\mu^\#} & (\wedge sV \otimes \wedge sV)^\# \\ \cong \downarrow \gamma_L & & \cong \downarrow \gamma_{L\oplus L} \\ \widehat{UL} & \xrightarrow{\Delta} & \widehat{UL}\widehat{\otimes}\widehat{UL}\,. \end{array}$$

which proves (ii).

□

## 2.7 The fundamental group of a 1-finite minimal Sullivan algebra

Recall from §1.5 that a minimal Sullivan algebra $(\wedge V, d)$ is 1-finite if $H^1(\wedge V, d) = H^1(\wedge V^1, d)$ is finite dimensional. Thus according to Corollary 2.2 (p. 51) the fundamental Lie algebra, $L_0$, of a 1-finite minimal Sullivan algebra is pronilpotent. Hence, by Proposition 2.3, $L_0$ maps isomorphically to $P_{L_0}$ and $\exp_{L_0}$ and $\log_{L_0}$ are inverse bijections connecting $L_0$ with the exponential group $G_L$.

**Definition.** The exponential group of the fundamental Lie algebra of a 1-finite minimal Sullivan algebra $(\wedge V, d)$ is called the *fundamental group* of $(\wedge V, d)$ and is denoted by $G_L$ (where $L$ is the homotopy Lie algebra of $(\wedge V, d)$). We write $\pi_1(\wedge V, d) = G_L$.

**Theorem 2.4.** *If $L_0$ is the fundamental Lie algebra of a 1-finite minimal Sullivan algebra $(\wedge V, d)$, then the composite*

$$\exp_{L_0} \circ \iota_1 : \pi_1(|\wedge V, d|, *) \to G_L$$

*is a natural isomorphism of groups.*

**Remark.** Let $(X,*)$ be a path connected CW complex, and suppose $\varphi : (\wedge V, d) \to A_{PL}(X)$ is a morphism from a minimal Sullivan algebra. Then as observed in §1.8, $\varphi$ induces a continuous map $|\varphi| : X \to |\wedge V, d|$ and thus, if $(\wedge V, d)$ is 1-finite, it induces a homomorphism of groups

$$\exp_{L_0} \circ \iota_1 \circ \pi_1(|\varphi|) : \pi_1(X,*) \to G_L \,.$$

For the proof of Theorem 2.4 we first need to set our convention for the product in the fundamental group of a pointed topological space $(X,*)$, as there are two possibilities. We make the following choice: let $\alpha, \beta \in \pi_1(X,*)$ be represented by loops $f : (I,0,1) \to (X,*,*)$ and $g : (I,0,1) \to (X,*,*)$. Then $\alpha \star \beta \in \pi_1(X,*)$ is the element represented by the loop $f \star g$ defined by

$$(f \star g)(t) = \begin{cases} f(2t)\,, & 0 \le t \le 1/2\,, \\ g(2t-1)\,, & 1/2 \le t \le 1\,. \end{cases}$$

Before starting the proof we also need to establish some preliminary machinery. Let $L_0(s,u) = L_0 \otimes \wedge(s,u)$, where $s$ and $u$ are variables of degree zero. Then denote by $F$ the completion of the Lie algebra $L_0(s,u)$,

$$F = \varprojlim_n L_0(s,u)/L_0(s,u)^n \,.$$

The derivations $\frac{d}{ds}$ and $\frac{d}{du}$ in $L_0(s,u)$ then extend uniquely to derivations in $\widehat{UF}$, and we denote these also by $\frac{d}{ds}$ and $\frac{d}{du}$. In particular, we may define a cdga,

$$(A, d_A) = (\widehat{UF} \otimes \wedge(ds, du), d_A)\,,$$

by setting $d_A(\Phi \otimes 1) = \frac{d\Phi}{ds} \otimes ds + \frac{d\Phi}{du} \otimes du$, since clearly $\frac{d^2}{dsdu}\Phi = \frac{d^2}{duds}\Phi\,.$

**Lemma 2.18.** *If $g \in G_F$ then $g^{-1}\frac{dg}{ds} = \frac{d}{ds}\log_F g$ and $g^{-1}\frac{dg}{du} = \frac{d}{du}\log_F g$. In particular, these are both elements in $F$.*

*proof:* Immediate from the definitions.

□

*proof of Theorem 2.4:* For simplicity, we denote the multiplication in $\widehat{UF}$ by $a \otimes b \mapsto ab$ and suppress the $*$ in the notation for the product in $G_F$ (p. 59). Suppose now that $x, y \in L_0$ and let $z = \log_{L_0}(\exp_{L_0} x \exp_{L_0} y)$. Define $a \in G_F$ by

$$a = (\exp_F uy)^{-1}(\exp_F ux)^{-1}\exp_F uz\,.$$

Then we write

$$a = 1 + \sum_{k\geq 1} a_k u^k$$

with $a_k \in I^k_{L_0}$. Since by Lemma 2.10 (p. 60), $L_0^k = L_0 \cap I_L^k$ it follows that

$$\log_F a = \sum_{k\geq 1} q_k u^k$$

with $q_k \in L_0^k$. Moreover,

$$\begin{aligned} 1 + \sum_{k\geq 1} a_k &= (\exp_{L_0} y)^{-1} (\exp_{L_0} x)^{-1} \exp_{L_0} z \\ &= (\exp_{L_0} y)^{-1} (\exp_{L_0} x)^{-1} \exp_{L_0} x \exp_{L_0} y = 1 \,. \end{aligned}$$

It follows that $\log_{L_0}(1 + \sum_{n\geq 1} a_n) = 0$; i.e. $0 = \sum_{k\geq 1} q_k$. In particular, since $\sum_{i=1}^k q_i = -\sum_{i>k} q_i \in L_0^{k+1}$ we may form the element

$$\sum_{k\geq 1} (q_1 + \cdots + q_k) u^{k-1} \in F \,.$$

A straightforward check shows that

$$u(1-u) \sum_{k=1}^n (q_1 + \cdots + q_k) u^{k-1} = \sum_{i=1}^n q_i u^i - u^{n+1} (\sum_{i=1}^n q_i) \,.$$

Since $\sum_{i=1}^n q_i \in L_0^{n+1}$ we obtain that

$$\log_F a = u(1-u) \sum_{k\geq 1} (q_1 + \cdots q_k) u^{k-1} \,.$$

Now define $g \in G_F$ by

$$g = \exp_F sx \exp_F ux \exp_F uy \exp_F \big(u(1-u-s) \sum_{k\geq 1} (q_1 + \cdots + q_k) u^{k-1}\big) \,.$$

Then set $\ell_s = g^{-1} \frac{dg}{ds}$ and $\ell_u = g^{-1} \frac{dg}{du}$; both $\ell_s$ and $\ell_u$ are elements in $F$. Thus a morphism of graded algebras, $\varphi : \wedge V^1 \to \wedge(s, u, ds, du)$ is given by

$$\varphi w = -\langle w, \ell_s \rangle \otimes ds - \langle w, \ell_u \rangle \otimes du \,, \qquad w \in V^1 \,.$$

(Note that because $w$ vanishes on $L_0^n$ for sufficiently large $n$ it follows that $\varphi$ does map into $\wedge(s, u, ds, du)$.)

We now show that $\varphi$ is a cdga morphism: $\varphi d = d\varphi$. First observe that since $\frac{d}{du}(gg^{-1}) = 0$ it follows that $\frac{dg^{-1}}{du} = -g^{-1} \frac{dg}{du} g^{-1}$. Thus

$$\frac{d}{du}(\ell_s) = -\ell_u \ell_s + g^{-1} \frac{d^2 g}{duds} \,.$$

Reversing the roles of $s$ and $u$ and then adding, we find that for $w \in V^1$,

$$d\varphi w = \left\langle w, \frac{d\ell_s}{du}\right\rangle \otimes ds \wedge du - \left\langle w, \frac{d\ell_u}{ds}\right\rangle \otimes ds \wedge du = \langle w, [\ell_s, \ell_u]\rangle \otimes ds \wedge du .$$

On the other hand, writing $dw = \sum w_i \wedge w_j$ we use the definition of the Lie bracket in $L$ at the start of this chapter and we find that

$$\begin{aligned}
\varphi dw &= \sum \varphi w_i \wedge \varphi w_j \\
&= \sum (\langle w_i, \ell_s\rangle\langle w_j, \ell_u\rangle - \langle w_i, \ell_u\rangle\langle w_j, \ell_s\rangle) \otimes ds \wedge du \\
&= -\sum \langle w_i \wedge w_j, \ell_s, \ell_u\rangle \otimes ds \wedge du = -\langle dw, \ell_s, \ell_u\rangle \otimes ds \wedge du \\
&= \langle w, [\ell_s, \ell_u]\rangle \otimes ds \wedge du = d\varphi w .
\end{aligned}$$

Next, recall from the Example of §1.4 that the correspondences $s \mapsto t_1, u \mapsto t_2$ define an isomorphism $\wedge(s, u, ds, du) \xrightarrow{\cong} (A_{PL})_2$ while the correspondence $t \mapsto t_1$ defines an isomorphism $\wedge(t, dt) \xrightarrow{\cong} (A_{PL})_1$. In particular, the first isomorphism identifies $\varphi$ as a 2-simplex in $\langle \wedge V^1, d\rangle$. Moreover, let $\lambda_0$, $\lambda_1$ and $\lambda_2$ denote the respective inclusions $\Delta[1] \to \Delta[2]$ corresponding to the edges $\langle e_1, e_2\rangle$, $\langle e_0, e_2\rangle$ and $\langle e_0, e_1\rangle$ (with the notation of (1.1), p. 9, $\lambda_i = \partial_i(id_{\Delta[1]})$). We may identify $\rho_i = A_{PL}(\lambda_i)$ as the morphisms

$$\rho_0 : s, u \mapsto 1 - t, t , \qquad \rho_1 : s, u \mapsto 0, t \quad \text{and } \rho_2 : s, u \mapsto t, 0 .$$

Now we set

$$E = \varprojlim_n L_0 \otimes \wedge t/(L_0 \otimes \wedge t)^n .$$

The morphisms $id \otimes \rho_i : L_0 \otimes \wedge(s, u) \to L_0 \otimes \wedge t$ complete to morphisms $p_i : \widehat{UF} \otimes \wedge(ds, du) \to \widehat{UE} \otimes \wedge dt$. Elementary calculus gives

$$p_i(-\ell_s \otimes ds - \ell_u \otimes du) = -p_i(g^{-1}\frac{dg}{ds} \otimes ds + g^{-1}\frac{dg}{du} \otimes du) = -p_i(g^{-1})\frac{d\rho_i(g)}{dt} \otimes dt .$$

Since for $w \in V^1$, $\rho_i(\varphi w) = \langle w, p_i(-\ell_s \otimes ds - \ell_u \otimes du)\rangle$ it follows from a simple computation that

$$\rho_2\varphi w = -\langle w, x\rangle dt , \qquad \rho_0\varphi w = -\langle w, y\rangle dt , \quad \text{and } \rho_1\varphi w = -\langle w, z\rangle dt .$$

Then note that we may regard $\varphi$ as a morphism $(\wedge V^1, d) \to A_{PL}(\Delta[2])$ whose adjoint $\langle\varphi\rangle : \Delta[2] \to \langle \wedge V^1, d\rangle$ realizes to a continuous map $|\langle\varphi\rangle| : \Delta^2 \to | \wedge V^1, d|$. The restrictions of $|\langle\varphi\rangle|$ to the edges $[e_0, e_1]$, $[e_1, e_2]$ and $[e_0, e_2]$ of $\Delta^2$ are then given respectively by $|\langle\rho_2\varphi\rangle|$, $|\langle\rho_0\varphi\rangle|$ and $|\langle\rho_1\varphi\rangle|$. All these maps send the vertices of $\Delta^2$ to the base point $*$ of $| \wedge V^1, d|$, and so $|\langle\rho_0\varphi\rangle|$, $|\langle\rho_1\varphi\rangle|$ and $|\langle\rho_2\varphi\rangle|$ are based loops in $| \wedge V^1, d|$ and $|\langle\varphi\rangle|$ is a based homotopy

$$|\langle\rho_2\varphi\rangle| * |\langle\rho_0\varphi\rangle| \sim_* |\langle\rho_1\varphi\rangle| .$$

Finally, let $\alpha,\ \beta \in \pi_1(|\wedge V^1, d|, *)$ be arbitrary. Set $\iota_1\alpha = x$ and $\iota_1\beta = y$. Then define $z \in L_0$ by $z = \log_{L_0}(\exp_{L_0} x \exp_{L_0} y)$, Now Theorem 1.3 (p. 37) asserts that there is a unique $\gamma \in \pi_1(|\wedge V^1, d|, *)$ such that $\iota_1\gamma = z$. It follows from the definition of $\iota_1$ in §1.8 that $\alpha, \beta, \gamma$ are represented respectively by $|\langle\rho_2\varphi\rangle|$, $|\langle\rho_0\varphi\rangle|$ and $|\langle\rho_1\varphi\rangle|$. Thus the equation above shows that $\alpha * \beta = \gamma$, i.e. that

$$\exp_{L_0}(\iota_1\alpha)\exp_{L_0}(\iota_1\beta) = \exp_{L_0} x \exp_{L_0} y = \exp_{L_0} z = \exp_{L_0} \iota_1(\alpha * \beta)\,.$$

□

Again, let $(\wedge V, d)$ be a minimal Sullivan algebra with $L_0$ its fundamental Lie algebra. Let $x, y$ and $z$ be the elements in $L_0$ corresponding respectively to $\alpha$, $\beta$ and $\alpha \star \beta$ in $\pi_1(|\wedge V, d|, *)$ under the bijection $\iota_1$.

**Corollary 2.6.** *With the hypotheses and notation above there is a morphism*

$$\varphi : (\wedge V, d) \to \wedge(s, u, ds, du)$$

*such that for* $w \in V^1$,

$$\rho_2\varphi w = -\langle w, x\rangle dt\,, \quad \rho_0\varphi w = -\langle w, y\rangle dt \quad \textit{and} \quad \rho_1\varphi w = -\langle w, z\rangle dt\,.$$

*proof:* It is sufficient to observe that the morphism $\varphi : (\wedge V^1, d) \to \wedge(s, u, ds, du)$ constructed in the proof of Theorem 2.4 extends to $(\wedge V, d)$ via the requirement $\varphi v = 0$, $v \in V^{\geq 2}$. This holds because for $v \in V^{\geq 2}$, $\deg dv \geq 3$ and $(A_{PL})_2$ vanishes in those degrees.

□

Now suppose $\varphi : (\wedge V, d) \to A_{PL}(X)$ and $\psi : (\wedge V, d) \to (\wedge U, d)$ are morphisms from a 1-finite minimal Sullivan model respectively to the PL forms on a based topological space $(X, *)$ and a minimal Sullivan algebra $(\wedge U, d)$. Again denote the fundamental Lie algebra of $(\wedge V, d)$ by $L_0$.

Recall from §1.8 that $|\langle\varphi\rangle| : (|Sing\, X|, *) \to (|\wedge V, d|, *)$ is the spatial realization of the adjoint of $\varphi$ and that we defined $\pi_1(\wedge V, d) = G_L$ and $\pi_1(\varphi) = \exp_{L_0} \circ \iota_1 \circ \pi_1(|\langle\varphi\rangle|)$. Thus, as promised in §1.8 we have

**Corollary 2.7.** $\pi_1(\varphi)$ *and* $\pi_1(\psi)$ *are homomorphisms of groups.*

Finally, suppose $(\wedge W, d)$ is a Sullivan 1-algebra with homotopy Lie algebra $L$ and $(X, *)$ is a pointed topological space with fundamental group $G_X = \pi_1(X, *)$.

**Proposition 2.6.** *Suppose $\varphi : (\wedge W, d) \to A_{PL}(X)$ is a morphism. If $(\wedge W, d)$ is 1-finite and $H^1(\varphi)$ is injective then $\pi_1(\varphi)$ induces linear surjections*

$$G_X^n/G_X^{n+1} \otimes_{\mathbb{Z}} k\!\!k \longrightarrow G_L^n/G_L^{n+1}\,, \qquad n \geq 1\,.$$

*proof:* First observe by Corollary 2.4 (p. 69) that $\exp_{L_0}$ induces isomorphisms of abelian groups $\overline{\exp_{L_0}} : L_0^n/L_0^{n+1} \xrightarrow{\cong} G_L^n/G_L^{n+1}$, thereby endowing $G_L^n/G_L^{n+1}$ with the structure of a $k\!\!k$-vector space. On the other hand, since $\pi_1(\varphi)$ is a homomorphism (Theorem 2.4) it factors to give homomorphisms of abelian groups $G_X^n/G_X^{n+1} \to G_L^n/G_L^{n+1}$, which then extend to linear maps

$$\pi_1(\varphi)^n : G_X^n/G_X^{n+1} \otimes_{\mathbb{Z}} k\!\!k \to G_L^n/G_L^{n+1}\,.$$

It remains to show that the $\pi_1(\varphi)^n$ are surjective.

We first show that

$$\pi_1(\varphi)^1 : G_X/[G_X, G_X] \otimes_{\mathbb{Z}} k\!\!k \to G_L/[G_L, G_L]$$

is surjective. For this note that $H^1(\varphi)$ dualizes to a surjective linear map $(H^1(\varphi))^\# : H_1(X; k\!\!k) \to (H^1(\wedge W, d))^\#$. Further, as described in §1.5, the surjection $\wedge^+ W \to W$ with kernel $\wedge^{\geq 2} W$ induces a linear map $\zeta : H^1(\wedge W, d) \to W$, which then dualizes to $\mathrm{Hom}(\zeta) : L_0 \to (H^1(\wedge W, d))^\#$.

Now Proposition 1.19 (p. 42) asserts that the diagram

$$\begin{array}{ccc} G_X & \xrightarrow{\log_L \circ \pi_1(\varphi)} & L_0 \\ \Big\downarrow{\scriptstyle hur} & & \Big\downarrow{\scriptstyle \zeta^\#} \\ H_1(X; k\!\!k) & \xrightarrow[(H^1(\varphi))^\#]{} & (H^1(\wedge W, d))^\# \end{array}$$

commutes. It is a standard fact (cf. [66]) that *hur* factors to yield an isomorphism $G_X/[G_X, G_X] \otimes_{\mathbb{Z}} k\!\!k \xrightarrow{\cong} H_1(X; k\!\!k)$. On the other hand, since $H^1(\wedge W, d) = W_0$, it follows from Theorem 2.1 (p. 50) that $\zeta^\#$ factors to give an isomorphism $L_0/[L_0, L_0] \xrightarrow{\cong} (H^1(\wedge W, d))^\#$. Finally, $\log_{L_0} \circ \pi_1(\varphi)$ factors to $\overline{\exp_{L_0}}^{-1} \circ \pi_1(\varphi)^1$, and the diagram

$$\begin{array}{ccc} G_X/[G_X, G_X] \otimes_{\mathbb{Z}} k\!\!k & \xrightarrow{\overline{\exp_L}^{-1} \circ \pi_1(\varphi)^1} & L_0/[L_0, L_0] \\ \Big\downarrow{\scriptstyle \cong} & & \Big\downarrow{\scriptstyle \cong} \\ H_1(X; k\!\!k) & \xrightarrow[(H^1(\varphi))^\#]{} & (H^1(\wedge W, d))^\# \end{array}$$

commutes. Since $(H^1(\varphi))^\#$ is surjective so is $\pi_1(\varphi)^1$.

Finally, a classical identity of Hall ( [45], Theorem 5.3) implies that the maps $(a,b) \mapsto aba^{-1}b^{-1}$ factor to natural surjections of abelian groups $G/[G,G] \otimes_{\mathbb{Z}} G^n/G^{n+1} \to G^{n+1}/G^{n+2}$. These give surjections of vector spaces $(G/[G,G] \otimes_{\mathbb{Z}} k) \otimes_k (G^n/G^{n+1} \otimes_{\mathbb{Z}} k) \to G^{n+1}/G^{n+2} \otimes_{\mathbb{Z}} k$ and the Proposition follows by induction on $n$. □

## 2.8 The homology Hopf algebra of a 1-finite minimal Sullivan algebra

Fix a 1-finite minimal Sullivan algebra $(\wedge V, d)$, denote its homotopy Lie algebra by $L$ and its fundamental Lie algebra by $L_0$. The *adjoint representation* of $L$ in $L$ is the representation by derivations given by

$$(ad\, x)(y) = [x,y]\,, \qquad x \in L\,, y \in L\,.$$

**Lemma 2.19.** *The adjoint representation of $L_0$ in $L$ determines a representation of $G_L$ by automorphisms of $L$ given by*

$$Ad(\exp_{L_0} x)(y) = e^{ad\, x}(y)\,, \qquad x \in L_0\,, y \in L\,.$$

*proof:* It follows from Lemma 2.1 that $V$ is the increasing union of subspaces $V(r)$ such that for any $r$ there is a number $q = q(r)$ such that for all $x_1, \cdots, x_q \in L_0$ and all $y \in L$, $ad\, x_1 \circ \cdots \circ ad\, x_q(y)$ vanishes on $V(r)$. This allows us to extend $ad$ to a representation of $\widehat{UL_0}$ in $L$.

Now, recall that $UL$ acts on $L$ by adjoint action

$$(x_1 \cdots x_n) \star \ell = [x_1, [x_2, \cdots [x_n, \ell] \cdots ]]\,.$$

We deduce that $(\exp_{L_0} x) \star \ell = e^{ad\, x}(\ell)$. Suppose $z = xy$ in $G_L$, then $\exp_{L_0} z = \exp_{L_0} x \cdot \exp_{L_0} y$, and

$$e^{ad\, z}(\ell) = (\exp_{L_0} z) \star \ell = (\exp_{L_0} x)(\exp_{L_0} y) \star \ell = e^{ad\, x} e^{ad\, y}(\ell)\,.$$

Thus $Ad$ restricts to a representation $Ad$ of $G_L$ in $L$ which by definition satisfies the equation of the Lemma. Because $ad\, x$ is a derivation it is then immediate that each $Ad(\exp_{L_0}(x))$ is an automorphism.

□

The adjoint representation of $L_0$ in $L$ extends uniquely to a representation (again denoted by $x \mapsto ad\, x$) by derivations of $L_0$ in $\widehat{UL}$. Thus

$$(ad\, x)(\Phi) = x\Phi - \Phi x\,, \qquad \Phi \in \widehat{UL}, x \in L_0\,.$$

As in Lemma 2.19 the adjoint representation $a \mapsto Ad\, a$ of $G_L$ in $L$ extends to the representation $a \mapsto Ad\, a$ of $G_L$ by automorphisms of $\widehat{UL}$ given by $Ad(\exp_{L_0} x) = e^{ad\, x}$. A straightforward induction shows that $\frac{(ad\, x)^n}{n!}\Phi = \sum_{s=0}^{n} \frac{x^s}{s!}\,\Phi\,\frac{(-x)^{n-s}}{n-s!}$, and hence that

$$(Ad\, a)\Phi = a\Phi a^{-1}\,, \qquad a \in G_L, \Phi \in \widehat{UL}\,.$$

**Corollary 2.8.** *For $a \in G_L$ and $y \in L_0$,*

$$\exp_{L_0}(Ad\, a(y)) = a\,(\exp_{L_0} y)\, a^{-1}\,.$$

*proof:* This follows directly from the formula $(Ad\, a)y = aya^{-1}$.

□

Now by construction, the adjoint representation of $G_L$ in $\widehat{UL}$ restricts to a representation of $G_L$ in $UL_{\geq 1}$. Denote by $\mathbb{k}[G_L]$ the Hopf algebra with $G_L$ as a basis and with multiplication and diagonal defined by multiplication in $G_L$ and by $g \mapsto g \otimes g$. The formula above then shows that $\mathbb{k}[G_L] \otimes UL_{\geq 1}$ is a cocommutative graded Hopf algebra, with multiplication given by

$$(a \otimes \Phi) \cdot (b \otimes \Psi) = ab \otimes (b^{-1}\Phi b)\Psi\,,$$

and with diagonal the tensor product of the diagonals.

**Definition.** The graded Hopf algebra, $\mathbb{k}[G_L] \otimes UL_{\geq 1}$ is the *homology Hopf algebra* of the 1-finite minimal Sullivan algebra $(\wedge V, d)$.

Note that the inclusion of $G_L$ and $UL_{\geq 1}$ in $\widehat{UL}$ extend uniquely to a morphism of graded algebras

$$\mathbb{k}[G_L] \otimes UL_{\geq 1} \to \widehat{UL}$$

which is compatible with the diagonals. Recall as well the classical definitions: the adjoint representation of $L_0$ in $L_{\geq 1}$ is *nilpotent* if for each $n \geq 1$ there is an integer $k = k(n)$ such that for all $x_1, \cdots, x_k \in L_0$, the composite $ad(x_1) \circ \cdots \circ ad(x_k)$ vanishes in $L_n$.

**Lemma 2.20.** *Suppose $(\wedge V, d)$ is a 1-finite minimal Sullivan algebra.*

*(i) Multiplication defines an injection*

$$\widehat{UL_0} \otimes UL_{\geq 1} \to \widehat{UL}$$

*and the morphism $\mathbb{k}[G_L] \to \widehat{UL_0}$ is also injective.*

*(ii) If $V^{\geq 2}$ has finite type then the adjoint representation of $L_0$ in $L_{\geq 1}$ is nilpotent.*

*(iii) If the adjoint representation of $L_0$ in $L$ is nilpotent then multiplication defines an isomorphism $\widehat{UL_0} \otimes UL_{\geq 1} \xrightarrow{\cong} \widehat{UL}$.*

*proof:* (i) For any finite dimensional subspace $S \subset L_{\geq 1}$ we may find a sub Sullivan algebra of $(\wedge V, d)$ of the form $(\wedge W, d) = (\wedge V^1 \otimes \wedge W^{\geq 2}, d)$ with homotopy Lie algebra $E = L_0 \oplus E_{\geq 1}$ and with these properties: $\dim W^{\geq 2} < \infty$ and the surjection $L_{\geq 1} \to E_{\geq 1}$ is injective on $S$. Thus if $a \in \widehat{UL_0} \otimes UL_{\geq 1}$ we may choose $(\wedge W, d)$ so that the image $b$ of $a$ in $\widehat{UL_0} \otimes UE_{\geq 1}$ is non-zero. By (iii) below, the image of $b$ in $\widehat{UE}$ is non-zero and so the image of $a$ in $\widehat{UL}$ is non-zero.

To complete the proof of (i) it is sufficient to show that the morphism $k[G_L] \to \widehat{UL_0}$ is injective. If not there are distinct elements $a_0, \cdots, a_n \in G_L$ such that $a_1, \cdots, a_n$ are linearly independent in $\widehat{UL_0}$, and

$$a_0 = \sum_{i=1}^{n} \lambda_i a_i \,,$$

with all the $\lambda_i \neq 0$. Since $\Delta_L a = a \otimes a$ for all $a \in G_L$ it follows that

$$(\sum \lambda_i a_i) \otimes (\sum \lambda_j a_j) = \sum \lambda_i a_i \otimes a_i \,.$$

Since the elements $a_i \otimes a_j$ are linearly independent in $\widehat{UL_0} \otimes \widehat{UL_0}$ this implies that $n = 1$ and that $\lambda_1^2 a_1 \otimes a_1 = \lambda_1 a_1 \otimes a_1$. Thus $\lambda_1 = 1$ and $a_1 = a$, contrary to our hypothesis.

(ii) This is immediate from the Sullivan condition in §1.5 (p. 18) and the definition of the Lie bracket in §2.1.

(iii) We have to show that for each $n \geq 1$, $\widehat{UL_0} \otimes (UL_{\geq 1})_n \xrightarrow{\cong} \left(\widehat{UL}\right)_n$. By hypothesis, each $L_n$ admits a finite filtration

$$0 = L_n(0) \subset \cdots \subset L_n(k)$$

for which $[L_0, L_n(q)] \subset L_n(q-1)$.

Now a straightforward computation shows that for $r \geq 0$, $\left(I_L^{r+nk}\right)_n \subset I_{L_0}^r \otimes (UL_{\geq 1})_n \subset (I_L^r)_n$. This gives surjections

$$(UL/I_L^{r+nk})_n \twoheadrightarrow UL_0/I_L^r \otimes (UL_{\geq 1})_n \twoheadrightarrow (UL/I_L^r)_n \,.$$

It follows that $\left(\widehat{UL_0} \otimes UL_{\geq 1}\right)_n \xrightarrow{\cong} \left(\widehat{UL}\right)_n$.

Thus the isomorphism $\varprojlim_r \left(UL/I_L^{r+nk}\right)_n \xrightarrow{\cong} \varprojlim_r (UL/I_L^r)_n$ factors through $\varprojlim_r \left[(UL_0/I_{L_0}^r) \otimes (UL_{\geq 1})_n\right]$ and since $(UL_{\geq 1})_n$ is finite dimensional,

$$\varprojlim_r \left[UL_0/I_{L_0}^r \otimes (UL_{\geq 1})_n\right] = \left(\varprojlim_r UL_0/I_{L_0}^r\right) \otimes (UL_{\geq 1})_n\,.$$

Thus $\widehat{UL_0} \otimes (UL_{\geq 1})_n \to \widehat{UL}$ is surjective.

On the other hand, suppose $(a_r \in \widehat{UL_0}/I_{L_0}^r \otimes (UL_{\geq 1})_n)$ is an element of $\varprojlim_r \left[UL_0/I_{L_0}^r \otimes (UL_{\geq 1})_n\right]$ which maps to zero in $\widehat{UL}$. Then the image of each $a_r$ in $UL/I_L^r$ is zero. If some $a_r \neq 0$ choose $b \in UL_0/I_{L_0}^{r+nk+s} \otimes (UL_{\geq 1})_n$ which maps to $a_{r+s}$, and so to $a_r$. Then $b \in I_L^{r+s}$. If $s > nk$ then $b \in I_{L_0}^r \otimes (UL_{\geq 1})_n$ and so its image $a_r$ is zero. Thus each $a_r = 0$ and so $\widehat{UL_0} \otimes UL_{\geq 1} \to \widehat{UL}$ is injective.

□

## 2.9 The action of $G_L$ on $\pi_n(|\wedge V, d|, *)$

For any based topological space $(X,*)$ the Whitehead product defines a natural right action of the fundamental group, $\pi_n(X,*) \times \pi_1(X,*) \to \pi_n(X,*)$, $n \geq 1$, which we denote by $(\beta,\alpha) \mapsto \beta \bullet \alpha$, and which is defined as follows: Let $D \subset \Delta[n+1]$ denote the subsimplicial set of all $n$-faces containing $e_{n+1}$. Thus $\partial D$ is the union of the $(n-1)$ faces of $D$ not containing $e_{n+1}$. Then denote by $K$ the simplicial set obtained from two copies, $D^+$ and $D^-$, of $D$ by identifying $\partial D^+$ with $\partial D^-$ through the simplicial isomorphism identifying $e_i^+$ with $e_i^-$, $0 \leq i \leq n$. Thus the vertices of $K$ are just $e_{n+1}^-$, $e_0, \cdots, e_n$, and $e_{n+1}^+$ and we take $e_0$ as the base vertex.

Next, denote by $\rho : D^- \to \Delta[1]$ the simplicial map sending $e_{n+1}^-$ to $e_0$ and $e_i$ to $e_1$, $0 \leq i \leq n$. (Thus $\rho$ does *not* preserve base points.) Then $\rho$ extends to a simplicial map (which does preserve base points)

$$\eta : K \to \left(D^+/\partial D^+\right) \vee \Delta[1]/\partial\Delta[1]$$

whose spatial realization, as usual, is written $|\eta|$.

Now suppose $f : (|\Delta[1]/\partial\Delta[1]|, *) \to (X,*)$ and $g : (|D^+/\partial D^+|, *) \to (X,*)$ represent respectively $\alpha$ and $\beta$. Then the composite $(g \vee f) \circ |\eta| : (|K|,*) \to (X,*)$ represents $\beta \bullet \alpha$, where we identify $(|K|,*) \simeq (S^n,*)$ via the base point preserving isomorphism $K/D^- \xrightarrow{\cong} D^+/\partial D^+$.

Now let $L$ be the homotopy Lie algebra of a 1-finite minimal Sullivan algebra $(\wedge V, d)$. In §2.8 we defined right adjoint actions of $G_L$ in $L_n$, $n \geq 1$

and in $G_L$; by definition (cf §1.8) these are actions of $G_L$ in $\pi_{n+1}(\wedge V, d)$, $n \geq 1$.

Then recall the isomorphisms (Theorem 1.3, p. 37)

$$\exp_L \circ \iota_1 : \pi_1(|\wedge V, d|, *) \cong G_L$$

and

$$\iota_n : \pi_n(|\wedge V, d|, *) \cong L_{n-1}\,, \qquad n \geq 2\,.$$

**Theorem 2.5.** *Let $L$ be the homotopy Lie algebra of a 1-finite minimal Sullivan algebra, $(\wedge V, d)$, and denote $\exp_{L_0} \circ \iota_1$ by $\iota$. Then for all $n \geq 1$ and all $\alpha \in \pi_1(|\wedge V, d|, *)$ and $\beta \in \pi_n(|\wedge V, d|, *)$*

$$\iota_n(\beta \bullet \alpha) = Ad(\iota(\alpha))^{-1}(\iota_n \beta)\,.$$

*In particular, if $n = 1$ then*

$$\iota(\beta \bullet \alpha) = \iota(\alpha^{-1})\iota(\beta)\iota(\alpha)\,.$$

*proof:* When $n = 1$ it is classical and straightforward that $\beta \bullet \alpha = \alpha^{-1}\beta\alpha$. Since (Theorem 2.4, p. 75) $\iota$ is a group homomorphism, the second equation of Theorem 2.5 follows. The first equation then follows from Corollary 2.8 (p. 82).

For the case $n \geq 2$, fix $\alpha$ and $\beta$ and write $x = \iota_1\alpha$ and $y = \iota_n\beta$. Thus $x \in L_0$ and $y \in L_{n-1}$. We adopt the following conventions: First, given any simplex $\sigma$ in any simplicial set $S$ and any morphism $\varphi : (\wedge V, d) \to A_{PL}(S)$ then $\varphi_\sigma : (\wedge V, d) \to A_{PL}(\sigma)$ is the restriction of $\varphi$. Second, if $v \in V^m$ and $z \in L_k$ then

$$\langle v, z\rangle = 0 \qquad \text{if } k \neq m-1\,.$$

Finally, we adopt the notation that $t^+_{n+1}$ and $t^-_{n+1}$ refer respectively to the coordinate functions corresponding to $e^+_{n+1}$ and $e^-_{n+1}$.

In particular, we may define morphisms

$$\tau : (\wedge V, d) \to A_{PL}(\Delta[1]/\partial\Delta[1]) \quad \text{and} \quad \varphi : (\wedge V, d) \to A_{PL}(D^+/\partial D^+)$$

by setting for $v \in V$:

$$\tau v = \langle v, x\rangle\, dt_1$$

and

$$\varphi_\sigma v = \begin{cases} (-1)^n \langle v, y\rangle\, dt_2 \wedge \cdots \wedge dt^+_{n+1}\,, & \text{if } \sigma = \langle e_1, \cdots, e_{n+1}\rangle \\ 0 & \text{otherwise}\,. \end{cases}$$

It follows from the definition of $\iota_1$ and $\iota_n$, and the base-point preserving homotopy equivalence $|K| \to |D^+/\partial D^+|$ that the spatial realizations $|\langle \tau \rangle|$ and $|\langle\varphi\rangle|$ respectively represent $\alpha^{-1}$ and $\beta$. Thus the composite

$$(|K|, *) \xrightarrow{|\eta|} (|D^+/\partial D^+|, *) \vee (|\Delta[1]/\partial\Delta[1]|, *) \xrightarrow{|\langle\varphi\rangle| \vee |\langle\tau\rangle|} (X, *)$$

represents $\beta \bullet \alpha$.

But $(|\langle\varphi\rangle| \vee |\langle\tau\rangle|) \circ |\eta| = |\langle\psi\rangle|$ where $\psi : (\wedge V, d) \to A_{PL}(K)$ is given by (for $v \in V$):

$$\psi_\sigma v = \begin{cases} \langle v, x\rangle \, dt^-_{n+1}\,, & \sigma \text{ an } n\text{-simplex of } D^- \\ (-1)^n \langle v, y\rangle \, dt_2 \wedge \cdots \wedge dt^+_{n+1}\,, & \sigma = \langle e_1, \cdots, e^+_{n+1}\rangle \\ 0\,, & \sigma = \langle e_0, \cdots, \widehat{e_i}, \cdots, e^+_{n+1}\rangle\,, \quad 1 \leq i \leq n\,. \end{cases}$$

On the other hand, set $z = Ad\,(\alpha)^{-1}(\iota_n\beta) = Ad\,(\exp_L(-x))(y)$, and define $\chi : (\wedge V, d) \to A_{PL}(K)$ by requiring for $v \in V$ that

$$\chi_\sigma v = \begin{cases} (-1)^n \langle v, z\rangle \, dt_2 \wedge \cdots \wedge dt^+_{n+1}\,, & \text{if } \sigma = \langle e_1, \cdots, e^+_{n+1}\rangle \\ 0\,, & \text{otherwise}\,. \end{cases}$$

Then $|\langle\chi\rangle|$ represents $\iota_n^{-1}(z)$, so that we need to prove that $|\langle\chi\rangle|$ and $|\langle\psi\rangle|$ are based homotopic with respect to some base point of $|K|$, and since $|K|$ is simply connected we are free to choose the base point. Thus by Theorem 1.1 (p. 33) it is sufficient to prove that $\chi$ and $\psi$ are based homotopic with respect to the augmentation $\varepsilon$ of $A_{PL}(K)$ induced by the vertex $e^-_{n+1}$.

First, define $\Phi : (\wedge V, d) \to \wedge(t, dt) \otimes A_{PL}(K)$ as follows: If $\sigma$ is an $n$-simplex of $D^-$ set

$$\Phi_\sigma v = (1-t)\, \langle v, x\rangle \, dt^-_{n+1} + dt\, \langle v, x\rangle \, (1 - t^-_{n+1})\,, \qquad v \in V\,.$$

Evidently $d\Phi_\sigma(v) = 0$. Since $\Phi_\sigma$ vanishes on $\wedge^{\geq 3}V$ and on $(\wedge^2 V)^{\geq 3}$, $\Phi_\sigma(dv) = 0$ unless $v \in V^1$. Then writing $dv = \sum w_i \wedge w_i'$ we have

$$\Phi_\sigma(dv) = (1-t)dt\, \langle v, [x, x]\rangle \, dt^-_{n+1} = 0\,.$$

It follows that $\Phi_\sigma$ is a morphism.

Then, if $\sigma$ is an $n$-simplex of $D^+$ and $v \in V$, set

$$\Phi_\sigma v = \begin{cases} dt\, \langle v, x\rangle + (-1)^n \langle v, e^{-tx} \bullet y\rangle \, dt_2 \wedge \cdots \wedge dt^+_{n+1}\,, \\ \qquad\qquad \text{if } \sigma = \langle e_1, \cdots, e^+_{n+1}\rangle \\ dt\, \langle v, x\rangle\,, \qquad \text{otherwise}\,. \end{cases} \tag{2.9}$$

Here $e^{-tx} \bullet y$ denotes $\sum_0^\infty \frac{(-t)^k}{k!} (ad\, x)^k(y)$. To check that $\Phi_\sigma dv = d\Phi_\sigma v$ it is sufficient to consider the case $v \in V^n$ and $\sigma = \langle e_1, \cdots, e_{n+1}^+\rangle$. In this case

$$d\Phi_\sigma v = (-1)^{n+1} dt\, \langle v, e^{-tx} \bullet [x, y]\rangle\, dt_2 \wedge \cdots \wedge dt_{n+1}^+ .$$

On the other hand, write $dv = \sum_i w_i \otimes v_i + \Omega$ with $w_i \in V^1$, $v_i \in V^n$ and $\Omega \in \wedge V^{<n}$. Then

$$\begin{aligned}
\Phi_\sigma(dv) &= \textstyle\sum_i \Phi_\sigma(w_i)\, \Phi_\sigma(v_i) \\
&= (-1)^n \textstyle\sum_i dt\, \langle w_i, x\rangle\, \langle v_i, e^{-tx} \bullet y\rangle\, dt_2 \wedge \cdots \wedge dt_{n+1}^+ \\
&= dt\, \langle dv, x, e^{-tx} \bullet y\rangle\, dt_2 \wedge \cdots \wedge dt_{n+1}^+ \\
&= (-1)^{n+1} dt\, \langle v, [x, e^{-tx} \bullet y]\rangle\, dt_2 \wedge \cdots \wedge dt_{n+1}^+ \\
&= d\Phi_\sigma v .
\end{aligned}$$

To check that the $\Phi_\sigma$ defines a morphism into $(\wedge t, dt) \otimes A_{PL}(K)$ it is sufficient to check that the two definitions restrict to the same morphism in the $(n-1)$ simplices of $D^- \cap D^+$. But in fact in this case both reduce to the morphism $v \mapsto dt\, \langle v, x\rangle$.

Finally, it is immediate from the definition that $\Phi$ reduces to $\psi$ when $t = 0$ and to $\chi$ when $t = 1$ and that $(id \otimes \varepsilon)\Phi = 0$, $\varepsilon$ denoting the augmentation at $e_{n+1}^-$. Thus $\Phi$ is a based homotopy from $\psi$ to $\chi$.

□

## 2.10 Formal Sullivan 1-algebras

Recall from ([18], p. 156) that a commutative cochain algebra $(A, d_A)$ for which $H^0(A, d_A) = \Bbbk$ is *formal* if $(A, d_A)$ and $(H(A), 0)$ are connected by quasi-isomorphisms. It follows from Proposition 1.10 (p. 21) that $(A, d_A)$ is formal if and only if there is a quasi-isomorphism $\rho : (\wedge V, d) \to (H(A), 0)$ from the minimal Sullivan model of $(A, d_A)$ such that $H(\rho) = id_{H(A)}$.

More generally, we have

**Definition.** A Sullivan minimal algebra $(\wedge V, d)$ is 1-*formal* if for some graded commutative algebra $A$ with differential 0 there is a cdga map

$$\varphi : (\wedge V, d) \to (A, 0)$$

such that $H^1(\varphi)$ is an isomorphism and $H^2(\varphi)$ is injective.

**Lemma 2.21.** *A Sullivan minimal algebra* $(\wedge V, d)$ *is* 1*-formal if and only if* $(\wedge V^1, d)$ *is 1-formal.*

*proof:* Since the injection $(\wedge V^1, d) \to (\wedge V, d)$ induces an iso on $H^1$ and an injection on $H^2$, if $(\wedge V, d)$ is 1-formal, then the same is true for $(\wedge V^1, d)$.

Conversely, suppose $\varphi : (\wedge V^1, d) \to (A, 0)$ satisfies $H^1(\varphi)$ is an isomorphism and $H^2(\varphi)$ is injective. By dividing by $A^{\geq 3}$ we can suppose $A = A^{\leq 2}$. Then $\varphi$ can be extended to $(\wedge V, d)$ by setting $\varphi(V^{\geq 2}) = 0$. □

Recall also from §2.2 that associated with a Sullivan 1-algebra $(\wedge W, d)$ is the filtration $\{W_n\}_{n\geq 0}$ of $W$ defined by

$$W_0 = \ker d \cap W \qquad \text{and} \quad W_{n+1} = d^{-1}(\wedge^2 W_n) \cap W\,.$$

**Proposition 2.7.** *A Sullivan 1-algebra* $(\wedge W, d)$ *is formal (resp. 1-formal) if and only if the following conditions hold with respect to the filtration* $W_n$ *defined in §2.1:*

(i) $W = \bigoplus_{n\geq 0} M_n$ *with* $W_0 = M_0$ *and* $W_{n+1} = M_{n+1} \oplus W_n$, $n \geq 0$;
(ii) $d : M_n \to \bigoplus_{i\leq n-1-i} M_i \wedge M_{n-1-i}$, $n \geq 1$;
(iii) *The morphism* $\rho : (\wedge W, d) \to H(\wedge W, d)$ *defined by*

$$\rho(w) = [w]\,, w \in W_0 \qquad \text{and} \quad \rho(M_n) = 0\,, \qquad n \geq 1$$

*is a quasi-isomorphism (resp. satisfies* $H^1(\rho)$ *and* $H^2(\rho)$ *are isomorphisms).*

To prove the Proposition we need

**Lemma 2.22.** *Suppose* $(\wedge W, d)$ *is a Sullivan 1-algebra. Then for* $n \geq 0$

(i) $d : W_{n+1} \to \sum_{i+j\leq n} W_i \wedge W_j$;
(ii) *If* $\Phi \in \wedge^2 W_n$ *is a boundary then* $\Phi = dw$ *for some* $w \in W_{n+1}$.

*proof:* (i) Let $M_k$ denote any direct summand for $W_{k-1}$ in $W_k$. If, for some $w \in W_{n+1}$, $w$ has a non-zero component in $M_i \wedge M_j$ with $i + j > n$ then there are elements $x \in L^{i+1}$ and $y \in L^{j+1}$ such that

$$\langle w, [x, y]\rangle = \pm\langle dw, x \wedge y\rangle \neq 0\,.$$

This contradicts (2.1), which asserts that $\langle W^{n+1}, L^{n+3}\rangle = 0$.

(ii) By hypothesis, $\Phi = dw$ for some $w \in W$. But then $w \in W \cap d^{-1}(\wedge W_n) = W_{n+1}$. □

*proof of Proposition 2.7*: We first give the proof in the formal case. If conditions (i)-(iii) hold then $(\wedge W, d)$ is formal by definition. Conversely, suppose $(\wedge W, d)$ is formal and let

$$\rho : (\wedge W, d) \to (H(\wedge W, d), 0)$$

be a quasi-isomorphism for which $H(\rho) = id_{H(\wedge W,d)}$. Then for $w \in W_0$, $\rho w = [w]$.

Suppose now by induction on $n$ that the spaces $M_k$, $k \leq n$, have been constructed to satisfy (i) and (ii), and so that $\rho(M_k) = 0$, $1 \leq k \leq n$. Define a direct decomposition $\wedge^2 W_n = \oplus_{k=0}^{2n} (\wedge^2 W_n)_k$ by assigning $M_i \wedge M_j$ lower degree $i + j$, and let $\{w_\ell\}$ be a basis for a direct summand of $W_n$ in $W_{n+1}$. Fix $w_\ell$. Then by Lemma 2.22(i),

$$dw_\ell = \sum_{i=0}^{n} \Phi_i$$

with $\Phi_i \in (\wedge^2 W_n)_i$.

By our induction hypothesis, each $\Phi_i$ is a cycle. Since $\rho \circ d = 0$, $\rho\Phi_0 = 0$ and so $\Phi_0 = du_0$ for some $u_0 \in M_1$. Since $\rho\Phi_i = 0$, $i \geq 1$, each $\Phi_i$, $i \geq 1$, is a boundary. Since $\Phi_i \in (\wedge^2 W_n)_i \subset \wedge^2 W_i$, it follows from Lemma 2.22(ii) that $\Phi_i = dv_i$ for $v_i \in W_{i+1}$. If $i < n$ our induction hypothesis gives $\Phi_i = du_i$ with $u_i \in M_{i+1}$. Thus $\sum_{i=0}^{n-1} \Phi_i = d(\sum_{i=0}^{n-1} u_i)$ with $\sum_{i=1}^{n-1} u_i \in W_n$. Replace $w_\ell$ by $w_\ell - \sum_{i=1}^{n-1} u_i$ to arrange that $dw_\ell \in (\wedge^2 W)_n$. Finally, since $\rho : W_0 \xrightarrow{\cong} H^1(\wedge W, d)$ it follows that $\rho w_\ell = \rho w'_\ell$ for some $w'_\ell$. Replace $w_\ell$ by $w_\ell - w'_\ell$ to arrange that $\rho w_\ell = 0$. This closes the induction and completes the proof when $(\wedge W, d)$ is formal.

Now suppose $(\wedge W, d)$ is 1-formal. As observed in the proof of Lemma 2.21, there is a morphism $\varphi : (\wedge W, d) \to (A, 0)$ for which $H^1(\varphi) : W_0 \xrightarrow{\cong} A^1$, $H^2(\varphi)$ maps $H^2(\wedge W, d)$ injectively to $A^2$, and $A^{\geq 3} = 0$. In particular since $\varphi(W) \subset A^1$, $\varphi(\wedge W)$ is contained in the algebra generated by $A^1$, which must then be $H(\varphi)\left(H^2(\wedge W, d)\right)$ in degree 2. In other words, we may regard $\varphi$ as a morphism $(\wedge W, d) \to \mathbb{k} \oplus H^1(\wedge W, d) \oplus H^2(\wedge W, d)$.

The restriction of $\varphi$ to $W$ then trivially extends to a morphism

$$\rho : (\wedge W, d) \to H(\wedge W, d)$$

for which $H^1(\rho)$ and $H^2(\rho)$ are isomorphisms.

The rest of the proof proceeds as in the formal case. □

**Corollary 2.9.** *(i) If $(\wedge W, d)$ is a formal Sullivan 1-algebra then $H(\wedge W, d) = \wedge W_0 / I$, where $I$ is the ideal generated by $d(M_1) \subset \wedge^2 W_0$.*

*(ii) If a Sullivan 1-algebra $(\wedge W, d)$ is 1-formal then $H^2(\wedge W, d)$ is generated by $H^1(\wedge W, d)$.*

**Remark.** Proposition 2.7 provides a "bigraded Sullivan model" for $H(\wedge W, d)$. This is a special case of the bigraded model for $(H(A, d_A), 0)$ where $(A, d_A)$ is a formal commutative cochain algebra ( [30]) but also has the interesting property of corresponding to the filtration $\{W_n\}$ of §2.2.

**Proposition 2.8.** *Let $(\wedge V, d)$ be a formal Sullivan 1-algebra. Let $P(t) = \sum_p \dim H^p(\wedge V, d)\, t^p$ and $\varphi_n = \dim M_{n-1}$. Then*

$$\prod_{n\geq 1}(1-t^n)^{\varphi_n} = P(-t)\,.$$

This formula is called the LCS formula (LCS = Lower Central Series) (see for instance [56]). As we will see later, for a space $X$ with $\dim H^1(X) < \infty$, fundamental group $G$ and minimal Sullivan model a Sullivan 1-algebra $(\wedge V, d)$, $\varphi_n = \dim G^n/G^{n+1} \otimes \mathbb{Q}$. Thus if $X$ is formal and its minimal Sullivan model $(\wedge V, d)$ satisfies $V = V^1$, then the LCS formula connects the cohomology of $X$ to the series of ranks of the quotients $G^n/G^{n+1}$.

*proof:* The graded algebra $\wedge V$ is bigraded, and for each $r$, $\bigoplus_{p+q=r}(\wedge^p V)_q$ is a subcomplex whose homology is $H^r(\wedge V, d)$. Thus $(-1)^r \dim H^r(\wedge V, d) = \sum_0^r (-1)^p \dim(\wedge^p V)_{r-p}$.

For a bigraded vector space $W = \oplus_n W_n^1$, write $Q_W = \sum_p (-1)^p \dim(\wedge^p W)_q t^{t+q}$. Then for bigraded vector spaces $W$ and $W'$, we have $Q_{W\oplus W'} = Q_W \cdot Q_{W'}$. Moreover, if $W = W_n^1$ with $\dim W_n^1 = d$, then $Q_W = (1-t^{n+1})^d$. Therefore

$$P(-t) = \sum_{p,q}(-1)^{p+q}\dim(\wedge^p M)_q t^{p+q} = \prod_n (1-t^{n+1})^{\dim M_n}\,.$$

□

# Chapter 3

# Fibrations and $\Lambda$-extensions

Recall from §1.4 that for any topological space, $X$, $H^*(X; \Bbbk) = H(A_{PL}(X))$ is denoted by $H(X)$.

## 3.1 Fibrations, Serre fibrations and homotopy fibrations

Recall first that a continuous map $p : X \to Y$ has the *lifting property* with respect to a pair of topological spaces $(Z, A)$ if for every diagram of the form

$$\begin{array}{ccc} A & \xrightarrow{f} & X \\ {\scriptstyle i}\big\downarrow & & \big\downarrow{\scriptstyle \pi} \\ Z & \xrightarrow{g} & Y \end{array}$$

there exists a continuous map $h : Z \to X$ such that $h$ lifts $g$ and extends $f$; i.e. $\pi \circ h = g$ and $h \circ i = f$.

**Definition.** A sequence of continuous maps

$$F \xrightarrow{j} X \xrightarrow{\pi} Y$$

is a *fibration* (respectively a *Serre fibration*) if $\pi$ is surjective, $j$ is the inclusion of $\pi^{-1}(*)$ for some base point $* \in Y$, and if $\pi$ has the lifting property with respect to any pair $(I \times A, 0 \times A)$, (respectively with respect to any pair $(Z, A)$ in which $A$ is a sub CW complex of a CW complex $Z$ and the inclusion $A \hookrightarrow Z$ is a homotopy equivalence).

A *fibre bundle* $\pi : X \to Y$ with fibre $F$ is a continuous map such that there exists an open covering $\{O_\alpha\}_\alpha$ of $Y$ for which there are commutative diagrams

$$\begin{array}{ccc} \pi^{-1}(O_\alpha) & \xrightarrow{\cong} & O_\alpha \times F \\ & \searrow^{\pi} \quad \swarrow_{\pi_L} & \\ & O_\alpha\,, & \end{array}$$

$\pi_L$ denoting projection on the left factor.

**Proposition 3.1.** ([18], Proposition 2.6). *Both a fibration and a fibre bundle are Serre fibrations.*

We recall now the classical construction of Moore which converts a continuous map into a fibration. A *free Moore path* of length $s$ is a pair $(f, s)$ in which $f : \mathbb{R} \to X$ is a continuous map satisfying $f(t) = f(0)$, $t \leq 0$ and $f(t) = f(s)$ for $t \geq s$. We denote by $MX \subset X^{\mathbb{R}} \times \mathbb{R}$ the space of free Moore paths while the subspaces of $MX$,

$$P(X, x) := \{(f, s) \in MX \mid f(0) = x\} \tag{3.1}$$

$$P'(X, x) = \{(f, s) \in MX \mid f(t) = x\,, t \geq s\} \tag{3.2}$$

$$\Omega(X, x) := \{(f, s) \in MX \mid f(0) = x = f(t)\,, t \geq s\} \tag{3.3}$$

are called respectively the *Moore path space*, the *reverse Moore path space*, and the *Moore loop space*. We call $s$ the *length* of $f$ and denote it by $|f|$.

We also usually abuse notation and, given a base point $* \in X$, write simply $PX, P'X$ and $\Omega X$ for $P(X, *)$, $P'(X, *)$ and $\Omega(X, *)$. All three spaces are assigned the base point $c_*$, which denotes the loop of length 0 at $*$.

Associated with any based continuous map $h : (X, *) \to (Y, *)$ is the commutative diagram

$$\begin{array}{ccc} X & \xrightarrow{\ell} & X \times_Y MY \\ & \searrow^{h} \quad \swarrow_{q} & \\ & Y & \end{array}$$

of based maps, with

$$\begin{cases} X \times_Y MY := \{(x, f) \in X \times MY \mid h(x) = f(0)\} \\ \ell(x) = (x, c_{h(x)})\,, \quad c_{h(x)} \text{ the path of length 0 at } h(x) \\ q(x, f) = f(|f|)\,, \end{cases}$$

and $(*, c_*)$ the base point of $X \times_Y MY$. The following Proposition is a classical and straightforward consequence of the definitions.

**Proposition 3.2.** *Suppose $(Y, *)$ is a based topological space,*

*(i)* $q : X \times_Y MY \to Y$ *is a fibration with fibre* $X \times_Y P'Y$ *at* $*$.
*(ii) In the diagram above $\ell$ is a based homotopy equivalence with based homotopy inverse the left projection* $\pi_L : X \times_Y MY \to X$.
*(iii)* $q \circ \ell = h$.
*(iv) If $h$ is a fibration (resp., a Serre fibration) with fibre $F$ at a base point $* \in Y$ then $\ell$ restricts to a homotopy equivalence (resp., a weak homotopy equivalence)* $\ell_0 : F \to q^{-1}(*)$.

**Definition.** (i) The fibration $q$ is called the *homotopy fibration* associated to $h$ and the fibre $X \times_Y P'Y$ is the *homotopy fibre* of $h$.

(ii) In particular, the homotopy fibration associated to the inclusion of a base point $* \to Y$ is the *path space fibration* $\Omega Y \to PY \to Y$, with homotopy fibre the *loop space* $\Omega Y$.

## 3.2 The classifying space fibration and Postnikov decompositions of a connected CW complex

**Definition.** The *classifying space* of a discrete group $G$ is a connected CW complex $(B, *)$, together with an isomorphism $G \overset{\cong}{\to} \pi_1(B, *)$, and such that $\pi_n(B, *) = 0$, $n \geq 2$.

Any group $G$ has a classifying space $(B, *)$, and if $G$ is the fundamental group of a connected CW complex $(X, *)$, then there is a continuous map

$$h : (X, *) \to (B, *)$$

such that $\pi_1(h) = id_G$. (The constructions are outlined just below.) In particular, we may apply the construction of §3.1 to convert $h$ to a fibration

$$B \xleftarrow{q} X \times_B MB \xleftarrow{j} F$$

for which there is a based homotopy equivalence $\ell : X \to X \times_B MB$ such that $q \circ \ell = h$, and $\pi_1(q)$ is an isomorphism. Then it follows from the long exact homotopy sequence for this fibration that $F$ is simply connected and $\pi_i(j)$ is an isomorphism for $i \geq 2$. This is the *classifying space fibration* for $X$ and we call $F$ the *homotopy universal cover* for $X$.

As we shall see in this and subsequent chapters, Sullivan's theory is particularly well adapted to the analysis of fibrations, and so the classifying space fibrations are a principal tool in the application of rational homotopy theory to non-simply connected spaces.

We now outline the promised constructions, noting that in §7.2 we shall provide a functorial simplicial construction of the classifying space of a group.

First, for a group $G$ let $(B(1), *) = (\vee S^1_\alpha, *)$ be a wedge of circles indexed by the elements $\alpha \in G$. By the Van Kampen theorem ( [32], Theorem 1.20) $\pi_1(\vee S^1_\alpha, *)$ is the free group on the set $\{g_\alpha\}_{\alpha\in G}$ where $g_\alpha$ is represented by the inclusion $S^1_\alpha \to \vee_\alpha S^1_\alpha$. In particular $g_\alpha \mapsto \alpha$ defines a surjection $\varphi : \pi_1(\vee_\alpha S^1_\alpha) \to G$. Next, adjoin 2-cells to $B(1)$ to kill $\operatorname{Ker}\varphi$ and denote the resulting space by $B(2)$. The Van Kampen Theorem ( [32], Proposition 1.26) asserts that $\pi_1(B(2)) = G$.

Now construct $B$ as the union of an increasing sequence of CW complexes $B(k)$ in which $B(k+1)$ is constructed from $B(k)$ by attaching $(k+1)$-cells to kill $\pi_k(B(k), *)$. Since for $n \geq 2$, any based map $(S^n, *) \to (B, *)$ is based homotopic to a map into $(B(n), *)$ it follows that it represents 0 in $\pi_n(B(n+1), *)$. Thus $\pi_{\geq 2}(B, *) = 0$ and $(B, *)$ is a classifying space for $G$.

Next, suppose $G$ is the fundamental group of a connected CW complex $(X, *)$ which, without loss of generality, we may suppose to have a single 0-cell, which then serves as base point. Let $\{\gamma\}$ be the 1-cells of $X$ and define $h : (\vee_\gamma S^1_\gamma, *) \to (B, *)$ by requiring that $h_{| S^1_\gamma}$ represents the class $[S^1_\gamma] \in G$. It is then immediate that $h$ extends to $X$ and that $\pi_1(h) = id_G$. In particular this construction shows that two classifying spaces for $G$ have the same homotopy type.

Classifying space fibrations generalize as follows. Suppose $h : (X, *) \to (Y, *)$ is a continuous map such that $\pi_1(h)$ is an isomorphism. Then, given $n \geq 1$, a *Postnikov decomposition* of $h$ is a fibration

$$X(n) \stackrel{q(n)}{\leftarrow} E \stackrel{j(n)}{\leftarrow} F(n)\,,$$

together with a homotopy equivalence $X \xrightarrow{\simeq} E$ such that $\pi_k(q(n))$ is an isomorphism for $k \leq n$ and $\pi_k(X(n)) = 0$ for $k > n$. It follows from the long exact homotopy sequence that $\pi_k(F(n)) = 0$ when $k \leq n$ and $\pi_k(j(n))$ is an isomorphism for $k > n$. Clearly when $n = 1$ these are just classifying space fibrations.

Postnikov decompositions are constructed in a similar way to classifying space fibrations: adjoin cells of dimension $\geq n+2$ to $X$ to successively kill homotopy groups of degree $k$ for $k = n+1, n+2, \cdots$. This produces a map $h(n) : X \to Y$ for which $\pi_k(Y,*) = 0$, $k \geq n+1$ and $\pi_k(h(m))$ is an isomorphism for $k \leq n$. Converting $h(m)$ to a fibration as described in §3.1 then yields the desired Postnikov decomposition.

### 3.3 Λ-extensions

The analogue in Sullivan's model theory of a (Serre) fibration $X \to Y$ is a Λ-extension of commutative cochain algebras, as described below.

**Definition.** A commutative cochain algebra morphism $(B,d) \to (B \otimes \wedge Z, d)$, $b \mapsto b \otimes 1$, satisfies the *Sullivan condition* if

(i) $H^0(B,d) = \Bbbk$,

(ii) $d : 1 \otimes Z \to (\Bbbk \oplus B^{\geq 1}) \otimes \wedge Z$, and

(iii) $Z$ is the increasing union of subspaces $0 = Z(0) \subset \cdots \subset Z(r) \subset \cdots$ for which

$$d : 1 \otimes Z^p(r) \to B \otimes \wedge(Z^{<p} + Z(r-1))\,, \qquad r \geq 1\,. \tag{3.4}$$

If instead of (3.4) the stronger condition

$$d : 1 \otimes Z^p(r) \to B \otimes \wedge(Z^{<p} \oplus Z^{\leq p}(r-1))\,, \qquad r \geq 1\,, \tag{3.5}$$

is satisfied then the morphism satisfies the *minimal Sullivan condition.*

**Remark.** A trivial induction shows that (3.4) is equivalent to the following condition: $Z$ is the increasing union of subspaces $0 = Z(0) \subset \cdots \subset Z(r) \subset \cdots$ with

$$d : 1 \otimes Z(r) \to B \otimes \wedge Z(r-1)\,, \qquad r \geq 1\,. \tag{3.6}$$

In particular, if (3.4) holds then $(B \otimes \wedge Z, d)$ is a semifree $(B,d)$-module, as follows from Lemma 14.1 in [18].

**Definition.** (i) A *(minimal) Λ-extension* of a commutative cochain algebra $(B,d)$ is a morphism of commutative cochain algebras $(B,d) \to (C,d)$ which may be written in the form $(B,d) \to (B \otimes \wedge Z, d)$, $b \mapsto b \otimes 1$, and which satisfies the (minimal) Sullivan condition.

(ii) A (minimal) Λ-extension $(B \otimes \wedge Z, d)$ is a *(minimal) Sullivan extension* if $Z = Z^{\geq 1}$.

**Note:** In [18] (minimal) Sullivan extensions are called *(minimal) relative Sullivan algebras.*

If $(B\otimes\wedge Z,d)$ is a minimal $\Lambda$-extension of $(B,d)$ then division by $B^+\otimes\wedge Z$ yields a cochain algebra of the form $B^0\otimes(\wedge Z,\overline{d})$. The cochain algebra $(\wedge Z,\overline{d})$ is a $\Lambda$-algebra (cf. §1.5):

$$\overline{d}: Z^p(r)\to\wedge(Z^{<p}\oplus Z^p(r-1))\,,\qquad p\geq 0, r\geq 1\,. \tag{3.7}$$

If $\varepsilon_B$ is an augmentation for $(B,d)$ then it induces an isomorphism $(\wedge Z,\overline{d})\overset{\cong}{\to}\Bbbk\otimes_B(B\otimes\wedge Z,d)$.

**Definition.** The $\Lambda$-algebra, $(\wedge Z,\overline{d})$, is called the *fibre* of $(B\otimes\wedge Z,d)$.

**Proposition 3.3.** *(i) If $\alpha:(\wedge Z_1,d)\to(\wedge Z_2,d)$ is a quasi-isomorphism between commutative cochain algebras satisfying (3.7) (with $B=\Bbbk$) then $\alpha$ is an isomorphism.*

*(ii) In particular, if $H(\wedge Z,d)=\Bbbk$ then $Z=0$.*

*proof:* Since the restriction of $\alpha$ to $\wedge Z_1^0$ coincides with $H^0(\alpha)$, it follows that $\alpha:\wedge Z_1^0\overset{\cong}{\to}\wedge Z_2^0$. Thus $\alpha$ extends to an isomorphism $\beta: K_1\overset{\cong}{\to}K_2$ between the respective fields of fractions of these polynomial algebras. In particular, $\beta\otimes\alpha: K_1\otimes_{\wedge Z_1^0}\wedge Z_1\to K_2\otimes_{\wedge Z_2^0}\wedge Z_2$ is a quasi-isomorphism between minimal Sullivan algebras defined over $K_1$ and $K_2$, and hence $\beta\otimes\alpha$ is an isomorphism (Corollary 1.5, p. 23). It follows that $\alpha$ is injective.

To show that $\alpha$ is surjective assume as an inductive hypothesis that for some $p\geq 1$ and some $r\geq 1$

$$Z_2^{<p}\oplus Z_2^p(r-1)\subset\operatorname{Im}\alpha\,.$$

If $z\in Z_2^p(r)$ then $dz\in\operatorname{Im}\alpha$ and we may set $dz=\alpha\Phi$. Thus $\alpha d\Phi=0$, and since $\alpha$ is injective, $d\Phi=0$. But $H(\alpha)[\Phi]=[dz]=0$ and so, since $\alpha$ is a quasi-isomorphism, $\Phi=d\Psi$. Now $z-\alpha\Psi$ is a cycle and so $z-\alpha\Psi=\alpha\Psi_1+d\gamma$ where $\Psi_1$ is a cycle in $\wedge Z_1$ and $\gamma\in(\wedge Z)^{p-1}$. By the induction hypothesis, $\gamma\in\operatorname{Im}\alpha$.

Since the induction hypothesis holds for $p=r=1$, the result follows by a double induction, first on $r$ and then on $p$.

□

**Proposition 3.4.** *Suppose $(B\otimes\wedge Z,d)$ and $(B\otimes\wedge Z\otimes\wedge W,d)$ are respectively $\Lambda$-extensions of $(B,d)$ and of $(B\otimes\wedge Z,d)$. Then $(B\otimes\wedge Z\otimes\wedge W,d)$ is a $\Lambda$-extension of $(B,d)$. In particular, $(B\otimes\wedge Z\otimes\wedge W,d)$ is $(B,d)$-semifree.*

*proof:* Define an increasing sequence $0=F_0'\subset\cdots\subset F_r'\subset\cdots$ of subspaces of $Z\oplus W$ as follows:

$$F_r'=\{\,y\in Z\oplus W\mid d(1\otimes y)\in B\otimes\wedge F_{r-1}'\,\}\,.$$

We only have to show that $Z \oplus W = \cup_r F'_r$.

Let $Z(q)$ be the filtration of (3.4). We first note that $Z \subset \cup_r F'_r$. For otherwise there is some $z \in Z^p(q)$ such that $z \notin \cup_r F'_r$ but $Z^{<p} + Z^p(< q) \subset \cup_r F'_r$. But then by (3.4) $d(1 \otimes z) = \sum_i b_i \otimes \Phi_i$ with $\Phi_i \in \wedge F'_{r_i}$. If $r = \max r_i$, $d(1 \otimes z) \in B \otimes \wedge F'_r$ and $z \in F'_{r+1}$.

Finally, denote by $W(p)$ the filtration of (3.4) for the Λ-extension of $(B \otimes \wedge Z, d)$. If $W \not\subset \cup_r F'_r$ then for some $w \in W^p(q)$, such that $w \notin \cup_r F'_r$ but $W^{<p} + W^p(< q) \subset \cup_r F'_r$. Since $Z \subset \cup_r F^r$ the same argument as above leads to a contradiction.

□

Generalizing the Lifting Lemma 1.2 we have

**Lemma 3.1.** *(Lifting Lemma) Suppose given a commutative diagram of commutative cochain algebras,*

$$\begin{array}{ccc} (B,d) & \xrightarrow{\quad\alpha\quad} & (A,d) \\ \big\downarrow & & \simeq\big\downarrow\eta \\ (B \otimes \wedge Z, d) & \xrightarrow{\quad\psi\quad} & (C,d) \end{array}$$

*in which $(B \otimes \wedge Z, d)$ is a Λ-extension of $(B,d)$ and $\eta$ is a surjective quasi-isomorphism. Then $\alpha$ extends to a morphism $\varphi : (B \otimes \wedge Z, d) \to (A,d)$ such that $\eta \circ \varphi = \psi$.*

*proof:* The argument in the proof of Lemma 12.4 of [18] applies verbatim.

□

We now introduce the relative version of homotopy. Recall that the cochain algebra $\wedge(t, dt)$ has two augmentations $\varepsilon_0, \varepsilon_1 : \wedge(t, dt) \to k$ defined by $\varepsilon_0 : t \mapsto 0$ and $\varepsilon_1 : t \mapsto 1$.

**Definition.** Suppose $\varphi_0, \varphi_1 : (B \otimes \wedge Z, d) \to (A,d)$ are morphisms from a Λ-extension of $(B,d)$ to a commutative cochain algebra. If $\varphi_0$ and $\varphi_1$ restrict to the same $\alpha$ in $B$ then they are *homotopic rel B* if there is a morphism

$$\Phi : (B \otimes \wedge Z, d) \to \wedge(t, dt) \otimes (A,d)$$

such that $(\varepsilon_i \otimes id) \circ \Phi = \varphi_i$, $i = 0, 1$, and $\Phi(b) = \alpha(b)$ for $b \in B$. The morphism $\Phi$ is a *homotopy rel B* from $\varphi_0$ to $\varphi_1$ and we write $\varphi_0 \sim_B \varphi_1$.

**Proposition 3.5.** *Suppose $(A,d)$ and $(B\otimes\wedge Z,d)$ are respectively a commutative cochain algebra and a $\Lambda$-extension of $(B,d)$. Suppose also that $\alpha:(B,d)\to(A,d)$ is a cdga morphism. Then*

*(i) Homotopy rel $B$ is an equivalence relation in the set of morphisms $(B\otimes\wedge Z,d)\to(A,d)$ that restrict to $\alpha$ in $B$.*
*(ii) If $\varphi_0\sim_B\varphi_1$, then $H(\varphi_0)=H(\varphi_1)$.*

*proof:* The proof of (i) is identical to the proof of the absolute case in [18], Proposition 12.7, except that in [18] $\wedge(t,dt)$ is written on the right and that here we use the Lifting Lemma 3.1. Assertion (ii) follows in the same way as Lemma 14.5 in [18].

□

As in Proposition 1.10 (p. 21) we can now extend the Lifting Lemma 3.1 to a homotopy lifting proposition. Let

$$\begin{array}{ccc} (B,d) & \xrightarrow{\alpha} & (A,d) \\ {\scriptstyle i}\downarrow & & \downarrow{\scriptstyle \simeq\ \eta} \\ (B\otimes\wedge Z,d) & \xrightarrow{\psi} & (C,d) \end{array}$$

be a commutative diagram of commutative cochain algebra morphisms in which $(B\otimes\wedge Z,d)$ is a $\Lambda$-extension of $(B,d)$ and $\eta$ is a quasi-isomorphism.

**Proposition 3.6.** *With the hypotheses and notation above, there is a morphism $\varphi:(B\otimes\wedge Z,d)\to(A,d)$ such that $\varphi\circ i=\alpha$ and $\eta\circ\varphi\sim_B\psi$. Moreover, $\varphi$ is unique up to homotopy rel $B$.*

*proof:* The argument in the proof of Lemma 14.6 in [18] applies verbatim.

□

Every cochain algebra of the form $(\wedge Z,d)$ admits an augmentation and, given an augmentation $\varepsilon$, may be written $(\wedge Z,d)$ with $\varepsilon(Z)=0$. With this choice a $\Lambda$-extension of a Sullivan algebra $(\wedge V,d)$ may be written $(\wedge V\otimes\wedge Z,d)$, augmented by $\varepsilon:V\oplus Z\to 0$. As in §1.5, if $(A,d)$ is a commutative cochain algebra augmented by $\varepsilon_A$ then a *based homotopy* $\Phi:(\wedge V\otimes\wedge Z,d)\to\wedge(t,dt)\otimes(A,d)$ is a homotopy satisfying $\Phi:\ker\varepsilon\to\wedge(t,dt)\otimes\ker\varepsilon_A$.

Next consider a commutative diagram of commutative cochain algebras in which $i$ is the inclusion of a $\Lambda$-extension of a Sullivan algebra,

$$\begin{array}{ccc} (\wedge V, d) & \xrightarrow{\;i\;} & (\wedge V \otimes \wedge Z, d) \\ \big\downarrow{\scriptstyle \varphi_0} & & \big\downarrow{\scriptstyle \psi_0} \\ (B, d) & \xrightarrow[\;\chi\;]{} & (C, d)\,. \end{array}$$

**Corollary 3.1.** *With the hypotheses and notation above, a (based) homotopy $\Phi$ from $\varphi_0$ to a second morphism $\varphi_1$ extends to a (based) homotopy $\Psi$ from $\psi_0$ to a second morphism $\psi_1 : (\wedge V \otimes \wedge Z, d) \to (C, d)$ and satisfying $\Psi \circ i = (id \otimes \chi) \circ \Phi$.*

*proof:* Apply Lemma 3.1 to the diagram

$$\begin{array}{ccc} (\wedge V, d) & \xrightarrow{\;(id\otimes\chi)\circ\Phi\;} & \wedge(t, dt) \otimes (C, d) \\ \big\downarrow{\scriptstyle i} & & \big\downarrow{\scriptstyle \varepsilon_0 \otimes id} \\ (\wedge V \otimes \wedge Z, d) & \xrightarrow[\;\psi_0\;]{} & (C, d)\,. \end{array}$$

□

Now consider a commutative diagram of cdga's:

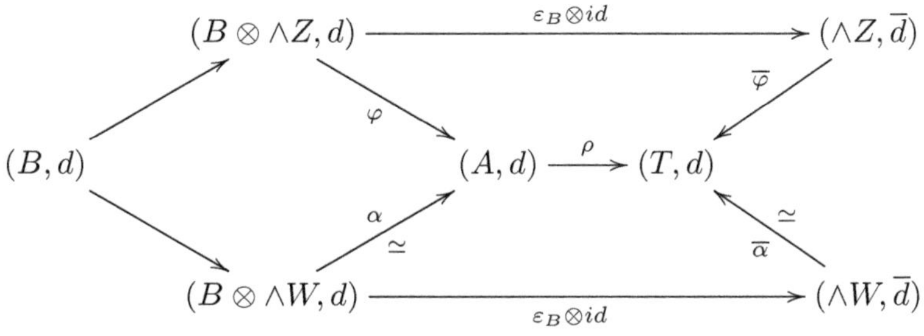


in which $(B, d)$ is augmented by $\varepsilon_B$, the left hand morphisms are $\Lambda$-extensions, and $\alpha$ and $\overline{\alpha}$ are quasi-isomorphisms.

**Corollary 3.2.** *With the hypotheses and notation above, $id_B$ extends to a morphism $\psi : (B \otimes \wedge Z, d) \to (B \otimes \wedge W, d)$ such that $\alpha \circ \psi \sim_B \varphi$, and $\overline{\psi} = \mathbb{k} \otimes_B \psi : (\wedge Z, \overline{d}) \to (\wedge W, \overline{d})$ satisfies $\overline{\alpha} \circ \overline{\psi} \sim \overline{\varphi}$.*

*proof:* By Proposition 3.6 there is such a morphism $\psi$ and a homotopy rel $B$, $H : (B \otimes \wedge Z, d) \to \wedge(t, dt) \otimes (A, d)$. Since $\psi$ is the identity on $B$, $\psi$

induces $\overline{\psi}$ with $\overline{\psi}(\varepsilon_B \otimes id) = (\varepsilon_B \otimes id)\psi$. Composition with $id \otimes \rho$ gives the homotopy between $\overline{\alpha} \circ \overline{\psi}$ and $\overline{\varphi}$.

□

Next consider a commutative diagram of commutative cochain algebras of the form

$$\begin{array}{ccc} & & (C,d) \\ & \overset{\varphi}{\nearrow} & \big\downarrow \eta \\ (B \otimes \wedge Z, d) & \xrightarrow[\psi]{} & (A,d) \end{array}$$

in which $(B \otimes \wedge Z, d)$ is a $\Lambda$-extension of $(B,d)$ and $\eta$ is surjective.

**Proposition 3.7.** *With the hypotheses and notation above, suppose*

$$\Psi : (B \otimes \wedge Z, d) \to \wedge(t, dt) \otimes (A, d)$$

*is a homotopy (rel $B$) from $\psi$ to a morphism $\psi' : (B \otimes \wedge Z, d) \to (A, d)$. Then there is a homotopy (rel $B$)*

$$\Phi : (B \otimes \wedge Z, d) \to \wedge(t, dt) \otimes (C, d)$$

*from $\varphi$ to a morphism $\varphi' : (B \otimes \wedge Z, d) \to (C, d)$ such that $(id \otimes \eta) \circ \Phi = \Psi$.*

*proof:* Let $Z(q)$ be a filtration of $Z$ satisfying the Sullivan condition (3.6). The Proposition will follow if we show that if $\Phi$ can be constructed in $B \otimes \wedge Z(q)$ then it can be extended to $B \otimes \wedge Z(q+1)$.

For this, fix a basis $z_i$ for a direct summand of $Z(q)$ in $Z(q+1)$, and as usual let $\varepsilon_0 : \wedge(t, dt) \to k\!\!k$ be the augmentation $t \mapsto 0$. Since $\eta\varphi z_i = \psi z_i$ we may find $w_i \in \ker \varepsilon_0 \otimes C$ such that $(id \otimes \eta)(\varphi z_i + w_i) = \Psi(z_i)$. On the other hand, since $dz_i \in B \otimes \wedge Z(q)$, $\Phi(dz_i)$ is a cycle in $\wedge(t, dt) \otimes C$. A simple computation then shows that

$$\Phi(dz_i) - d(\varphi z_i + w_i) \in \ker \varepsilon_0 \otimes \ker \eta\,.$$

But $H(\ker \varepsilon_0 \otimes \ker \eta) = H(\ker \varepsilon_0) \otimes H(\ker \eta) = 0$, because $\varepsilon_0$ is a surjective quasi-isomorphism. Thus

$$\Phi(dz_i) = d(\varphi z_i + w_i + u_i)$$

for some $u_i \in \ker \varepsilon_0 \otimes \ker \eta$. Extend $\Phi$ by setting $\Phi(z_i) = \varphi(z_i) + w_i + u_i$.

□

Finally, we have

**Proposition 3.8.** *If $(\wedge V \otimes \wedge Z, d)$ is a $\Lambda$-extension of a Sullivan algebra in which $H(\wedge V, d)$ and $H(\wedge Z, \overline{d})$ have finite type then $H(\wedge V \otimes \wedge Z, d)$ has finite type.*

*proof:* Filter $\wedge V \otimes \wedge Z$ by the ideals $(\wedge V)^{\geq k} \otimes \wedge Z$. The resulting spectral sequence $(E_i, d_i)$ converges to $H(\wedge V \otimes \wedge Z, d)$ from

$$E_1^{p,q} = (\wedge V)^p \otimes H^q(\wedge Z, \overline{d})\,.$$

The Sullivan condition (3.4) gives an increasing filtration $F_k = F_k(H^p(\wedge Z))$ such that

$$d_1 : F_k(H^p(\wedge Z)) \to V^1 \otimes F_{k-1}(H^p(\wedge Z))\,.$$

This filtration is necessarily finite because $H^p(\wedge Z)$ is finite dimensional, and in the quotient spaces $\wedge V \otimes F_k/F_{k-1}$ the differential $d_1$ reduces to $d_1 \otimes id$. It follows that $E_2 = H(\wedge V \otimes H(\wedge Z), d_1)$ has finite type.

□

**Lemma 3.2.** *Suppose $(B \otimes \wedge Z, d)$ is a $\Lambda$-extension in which $Z$ and $H(B \otimes \wedge Z, d)$ both have finite type. Then $H(B, d)$ has finite type.*

*proof:* For any $p$, $H^{\leq p}(B \otimes \wedge Z, d) \cong H^{\leq p}(B \otimes \wedge W, d)$ with $\dim W < \infty$. Thus it is sufficient to show that if $\dim H^{\leq p}(B \otimes \wedge W, d) < \infty$ and $\dim W < \infty$ then $\dim H^{\leq p}(B, d) < \infty$, and we do so by induction on $\dim W$. Write $(B \otimes \wedge W, d)$ as a $\Lambda$-extension of the form $((B \otimes \wedge U) \otimes \wedge w, d)$. Then define a $\Lambda$-extension

$$(\wedge w', 0) \to (\wedge w' \otimes B \otimes \wedge U \otimes \wedge w, D)$$

by setting $Dw = dw - w'$. Filtering by the degree of $B \otimes \wedge U$ gives a spectral sequence, which shows that $(B \otimes \wedge U, d) \simeq (\wedge w' \otimes B \otimes \wedge U \otimes \wedge w, D)$. Filtering by the degree in $\wedge w'$ gives a spectral sequence converging from $\wedge w' \otimes H(B \otimes \wedge W, d)$. In particular, $\dim H^{\leq p}(B \otimes \wedge U, d) < \infty$ and the Lemma follows by induction.

□

## 3.4 Existence of minimal Sullivan models

Let $\varphi : (B, d) \to (C, d)$ be a morphism of commutative cochain algebras with $H^0(B) = H^0(C) = \Bbbk$.

**Definition.** (i) A *minimal Sullivan model* for $\varphi$ is a quasi-isomorphism,

$$\gamma : (B \otimes \wedge Z, d) \xrightarrow{\simeq} (C, d)$$

from a minimal $\Lambda$-extension of $(B, d)$ such that $\gamma(b \otimes 1) = \varphi(b)$ for $b \in B$.

(ii) A minimal Sullivan model $\chi : (B \otimes \wedge U, d) \xrightarrow{\simeq} k\!\!k$ for an augmentation $\varepsilon_B : (B, d) \to k\!\!k$ is called an *acyclic closure* for $(B, d)$.

**Theorem 3.1.** *(i) Every morphism $\varphi : (B, d) \to (C, d)$ of commutative cochain algebras in which $H^0(B) = H^0(C) = k\!\!k$ has a minimal Sullivan model, $\gamma : (B \otimes \wedge Z, d) \xrightarrow{\simeq} (C, d)$.*

*(ii) If for some $k \geq 0$, $H^i(\varphi)$ is an isomorphism for $i \leq k$ and injective for $i = k + 1$ then $Z = Z^{\geq k+1}$.*

**Remark.** The hypotheses $H^0(B) = H^0(C) = k\!\!k$ are necessary. For instance for the injection $(k\!\!k, 0) \to (k\!\!k[x]/x^2, 0)$, $\deg x = 0$, there is no $\Lambda$-extension $(k\!\!k, 0) \to (\wedge Z, d)$ with a quasi-isomorphism $\gamma : (\wedge Z, d) \to (k\!\!k[x]/x^2, 0)$.

On the other hand, the morphism $\varphi : (k\!\!k, 0) \to (\wedge(u, t_0, t_1, \cdots), d)$ with $\deg u = 1$, $\deg t_n = 0$, $du = 0$, $dt_0 = 0$ and $dt_n = ut_{n-1}$, $n > 0$, is a $\Lambda$-extension that does not admit a minimal Sullivan model, because for each minimal Sullivan model $(B \otimes \wedge Z, d)$, in the fibre $\overline{d} = 0$ in $Z^0$.

**Corollary 3.3.** *If in Theorem 3.1 $\varphi$ preserves augmentations $\varepsilon_B$ and $\varepsilon_C$ then the minimal Sullivan model $\gamma : (B \otimes \wedge Z, d) \to (C, d)$ can be chosen to preserve augmentations $\varepsilon_C$ and $\varepsilon$ with $\varepsilon_{|B} = \varepsilon_B$ and $\varepsilon(Z) = 0$.*

*proof:* Let $\gamma' : (B \otimes \wedge Z, d') \xrightarrow{\simeq} (C, d)$ be a minimal Sullivan model, and define a differential $d$ in $B \otimes \wedge Z$ and an isomorphism $(B \otimes \wedge Z, d) \xrightarrow{\cong} (B \otimes \wedge Z, d')$ extending $id_B$, by sending $1 \otimes z \mapsto 1 \otimes z - \varepsilon_C \gamma(1 \otimes z)$. Since, for the filtration $Z(q)$ in (3.4), this isomorphism sends $1 \otimes Z^p(q) \to k\!\!k \oplus Z^p(q)$, $(B \otimes \wedge Z, d)$ is a $\Lambda$-extension. Composition with $\gamma'$ gives an augmentation preserving quasi-isomorphism $\gamma : (B \otimes \wedge Z, d) \xrightarrow{\simeq} (C, d)$.

□

**Remark.** In Theorem 3.2 (p. 110) we shall show that minimal Sullivan models are unique up to isomorphism.

Before proving the theorem in general we establish the special case when $\varphi : (\wedge V, d) \to (\wedge W, d)$ is a surjection of minimal Sullivan algebras. In this case $\varphi$ induces a surjection $\wedge^+ V / \wedge^{\geq 2} V \to \wedge^+ W / \wedge^{\geq 2} W$ and so we may and do choose the generating space $V$ so that $\varphi$ restricts to a surjection $V \to W$.

**Proposition 3.9.** *Suppose* $\rho : (\wedge V, d) \to (\wedge Z, d)$ *is a morphism of minimal Sullivan algebras that restricts to a surjection* $V \to Z$. *Then* $\rho$ *extends to a quasi-isomorphism from a minimal* Λ*-extension of* $(\wedge V, d)$,

$$\chi : (\wedge V \otimes \wedge U, d) \xrightarrow{\cong} (\wedge Z, d)\,,$$

*in which,* $\chi(U) = 0$, *and for* $z \in U$, $d(1 \otimes z) = \alpha(z) \otimes 1 + \beta(z)$ *with*

$$\begin{cases} \alpha : U^p \xrightarrow{\cong} (\mathit{ker}\,\rho \cap V)^{p+1}\,, \qquad p \geq 0\,, \qquad \textit{and,} \\ \beta : U \to \mathit{ker}\,\chi \cap (V \wedge \wedge^+ (V \oplus U))\,. \end{cases}$$

*proof:* We first exhibit $\ker \rho \cap V$ as the increasing union of a sequence of subspaces:

$$0 = K(0) \subset K(1) \subset \cdots \subset K(q) \subset \cdots\,,$$

with $d : K(q+1) \to K(q) \wedge \wedge^+ V$. In fact, given $K(q)$, note that the ideal $K(q) \wedge \wedge^+ V$ is preserved by $d$ and division by this ideal gives a commutative diagram

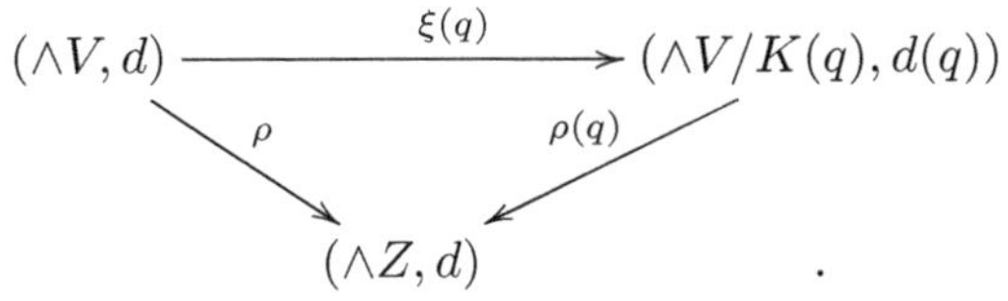


.

Choose $K(q+1)$ to be the subspace of $\ker \varphi$ such that

$$K(q+1)/K(q) = (V/K(q)) \cap \ker \rho(q) \cap \ker d(q)\,.$$

We show now that $\ker \rho \cap V = \cup_q K(q)$. Indeed, set $K = \cup_q K(q)$, so that $K \subset \ker \rho \cap V$ by construction. To show the reverse inclusion recall from (3.3) with $B = \Bbbk$ that $V = \cup_r V(r)$ where $0 = V(0) \subset V(1) \subset \cdots \subset V(r) \subset \cdots$ and $d : V(r) \to \wedge V(r-1)$. If $K \neq \ker \rho$ choose the least $r$ for which for some $x \in \ker \varphi \cap V(r)$, $x \notin K$. Then $dx \in \wedge V(r-1)$. Since $(\wedge V, d)$ is a minimal Sullivan algebra, $d(V) \subset \wedge^{\geq 2} V$ and $dx$ is also in the ideal $\ker \rho \cap \wedge^{\geq 2} V$. Simple algebra shows that $\ker \rho \cap \wedge^{\geq 2} V = (\ker \rho \cap V) \wedge \wedge^+ V$. But

$$\left[(\ker \rho \cap V) \wedge \wedge^+ V\right] \cap \wedge V(r-1) \subset \left[\ker \rho \cap V(r-1)\right] \wedge \wedge^+ V = K \wedge \wedge^+ V\,.$$

It follows that $dx \in K(q)\wedge \wedge^+ V$ for some $q$, and thus $x \in K(q+1)$. This is a contradiction, and hence establishes $K = \ker \rho \cap V$. In particular, $\rho$ factors to give an isomorphism

$$(\wedge(V/K), \overline{d}) \xrightarrow{\cong} (\wedge Z, d)\,.$$

Now suppose by induction that we have constructed $U(q)$, and a commutative cochain algebra $(\wedge V \otimes \wedge U(q), d)$ together with a quasi-isomorphism

$$\chi(q) : (\wedge V \otimes \wedge U(q), d) \xrightarrow{\simeq} (\wedge(V/K(q)), d(q))$$

extending the surjection $\xi(q) : V \to V/K(q)$ and sending $U(q) \to 0$. Suppose further that for $z \in U(q)$, $dz = \alpha(z)\otimes 1 + \beta(z)$ with $\alpha : U(q) \xrightarrow{\cong} K(q)$ and

$$\beta : U(q) \to \ker \chi(q) \cap \left(V\wedge \wedge^+ (V \oplus U(q-1))\right)\,.$$

Then division by $\wedge^{\geq 2}V \otimes \wedge U(q)$ defines a cochain algebra $((V \oplus \Bbbk) \otimes \wedge U(q), \overline{d})$ in which $\overline{d}(V) = 0$ and

$$\overline{d}z = \alpha(z) \otimes 1 + \omega(z)\,, \qquad z \in U(q)$$

with $\omega(z) \in V \otimes \wedge^+ U(q-1)$. The linear maps $z \mapsto \alpha(z) \otimes 1$, $z \in U(q)$ and $v \mapsto 0$, $v \in V$ extend uniquely to a derivation $\theta$ in $(V \oplus \Bbbk) \otimes \wedge U(q)$, and it is straightforward to check that $\theta : \wedge U(q) \to V \otimes \wedge U(q)$ is injective. But

$$\theta : (V \oplus \Bbbk) \otimes \wedge^k U(q) \to (V \oplus \Bbbk) \otimes \wedge^{k-1} U(q)$$

while $\overline{d}$ vanishes in $V \otimes \wedge U(q)$ and

$$\overline{d} - \theta : \Bbbk \otimes \wedge^k U(q) \to V \otimes \wedge^{\geq k} U(q)\,.$$

It follows that $\overline{d} : \wedge U(q) \to V \otimes \wedge U(q)$ is injective and $\ker \overline{d} = V \otimes \wedge U(q)\,.$

Choose a basis $y_i$ of $K(q+1)/K(q)$. Since $\chi(q)$ is a surjective quasi-isomorphism there are cycles $\Phi_i \in \wedge V \otimes \wedge U(q)$ such that $\chi(q)\Phi_i = y_i$. The image of $\Phi_i$ in $(V \oplus \Bbbk) \otimes \wedge U(q)$ is then a cycle, necessarily of the form $x_i + \Psi_i$, with $x_i \in V$ and $\Psi_i \in \wedge U(q) \oplus (V \otimes \wedge^+ U(q))$. Extend $\alpha$ to a derivation $d_\alpha$ from $\wedge U(q)$ to $V \otimes \wedge U(q)$ and let $\Omega_i$ be the component of $\Psi_i$ in $\wedge U(q)$. Then $d_\alpha \Omega_i$ is the component of $d\Phi_i$ in $V \otimes \wedge U(q)$. It follows that $d_\alpha \Omega_i = 0$. But $d_\alpha$ is injective, and so $\Omega_i = 0$ and $\Psi_i \in V \otimes \wedge^+ U(q)$. Thus $\Psi_i \in \ker \chi(q)$ and so $y_i = \chi(q)\Phi_i = \chi(q)x_i + \Gamma_i$, with $\Gamma_i \in \wedge^{\geq 2} V/K(q)$. But $\chi(q)x_i = \xi(q)x_i \in V/K(q)$ and therefore $\chi(q)x_i = y_i$. Thus the $x_i$ are a basis for a direct summand of $K(q)$ in $K(q+1)$.

Now extend $U(q)$ to $U(q+1)$ by adjoining a basis $z_i$ and setting $dz_i = \Phi_i$. Since $\chi(q)\Phi_i \in K(q+1)$ it follows that the image of $\chi(q)\Phi_i \in \wedge V/K(q+1)$ is zero. Thus we may define $\chi(q+1)$ to be the composite

$$\begin{array}{ccc}(\wedge V \otimes \wedge U(q) \otimes \wedge(z_i), d) & \xrightarrow[\simeq]{\chi(q)\otimes id} & (\wedge\,(V/K(q)) \otimes \wedge(z_i), d(q)) \\ & & \Big\downarrow \gamma \\ & & \wedge(V/K(q+1))\,,\end{array}$$

where $\gamma$ restricts to the surjection $V/K(q) \to V/K(q+1)$ and $\gamma z_i = 0$.

Finally, write $(\wedge(V/K(q)) \otimes \wedge(z_i), d(q))$ as a minimal $\Lambda$-extension of $(\wedge(K(q+1)/K(q)) \otimes \wedge(z_i), d)$ with fibre $(\wedge(V/K(q+1)), d)$. Since $d(q)z_i = x_i$ is a basis for $K(q+1)/K(q)$, an easy calculation shows that $(\wedge(K(q+1)/K(q)) \otimes \wedge(z_i), d)$ is acyclic. It follows that $\gamma$ is also a quasi-isomorphism and hence so is $\chi(q+1)$. Now set $U = \cup_q U(q)$; then

$$\chi = \varinjlim \chi(q) : (\wedge V \otimes \wedge U, d) \xrightarrow{\simeq} (\wedge V/K, d) \xrightarrow{\cong} (\wedge W, d)$$

is the desired minimal Sullivan model of $\varphi$. □

**Corollary 3.4.** *Suppose in Proposition 3.9 that $d : V \to \wedge^2 V$ and $d : Z \to \wedge^2 Z$. Then the Sullivan model for $\rho$ may be constructed so that $d : U \to V \otimes \wedge U$, and so*

$$d : U \to (ker\,\rho \otimes 1) \oplus \left(V \otimes \wedge^+ U\right)\,.$$

*proof:* Suppose by induction that $d : U(q) \to V \otimes \wedge U(q-1)$. Set

$$[\wedge V \otimes \wedge U(q)]_k = \wedge^k V \otimes \wedge U(q) \qquad \text{and} \quad (\wedge W)_k = \wedge^k W\,.$$

Then $d$ increases lower degrees by 1, and $\chi(q)$ preserves them. It follows that in the construction in the proof of the Proposition we may take $\Phi_i \in V \otimes \wedge U(q)$ and therefore for the adjoined elements $z_i$ have $d(z_i) \in V \otimes \wedge U(q)$. □

*proof of Theorem 3.1:* (i) Let $m : (\wedge V, d) \xrightarrow{\simeq} (B, d)$ and $n : (\wedge W, d) \xrightarrow{\simeq} (C, d)$ be minimal Sullivan models (Proposition 1.8, p. 19). Then Proposition 1.10 (p. 21) provides a morphism $\psi : (\wedge V, d) \to (\wedge W, d)$ and a homotopy $\Phi : n \circ \psi \sim \varphi \circ m : (\wedge V, d) \to (C, d)$.

In particular, $\psi$ restricts to a morphism $(\wedge V^1, d) \to (\wedge W^1, d)$ whose image is necessarily a Sullivan algebra, namely $(\wedge \psi V^1, d)$. Proposition 3.9 (p. 103) then provides a quasi-isomorphism from a $\Lambda$-extension,

$$(\wedge V^1 \otimes \wedge Z^0, d) \xrightarrow{\simeq} (\wedge \psi V^1, d)\,,$$

restricting to $\psi$ in $\wedge V^1$. Since $(\wedge V, d)$ is a semifree $(\wedge V^1, d)$-module (Proposition 3.4, p. 96) it follows from Proposition 1.1 (p. 7) that

$$\big((\wedge V^1 \otimes \wedge Z^0) \otimes_{\wedge V^1} \wedge V, d\big) \xrightarrow{\simeq} (\wedge \psi V^1 \otimes_{\wedge V^1} \wedge V, d)$$

is a quasi-isomorphism, which we may write as

$$(\wedge V \otimes \wedge Z^0, d) \xrightarrow{\simeq} (\wedge \psi V^1 \otimes \wedge V^{\geq 2}, d)\,.$$

Compose this with $id \otimes \psi : (\wedge \psi V^1 \otimes \wedge V^{\geq 2}, d) \to (\wedge W, d)$ to obtain a morphism

$$\chi(1) : (\wedge V \otimes \wedge Z^0, d) \to (\wedge W, d)$$

in which $H^1(\chi(1))$ is the composite $H^1(\wedge V \otimes \wedge Z^0, d_1) \overset{\cong}{\to} H^1(\wedge \psi V^1, d_1) \to H^1(\wedge W)$. In particular, $H^1(\chi(1))$ is injective. Now it follows from Proposition 14.3 in [18] that $\chi(1)$ extends to a quasi-isomorphism,

$$\chi : \big((\wedge V \otimes \wedge Z^0) \otimes \wedge Z^{\geq 1}, d\big) \xrightarrow{\simeq} (\wedge W, d)\,,$$

from a minimal Sullivan extension of $(\wedge V \otimes \wedge Z^0, d)$. But by Proposition 3.4 (p. 96) $(\wedge V \otimes \wedge Z, d)$ is then a $\Lambda$-extension of $(\wedge V, d)$, which is trivially minimal, and $\chi$ restricts to $\psi$ in $\wedge V$.

Now it follows from Corollary 3.1 (p. 99) that the homotopy $\Phi$ extends to a homotopy $\Psi : \eta \circ \chi \sim \xi : (\wedge V \otimes \wedge Z, d) \to (C, d)$ where $\xi$ restricts to $\varphi \circ m$ in $(\wedge V, d)$. Since $\chi$ is a quasi-isomorphism so is $\xi$ (Proposition 1.9(iv), p. 20). Moreover, set $B \otimes \wedge Z = B \otimes_{\wedge V} (\wedge V \otimes \wedge Z)$. Then $\xi$ factors as

$$(\wedge V \otimes \wedge Z, d) \longrightarrow (B \otimes \wedge Z, d) \xrightarrow[\gamma]{} (C, d)\,,$$

with $\gamma_{|B} = \varphi$. Since $m$ is a quasi-isomorphism and $(\wedge V \otimes \wedge Z, d)$ is $(\wedge V, d)$-semifree, Proposition 1.1 (p. 7) asserts that $B \otimes_{\wedge V} -$ is a quasi-isomorphism. It follows that $\gamma$ is a quasi-isomorphism, and that $H^0(B \otimes \wedge Z, d) = \Bbbk$. Moreover, by construction $(B \otimes \wedge Z, d)$ is a minimal $\Lambda$-extension of $(B, d)$.

In summary, $\gamma$ is a quasi-isomorphism which extends $\varphi$ from a minimal $\Lambda$-extension of $(B, d)$, and thus is a minimal Sullivan model for $\varphi$.

(ii) If $p$ is the least integer such that $Z^p \neq 0$ then the Sullivan condition implies that for some $0 \neq z \in Z^p$, $d(1 \otimes z) \in B$. If $p < k+1$ then $H^{p+1}(B) \to H^{p+1}(B \otimes \wedge Z)$ is injective and so $dz = dx$ for some $x \in B$. But then $1 \otimes z - x \otimes 1$ is a cycle and since $H^p(B) \to H^p(B \otimes \wedge Z)$ is surjective, $1 \otimes z - x \otimes 1 = y \otimes 1 + dw$ for some $w \in (B \otimes \wedge Z)^{p-1}$. But since $(B \otimes \wedge Z, d)$ is a relative $\Lambda$-algebra, $dw \in B \otimes \wedge Z^{\leq p-1}$ and so $1 \otimes z \in B \otimes \wedge Z^{\leq p-1}$, which is impossible. It follows that $p \geq k+1$.

□

### 3.5 Uniqueness of minimal Sullivan models

**Proposition 3.10.** *Suppose*

$$\begin{array}{ccccc} (B,d) & \xrightarrow{\ i\ } & (B\otimes \wedge Z,d) & \xrightarrow{\varepsilon_B\otimes id} & (\wedge Z,\overline{d}) \\ {\scriptstyle\psi}\big\downarrow & & {\scriptstyle\varphi}\big\downarrow & & {\scriptstyle\chi}\big\downarrow \\ (B',d') & \xrightarrow[i']{} & (B'\otimes \wedge Z',d') & \xrightarrow[\varepsilon'_B\otimes id]{} & (\wedge Z',\overline{d'}) \end{array}$$

*is a commutative diagram in which $\varepsilon_B$ and $\varepsilon'_B$ are augmentations, $(B \otimes \wedge Z, d)$ and $(B' \otimes \wedge Z', d')$ are Λ-extensions of $(B, d)$ and $(B', d')$, and $H^0(B \otimes \wedge Z, d) = \Bbbk$ $H^0(B \otimes \wedge Z', d) = \Bbbk$. Then if any two of the three morphisms $\psi$, $\varphi$ and $\chi$ are quasi-isomorphisms, so is the third.*

*In this case, if the Λ-extensions are minimal then $\chi$ is an isomorphism.*

*proof:* The proof is in a number of steps.

*Step One: If $\varphi$ and $\psi$ are quasi-isomorphisms then $\chi$ is a quasi-isomorphism.*

*proof:* Let $(B \otimes \wedge U, d)$ be an acyclic closure for $(B, d)$. Then we have the commutative diagram

$$\begin{array}{ccccc} (B\otimes \wedge U,d) & \longrightarrow & (B\otimes \wedge Z\otimes \wedge U,d) & \xrightarrow{\varepsilon_B\otimes id\otimes\varepsilon} & (\wedge Z,\overline{d}) \\ {\scriptstyle\psi\otimes id}\big\downarrow & & {\scriptstyle\varphi\otimes id}\big\downarrow & & {\scriptstyle\chi}\big\downarrow \\ (B'\otimes \wedge U,d) & \longrightarrow & (B'\otimes \wedge Z'\otimes \wedge U,d') & \xrightarrow{\varepsilon'_B\otimes id\otimes\varepsilon} & (\wedge Z',\overline{d'})\,. \end{array}$$

Since $(B \otimes \wedge U, d)$ is $(B, d)$-semifree it follows from Proposition 1.1 (p. 7) that because $\psi$ is a quasi-isomorphism then so is $\psi \otimes id$. Similarly,

$$\varepsilon_B \otimes \varepsilon \otimes id : (B \otimes \wedge U)_{B\otimes\wedge U}(B \otimes \wedge U \otimes \wedge Z) \xrightarrow{\simeq} \Bbbk \otimes_{B\otimes\wedge U} (B \otimes \wedge U \otimes \wedge Z)\,,$$

because $H(B \otimes \wedge U, d) = \Bbbk$. Since $H(B' \otimes \wedge U) \cong H(B \otimes \wedge U) = \Bbbk$ and since $\varphi$ is a quasi-isomorphism it follows in the same way that $\varepsilon_{B'} \otimes \varepsilon \otimes id$ and $\varphi \otimes id$ are quasi-isomorphisms. Thus $\chi$ is a quasi-isomorphism and so Proposition 3.3 (p. 96) asserts that it is an isomorphism.

*Step Two: If $\psi$ and $\chi$ are quasi-isomorphisms so is $\varphi$.*

*proof:* First observe that since $Z = Z^{\geq 0}$ and $H^0(B \otimes \wedge Z, d) = \Bbbk$, then $H^0(B) = \Bbbk$. Fix a minimal Sullivan model $m : (\wedge V, d) \xrightarrow{\simeq} (B, d)$. By

Theorem 3.1 there is a minimal Sullivan model $\gamma : (\wedge V \otimes \wedge W, d) \xrightarrow{\simeq} (B \otimes \wedge Z, d)$ for $i \circ m : (\wedge V, d) \to (B \otimes \wedge Z, d)$. Then let $\eta : (\wedge V \otimes \wedge W \otimes \wedge U, d) \xrightarrow{\simeq} (B' \otimes \wedge Z', d')$ be a minimal Sullivan model for $\varphi \circ \gamma$. Applying $k\!\!k \otimes_{\wedge V} -$ and $k\!\!k \otimes_B -$ we obtain the commutative diagram

$$\begin{array}{ccc} \wedge W & \xrightarrow{\overline{\gamma}} & \wedge Z \\ \downarrow & & \downarrow \chi \\ \wedge W \otimes \wedge U & \xrightarrow[\overline{\eta}]{} & \wedge Z'. \end{array}$$

It follows from Step One that $\overline{\gamma}$ and $\overline{\eta}$ are quasi-isomorphisms, and hence so is $(\wedge W, \overline{d}) \to (\wedge W \otimes \wedge U, \overline{d})$. Therefore $U = 0$, and so $\varphi \circ \gamma$ is a quasi-isomorphism, hence so is $\varphi$.

*Step Three: If $\varphi$ and $\chi$ are quasi-isomorphisms, so is $\psi$.*

*proof:* Let $(B \otimes \wedge V, d) \xrightarrow[\simeq]{\eta} (B', d')$ be a minimal Sullivan model for $\psi$. Then, as above we can form

$$\eta \otimes_B \varphi : (B \otimes \wedge V \otimes \wedge Z, d) = (B \otimes \wedge V, d) \otimes_B (B \otimes \wedge Z, d) \to (B' \otimes \wedge Z', d')\,.$$

Since $\eta$ is a quasi-isomorphism and $(B \otimes \wedge Z, d)$ is $B$-semifree if follows from Proposition 1.1 (p. 7) that

$$\eta \otimes_B id : (B \otimes \wedge V, d) \otimes_B (B \otimes \wedge Z, d) \to (B' \otimes_B (B \otimes \wedge Z), d)$$

is a quasi-isomorphism. On the other hand, the morphism

$$(B' \otimes_B (B \otimes \wedge Z), d) \xrightarrow{id \otimes_B \varphi} (B' \otimes_B (B' \otimes \wedge Z'), d') \xrightarrow{mult \otimes id} (B' \otimes \wedge Z', d')$$

can be written as

$$\varphi' : (B' \otimes \wedge Z, d) \to (B' \otimes \wedge Z', d')\,.$$

Since $\chi$ is a quasi-isomorphism it follows from Step Two that $\varphi'$ is a quasi-isomorphism. But $\eta \otimes_B \varphi = \varphi' \circ (\eta \otimes_B id)$, and so $\eta \otimes_B \varphi$ is a quasi-isomorphism.

Since $\varphi$ is also a quasi-isomorphism it follows from Step One that applying $k\!\!k \otimes_{B \otimes \wedge Z} -$ and $k\!\!k \otimes_{B' \otimes \wedge Z'} -$ gives a quasi-isomorphism $(\wedge V, \overline{d}) \xrightarrow{\simeq} k\!\!k$. Now Proposition 3.3 (p. 96) gives $V = 0$, and $\psi$ is a quasi-isomorphism.

□

**Proposition 3.11.** *Suppose $\varphi : (B\otimes\wedge Z, d) \to (B\otimes\wedge W, d)$ is a morphism between Λ-extensions of $(B, d)$ which restricts to $id_B$ in $B$. Suppose $H^0(B\otimes\wedge Z) = H^0(B\otimes\wedge W) = \Bbbk$. If $\varepsilon_B : (B, d) \to \Bbbk$ is an augmentation, and if $H^i(\varphi)$ is an isomorphism for $i \le k$ and injective for $i = k+1$, then*

$$H^i(\Bbbk\otimes_B\varphi) : H^i(\wedge Z, \overline{d}) \to H^i(\wedge W, \overline{d})$$

*is an isomorphism for $i \le k$ and injective for $i = k+1$.*

*proof:* Let $(B\otimes\wedge U, d)$ be an acyclic closure, and denote by $\varepsilon : B\otimes\wedge U \to \Bbbk$ the augmentation $\varepsilon_B\otimes\varepsilon$ where $\varepsilon(U) = 0$. But $(B\otimes\wedge Z, d)$ and $(B\otimes\wedge W, d)$ are $(B, d)$-semifree and hence in

$$\begin{array}{ccc} (B\otimes\wedge U)\otimes_B(B\otimes\wedge Z, d) & \xrightarrow{(B\otimes\wedge U)\otimes_B\varphi} & (B\otimes\wedge U)\otimes_B(B\otimes\wedge W, d) \\ {\scriptstyle \varepsilon\otimes\varepsilon_B}\big\downarrow & & \big\downarrow{\scriptstyle \varepsilon\otimes\varepsilon_B} \\ (\wedge Z, \overline{d}) & \xrightarrow{\Bbbk\otimes_B\varphi} & (\wedge W, \overline{d}) \end{array}$$

the vertical arrows are quasi-isomorphisms. Moreover, since $(B\otimes\wedge U, d)$ is $(B, d)$-semifree $\wedge U$ is the increasing union of subspaces $0 = F_0 \subset F_1 \subset \cdots \subset F_r \subset \cdots$ such that the differential in $B\otimes F_r/F_{r-1}$ is just $d\otimes id$. It follows that $H^i((B\otimes F_r/F_{r-1})\otimes_B\varphi)$ is a quasi-isomorphism for $i \le k$ and injective for $i = k+1$. A standard spectral sequence argument then implies that $H^i(\Bbbk\otimes_B\varphi)$ is an isomorphism for $i \le k$ and injective for $i = k+1$.

□

**Definition.** Let $(C, d)$ be a commutative cochain algebra for which $H^0(C) = \Bbbk$. A *minimal Sullivan $k$-model* for $(C, d)$ is a morphism $\varphi : (\wedge V, d) \to (C, d)$ from a minimal Sullivan algebra in which $V = V^{\le k}$ and $H^i(\varphi)$ is an isomorphism for $i \le k$ and injective for $i = k+1$. If $X$ is a path connected topological space then a $k$-minimal Sullivan model for $A_{PL}(X)$ is called a *minimal Sullivan $k$-model for $X$*.

**Corollary 3.5.** *Suppose $(C, d)$ is a commutative cochain algebra and $H^0(C) = \Bbbk$. Then for each $k \ge 1$, $(C, d)$ has a minimal Sullivan $k$-model. If*

$$(\wedge V, d) \xrightarrow{\varphi} (C, d) \quad \text{and} \quad (\wedge W, d) \xrightarrow{\psi} (C, d)$$

*are two minimal Sullivan $k$-models then there is an isomorphism $\chi : (\wedge V, d) \xrightarrow{\cong} (\wedge W, d)$ such that $\psi\circ\chi \sim \varphi$. If $C$ is augmented then $\chi$ may be chosen so that $\psi\circ\chi \sim_* \varphi$.*

*proof:* If $\alpha : (\wedge Z, d) \xrightarrow{\simeq} (C, d)$ is a minimal Sullivan model then the minimality condition ensures that each $\wedge Z^{\leq k}$ is preserved by $d$ and hence that the restriction of $\alpha$ to $(\wedge Z^{\leq k}, d)$ is a minimal Sullivan $k$-model.

Now suppose $(\wedge V, d) \xrightarrow{\varphi} (C, d)$ and $(\wedge W, d) \xrightarrow{\psi} (C, d)$ are both minimal Sullivan $k$-models. Then Theorem 3.1(ii) provides quasi-isomorphisms

$$(\wedge V \otimes \wedge U, d) \xrightarrow{\simeq} (C, d) \quad \text{and} \quad (\wedge W \otimes \wedge Y, d) \xrightarrow{\simeq} (C, d)$$

from minimal relative Sullivan algebras in which $U = U^{\geq k+1}$ and $Y = Y^{\geq k+1}$. It follows that $(\wedge V \otimes \wedge U, d)$ and $(\wedge W \otimes \wedge Y, d)$ are minimal Sullivan algebras, and so the Corollary follows from Proposition 1.11 (p. 23).

□

**Theorem 3.2.** *Let $\varphi : (B, d) \to (C, d)$ be a morphism of commutative cochain algebras in which $H^0(B) = H^0(C) = \Bbbk$. If $\gamma : (B \otimes \wedge Z, d) \xrightarrow{\simeq} (C, d)$ and $\gamma' : (B \otimes \wedge Z', d') \xrightarrow{\simeq} (C, d)$ are minimal Sullivan models of $\varphi$, then there is an isomorphism*

$$m : (B \otimes \wedge Z, d) \xrightarrow{\cong} (B \otimes \wedge Z', d')$$

*such that $m_{|B} = id_B$ and $\gamma' \circ m \sim_B \gamma$.*

*In particular, acyclic closures are unique up to isomorphism.*

**Corollary 3.6.** *With the hypotheses and notation of Theorem 3.2, suppose $B$ is augmented by $\varepsilon_B$. Then $m$ induces an isomorphism*

$$\overline{m} : (\wedge Z, \overline{d}) = \Bbbk \otimes_B (B \otimes \wedge Z) \xrightarrow{\cong} \Bbbk \otimes_B (B \otimes \wedge Z') = (\wedge Z', \overline{d'})\,.$$

*proof of Theorem 3.2:* Choose a subcochain algebra $(\widehat{B}, d)$ of $(B, d)$ such that

$$\widehat{B}^0 = \Bbbk\,, \qquad \widehat{B}^1 \oplus d(B^0) = B^1 \qquad \text{and } \widehat{B}^i = B^i \quad \text{for } i \geq 2\,.$$

By definition of a $\Lambda$-extension for each $z \in Z^n$ there are linearly independent elements $b_i \in B^0$, linearly independent elements $u_j$ in $\widehat{B}^1$ and elements $w_i$ and $v_j$ in $(\wedge Z)^n$ such that

$$dz - \sum (db_i) \otimes w_i - \sum u_j \otimes v_j \in B^{\geq 2} \otimes \wedge Z\,.$$

Then $d^2 = 0$ implies $\overline{d} w_i = 0$. Define an isomorphism $\sigma : B \otimes \wedge Z \to B \otimes \wedge Z$ extending $id_B$ by setting $\sigma(z) = z - \sum b_i \otimes w_i$, and let $\widetilde{d}$ be the differential in $B \otimes \wedge Z$ defined by $\sigma \circ \widetilde{d} = d \circ \sigma$. Now $\gamma \circ \sigma : (B \otimes \wedge Z, \widetilde{d}) \to (C, d)$ is a minimal Sullivan model such that $\widehat{B} \otimes \wedge Z$ is preserved by $\widetilde{d}$. Do the same for $(B \otimes \wedge Z', d')$. Thus we may and do suppose that $\widehat{B} \otimes \wedge Z$ and $\widehat{B} \otimes \wedge Z'$ are preserved by $d$ and $d'$.

Moreover, since the inclusion $(\widehat{B}, d) \to (B, d)$ is a quasi-isomorphism, it follows from Proposition 1.1 (p. 7) and (3.3) that

$$(\widehat{B} \otimes \wedge Z, d) = \widehat{B} \otimes_{\widehat{B}} (\widehat{B} \otimes \wedge Z, d) \to B \otimes_{\widehat{B}} (\widehat{B} \otimes \wedge Z, d) = (B \otimes \wedge Z, d)$$

is a quasi-isomorphism. Similarly, $(\widehat{B} \otimes \wedge Z', d') \xrightarrow{\simeq} (B \otimes \wedge Z', d')$. Thus

$$\gamma : (\widehat{B} \otimes \wedge Z, d) \xrightarrow{\simeq} (C, d) \quad \text{and} \quad \gamma' : (\widehat{B} \otimes \wedge Z', d') \xrightarrow{\simeq} (C, d).$$

Now Proposition 3.6 (p. 98) gives a quasi-isomorphism

$$n : (\widehat{B} \otimes \wedge Z, d) \xrightarrow{\simeq} (\widehat{B} \otimes \wedge Z', d')$$

such that $n_{|\widehat{B}} = id_{\widehat{B}}$ and $\gamma' \circ n \sim_{\widehat{B}} \gamma$. Dividing by $\widehat{B}^+$ gives a morphism $\overline{n} : (\wedge Z, \overline{d}) \to (\wedge Z', \overline{d})$, which is an isomorphism by Proposition 3.10 (p. 107). Filter both sides by the ideals $\widehat{B}^{\geq k} \otimes -$ and note that the morphism of associated bigraded algebras is the isomorphism $id \otimes \overline{n}$. Since $\widehat{B} = \Bbbk \oplus \widehat{B}^{\geq 1}$ it follows that $n$ is an isomorphism and hence so is $m = B \otimes_{\widehat{B}} (\widehat{B} \otimes \wedge Z) \xrightarrow{\cong} B \otimes_{\widehat{B}} (\widehat{B} \otimes \wedge Z')$. Clearly $\gamma' \circ m \sim_B \gamma$.

□

## 3.6 The acyclic closure of a minimal Sullivan algebra

**Proposition 3.12.** *A minimal Λ-extension $(\wedge V \otimes \wedge U, d)$ of a minimal Sullivan algebra $(\wedge V, d)$ is an acyclic closure if and only if there is a linear isomorphism of degree one, $\alpha : U \xrightarrow{\cong} V$ such that for $u \in U$*

$$d(1 \otimes u) - \alpha(u) \otimes 1 \in V \wedge \wedge^+ (V \oplus U).$$

*proof:* Suppose $(\wedge V \otimes \wedge U, d)$ is an acyclic closure. It follows from Proposition 3.9 (p. 103) applied to the unique augmentation $(\wedge V, d) \to \Bbbk$ that there is an acyclic closure $(\wedge V \otimes \wedge U', d)$ satisfying the condition of the Proposition. Now Theorem 3.2 provides a commutative diagram

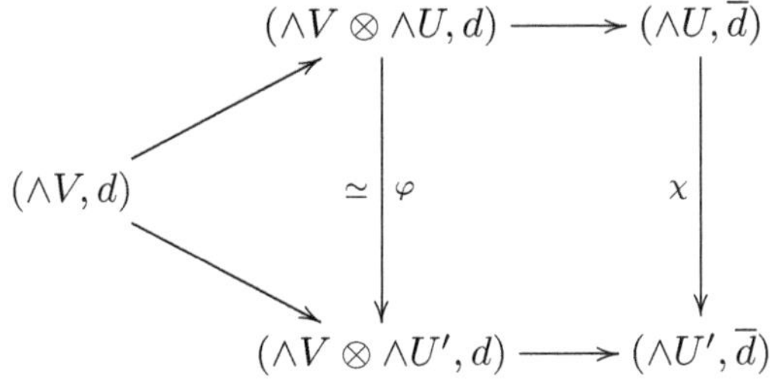

where $\varphi$ is a quasi-isomorphism, and Proposition 3.10 (p. 107) asserts that $\chi$ is an isomorphism. It follows that $\varphi$ is an isomorphism.

Conversely, suppose the $\Lambda$-extension satisfies the condition of the Proposition, and let $0 = U(0) \subset \cdots \subset U(r) \subset \cdots$ be an increasing filtration of $U$ such that

$$d : 1 \otimes U^p(r) \to \wedge V \otimes \wedge(U^{<p} \oplus U^p(r-1)) .$$

It follows that $d(1 \otimes u) = \alpha(u) \otimes 1$ for $u \in U^0(1)$. Thus $(\wedge\alpha(U^0(1)) \otimes \wedge U^0(1), d)$ is an acyclic closure of $(\wedge\alpha(U^0(1)), d)$, and the resulting surjection

$$(\wedge V^1 \otimes \wedge U^0, d) \to \left(\wedge(V^1/\alpha(U^0(1))) \otimes \wedge(U^0/U^0(1)), \overline{d}\right)$$

is a quasi-isomorphism.

Iterating the observation yields quasi-isomorphisms

$$(\wedge V^1 \otimes \wedge U^0, d) \xrightarrow{\simeq} \left(\wedge(V^1/\alpha(U^0(r))) \otimes \wedge(U^0/U^0(r)), \overline{d}\right) , \qquad r \geq 1 .$$

Any cycle in $(\wedge V^1 \otimes \wedge U^0)^+$ will be killed by one of these surjections and so $H(\wedge V^1 \otimes \wedge U^0, d) = \Bbbk$. This in turn yields a quasi-isomorphism $(\wedge V \otimes \wedge U, d) \xrightarrow{\simeq} (\wedge V^{\geq 2} \otimes \wedge U^{\geq 1}, \overline{d})$, and iterating yields quasi-isomorphisms $(\wedge V \otimes \wedge U, d) \xrightarrow{\simeq} (\wedge V^{\geq p+1} \otimes \wedge U^{\geq p}, \overline{d})$. As above for $(\wedge V^1 \otimes \wedge U^0, d)$ it follows that $H(\wedge V \otimes \wedge U, d) = \Bbbk$.

□

**Corollary 3.7.** *If $(\wedge V \otimes \wedge U, d)$ is an acyclic closure of a minimal Sullivan algebra $(\wedge V, d)$ then each $(\wedge V^{\leq p+1} \otimes \wedge U^{\leq p}, d)$ is an acyclic closure of $(\wedge V^{\leq p+1}, d)$.*

Now suppose $\psi : (\wedge V, d) \to (\wedge Z, d)$ is a surjection of minimal Sullivan algebras and extend $\psi$ to the quasi-isomorphism $\varphi : (\wedge V \otimes \wedge S, d) \xrightarrow{\simeq} (\wedge Z, d)$ constructed in Proposition 3.9 (p. 103). Then $\varphi$ has a right inverse, $\sigma : (\wedge Z, d) \xrightarrow{\simeq} (\wedge V \otimes \wedge S, d)$. Thus if $(\wedge Z \otimes \wedge U, \overline{d})$ is an acyclic closure for $(\wedge Z, d)$ we may use $\sigma$ to form the cdga,

$$(\wedge V \otimes \wedge S, d) \otimes_{\wedge Z} (\wedge Z \otimes \wedge U, d) = (\wedge V \otimes \wedge S \otimes \wedge U, d) .$$

**Proposition 3.13.** *$(\wedge V \otimes \wedge S \otimes \wedge U, d)$ is an acyclic closure of $(\wedge V, d)$.*

*proof*: It follows from Proposition 1.1 (p. 7) that

$$(\wedge V \otimes \wedge S \otimes \wedge U, d) \xrightarrow{\varphi \otimes id} (\wedge Z \otimes \wedge U, \overline{d})$$

is a quasi-isomorphism, and so $H(\wedge V \otimes \wedge S \otimes \wedge U) = \Bbbk$. It remains to show that this is a minimal Λ-extension of $(\wedge V, d)$. Since $\wedge V \to \wedge V \otimes \wedge S$ and $\wedge V \otimes \wedge S \to \wedge V \otimes \wedge S \otimes \wedge U$ are Λ-extensions it follows from Proposition 3.4 (p. 96) that so is $\wedge V \to \wedge V \otimes \wedge S \otimes \wedge U$. Thus we need only show this is a minimal Λ-extension; i.e. that if $u \in S^p \oplus U^p$ then $d(1 \otimes u) \in \wedge V \otimes \wedge(S^{\leq p} \oplus U^{\leq p})$. But this is true for $u \in S^p$ since $(\wedge V \otimes \wedge S, d)$ is a minimal Λ-extension of $(\wedge V, d)$. It therefore remains to consider the case $u \in U^p$. In this case

$$d(1 \otimes u) \in \wedge V \otimes \wedge S^{\leq p+1} \otimes \wedge U^{\leq p}$$

and we have to show that $d(1 \otimes u) \in \wedge V \otimes \wedge S^{\leq p} \otimes \wedge U^{\leq p}$. But if not then $d(1 \otimes u)$ has a non-zero component $\Omega$ in $1 \otimes \wedge S^0 \otimes S^{p+1} \otimes \wedge U^0$. We assume this occurs and deduce a contradiction.

For this we tensor with the algebraic closure of $\Bbbk$ to reduce to the case that $\Bbbk$ is algebraically closed. Then we note that $\varphi$ restricts to a quasi-isomorphism $\wedge V^1 \otimes \wedge S^0 \xrightarrow{\simeq} \wedge Z^1$, and that the restriction of $\varphi$ to $\wedge S^0$ may be any augmentation $\varepsilon$. In particular the induced surjection

$$\varepsilon \otimes id : (\wedge V \otimes \wedge S \otimes \wedge U, d) \to (\wedge Z^1 \otimes \wedge V^{\geq 2} \otimes \wedge S^{\geq 1} \otimes \wedge U, d)$$

is a quasi-isomorphism.

Now $\Omega \in 1 \otimes \wedge T \otimes \wedge S^{p+1} \otimes \wedge U^0$ for some finite dimensional subspace $T \subset S^0$. Write $\Omega = A \otimes u_0 + \sum \omega_i \otimes u_i$ with $\omega_i \in \wedge T$ and $u_i \in \wedge S^{p+1} \otimes \wedge U^0$. By the Hilbert Nullstellensatz ([22]) there is an augmentation $\overline{\varepsilon} : \wedge T \to \Bbbk$ such that $\overline{\varepsilon}(\omega_i - 1) = 0$ for each $i$. It follows that we may choose $\varepsilon$ so that $(\varepsilon \otimes id)\Omega \neq 0$. Thus we may assume $V = Z^1 \oplus V^{\geq 2}$, $S = S^{\geq 1}$, and that

$$\Omega \in 1 \otimes S^{p+1} \otimes \wedge U^0 .$$

Next observe that $\varphi$ restricts to a quasi-isomorphism $\varphi' : (\wedge V^{\leq p+1} \otimes \wedge S^{\leq p}, d) \to (\wedge Z^{\leq p+1}, d)$. Apply $\varphi' \otimes id$ to reduce to the case $V^{\leq p+1} = Z^{\leq p+1}$ and $S = S^{\geq p+1}$. Then

$$d(1 \otimes u) = \Omega + 1 \otimes \overline{d}u .$$

Now $\overline{d}u \in Z^{p+1} \otimes \wedge U^0$, and since $(\wedge Z \otimes \wedge U, \overline{d})$ is a minimal Λ-extension, $\overline{d} : Z^{p+1} \otimes \wedge U^0 \to \wedge Z \otimes \wedge U^0$. Moreover, $d : S^{p+1} \xrightarrow{\cong} W^{p+2} \subset V$, where $W^{p+2} \oplus Z^{p+2} = V^{p+2}$. It follows that $d\Omega$ has a non-trivial component in $W^{p+2} \otimes \wedge U^0$ whereas $d(\overline{d}u) \in \wedge Z \otimes \wedge U^0$. Thus $d(\Omega + 1 \otimes \overline{d}u) \neq 0$, which contradicts $d^2 = 0$.

□

**Corollary 3.8.** *An acyclic closure for a minimal Sullivan extension* $(\wedge W \otimes \wedge Z, d)$ *of a minimal Sullivan algebra* $(\wedge W, d)$ *decomposes as a minimal* $\Lambda$-*extension*

$$(\wedge W \otimes \wedge U_W, d) \to (\wedge W \otimes \wedge U_W \otimes \wedge Z \otimes \wedge U_Z, d) \to (\wedge Z \otimes \wedge U_Z, \overline{d})$$

*of* $(\wedge W \otimes \wedge U_W, d)$ *in which the initial and terminal cochain algebras are acyclic closures of* $(\wedge W, d)$ *and* $(\wedge Z, \overline{d})$.

*proof:* Let $(\wedge W \otimes \wedge U_W, d)$ be an acyclic closure for $(\wedge W, d)$ and write

$$(\wedge W \otimes \wedge Z \otimes \wedge U_W, d) = (\wedge W \otimes \wedge Z, d) \otimes_{\wedge W} (\wedge W \otimes \wedge U_W, d)\,.$$

Thus an augmentation $\varepsilon$ in $(\wedge W \otimes \wedge U_W, d)$ defines a quasi-isomorphism

$$\varepsilon \otimes id : (\wedge W \otimes \wedge Z \otimes \wedge U_W, d) \xrightarrow{\simeq} (\wedge Z, \overline{d})\,.$$

Now apply Proposition 3.13 with $\wedge V = \wedge W \otimes \wedge Z$, and $\psi$ the projection $\Bbbk \otimes_{\wedge W} - : \wedge W \otimes \wedge Z \to \wedge Z$.

□

## 3.7 Sullivan extensions and fibrations

We turn now to the connection between Sullivan extensions and fibrations. Let

$$Y \xleftarrow{p} X \xleftarrow{j} F$$

be a Serre fibration in which $X$ and $Y$ are path connected spaces. Then $A_{PL}$ converts the diagram

$$\begin{array}{ccc} F & \xrightarrow{j} & X \\ \downarrow & & \downarrow p \\ * & \longrightarrow & Y \end{array} \qquad \text{to} \qquad \begin{array}{ccc} A_{PL}(F) & \xleftarrow{A_{PL}(j)} & A_{PL}(X) \\ \uparrow & & \uparrow A_{PL}(p) \\ \Bbbk & \longleftarrow & A_{PL}(Y) \end{array}$$

By Theorem 3.1 (p. 102), a minimal Sullivan model $m_Y : (\wedge W, d) \xrightarrow{\simeq} A_{PL}(Y)$ extends to a minimal Sullivan model for $A_{PL}(p) \circ m_Y$:

$$m_X : (\wedge W \otimes \wedge Z, d) \xrightarrow{\simeq} A_{PL}(X)\,.$$

The augmentation $\varepsilon$ in $(\wedge W, d)$ then gives the quotient $\Lambda$-algebra $(\wedge Z, \overline{d})$. Moreover, since $A_{PL}(j) \circ A_{PL}(p) = \varepsilon$, the morphism $m_X$ factors to yield a quotient morphism $m_F : (\wedge Z, \overline{d}) \to A_{PL}(F)$ which makes commutative the diagram

$$\begin{array}{ccccc}
(\wedge W, d) & \longrightarrow & (\wedge W \otimes \wedge Z, d) & \longrightarrow & (\wedge Z, \overline{d}) \\
\simeq \downarrow m_Y & & \simeq \downarrow m_X & & \downarrow \overline{m} \\
A_{PL}(Y) & \xrightarrow{A_{PL}(p)} & A_{PL}(X) & \xrightarrow{A_{PL}(j)} & A_{PL}(F)\,.
\end{array} \tag{3.8}$$

In Chapter 5, §5.3 we shall apply this construction to the classifying fibration

$$B \leftarrow X \times_B MB \leftarrow F$$

of a connected CW complex $X$.

Note that if $F$ is path connected then $\pi_1(p)$ is surjective and $H^1(p : \Bbbk)$ is injective. In this case it follows from Theorem 3.1(ii) (p. 102) that $Z = Z^{\geq 1}$, i.e. $(\wedge Z, \overline{d})$ is a minimal Sullivan algebra.

In §5.1, Theorem 5.1 (p. 145) we shall show that under very general hypotheses $\overline{m}$ is a quasi-isomorphism and hence a minimal Sullivan model for $F$. Thus Sullivan models become a valuable tool in the study of fibrations.

In general, the proof of ( [18], Proposition 17.9) shows that if $(\wedge W, d) \xrightarrow{i} (\wedge W \otimes \wedge Z, d) \xrightarrow{\rho} (\wedge Z, \overline{d})$ is a Λ-extension then the spatial realization (§1.6)

$$|\wedge W, d| \xleftarrow{|i|} |\wedge W \otimes \wedge Z, d| \xleftarrow{|\rho|} |\wedge Z, \overline{d}|$$

is a fibre bundle of CW complexes, and hence ([18], Proposition 2.6) is a Serre fibration.

**Proposition 3.14.** *With the above notation, we have the commutative diagram*

$$\begin{array}{ccccc}
(\wedge W, d) & \xrightarrow{i} & (\wedge W \otimes \wedge Z, d) & \xrightarrow{\rho} & (\wedge Z, \overline{d}) \\
\downarrow m_{|\wedge W, d|} & & \downarrow m_{|\wedge W \otimes \wedge Z, d|} & & \downarrow m_{|\wedge Z, \overline{d}|} \\
A_{PL}(|\wedge W, d|) & \xrightarrow{A_{PL}(|i|)} & A_{PL}(|\wedge W \otimes \wedge Z, d|) & \xrightarrow{A_{PL}(|\rho|)} & A_{PL}(|\wedge Z, \overline{d}|)\,.
\end{array} \tag{3.9}$$

*proof:* Recall the commutative diagram from formula (1.12) (p. 28)

$$\begin{array}{ccccc}
(\wedge W,d) & \xrightarrow{\quad i \quad} & (\wedge W\otimes\wedge Z,d) & \xrightarrow{\quad \rho \quad} & (\wedge Z,\overline{d}) \\
\Big\downarrow{\scriptstyle m_{\langle\wedge W,d\rangle}} & & \Big\downarrow{\scriptstyle m_{\langle\wedge W\otimes\wedge Z,d\rangle}} & & \Big\downarrow{\scriptstyle m_{\langle\wedge Z,\overline{d}\rangle}} \\
A_{PL}(\langle\wedge W,d\rangle) & \xrightarrow{A_{PL}(\langle i\rangle)} & A_{PL}(\langle\wedge W\otimes\wedge Z,d\rangle) & \xrightarrow{A_{PL}(\langle\rho\rangle)} & A_{PL}(\langle\wedge Z,\overline{d}\rangle) \\
{\scriptstyle\beta}\Big\uparrow{\scriptstyle\simeq} & & \Big\uparrow{\scriptstyle\simeq} & & \Big\uparrow{\scriptstyle\simeq} \\
A_{PL}(|\wedge W,d|) & \xrightarrow{A_{PL}(|i|)} & A_{PL}(|\wedge W\otimes\wedge Z,d|) & \xrightarrow{A_{PL}(|\rho|)} & A_{PL}(|\wedge Z,\overline{d}|)\,.
\end{array}$$

By Proposition 1.10 (p. 21) we may find a quasi-isomorphism $m_{|\wedge W,d)|}$ : $(\wedge W,d) \xrightarrow{\simeq} A_{PL}(|\wedge W,d|)$ such that there is a homotopy $m_{\langle\wedge W,d\rangle} \sim \beta\circ m_{|\wedge W,d|}$. By Corollary 3.1 (p. 99), this homotopy extends to a homotopy

$$m_{\langle\wedge W\otimes\wedge Z,d\rangle} \sim m' : (\wedge W\otimes\wedge Z,d) \to A_{PL}(\langle\wedge W\otimes\wedge Z,d\rangle)\,.$$

This yields the commutative diagram

$$\begin{array}{ccc}
(\wedge W,d) & \xrightarrow{A_{PL}(|\,i|)(\circ m_{|\wedge W,d|}} & A_{PL}(|\wedge W\otimes\wedge Z,d|) \\
\Big\downarrow & & \Big\downarrow{\scriptstyle\simeq} \\
(\wedge W\otimes\wedge Z,d) & \xrightarrow{\quad m' \quad} & A_{PL}(\langle\wedge W\otimes\wedge Z,d\rangle)
\end{array}$$

and so (3.9) follows directly from Proposition 3.6 and formula (3.7).

□

# Chapter 4

# Holonomy

Recall from §1.4 (p. 17) that for any topological space $X$, $H^*(X;\Bbbk) = H(A_{PL}(X))$ is denoted by $H(X)$.

## 4.1 Holonomy of a fibration

**Definition.** A sequence of maps $\xi : F \stackrel{j}{\to} X \stackrel{p}{\to} Y$ *satisfies the fibration condition* if:

$$\left.\begin{array}{l} \text{- } Y \text{ is path connected and pointed by } * \in Y, \\ \text{- } X \text{ is path connected,} \\ \text{- } j \text{ is the inclusion of } p^{-1}(*), \text{ and} \\ \text{- either } \xi \text{ is a fibration, or else it is a Serre fibration in} \\ \quad \text{which } F \text{ is a CW complex.} \end{array}\right\} \qquad (4.1)$$

Suppose $F \stackrel{j}{\to} X \stackrel{p}{\to} Y$ satisfies the fibration condition and let $f : (I, \partial I) \to (Y, *)$ be a loop in $Y$. Then, because of our hypotheses above, the diagram

$$\begin{array}{ccc} \{0\} \times F & \xrightarrow{j} & X \\ \downarrow & & \downarrow{\scriptstyle p} \\ I \times F & \xrightarrow[f_F]{} & Y \end{array}$$

with $f_F(t, z) = f(t)$, admits a map $\Phi_f : I \times F \to X$ such that

$$\Phi_f(0, z) = j(z) \quad \text{and} \quad p \circ \Phi_f = f_F\,.$$

The condition $p \circ \Phi_f = f_F$ is equivalent to the commutative diagram

$$\begin{array}{ccc} I \times F & \xrightarrow{\Phi_f} & X \\ {\scriptstyle \pi_I}\downarrow & & \downarrow{\scriptstyle p} \\ I & \xrightarrow[f]{} & Y \end{array}$$

and a homotopy $I \times F \to X$ for which this diagram commutes is said to *cover* $f$.

In particular it follows from the diagram that $\Phi_f$ restricts to a continuous map $\{1\} \times F \to F$, which we denote by $\lambda_f : F \to F$. Thus $\Phi_f$ is a homotopy from $j$ to $j \circ \lambda_f$. Moreover, if $j_0$ and $j_1$ denote the inclusions of $F$ in $I \times F$ opposite $\{0\}$ and $\{1\}$ then

$$\Phi_f \circ j_0 = j \qquad \text{and} \qquad \Phi_f \circ j_1 = j \circ \lambda_f \tag{4.2}$$

**Lemma 4.1.** *Suppose $F \xrightarrow{j} X \xrightarrow{p} Y$ satisfies the fibration condition and that $\alpha \in \pi_1(Y, *)$, and suppose $f, g : (I, \partial I) \to (Y, *)$ are representatives of $\alpha$. If $\Phi_f, \Phi_g : I \times F \to X$ are homotopies starting at $j$, and covering $f$, $g$, then the corresponding maps $\lambda_f, \lambda_g : F \to F$ are homotopic.*

*proof:* Set $J = I \times \{0\} \cup \{0,1\} \times I \subset I \times I$ and let $H : (I \times I, I \times \partial I) \to (Y, *)$ be a based homotopy from $f$ to $g$. Then we have the commutative diagram

$$\begin{array}{ccc} J \times F & \xrightarrow{\quad\Phi\quad} & X \\ \downarrow & & \downarrow{\scriptstyle p} \\ (I \times I) \times F & \xrightarrow{\quad\Psi\quad} & Y \end{array}$$

in which $\Psi(s,t,x) = H(s,t)$, $\Phi(s,0,x) = j(x)$, $\Phi(0,t,x) = \Phi_f(t,x)$ and $\Phi(1,t,x) = \Phi_g(t,x)$. Because (4.1) is satisfied it follows that $\Phi$ extends to a map $\Omega : (I \times I) \times F \to X$ such that $p \circ \Omega = \Psi$. In particular, the restriction of $\Omega$ to $I \times \{1\} \times F$ is a homotopy $\lambda_f \sim \lambda_g$.

□

Now suppose $f, g : (I, \partial I) \to (Y, *)$ are loops representing $\alpha, \beta \in \pi_1(Y, *)$ recall from §2.7 (p. 76) that the loop $f \star g$ defined by

$$(f \star g)(t) = \begin{cases} f(2t)\,, & 0 \le t \le 1/2 \\ g(2t-1)\,, & 1/2 \le t \le 1 \end{cases}$$

represents the product $\alpha \star \beta$. We denote by $f^{-1}$ the loop $f^{-1}(t) = f(1-t)$, and by $c$ the constant loop $I \to *$, and note that $f^{-1} \star f \sim_* c \sim_* f \star f^{-1}$.

**Lemma 4.2.** *With the hypotheses and notation above,*

*(i)* $\lambda_{f\star g} \sim \lambda_g \circ \lambda_f$.
*(ii)* $\lambda_c \sim id_F$.
*(iii)* $\lambda_f \circ \lambda_{f^{-1}} \sim id_F \sim \lambda_{f^{-1}} \circ \lambda_f$.

*proof:* (i) Let $\Phi_f$ and $\Phi_g$ satisfy the conditions above, and define $\Phi_{f\star g} : I \times F \to X$ by

$$\Phi_{f\star g}(t,x) = \begin{cases} \Phi_f(2t,x)\,, & 0 \le t \le 1/2 \\ \Phi_g(2t-1, \lambda_f(x))\,, & 1/2 \le t \le 1\,. \end{cases}$$

Then $\Phi_{f\star g}$ is a homotopy covering $f\star g$, $\Phi_{f\star g}(0,x) = j(x)$ and $\Phi_{f\star g}(1,x) = j \circ \lambda_g \circ \lambda_f(x)$. Thus $\lambda_{f\star g} \sim \lambda_g \circ \lambda_f$. Assertions (ii) and (iii) follow from Lemma 4.1.

□

It follows from Lemmas 4.1 and 4.2 that for $\xi : F \to X \to Y$, the map $\lambda_f : F \to F$ is a homotopy equivalence with homotopy inverse $\lambda_{f^{-1}}$, and that the homotopy classes $[\lambda_f]$ and $[\lambda_{f^{-1}}]$ depend only on the element $\alpha \in \pi_1(Y,*)$ represented by $f$. We write

$$hol_\xi(\alpha) = [\lambda_f]\,; \tag{4.3}$$

where no ambiguity is possible we write simply $hol\,(\alpha)$.

Now suppose $F \to X \to Y$ satisfies the fibration condition (4.1) and denote by $\mathcal{G}_F$ the group of homotopy classes of homotopy equivalences of a space $F$. Then it follows from Lemma 4.2 that the map

$$hol : \pi_1(Y,*) \to \mathcal{G}_F$$

defined by $\alpha \mapsto hol\,\alpha$ preserves the identity and satisfies $hol(\alpha \star \beta) = hol(\beta)\,hol(\alpha)$. Such a map is called an *anti-homomorphism* of groups. Similarly, a right *representation* of a group $G$ in a graded vector space $M$ is an anti-homomorphism $G \to \operatorname{Aut} M$ of $G$ into the group of linear automorphisms of $M$.

**Definition.** The anti-homomorphism $hol : \pi_1(Y, y_0) \to \mathcal{G}_F$ is called the *holonomy action* of $\pi_1(Y,*)$ on $F$.

In particular, associated with the homotopy class $[\psi]$ of a homotopy equivalence $\psi$ of $F$ are the corresponding automorphisms of $H_*(F;\Bbbk)$ and

$H^*(F;\Bbbk)$ which we denote respectively by $H_*([\psi])$ and $H^*([\psi])$. This defines maps

$$\mathcal{G}_F \to \operatorname{Aut} H_*(F;\Bbbk) \quad \text{and} \quad \mathcal{G}_F \to \operatorname{Aut} H^*(F;\Bbbk)$$

which are, respectively, an anti-homomorphism and a homomorphism of groups. Thus if $F \to X \to Y$ satisfies the fibration condition (4.1) then the correspondences

$$\alpha \mapsto H_*(\operatorname{hol}\alpha) \quad \text{and} \quad \alpha \mapsto H(\operatorname{hol}\alpha)\,, \qquad \alpha \in \pi_1(Y,*)\,,$$

are respectively a right representation and a representation of $\pi_1(Y,*)$ in $H_*(F;\Bbbk)$ and $H(F)$. These are called the *holonomy representations* of $\pi_1(Y,*)$, and $H_*(F)$ and $H(F)$ are *holonomy $\pi_1(Y,*)$-modules.*

Now a representation of a group $G$ in a graded vector space $M$ is *locally nilpotent* if for each $z \in M$ there is an integer $n = n(z)$ such that for all $g_1, \cdots, g_n \in G$,

$$(g_1 - id) \circ \cdots \circ (g_n - id)z = 0\,. \tag{4.4}$$

If for each $k$ this holds for a single $n = n(k)$ and all $z \in M^k$, then the representation is *nilpotent.*

As in §2.5 define the lower central series for $G$ by $G^1 = G$ and $G^{n+1} = [G, G^n]$. It is straightforward to check that if (4.4) holds then $G^n.z = z$.

Finally, let $\xi : F \to X \to Y$ satisfy the fibration condition (4.1) and suppose $\varphi : (Z,*) \to (Y,*)$ is a continuous map from a CW complex $Z$. Then we can form the pull-back diagram

$$\begin{array}{lccccc}
\xi: & Y & \xleftarrow{\;p\;} & X & \longleftarrow & F \\
 & \uparrow{\scriptstyle\varphi} & & \uparrow{\scriptstyle\psi} & & \uparrow{\scriptstyle id} \\
\xi_\varphi: & Z & \longleftarrow & Z \times_Y X & \longleftarrow & F
\end{array}$$

and (cf. [18], §2) $\xi_\varphi$ again satisfies the fibration condition.

**Proposition 4.1.** *If the holonomy representation of $\pi_1(Y,*)$ in $H_*(F;\Bbbk)$ is locally nilpotent and if $H_*(\varphi;\Bbbk)$ is an isomorphism then $H_*(\psi;\Bbbk)$ is an isomorphism.*

*proof:* In ([66], Chap. VI-1, Example 3) the homology, $H_*(Y;\mathcal{H}_*(F;\Bbbk))$, of $Y$ in the local coefficient system $\mathcal{H}_*(F,\Bbbk) = \{\, H_*(p^{-1}(y);\Bbbk) \,\}$ of $\pi_1(Y,*)$-modules is constructed and in ([66], Chap. XIII) the Leray-Serre spectral sequence for $F \to X \to Y$ is constructed and shown to converge from $E^2_{p,q} =$

$H_p(Y; \mathcal{H}_q(F; \Bbbk))$ to $H_*(X; \Bbbk)$. The local nilpotent hypothesis implies that $H_*(F; \Bbbk)$ is the increasing union of sub $\pi_1(Y, *)$-modules

$$0 = \mathcal{U}_0 \subset \mathcal{U}_1 \subset \cdots \subset \mathcal{U}_n \subset \cdots$$

such that each $\mathcal{U}_{i+1}/\mathcal{U}_i$ is the trivial module. It follows that for the corresponding local coefficient systems $\mathcal{U}_{i+1}/\mathcal{U}_i$, the homology

$$H_*(Y; \mathcal{U}_{i+1}/\mathcal{U}_i) = H_*(Y) \otimes \mathcal{U}_{i+1}/\mathcal{U}_i \,.$$

On the other hand, by construction,

$$\mathrm{hol}_{\xi_\varphi}(\alpha) = \mathrm{hol}_\xi(\pi_1(\varphi)\alpha) \,, \qquad \alpha \in \pi_1(Y, *) \,.$$

Moreover the Leray-Serre spectral sequence is natural with respect to maps of fibrations, and so it follows from the identity above that $\varphi$ induces isomorphisms $H_*(Z, \mathcal{U}_{i+1}/\mathcal{U}_i) \xrightarrow{\cong} H_*(Y, \mathcal{U}_{i+1}/\mathcal{U}_i)$. Now a long exact sequence argument shows that $\psi$ induces isomorphisms $H_*(Z, \mathcal{U}_i) \xrightarrow{\cong} H_*(Y, \mathcal{U}_i)$ and a direct limit argument shows that $H_*(Z, \mathcal{H}_*(F; \Bbbk)) \xrightarrow{\cong} H_*(Y, \mathcal{H}_*(F; \Bbbk))$. Thus $E^2(\varphi)$ is an isomorphism and hence so is $H_*(\psi; \Bbbk)$.

□

In particular, suppose $h : (X, *) \to (Y, *)$ is a continuous map between connected CW complexes, and recall from §3.1 the fibrations

$$Y \xleftarrow{q} X \times_Y MY \longleftarrow X \times_Y P'Y \qquad \text{and} \quad Y \longleftarrow PY \longleftarrow \Omega Y$$

constructed respectively from $h$ and from the inclusion $* \hookrightarrow Y$. Then define a continuous map

$$(X \times_Y MY) \times_Y P'Y \to X \times_Y P'Y \,, \qquad ((x, g), f) \mapsto (x, g) \bullet f$$

by $(x, g) \bullet f = (x, g')$, where

$$g'(t) = \begin{cases} g(t) \,, & 0 \le t \le |g| \\ f(t - |g|) \,, & t \ge |g| \end{cases} \qquad \text{and} \quad |g'| = |f| + |g| \,.$$

This map restricts to a continuous map

$$\mu_X : (X \times_Y P'Y) \times \Omega Y \to X \times_Y P'Y \,,$$

which in the case of the path space fibration, is a map $\mu_* : \Omega Y \times \Omega Y \to \Omega Y$, $(f, g) \mapsto f \bullet g$. Moreover, if $f_1, f_2 \in \Omega Y$ have length 1 then $f_1 \bullet f_2$ is a loop of length 2 based homotopic to the loop $f_1 * f_2$ used in §2.7 to define the product in $\pi_1(Y, *)$, and then used in §4.1.

**Proposition 4.2.** *With the hypotheses and notation above, for* $(x, g) \in X \times_Y P'Y$ *and* $f_1, f_2 \in \Omega Y$,

*(i)* $((x, g) \bullet f_1) \bullet f_2) = (x, g) \bullet (f_1 \bullet f_2)$
*(ii)* $(x, g) \bullet c_x = (x, g)$.

*In particular* $\mu_*$ *makes* $\Omega Y$ *into a topological monoid acting via* $\mu_X$ *from the right on* $X \times_Y P'Y$.

*proof:* This is immediate from the definitions.

□

**Definition.** The map $\mu_X$ is the *holonomy action* of $\Omega Y$ on $X \times_Y P'Y$.

The constant loop, $c_*$, at $*$ serves as a base point and is the identity for the topological monoid $\Omega Y$. Its path component is denoted by $\Omega_0 Y$, and $\Omega_0 Y$ is a sub topological monoid.

It follows from Proposition 4.2 that the multiplication in $\Omega Y$ defines an associative product in the set $\pi_0(\Omega Y)$ of path components of $Y$, with unit element the path component containing $c_*$. Moreover, if $f \in \Omega Y$, denote by $f^{-1} \in \Omega Y$ the loop of length $|f|$ given by $f^{-1}(t) = f(|f| - t)$. Then the path component of $f^{-1}$ is inverse to the path component of $f$ and so $\pi_0(\Omega Y)$ is a group.

Now observe that (§2 in [18]) the long exact homotopy sequence for the path space fibration yields a bijection

$$\partial_1 : \pi_1(Y, *) \xrightarrow{\cong} \pi_0(\Omega Y)$$

which sends $\alpha \in \pi_1(Y, *)$ to the path component of a representative loop. It is immediate that this is an isomorphism of groups.

Recall from §4.1 that $\mathcal{G}_{X \times_Y P'Y}$ denotes the group of homotopy classes of homotopy equivalences of $X \times_Y P'Y$. Evidently right multiplication by $f$ is homotopy inverse to right multiplication by $f^{-1}$ as self maps of $X \times_Y P'Y$ and so right multiplication by $f$ is a homotopy equivalence whose homotopy class depends only on the path component containing $f$. In view of Proposition 4.2, right multiplication by $f$ thus defines an anti-homomorphism

$$\varphi : \pi_0(\Omega Y) \to \mathcal{G}_{X \times_Y P'Y} .$$

On the other hand, denote by

$$hol : \pi_1(Y, *) \to \mathcal{G}_{X \times_Y P'Y} ,$$

the holonomy action associated with the fibration $q$. We have

**Proposition 4.3.** *With the notation and hypotheses above, if $\alpha \in \pi_1(Y, *)$ is represented by a loop $f \in \Omega Y$ then $hol(\alpha)$ is represented by right multiplication by $f$ in $X \times_Y P'Y$. In particular,*

*(i) Right multiplication by $f$ induces the maps $H(hol\,\alpha)$ and $H_*(hol\,\alpha)$ respectively in $H(X \times_Y P'Y)$ and $H(X \times_Y P'Y; \Bbbk)$.*
*(ii) $\varphi(\partial_1\alpha) = hol\,(\alpha)$.*

*proof:* Suppose $\alpha \in \pi_1(Y, *)$ is represented by $f : (I, \partial I) \to (Y, *)$. Regard $f$ as a path of length 1 in $Y$. Then for $s \in I$ define $f_s \in PY$ to be the path of length $s$ in $Y$ given by

$$f_s(t) = \begin{cases} f(t)\,, & 0 \le t \le s\,, \\ f(s)\,, & t \ge s\,. \end{cases}$$

Then define

$$\Psi : I \times (X \times_Y P'Y) \to X \times_Y MY$$

by setting $\Psi : (s, (x, g)) \mapsto (x, g) \bullet f_s$. By definition, $(q \circ \Phi_f)(s, (x, g)) = f(s)$ and so, in the terminology of §4.1, $\Psi$ covers $f$. Moreover, the restriction of $\Psi$ to $\{1\} \times (X \times_Y P'Y)$ is multiplication by $f$ and so, by the definition of *hol* in §4.1, right multiplication by $f$ represents $hol(\alpha)$. Assertion (i) follows.

Finally, $s \mapsto f_s$ is a path in $PY$ covering $f$ and $f_0$ is the constant path $c_*$. Thus by definition $f_1$ is a loop in the path component $\partial_1\alpha$. But $f_1 = f$ and so $\varphi(\partial_1\alpha) = hol(\alpha)$.

□

**Example.** Let $(Y, *)$ be a pointed space and $LY \xrightarrow{q} Y$ be the free loop space fibration. Here $LY = \{(f, s) \in MY \mid f(0) = f(s)\}$ and $q(f, s) = f(0)$. The space $LY \times_Y P'Y$ has the homotopy type of $\Omega Y$ and we have homotopy inverse homotopy equivalences

$$\Omega Y \xrightarrow{i} LY \times_Y P'Y \xrightarrow{c} \Omega Y$$

with $i(f, r) = ((f, r), (c_*, 0))$ and $c((f, r), (g, s)) = (g^{-1} \bullet f \bullet g, r + 2s)$. Then the holonomy action is homotopic to $c \circ \mu_{LY} \circ (i \times id) : \Omega Y \times \Omega Y \to \Omega Y$ that is the conjugation of loops.

**Example: Covering spaces.** Suppose

$$B \xleftarrow{\;q\;} Y \times_B MB \xleftarrow{\;j\;} F$$

is a classifying space fibration for a connected CW complex $(Y, *)$ (see p. 93 for the definition). Denote the universal cover of $Y$ by $\rho : (\widetilde{Y}, *) \to (Y, *)$. Then each $g \in \pi_1(Y, *)$ determines a covering transformation $\mu_g$ of $\widetilde{Y}$. Because, consistent with our conventions, we want covering transformations to act on the right, $\mu_g$ is the unique covering map of $id_Y$ such that a path from $*$ to $\mu_{g^{-1}}(*)$ maps under $\rho$ to a loop representing $g$. The next Lemma shows that the covering transformations of $\widetilde{Y}$ represent the holonomy action of $\pi_1(Y, *)$ on $F$.

**Lemma 4.3.** *With the hypotheses and notation above, there is a commutative diagram*

$$\begin{array}{ccc} (Y, *) & \xleftarrow{\quad \rho \quad} & (\widetilde{Y}, *) \\ {\scriptstyle \pi_Y} \uparrow & & \uparrow {\scriptstyle \sigma} \\ (Y \times_B MB, *) & \xleftarrow[\quad j \quad]{} & (F, *) \end{array}$$

*in which $\sigma$ is a weak homotopy equivalence. Moreover, if $\mu_g$ denotes the covering transformation of $\widetilde{Y}$ corresponding to $g \in G$ and if $\lambda_g : F \to F$ represents the holonomy action of $g$, then*

$$\mu_g \circ \sigma \sim \sigma \circ \lambda_g .$$

*proof:* Since $F$ is simply connected, the standard properties of covering maps imply that there is a unique continuous map $\sigma : (F, *) \to (\widetilde{Y}, *)$ such that $\rho \circ \sigma = \pi_Y \circ j$. Since $\pi_k(j)$ is an isomorphism for $k \geq 2$, each $\pi_k(\sigma)$ is an isomorphism.

Next, since $\pi_1(Y \times_B MB, *) = \pi_1(Y, *) = \pi_1(B, *) = G$, for any $g \in G$ we may choose a loop $\beta_g : (I, \partial I) \to (Y \times_B MB, *)$ such that $\alpha_g = q \circ \beta_g$ represents $g \in \pi_1(B, *)$. As in §4.2 let

$$\Phi_g : I \times F \to Y \times_B MB$$

be a map satisfying $\Phi_g(0, x) = j(x)$ and $q \circ \Phi_g(t, x) = \alpha_g(t)$. Then $\Phi_g(1, x) = j \circ \lambda_g(x)$ and $\lambda_g : F \to F$ represents $hol(g)$. Because $q(\beta_g(t)) = \alpha_g(t)$, $t \in I$, we may choose $\Phi_g$ so that

$$\Phi_g(t, *) = \beta_g(t) , \qquad t \in I .$$

In particular, $\Phi_g$ and $\lambda_g$ are based maps.

Then, since $F$ is simply connected it follows from the standard properties of covering spaces that there is a unique continuous map

$$\Psi_g : I \times F \to \widetilde{Y}$$

such that $\rho \circ \Psi_g = \pi_Y \circ \Phi_g$ and $\Psi_g(0, *) = *$. Since $\rho\Psi_g(0, x) = \pi_Y \circ j(x) = \rho \circ \sigma(x)$, $x \in F$, it follows that $\Psi_g$ restricts to $\sigma$ in $\{0\} \times F$. Let $\psi_g$ be the restriction of $\Psi_g$ to $\{1\} \times F$. Thus for $x \in F$,

$$\rho\psi_g(x) = \pi_Y \Phi_g(1, x) = \pi_Y j(\lambda_g x) = \rho\sigma(\lambda_g x) = \rho\mu_{g^{-1}}\sigma\lambda_g(x)\,.$$

On the other hand, $\Psi_g(t, x)$ is a path in $\widetilde{Y}$ starting at $*$ and satisfying $\rho\Psi_g(t, *) = \pi_Y \Phi_g(t, *) = \pi_Y \beta_g(t)$. Since $\pi_Y \beta_g$ is a loop in $(Y, *)$ representing $g \in \pi_1(Y, *)$, and since $\psi_g(*) = \Psi_g(1, *)$ it follows that $\psi_g(x) = \mu_{g^{-1}}(x)$. Now the equation above gives

$$\psi_g = \mu_{g^{-1}}\sigma\lambda_g\,.$$

Since $\Psi_g$ is a homotopy from $\psi_g$ to $\sigma$ we obtain $\mu_g \circ \sigma \sim \sigma \circ \lambda_g$.

□

## 4.2 Holonomy of a Λ-extension

Recall from §1.5 that a minimal Sullivan algebra $(\wedge V, d)$ is *1-finite* if $\dim H^1(\wedge V, d) < \infty$. Now suppose that

$$(\wedge V, d) \xrightarrow{i} (\wedge V \otimes \wedge Z, d) \xrightarrow{\rho} (\wedge Z, \overline{d}) \tag{4.5}$$

is a minimal Λ-extension (cf §3.1) of a 1-finite minimal Sullivan algebra, $(\wedge V, d)$. Denote by $L$ the homotopy Lie algebra of $(\wedge V, d)$. Since $(\wedge V, d)$ is 1-finite, its fundamental group $G_L = \exp_{L_0}(L_0)$ is defined (§2.6).

Denote by $\mathcal{G}_{(\wedge Z, \overline{d})}$ the group of homotopy classes of quasi-isomorphisms of $(\wedge Z, \overline{d})$. We shall construct a homomorphism

$$hol : G_L \to \mathcal{G}_{(\wedge Z, \overline{d})}$$

and then in §4.5 establish its connection with the holonomy action defined by a fibration.

First recall that since $\mathrm{End}(\wedge V)$ is a ($\mathbb{Z}$-graded) algebra, a Lie bracket in $\mathrm{End}(\wedge V)$ is defined by

$$[\varphi, \psi] = \varphi \circ \psi - (-1)^{\deg\varphi \cdot \deg\psi}\psi \circ \varphi\,.$$

It is straightforward to check that if $\varphi$ and $\psi$ are derivations so is $[\varphi, \psi]$; and so the subspace of derivations is a sub Lie algebra.

**Definition.** The Lie algebra $\mathrm{Der}(\wedge V, d)$ is the sub Lie algebra of derivations $\theta$ satisfying

$$[\theta, d] = 0\,.$$

Two derivations $\theta$ and $\theta'$ of the same degree in $\mathrm{Der}(\wedge V, d)$ are *homotopic* ($\theta \sim \theta'$) if there is a derivation $\tau \in \mathrm{Der}(\wedge V)$ such that

$$\theta - \theta' = [d, \tau]\,.$$

Now recall the pairing $\langle\,,\,\rangle : V \times L \to \Bbbk$ defined in §2.1. Using the tensor product decomposition $\wedge V \otimes \wedge Z$ we write

$$d(1 \otimes \Phi) = 1 \otimes \overline{d}\Phi + \sum_i v_i \otimes \theta_i \Phi + \Omega\,, \qquad \Phi \in \wedge Z\,,$$

where $v_i$ is a basis of $V$ and $\Omega \in \wedge^{\geq 2} V \otimes \wedge Z$. Then for $x \in L$ we set

$$\theta(x)(\Phi) = -\sum \langle v_i, x\rangle\, \theta_i \Phi\,, \qquad \Phi \in \wedge Z\,. \tag{4.6}$$

An easy check shows that $\theta(x)$ is a derivation of degree $-\deg x$ in $\mathrm{Der}(\wedge V, d)$. Moreover the Sullivan condition (3.6) implies that for all $\Phi \in \wedge Z$ there is some $N$ (depending on $\Phi$) such that $\theta(x)^N(\Phi) = 0$; we say $\theta(x)$ acts *locally nilpotently* in $(\wedge Z, \overline{d})$. In particular, if $x \in L_0$ then $\exp \theta(x)$ is an automorphism of $(\wedge Z, \overline{d})$. As in §1.5, we denote the homotopy class of $\exp \theta(x)$ by $[\exp \theta(x)]$.

**Theorem 4.1.** *Suppose* $(\wedge V, d) \xrightarrow{i} (\wedge V \otimes \wedge Z, d) \xrightarrow{\rho} (\wedge Z, \overline{d})$ *is a $\Lambda$-extension of a minimal Sullivan algebra $(\wedge V, d)$ satisfying $\dim H^1(\wedge V, d) < \infty$. Then a homomorphism of groups*

$$hol : G_L \to \mathcal{G}_{(\wedge Z, \overline{d})}$$

*is defined by* $hol(\exp_{L_0} x) = [\exp \theta(x)]$, $x \in L_0$.

**Definition.** $hol : G_L \to \mathcal{G}_{(\wedge Z, \overline{d})}$ is the *holonomy homomorphism* of the $\Lambda$-extension.

For the proof of Theorem 4.1 we first need to show that $[\exp \theta(x)]$ is independent of the choices involved in the definition of $\theta(x)$. For this suppose $\psi : (\wedge V \otimes \wedge Z, d) \to (\wedge V \otimes \wedge Z, d')$ is a cdga isomorphism between $\Lambda$-extensions restricting to the identity in $\wedge V$ and inducing the identity on $\wedge Z$. As with $d$ above we write

$$\psi(1 \otimes \Phi) = 1 \otimes \Phi + \sum_i v_i \otimes \sigma_i \Phi + \Gamma\,, \qquad \Phi \in \wedge Z\,,$$

with $\Gamma \in \wedge^{\geq 2} V \otimes \wedge Z'$. Then for $x \in L$ we set

$$\sigma_x = -\sum_i \langle v_i, x\rangle \sigma_i : \wedge Z \to \wedge Z'\,.$$

A straightforward calculation then gives

**Lemma 4.4.** *With the hypotheses and notation above $\sigma_x$ is a derivation decreasing degrees by $\deg x + 1$, and satisfying*

$$\theta'(x) - \theta(x) = (-1)^{\deg x}[d, \sigma_x]\,.$$

*(Here $\theta'(x)$ is the derivation in $\wedge Z$ determined by the $\Lambda$-extension $(\wedge V \otimes \wedge Z', d')$.)*

□

**Lemma 4.5.** *With the hypotheses and notation above, if $x \in L_0$, then $\exp\theta(x) \sim \exp\theta'(x)$. In particular, the homotopy class $[\exp\theta(x)]$ is independent of the choice of the right inverse of $\rho$.*

*proof:* Denote by $\delta$ the differential $d \otimes id + id \otimes \overline{d}$ in $\wedge(t, dt) \otimes \wedge Z$, and define a derivation $\tau_x$ in $\wedge(t, dt) \otimes \wedge Z$ by setting

$$\tau_x = \delta(t \otimes \sigma_x) + (t \otimes \sigma_x) \circ \delta\,.$$

Since $\sigma_x$ decreases degrees and the $\Lambda$-extensions satisfy (3.5) it follows that $\tau_x : Z^p \to \wedge(t, dt) \otimes \wedge Z^{<p}$. Thus we may define

$$\gamma : (\wedge Z, \overline{d}) \to \wedge(t, dt) \otimes (\wedge Z, \overline{d})$$

by setting $\gamma(\Phi) = \exp(id \otimes \theta(x) - \tau_x)(1 \otimes \Phi)$. A quick check shows that $\gamma$ is a homotopy from $\exp\theta(x)$ to $\exp\theta'(x)$.

□

We have now showed that $[\exp\,\theta(x)]$ is well defined. A final construction and Lemma are required before we can complete the proof of Theorem 4.1.

Fix a vector $u$ of degree 1, and for $x \in L_0$ define $(\wedge V, d) \to (\wedge u, 0)$ by $v \mapsto \langle v, x\rangle u$. Then in the $\Lambda$-extension

$$(\wedge u \otimes \wedge Z, d_x) := \wedge u \otimes_{\wedge V} (\wedge V \otimes \wedge Z, d)$$

the differential satisfies

$$d(1 \otimes \Phi) = 1 \otimes \overline{d}\Phi \;-\; u \otimes \theta(x)\Phi\,,$$

where $\theta(x)$ is the derivation of degree zero in $(\wedge Z, \overline{d})$ defined in (4.6).

On the other hand, recall from §1.4 that the map $\Delta^1 \to \Delta^1/\partial\Delta^1$ induces a homomorphism $A_{PL}(\Delta^1/\partial\Delta^1) \to A_{PL}(\Delta^1)$ and, according to formula (1.4), p. 17, maps the fundamental cocycle $u$ in $\Delta^1/\partial\Delta^1$ to $-dt$. In analogy we use the morphism $\wedge u \to \wedge(t, dt)$, $u \mapsto -dt$, to construct the $\Lambda$-extension

$$(\wedge(t, dt) \otimes \wedge Z, d_x) := \wedge(t, dt) \otimes_{\wedge u} (\wedge u \otimes \wedge Z, d_x)\,, \tag{4.7}$$

which we denote by $A_x$, and in which $d_x(1 \otimes \Phi) = 1 \otimes \overline{d}\Phi + dt \otimes \theta(x)\Phi$.

A straightforward computation now shows that a cdga isomorphism,

$$\psi_x : (\wedge(t, dt) \otimes \wedge Z, d_x) \xrightarrow{\cong} \wedge(t, dt) \otimes (\wedge Z, \overline{d}),$$

is defined by $\psi_x(t \otimes 1) = t \otimes 1$ and

$$\psi_x(1 \otimes \Phi) = \sum_{n \geq 0} \frac{t^n}{n!} \otimes (\theta(x))^n \Phi, \quad \Phi \in \wedge Z. \tag{4.8}$$

In particular, we have the identities

$$(\varepsilon_0 \otimes id) \circ \psi_x = \varepsilon_0 \otimes id \quad \text{and} \quad (\varepsilon_1 \otimes id) \circ \psi_x = \varepsilon_1 \otimes \exp \theta(x), \tag{4.9}$$

where as usual $\varepsilon_0$, $\varepsilon_1$ denote the augmentations $t \mapsto 0, 1$ in $\wedge(t, dt)$.

**Lemma 4.6.** *Suppose in the $\Lambda$-extension (4.7) that $\varepsilon : \wedge(t, dt) \to lk$ is an augmentation and*

$$\sigma : (\wedge Z, \overline{d}) \to (\wedge(t, dt) \otimes \wedge Z, d_x)$$

*satisfies $(\varepsilon \otimes id) \circ \sigma = id$. Then*

$$(\varepsilon_1 \otimes id) \circ \sigma \sim \exp(-\theta(x)) \circ (\varepsilon_0 \otimes id) \circ \sigma.$$

*proof:* By replacing $t$ by $t - \varepsilon(t)$ we may reduce to the case $\varepsilon(t) = 0$ without affecting $d_x$ or $\theta(x)$. In this case $\varepsilon_0 = \varepsilon$. Moreover $\psi_x^{-1}$ is easily seen to satisfy

$$\psi_x^{-1}(1 \otimes \Phi) = \sum_{n=0}^{\infty} \frac{t^n}{n!} \otimes [-\theta(x)]^n \Phi.$$

In particular if we set $\tau(\Phi) = \psi_x^{-1}(1 \otimes \Phi)$ then $(\varepsilon_0 \otimes id) \circ \tau = id$. It follows from Proposition 1.10 (p. 21) that, since $\varepsilon_0 \otimes id$ is a quasi-isomorphism, $\sigma \sim \tau$. But then

$$(\varepsilon_1 \otimes id) \circ \sigma \sim (\varepsilon_1 \otimes id) \circ \tau = \exp(-\theta(x)) = \exp(-\theta(x)) \circ (\varepsilon_0 \otimes id) \circ \sigma.$$

□

*proof of Theorem 4.1:* We have to show that for $x, y \in L_0$, and $z = \log_{L_0}(\exp_{L_0} x \star \exp_{L_0} y)$

$$hol(\exp_{L_0} x) \circ hol(\exp_{L_0} y) = hol(\exp_{L_0} z);$$

i.e. that $\varphi_x \varphi_y \sim \varphi_z$ where $\varphi_x = \exp \theta(x)$, $\varphi_y = \exp \theta(y)$ and $\varphi_z = \exp \theta(z)$.

Recall from Corollary 2.6 (p. 79) the morphism,

$$\varphi : (\wedge V, d) \to \wedge(s, u, ds, du),$$

satisfying for $w \in V^1$ the equations

$$\begin{aligned} &\rho_2\varphi w = -\langle w, x\rangle\, dt\,, \qquad \rho_0\varphi w = -\langle w, y\rangle\, dt \\ &\text{and } \ \rho_1\varphi w = -\langle w, z\rangle\, dt\,, \end{aligned} \tag{4.10}$$

where $\rho_0, \rho_1, \rho_2 : \wedge(s, u, ds, du) \to \wedge(t, dt)$ are defined by

$$\begin{aligned} &\rho_0 : (s, u) \mapsto (1-t, t)\,, \qquad \rho_1 : (s, u) \mapsto (0, t)\,, \\ &\text{and } \ \rho_2 : (s, u) \mapsto (t, 0)\,. \end{aligned} \tag{4.11}$$

Apply $\wedge(s, u, ds, du) \otimes_\varphi -$ to $(\wedge V \otimes \wedge Z, d)$ to obtain the $\Lambda$-extension $(\wedge(s, u, ds, du) \otimes \wedge Z, d)$ of $(\wedge(s, u, ds, du))$. Then apply the constructions $\wedge(t, dt) \otimes_{\rho_i} -$ to obtain three $\Lambda$-extensions of the form $(\wedge(t, dt) \otimes \wedge Z, d_i)$. In view of the definition (4.6) of $\theta(x)$ and of (4.10), these coincide with the constructions (4.7):

$$\begin{aligned} &(\wedge(t, dt) \otimes \wedge Z, d_0) = (A_y, d_y)\,, \quad (\wedge(t, dt) \otimes \wedge Z, d_1) = (A_z, d_z) \\ &\text{and } \ (\wedge(t, dt) \otimes \wedge Z, d_2) = (A_x, d_x)\,. \end{aligned}$$

On the other hand, let $\varepsilon : \wedge(s, u, ds, du) \to I\!k$ denote the augmentation sending $s, u \mapsto 0$. Then Lemma 1.2 (p. 19) yields a quasi-isomorphism $\chi : (\wedge Z, \overline{d}) \to (\wedge(s, u, ds, du) \otimes \wedge Z, d)$ such that $(\varepsilon \otimes id) \circ \chi = id_{\wedge Z}$. Now for $i = 0, 1, 2$ set

$$j_i = (\rho_i \otimes id) \circ \chi : (\wedge Z, \overline{d}) \xrightarrow{\ \simeq\ } (\wedge(t, dt) \otimes \wedge Z, d_i)\,.$$

It follows immediately from (4.11) that $(\varepsilon_0 \otimes id) \circ j_2 = id_{\wedge Z}$, $(\varepsilon_1 \otimes id) \circ j_2 = (\varepsilon_0 \otimes id) \circ j_0$, $(\varepsilon_1 \otimes id) \circ j_0 = (\varepsilon_1 \otimes id) \circ j_1$ and $(\varepsilon_0 \otimes id) \circ j_1 = id_{\wedge Z}$. Write $\varphi_u = \exp(\theta(u))$. Thus Lemma 4.6 (p. 128) gives

$$\begin{aligned} &id_{\wedge Z} = (\varepsilon_0 \otimes id) j_2 \sim \varphi_x(\varepsilon_1 \otimes id) j_2 =\sim \varphi_x \circ (\varepsilon_0 \otimes id) j_0 \\ &\sim \varphi_x\varphi_y(\varepsilon_1 \otimes id) j_0 = \varphi_x\varphi_y(\varepsilon_1 \otimes id) j_1 \sim \varphi_x\varphi_y\varphi_z^{-1}(\varepsilon_0 \otimes id) j_1 = \varphi_x\varphi_y\varphi_z^{-1}\,. \end{aligned}$$

It follows that $\varphi_x\varphi_y \sim \varphi_z$, as desired.

□

**Corollary 4.1.** *The holonomy representation induces the homomorphism*

$$Hhol : G_L \to Aut(H(\wedge Z, \overline{d}))$$

*which sends* $\exp_{L_0} x \mapsto H(\exp\, \theta(x))$.

**Definition.** *Hhol* is the holonomy representation of $G_L$ in $H(\wedge Z, \overline{d})$.

We close this section with a final observation.

**Lemma 4.7.** *In the minimal $\Lambda$-extension (4.5), if $\Phi \in \wedge V \otimes \wedge Z$ is a cycle, then for all $x \in L$, $\theta(x)(\rho\Phi) \in \overline{d}(\wedge Z)$.*

*proof:* Write $\Phi = 1 \otimes \rho\Phi + \sum v_i \otimes \Phi_i + \Omega$, with $\Omega \in \wedge^{\geq 2} V \otimes \wedge Z$. Then $d(1 \otimes \rho\Phi)$ has the form

$$d(1 \otimes \rho\Phi) = 1 \otimes \overline{d}\rho\Phi + \sum v_i \otimes \Psi_i + \Psi$$

with $\Psi \in \wedge^{\geq 2} V \otimes \wedge Z$. From $d\Phi = 0$ we obtain $\overline{d}\rho\Phi = 0$ and

$$\sum v_i \otimes (\Psi_i - (-1)^{\deg v_i} \overline{d}\Phi_i) = 0\,.$$

Thus $\Psi_i = (-1)^{\deg v_i} \overline{d}\Phi_i$ and

$$\theta(x)(\rho\Phi) = \sum -\langle v_i, x\rangle \Psi_i = \overline{d}\left(\sum (-1)^{\deg v_i + 1} \langle v_i, x\rangle \Phi_i\right)\,.$$

□

## 4.3 Holonomy representations for a $\Lambda$-extension

Suppose $(\wedge V, d) \to (\wedge V \otimes \wedge Z, d) \to (\wedge Z, \overline{d})$ is a minimal $\Lambda$-extension of a 1-finite minimal Sullivan algebra $(\wedge V, d)$ with homotopy Lie algebra, $L$. Recall from §4.2 that the holonomy homomorphism, $hol : G_L \to \mathcal{G}_{(\wedge Z, \overline{d})}$, induces the representation $Hhol$ of $G_L$ in $H(\wedge Z, \overline{d})$.

Recall also from (4.6) that given a basis $\{v_i\}$ of $V$, and for $\Phi \in \wedge Z$, we write $d(1 \otimes \Phi) = 1 \otimes \overline{d}\Phi + \sum v_i \otimes \theta_i\Phi + \Omega$ with $\Omega \in \wedge^{\geq 2} V \otimes \wedge Z$ and for $x \in L$ set $\theta(x) = -\sum \langle v_i, x\rangle \theta_i$. Then $\theta(x)$ is a derivation commuting in the graded sense with $\overline{d}$, and so we may define a map

$$\overline{\theta} = H(\theta) : L \to \operatorname{Der} H(\wedge Z, \overline{d})\,.$$

On the other hand, filtering $(\wedge V \otimes \wedge Z, d)$ by the ideals $\wedge^{\geq k} V$ produces a spectral sequence whose $E_1$-term is a $(\wedge V, d_1)$-module,

$$(E_1, d_1) = (\wedge V \otimes H(\wedge Z, \overline{d}), d_1)\,,$$

in which $d_1$ maps $\wedge^k V \otimes H(\wedge Z, \overline{d})$ to $\wedge^{k+1} V \otimes H(\wedge Z, \overline{d})$. In particular, the restriction of $d_1$ to $\wedge V$ is defined by $d_1 : V \to \wedge^2 V$ and $d - d_1 : V \to \wedge^{\geq 3} V$. Then $d_1$ has the form

$$d_1(1 \otimes \alpha) = \sum_i v_i \otimes \overline{\theta}_i(\alpha)\,, \qquad \alpha \in H(\wedge Z, \overline{d})$$

and

$$\overline{\theta}(x) = -\sum_i \langle v_i, x\rangle\, \overline{\theta}_i\,.$$

The right-hand side is therefore independent of the choice of basis $\{v_i\}$.

**Lemma 4.8.** *$\overline{\theta}$ is a representation of $L$ in $H(\wedge Z, \overline{d})$.*

**Definition.** $\overline{\theta}$ is the *holonomy representation* of $L$ in $H(\wedge Z, \overline{d})$.

*proof of Lemma 4.8:* It is immediate from the description above that $\overline{\theta}$ is a linear map. Now fix $\alpha \in H(\wedge Z, \overline{d})$ and write

$$d_1\alpha = \sum_i v_i \otimes \alpha_i \qquad \text{and} \quad d_1(1\otimes \alpha_i) = \sum_j v_j \otimes \beta_{ij}\,.$$

From $d_1^2 = 0$ we obtain $\sum_i dv_i \otimes \alpha_i + \sum_{i,j} (-1)^{\deg v_i} v_i \wedge v_j \otimes \beta_{ij} = 0$. Then for $x, y \in L$,

$$\overline{\theta}([x,y])\alpha = -\sum_i \langle v_i, [x,y]\rangle\, \alpha_i = (-1)^{\deg y} \sum_i \langle dv_i, x, y\rangle\, \alpha_i$$

$$= \sum_{i,j} (-1)^{\deg y + \deg v_i + 1}\, \langle v_i \wedge v_j, x, y\rangle\, \beta_{ij}$$

$$= \textstyle\sum_{ij} \left((-1)^{\deg v_i \cdot \deg v_j + \deg v_i + \deg v_j}\, \langle v_i, x\rangle\langle v_j, y\rangle + \langle v_i, y\rangle\langle v_j, x\rangle\right)\, \beta_{ij}\,.$$

Since $\beta_{ij} = \overline{\theta}_j(\alpha_i)$ it follows that $\overline{\theta}[x,y] = [\overline{\theta}x, \overline{\theta}y]$. Thus $\overline{\theta}$ is a representation of $L$.

□

It follows from Lemma 4.7 (p. 130) that this homomorphism is independent of the choice of the tensor decomposition $\wedge V \otimes \wedge Z$ extending the inclusion of $(\wedge V, d)$.

**Remark.** This representation for $L$ differs slightly from that in the Introduction to §31 in [18]. Our representation here acts from the left while that in [18] acts from the right.

A representation $\theta$ of $L$ in a graded vector space $M$ is called *locally nilpotent* if for each $\Phi \in M$ there is an $n = n(\Phi)$ such that

$$\theta(x_1) \circ \cdots \circ \theta(x_n)\Phi = 0\,, \qquad x_i \in L\,.$$

The representation is called *nilpotent* if for each $k$, $n$ can be chosen independently of $\Phi \in M_k$.

Any representation $\theta$ of $L$ extends uniquely to a representation $U\theta$ of the universal enveloping algebra $UL$, and if the representation is locally nilpotent it extends to a representation $\widehat{UL(\theta)}$ of $\widehat{UL}$. Moreover, recall from §2.4 the definition of the group $G_L \subset \widehat{UL}$, and from (4.4) that a representation of a group $G$ in a vector space $M$ is locally nilpotent if for each $\Phi \in M$ there is an $n = n(\Phi)$ such that

$$(g_1 - id) \circ \cdots \circ (g_n - id)\Phi = 0\,, \qquad g_i \in G\,.$$

In particular, from the Sullivan condition (3.6) (p. 95) we obtain

**Proposition 4.4.** *The holonomy representation $\overline{\theta}$ of $L$ in $H(\wedge Z, \overline{d})$ is locally nilpotent, and hence the restriction of $\widehat{U\theta}$ to $G_L$ is a locally nilpotent representation of $G_L$.*

The representations $U\overline{\theta}$ and $\widehat{U\overline{\theta}}$ are also called *holonomy representations,* respectively of $UL$ and $\widehat{UL}$ in $H(\wedge Z, \overline{d})$.

Finally, recall from §2.8 that since $(\wedge V, d)$ is 1-finite there is a natural inclusion of the homology Hopf algebra

$$\Bbbk[G_L] \otimes UL_{\geq 1} \to \widehat{UL}\,.$$

Thus $\widehat{U\overline{\theta}}$ restricts to a representation $Hhol$ of $\Bbbk[G_L] \otimes UL_{\geq 1}$ in $H(\wedge Z, \overline{d})$; this is the *holonomy representation* of $\Bbbk[G_L] \otimes UL_{\geq 1}$.

Immediately from the definitions we have the

**Proposition 4.5.** *If $(\wedge V, d) \xrightarrow{i} (\wedge V \otimes \wedge Z, d) \xrightarrow{\rho} (\wedge Z, \overline{d})$ is a minimal $\Lambda$-extension of a 1-finite minimal Sullivan algebra $(\wedge V, d)$, then*

$$H\,hol(\exp_{L_0} x) = \exp \overline{\theta}(x)\,, \qquad x \in L_0\,,$$

*and so $\widehat{U\overline{\theta}}$ restricts to the holonomy representation $Hhol$ of $G_L$.*

*In particular $H\,hol$ is a locally nilpotent representation.*

□

The construction above generalizes as follows. Suppose $(\wedge V, d)$ is a 1-finite minimal Sullivan algebra and that $(\wedge V \otimes M, d_1)$ is a $(\wedge V, d_1)$-module in which $d_1 : 1 \otimes M \to V \otimes M$. Then define a linear map $\theta : L \to \mathrm{Hom}(M, M)$ by writing $d_1(1 \otimes \alpha) = \sum v_i \otimes \theta_i(\alpha)$ and setting

$$\theta(x) = -\sum \langle v_i, x\rangle \theta_i\,, \qquad x \in L\,.$$

The argument of Lemma 4.8 (p. 131) shows that $\theta$ is a representation of $L$ in $M$.

Next, suppose that $(\wedge V \otimes M, d_1)$ satisfies the following analogue of the Sullivan condition (3.6): $M$ is the increasing union of subspaces $0 = M(0) \subset \cdots \subset M(r) \subset \cdots$ such that $d_1(1 \otimes M(r)) \subset \wedge V \otimes M(r-1)$. It follows that $\theta$ is a locally nilpotent representation and so extends to representations $U\theta$ and $\widehat{U\theta}$ of $UL$ and $\widehat{UL}$ in $M$. Moreover $\widehat{U\theta}$ restricts to the representation $\exp_{L_0} x \mapsto \exp \theta(x)$ of $G_L$ in $M$.

**Lemma 4.9.** *If $(\wedge V \otimes M, d_1)$ satisfies this Sullivan-type condition then $\widehat{UL} \cdot \alpha$ is finite dimensional for each $\alpha \in M$.*

*proof:* Note that $d(1 \otimes \alpha) = \sum v_i \otimes \theta_i \alpha$ is a finite sum, and so a subspace $S \subset L$ of finite codimension annihilates $\alpha$. Thus $L \cdot \alpha$ is finite dimensional and hence so is each $L \cdot L \cdot \cdots \cdot L \cdot \alpha$. The Lemma follows from the local nilpotence of the representation.

□

## 4.4 Nilpotent and locally nilpotent representations

**Proposition 4.6.** *Suppose $\sigma : G \to \mathcal{G}_{(\wedge Z, d)}$ is a homomorphism of groups, with $(\wedge Z, d)$ a $\Lambda$-algebra. For $g \in G$, let $\varphi(g)$ be any choice of an automorphism of $(\wedge Z, d)$ representing $\sigma(g)$. If the representation $H(\sigma) : G \to Aut\, H(\wedge Z, d)$ is locally nilpotent then for each $\Phi \in \wedge Z$ there is an $n = n(\Phi)$ such that*

$$(\varphi(g_1) - id) \circ \cdots \circ (\varphi(g_n) - id)\Phi = 0\,, \qquad g_i \in G\,.$$

*proof:* The Sullivan condition for a $\Lambda$-algebra (§1.5) gives a filtration of $Z$ such that $Z = \cup_r Z(r)$ and $d : Z^p_r \to \wedge(Z^{<p} \oplus Z^p(r-1))$. Suppose then that the Proposition holds for $\Phi \in \wedge(Z^{<p} \oplus Z^p(r-1))$ and let $z \in Z^p(r)$. Then $dz \in \wedge(Z^{<p} \oplus Z^p(r-1))$ and so for some $n$,

$$\begin{aligned} & d\,(\,(\varphi(g_1) - id) \circ \cdots \circ (\varphi(g_n) - id)\,)\; z \\ &= (\,(\varphi(g_1) - id) \circ \cdots \circ (\varphi(g_n) - id)\,)\; dz = 0 \end{aligned}$$

for all representatives $\varphi(g_i)$ of $\sigma(g_i)$ and all $g_i \in G$. Since the representation $H(\sigma)$ is locally nilpotent this implies that for some $m \geq n$, and all $g_i \in G$ and all representatives $\varphi(g_i)$ of $\sigma(g_i)$,

$$(\,(\varphi(g_1) - id) \circ \cdots \circ (\varphi(g_m) - id)\,)\,(z) = d\Phi\,.$$

But necessarily $\deg\Phi = p-1$ and so $\Phi \in \wedge Z^{<p}$. Thus by our assumption, for some $q \geq m$

$$(\varphi(g_1) - id) \circ \cdots \circ (\varphi(g_q) - id)\, z = 0\,,$$

for all $g_i \in G$ and all representatives $\varphi(g_i)$ of $\sigma(g_i)$. Since this holds for all $z \in Z^p(r)$ it follows that the Proposition holds for $\Phi \in \wedge(Z^{<p} \oplus Z^p(r))$ and hence by induction on $r$ that it holds in $\wedge Z^{<p} \oplus Z^p$. Because the $\varphi(g)$ are homomorphisms it follows that the Proposition holds in $\wedge Z^{\leq p}$. Thus by induction on $p$ it holds in $\wedge Z$.

□

**Corollary 4.2.** *For $g \in G$, every representative $\varphi(g)$ of $\sigma(g)$ has the form*

$$\varphi(g) = \exp\theta(g)$$

*for some locally nilpotent $\theta(g) \in Der(\wedge Z, d)$.*

*proof:* By the Proposition $(\varphi(g) - id)^{n(\Phi)}\,\Phi = 0$ for every $\Phi \in \wedge Z$. Thus we may define $\theta(g) : (\wedge Z, d) \to (\wedge Z, d)$ by

$$\theta(g) = \sum_{n\geq 1} \frac{(-1)^{n+1}}{n}\,(\varphi(g) - id)^n\,.$$

A formal series argument shows that $\theta(g)$ is a derivation and that $\exp\theta(g) = \varphi(g)$.

□

Suppose now that $(\wedge Z, d)$ is a minimal Sullivan algebra and that $H(\wedge Z, d)$ has finite type.

**Proposition 4.7.** *Suppose $(\wedge V, d)$ is a minimal Sullivan algebra and denote by $D$ the Lie algebra of derivations in $(\wedge V, d)$ which preserve each $(\wedge V)^{\leq k}$. If $H(\wedge V, d)$ has finite type then there is a linear map $\psi : D \to D$ of degree zero such that*

*(i)* $\psi(\theta) \sim \theta$, $\theta \in D$,
*(ii)* $\dim\psi(D)(v) < \infty$, $v \in V$, *and*
*(iii)* $\psi(\theta) - \theta : V^p \to \wedge V^{<p}$.

*proof:* Fix a basis $\theta_i$ of $D$. We shall construct derivations $\tau_i$ in $\wedge V$ which preserve $\wedge^+ V$, and so that the linear map

$$\psi : \theta_i \mapsto \theta_i + [\tau_i, d]$$

has the desired property.

Since $H(\wedge V, d)$ has finite type, $V$ has a basis $v_1, v_2, \cdots$ such that $dv_1 = 0$ and $dv_{k+1} \in \wedge(v_1, \cdots, v_k)$. Moreover, choosing a direct summand $H$ of $\operatorname{Im} d$ in $\operatorname{Ker} d$ we have that $H$ is a graded vector space of finite type.

Now the $\tau_i$ will be constructed inductively on $k$ as maps $\wedge(v_1, \cdots, v_k) \to \wedge V$ satisfying $\tau_i(\Phi \wedge \Psi) = (\tau_i \Phi) \wedge \Psi + (-1)^{\deg \tau_i \cdot \deg \Phi} \Phi \wedge \tau_i \Psi$. In particular, since $dv_1 = 0$,

$$\theta_i v_1 = \Phi_i - d\Psi_i$$

with $\Phi_i \in H^{\leq \deg v_1}$ and $\Psi_i \in \wedge^+ V$. Set $\tau_i(v_1) = (-1)^{\deg \theta_i} \Psi_i$; then $\theta_i + [\tau_i, d] : \Bbbk v_1 \to H^{\leq |v_1|}$.

Next suppose by induction that the $\tau_i$ have been defined in $\wedge(v_1, \cdots, v_{k-1})$ and that for some finite dimensional subspace $F \subset \wedge V$, for all $i$

$$\theta_i' = \theta_i + [\tau_i, d] : \oplus_{j=1}^{k-1} \Bbbk v_j \to F .$$

Since the $\theta_i'$ are derivations from $\wedge(v_1, \cdots, v_{k-1})$ to $\wedge V$ it follows for some finite dimensional subspace $K \subset \wedge V$ that for all $i$,

$$\theta_i'(dv_k) \in K .$$

On the other hand, an easy calculation gives

$$d\theta_i(v_k) + d\tau_i(dv_k) = (-1)^{\deg \theta_i} \theta_i'(dv_k) \in K .$$

Since $K$ is finite dimensional it follows that $d^{-1}(K) = C \oplus \ker d = C \oplus H \oplus \operatorname{Im} d$, where $C$ is also finite dimensional. Thus we may choose $\tau_i(v_k) \in \wedge^+ V$ so that for all $i$,

$$(\theta_i + [\tau_i, d])(v_k) = \theta_i(v_k) + \tau_i(dv_k) + (-1)^{\deg \tau_i + 1} d\tau_i(v_k) \in C \oplus H .$$

Now, define $\tau : D \to \operatorname{Der}(\wedge V)$ by setting $\tau(\theta_i) = \tau_i$ and define $\psi : D \to D$ by setting $\psi(\theta) = \theta + [\tau(\theta), d]$. By construction, $\dim \psi(D)(v_k) < \infty$ for each $k$. It follows that $\dim \psi(D)(v) < \infty$ for all $v$.

Finally, since $(\wedge V, d)$ is a minimal Sullivan algebra, $d : V^p \to (V^1 \wedge V^p) + \wedge V^{<p}$. On the other hand, any $\theta \in D$ preserves each $(\wedge V)^{\leq p}$ and so $\tau(\theta)$ decreases degrees by at least one. Since $\tau(\theta)$ preserves $\wedge^+ V$ it vanishes in $V^1$ and thus

$$\psi(\theta) - \theta = [\tau(\theta), d] : V^p \to \wedge V^{<p} .$$

□

Given a representation $G \to \operatorname{Aut} M$ of a group $G$ in a graded vector space $M$ we say $M$ is a $G$-module and subspaces stable under the action of

$G$ are submodules. In particular a homomorphism $G' \to G$ of groups makes $M$ into a $G'$-module and $G$-submodules are automatically $G'$-submodules.

In particular suppose that

$$\begin{array}{ccccc} (\wedge V, d) & \xrightarrow{i} & (\wedge V \otimes \wedge Z, d) & \xrightarrow{\rho} & (\wedge Z, \overline{d}) \\ \downarrow{\scriptstyle \varphi_Y} & & \downarrow{\scriptstyle \varphi_X} & & \downarrow{\scriptstyle \varphi_F} \\ A_{PL}(Y) & \xrightarrow{A_{PL}(\pi)} & A_{PL}(X) & \xrightarrow{A_{PL}(j)} & A_{PL}(F) \end{array}$$

is a commutative diagram of commutative cochain algebras, in which $F \to X \to Y$ satisfies the fibration condition (4.1) and the upper row is a minimal $\Lambda$-extension (§3.1) of a 1-finite minimal Sullivan algebra $(\wedge V, d)$. As usual, denote the fundamental group of $(\wedge V, d)$ by $G_L$. Then, *H hol* makes $H(\wedge Z, \overline{d})$ into a locally nilpotent $G_L$-module (Proposition 4.5, p. 132) and (Corollary 2.7 (p. 79)) $\pi_1(\varphi_Y) : \pi_1(Y, *) \to G_L$ makes any $G_L$-module $M$ into a $\pi_1(Y, *)$-module.

**Proposition 4.8.** *With the hypotheses and notation above, suppose $H^1(\varphi_Y)$ is injective and $M$ is a locally nilpotent $G_L$-module.*

*(i) If $\pi_1(Y, *)$ acts trivially in a subspace of $M$ then $G_L$ acts trivially in that subspace.*
*(ii) Every $\pi_1(Y, *)$-submodule of $M$) is a $G_L$-submodule.*
*(iii) The representation of $\pi_1(Y, *)$ in each $H^k(\wedge Z, \overline{d})$ is locally nilpotent, and is nilpotent if and only if the representation of $G_L$ in $H^k(\wedge Z, \overline{d})$ is nilpotent.*

*proof:* (i) Denote the lower central series (§2.5) of $G_L$ by $G_L^n$. Suppose $\alpha \in M$ is fixed by $\pi_1(Y, *)$. Since the representation of $G_L$ in $M$ is locally nilpotent it follows that for some $n$, $G_L^n \cdot \alpha = \alpha$. Thus the abelian group $G_L^{n-1}/G_L^n$ acts on $\alpha$, and by Proposition 2.6 (p. 80) this vector space is generated by the image of $\pi_1(Y, *)$. It follows that $G_L^{n-1}$ acts trivially on $\alpha$ and by reverse induction on $n$, $G_L \cdot \alpha = \alpha$.

(ii) Since $G_L$ acts locally nilpotently in $M$ the same is true for $\pi_1(Y, *)$. Thus any $\pi_1(Y, *)$-submodule is the union of finite dimensional $\pi_1(Y, *)$-submodules, and so it is sufficient to prove that these are all $G_L$-modules. But a finite dimensional $\pi_1(Y)$-submodule contains a non-zero submodule in which $\pi_1(Y, *)$ acts trivially; this is then a trivial $G_L$-submodule by (i). Now (ii) follows by induction on dimension via the obvious exact sequence argument.

(iii) This follows from a standard exact sequence argument. □

**Lemma 4.10.** *Suppose $\varphi : G \to Aut\, M$ is a representation of a group $G$ in a vector space $M$ and the dual representation $\varphi^{\#} : f \mapsto \varphi(g)^{\#}(f) = f \circ \varphi(g^{-1})$ in $Hom(M; \Bbbk)$ is locally nilpotent. Then both $\varphi$ and its dual are nilpotent.*

*proof:* If $\varphi$ is not nilpotent then for each $k$ there are vectors $m_k \in M$ and elements $g_{k1}, \cdots, g_{kk} \in G$ such that the vectors $m'_k = (\varphi(g_{k1}) - id) \circ \cdots \circ (\varphi(g_{kk}) - id)m_k$ are linearly independent. Choose $f \in M^{\#}$ to satisfy $f(m'_k) = 1$, $k \geq 1$. Then

$$\left[(\varphi(g_{kk}^{-1})^{\#} - id) \circ \cdots \circ (\varphi(g_{k1}^{-1})^{\#} - id)(f)\right](m_k) = f(m'_k) = 1$$

and so the dual representation is not locally nilpotent. □

## 4.5 Connecting topological and Sullivan holonomy

The general connection between holonomy actions for Sullivan extensions and for fibrations can now be established. For this consider any commutative diagram of commutative cochain algebras

$$\begin{array}{ccccc} (\wedge V, d) & \xrightarrow{i} & (\wedge V \otimes \wedge Z, d) & \xrightarrow{\rho} & (\wedge Z, \overline{d}) \\ \Big\downarrow{\scriptstyle \varphi_Y} & & \Big\downarrow{\scriptstyle \varphi_X} & & \Big\downarrow{\scriptstyle \varphi_F} \\ A_{PL}(Y) & \xrightarrow{A_{PL}(p)} & A_{PL}(X) & \xrightarrow{A_{PL}(j)} & A_{PL}(F) \end{array} \tag{4.12}$$

in which the top row is a minimal $\Lambda$-extension, $(\wedge V, d)$ is a 1-finite minimal Sullivan algebra, $L_0$ is its fundamental Lie algebra, and $F \xrightarrow{j} X \xrightarrow{p} Y$ satisfies the fibration condition (4.1), p. 117.

Then recall from Corollary 2.7 (p. 79) that $\pi_1(\varphi_Y)$ is a homomorphism from $\pi_1(Y, *)$ to the fundamental group, $G_L = \exp_{L_0}(L_0)$ of $(\wedge V, d)$.

**Theorem 4.2.** *With the hypotheses and notation above, for $\alpha \in \pi_1(Y, *)$,*

$$\varphi_F \circ hol(\pi_1(\varphi_Y)\alpha) \sim A_{PL}(hol\, \alpha) \circ \varphi_F\,,$$

$$\begin{array}{ccc} (\wedge Z, \overline{d}) & \xrightarrow{hol(\pi_1(\varphi_Y)\alpha)} & (\wedge Z, \overline{d}) \\ \Big\downarrow{\scriptstyle \varphi_F} & & \Big\downarrow{\scriptstyle \varphi_F} \\ A_{PL}(F) & \xrightarrow{A_{PL}(hol\, \alpha)} & A_{PL}(F) \end{array}$$

*and so*

$$H hol\, \alpha \circ H(\varphi_F) = H(\varphi_F) \circ [\, H hol(\pi_1(\varphi_Y)\alpha)]\,.$$

**Remark.** In §6.4, Theorem 6.5 (p. 193), we shall extend Theorem 4.2 to the holonomy representation of the Hopf algebra $\mathbb{Q}[G_L] \otimes UL_{\geq 1}$, where $L$ is the homotopy Lie algebra of $(\wedge V, d)$.

**Note:** In the statement of the Theorem we have abused notation: $hol(\pi_1(\varphi_Y)\alpha)$ means "any isomorphism representing $hol(\pi_1(\varphi_Y)\alpha)$" and $A_{PL}(hol\,\alpha)$ means "$A_{PL}(\psi)$ for any continuous map $\psi : F \to F$ representing $hol\,\alpha$".

**Corollary 4.3.** *If $H(\varphi_F)$ is an isomorphism, then the holonomy representation of $\pi_1(Y,*)$ in $H(F)$ is nilpotent. If in addition $H^1(\varphi_Y)$ is injective, then the holonomy representation of $G_L$ in $H(\wedge Z, \overline{d})$ is nilpotent.*

*proof:* Suppose $H(\varphi_F)$ is an isomorphism. Since the holonomy representation of $G_L$ is locally nilpotent, it follows that the holonomy representation of $\pi_1(Y,*)$ in $H(\wedge Z, \overline{d})$ is locally nilpotent. Thus the holonomy representation of $\pi_1(Y,*)$ in $H(F)$ is locally nilpotent, and therefore nilpotent by Lemma 4.10. The final assertion follows from Proposition 4.8(iii).

□

*proof of Theorem 4.2:* Let $\psi : (S^1,*) \to (Y,*)$ represent $\alpha$, and let

$$S^1 \leftarrow X_{S^1} \leftarrow F$$

be the pullback over $\psi$ of $Y \leftarrow X \leftarrow F$. Then (4.12) yields the commutative diagram

$$\begin{array}{ccccc}
(\wedge V, d) & \longrightarrow & (\wedge V \otimes \wedge Z, d) & \longrightarrow & (\wedge Z, \overline{d}) \\
\downarrow & & \downarrow & & \downarrow{\scriptstyle \varphi_F} \\
A_{PL}(S^1) & \longrightarrow & A_{PL}(X_{S^1}) & \longrightarrow & A_{PL}(F)\,.
\end{array}$$

Thus by naturality it is sufficient to prove the theorem when $Y = S^1$ and $\alpha$ is represented by $id_{S^1}$. We make that assumption.

But in this case $\varphi_Y$ is homotopic to a morphism $\varphi'_Y$ which vanishes on $V^{\geq 2}$ and maps $V^1$ to $\Bbbk u$, where $u$ represents the fundamental class of $S^1$. According to Corollary 3.1 (p. 99) this homotopy extends to a homotopy $\varphi_X \sim \varphi'_X$; this then restricts to a homotopy $\varphi_F \sim \varphi'_F$. In summary, we are reduced to proving the theorem when the diagram,

$$\begin{array}{ccccc}
(\wedge V, d) & \longrightarrow & (\wedge V \otimes \wedge Z, d) & \longrightarrow & (\wedge Z, \overline{d}) \\
\downarrow{\scriptstyle \varphi_Y} & & \downarrow{\scriptstyle \varphi_X} & & \downarrow{\scriptstyle \varphi_F} \\
A_{PL}(S^1) & \longrightarrow & A_{PL}(X) & \longrightarrow & A_{PL}(F)
\end{array}$$

factors as

$$\begin{array}{ccccc}
(\wedge V, d) & \longrightarrow & (\wedge V \otimes \wedge Z, d) & \longrightarrow & (\wedge Z, \overline{d}) \\
\downarrow{\scriptstyle \wedge u \otimes_{\wedge V} -} & & \downarrow{\scriptstyle \wedge u \otimes_{\wedge V} -} & & \downarrow{\scriptstyle id} \\
(\wedge u, 0) & \longrightarrow & (\wedge u \otimes \wedge Z, d) & \longrightarrow & (\wedge Z, \overline{d}) \\
\downarrow{\scriptstyle \simeq} & & \downarrow{\scriptstyle \varphi} & & \downarrow{\scriptstyle \varphi_F} \\
A_{PL}(S^1) & \longrightarrow & A_{PL}(X) & \longrightarrow & A_{PL}(F)\,.
\end{array}$$

Again by naturality it is sufficient to prove the theorem for a diagram

$$\begin{array}{ccccc}
(\wedge u, 0) & \longrightarrow & (\wedge u \otimes \wedge Z, d) & \xrightarrow{\ p\ } & (\wedge Z, \overline{d}) \\
\simeq \downarrow{\scriptstyle \varphi_Y} & & \downarrow{\scriptstyle \varphi} & & \downarrow{\scriptstyle \varphi_F} \\
A_{PL}(S^1) & \longrightarrow & A_{PL}(X) & \xrightarrow[A_{PL}(j)]{} & A_{PL}(F)
\end{array} \tag{4.13}$$

in which $\varphi_Y(u)$ represents the fundamental class of $S^1$ and $\alpha = id_{S^1}$.

Since $\varphi_Y$ is a minimal model for $S^1$ it follows from the definition of $\pi_1(\varphi_Y)$ in §1.8 that $\pi_1(\varphi_Y)\alpha = \exp_{L_0} x$ where $x \in L_0$ satisfies $\langle u, x\rangle = 1$ and $L_0$ is the fundamental Lie algebra of $(\wedge u, 0)$. Now write the differential in $\wedge u \otimes \wedge Z$ as $d(1 \otimes \Phi) = 1 \otimes \overline{d}\Phi - u \otimes \theta(\Phi)$, where $\theta$ is a derivation of degree zero in $(\wedge Z, \overline{d})$. Then by (4.6) and the definition on p. 126, *hol* $(\pi_1(\varphi)\alpha)$ is represented by $\exp(\theta)$, and we have only to show that

$$\varphi_F \circ \exp(\theta) \sim A_{PL}(\lambda) \circ \varphi_F \tag{4.14}$$

where $\lambda : F \to F$ represents *hol* $\alpha$.

Next, recall from §4.1 that there is a commutative diagram

$$\begin{array}{ccc}
I \times F & \xrightarrow{\ \Phi_f\ } & X \\
\downarrow & & \downarrow{\scriptstyle \pi} \\
I & \xrightarrow[f]{} & S^1
\end{array}$$

in which $f : (I, \partial I) \to (S^1, *)$ is the identification map, and in which the following is satisfied: let $j_0, j_1 : F \to I \times F$ be the inclusions opposite $\{0\}$ and $\{1\}$ and recall that $j : F \to X$ is the inclusion. Then

$$\Phi_f \circ j_0 = j \qquad \text{and} \qquad \Phi_f \circ j_1 = j \circ \lambda_f\,,$$

where $\lambda_f$ represents *hol* $\alpha$.

Furthermore, according to the Example in §1.3,

$$I = |\,\Delta[1]\,| \qquad \text{and} \qquad S^1 = |\,\Delta[1]/\partial\Delta[1]\,|\,.$$

Thus with the notation of (1.3), p. 12, the inclusions $\lambda_{\Delta[1]} : \Delta[1] \to Sing\, I$ and $\Delta[1]/\partial\Delta[1] \to Sing\, S^1$ induce reverse morphisms of polynomial differential forms. As observed in §1.4, there is a single representative cycle of the fundamental class in $H^1(A_{PL}(\Delta[1]/\partial\Delta[1])$ and the morphism $A_{PL}(\Delta[1]/\partial\Delta[1]) \to A_{PL}(\Delta[1])$ maps that cycle to $-dt$.

Then consider the map

$$\Delta[1] \times Sing\, F \xrightarrow{\lambda_{\Delta[1]} \times id} Sing\, I \times Sing\, F = Sing(I \times F)\,,$$

and define

$$\Psi_f : A_{PL}(X) \to A_{PL}(\Delta[1] \times Sing\, F)$$

by setting $\Psi_f = A_{PL}(\lambda_{\Delta[1]} \times id) \circ A_{PL}(\Phi_f)$. The inclusions of the vertices $e_0$, $e_1$ in $\Delta[1]$ define morphisms

$$\rho_0, \rho_1 : A_{PL}(\Delta[1] \times Sing\, F) \to A_{PL}(F)$$

and it follows from the conditions on $\Phi_f$ above that

$$\begin{cases} \rho_0 \circ \Psi_f = A_{PL}(j)\,, & \text{and} \\ \rho_1 \circ \Psi_f = A_{PL}(\lambda_f) \circ A_{PL}(j)\,. \end{cases}$$

Additionally, by our discussion above, $\Psi_f \circ \varphi(u \otimes 1) = -dt \otimes 1$. Thus $\Psi_f \circ \varphi$ extends to a homomorphism

$$\chi : \wedge(t, dt) \otimes_{\wedge u} (\wedge u \otimes \wedge Z, d) \to A_{PL}(\Delta[1] \times Sing\, F)\,, \qquad \chi(u) = -dt\,,$$

in which $\wedge(t, dt) \otimes_{\wedge u} (\wedge u \otimes \wedge Z) = (\wedge(t, dt) \otimes \wedge Z, d_x)$ and

$$d_x(1 \otimes \Phi) = 1 \otimes \overline{d}\Phi + dt \otimes \theta(\Phi)\,.$$

On the other hand, in §4.2 (formulas (4.8) and (4.9), p. 128) we constructed an isomorphism

$$\psi_x : (\wedge(t, dt) \otimes \wedge Z, d_x) \xrightarrow{\cong} \wedge(t, dt) \otimes (\wedge Z, \overline{d})$$

extending the identity in $\wedge(t, dt)$ and satisfying

$$\begin{cases} (\varepsilon_0 \otimes id) \circ \psi_x = \varepsilon_0 \otimes id\,, & \text{and} \\ (\varepsilon_1 \otimes id) \circ \psi_X = \varepsilon_1 \otimes \exp\theta\,. \end{cases}$$

Compose $\psi_x$ with $id \otimes \varphi_F$ to obtain a morphism

$$\psi : (\wedge(t, dt) \otimes \wedge Z, d_x) \to A_{PL}(\Delta[1] \times Sing\, F)\,.$$

Then observe from (4.13) that

$$A_{PL}(j) \circ \varphi = \varphi_F \circ p\,.$$

Combining these facts yields first

$$\rho_0 \circ \chi = \varepsilon_0 \otimes \varphi_F \qquad \text{and} \qquad \rho_1 \circ \chi = \varepsilon_1 \otimes (A_{PL}(\lambda_f) \circ \varphi_F)\,,$$

and then that

$$\rho_0 \circ \psi = \varepsilon_0 \otimes \varphi_F \qquad \text{and} \qquad \rho_1 \circ \psi = \varepsilon_1 \otimes (\varphi_F \circ \exp\theta)\,.$$

Since $\rho_0$ is a quasi-isomorphism it follows from Proposition 1.10 (p. 21) that $\psi \sim \chi$, and so $\rho_1 \circ \chi \sim \rho_1 \circ \psi$. But $\varepsilon_1$ is also a quasi-isomorphism, and thus

$$\varphi_F \circ \exp\theta \sim A_{PL}(\lambda_f) \circ \varphi_F\,.$$

□

**Example.** Suppose

$$B \xleftarrow{\ q\ } Y \times_B MB \xleftarrow{\ j\ } F$$

is a classifying space fibration for a connected CW complex $(Y,*)$. Recall (Lemma 4.3, p. 124) that the natural weak homotopy equivalence $\sigma : F \to \widetilde{Y}$ transforms the holonomy action of $\pi_1(Y)$ on $H^*(F)$ into the action by covering transformations of $\pi_1(Y)$ on $H^*(\widetilde{Y})$.

Now let $\varphi_B : (\wedge V^1, d) \to A_{PL}(B)$ be a minimal Sullivan 1-model for $B$ (cf. p. 109 for the definition). Since $H^1(q)$ is an isomorphism and $H^2(q)$ is injective, $A_{PL}(q) \circ \varphi_B$ is a Sullivan 1-model for $Y \times_B MB$ that we extend (by Theorem 3.1(ii), p. 102) into a minimal Sullivan model $\varphi_Y : (\wedge V^1 \otimes \wedge V^{\geq 2}, d) \to A_{PL}(Y \times_B MB)$. This gives the commutative diagram

$$\begin{array}{ccccc}
(\wedge V^1, d) & \longrightarrow & (\wedge V, d) & \xrightarrow{\ \zeta\ } & (\wedge V^{\geq 2}, \overline{d}) \\
\big\downarrow {\scriptstyle \varphi_B} & & \simeq \big\downarrow {\scriptstyle \varphi_Y} & & \big\downarrow {\scriptstyle \varphi_F} \\
A_{PL}(B) & \xrightarrow{A_{PL}(q)} & A_{PL}(Y \times_B MB) & \xrightarrow{A_{PL}(j)} & A_{PL}(F)\,,
\end{array} \qquad (4.15)$$

It then follows from Theorem 4.2 that

$$H^*(\varphi_F) \circ H^*(\sigma) : H^*(\widetilde{Y}; \mathbb{Q}) \to H^*(\wedge Z^{\geq 2}, \overline{d})$$

is a $\pi_1(Y)$ equivariant map, where $\pi_1(Y)$ acts on $\widetilde{Y}$ by covering transformations and on $H^*(\wedge V^{\geq 2}, \overline{d})$ via the holonomy action.

## 4.6 The holonomy action on the homotopy groups of a fibre

Suppose $\xi : F \to X \to Y$ satisfies the fibration condition (4.1) and that $F$ is simply connected. Fix a base point, $* \in F$. Then if $(C, *)$ is a pointed CW complex any continuous map $C \to F$ is homotopic to a based map and if two maps $C \to F$ are homotopic, the corresponding based maps are based homotopic. (See [66], III(1.11), for a proof.)

Thus for any $\varphi : F \to F$ composition with $\varphi$ defines a map $\pi_n(\varphi) : \pi_n(F) \to \pi_n(F)$ which depends only on the homotopy class of $\varphi$, and so may be written $\pi_n([\varphi])$. In particular, right representations of $\pi_1(Y, *)$ in $\pi_n(F)$ are determined by

$$\alpha \mapsto \pi_n(\, hol\,(\alpha))\,, \qquad \alpha \in \pi_1(Y, *)$$

and these are called $\pi_*$-*holonomy representations* of $\pi_1(Y, *)$. It is immediate from the definitions that these are connected to the holonomy representations of §4.1 by the Hurewicz homomorphisms via the commutative diagrams, again for $\alpha \in \pi_1(Y, *)$,

$$\begin{array}{ccc} \pi_*(F) & \xrightarrow{\pi_*(\,hol\,\alpha)} & \pi_*(F) \\ \Big\downarrow{\scriptstyle hur} & & \Big\downarrow{\scriptstyle hur} \\ H_*(F; \Bbbk) & \xrightarrow{H_*(\,hol\,\alpha)} & H_*(F; \Bbbk)\,. \end{array} \tag{4.16}$$

On the other hand, suppose $(\wedge V, d) \to (\wedge V \otimes \wedge Z, d) \to (\wedge Z, \overline{d})$ is a Sullivan extension in which $(\wedge V, d)$ is minimal and $Z = Z^{\geq 2}$. Then the differential satisfies for $z \in Z^n$ that

$$d(1 \otimes z) = \sum v_i \otimes z_i + \Phi$$

with $v_i \in V^1$, $z_i \in Z^n$ and $\Phi \in \wedge V \otimes \wedge Z^{<n}$. It follows as in Lemma 4.8 that a locally nilpotent representation $\sigma$ of $L$ in $1 \otimes Z$ is defined by

$$\sigma(x)(z) = \sum -\langle v_i, x\rangle\, z_i\,, \qquad x \in L\,.$$

This representation is independent of the choice of subspace $1 \otimes Z$ mapping to $Z$, and so defines a representation of $L$ in $(\wedge Z, \overline{d})$. In particular, a representation of $G_L$ in $(\wedge Z, \overline{d})$ is defined by setting

$$a \bullet z = \exp(\sigma(x))z \qquad \text{if } a = \exp_L x\,,$$

and it dualizes to a right representation $(\alpha, a) \mapsto \alpha \bullet a$ of $G_L$ in $\pi_*(\wedge Z) = Z^{\#}$.

**Definition.** The correspondences $(\alpha, a) \mapsto \alpha \bullet a$ are the $\pi_*$-*holonomy representations* of $\pi_1(\wedge V, d)$ in $\pi_n(\wedge Z, d)$.

Now consider again the diagram (4.12)

$$\begin{array}{ccccc} (\wedge V, d) & \longrightarrow & (\wedge V \otimes \wedge Z, d) & \longrightarrow & (\wedge Z, \overline{d}) \\ \downarrow{\scriptstyle \varphi_Y} & & \downarrow{\scriptstyle \varphi_X} & & \downarrow{\scriptstyle \varphi_F} \\ A_{PL}(Y) & \longrightarrow & A_{PL}(X) & \longrightarrow & A_{PL}(F) \end{array}$$

in which $Z = Z^{\geq 2}$, $F$ is simply connected, and the hypotheses at the start of §4.2 are satisfied. Then it follows at once from Theorem 4.2 that

$$\pi_n(\varphi_F)(\beta \bullet \alpha) = \pi_n(\varphi_F\beta) \bullet \pi_1(\varphi_Y)(\alpha)\,, \qquad \alpha \in \pi_1(Y, *), \beta \in \pi_n(F, *)\,;$$

i.e., the diagram

$$\begin{array}{ccc} \pi_n(F) \otimes \pi_1(Y) & \xrightarrow{\quad\bullet\quad} & \pi_n(F) \\ \downarrow{\scriptstyle \pi_n(\varphi_F)\otimes\pi_1(\varphi_Y)} & & \downarrow{\scriptstyle \pi_n(\varphi_F)} \\ \pi_n(\wedge Z, \overline{d}) \otimes \pi_1(\wedge V, \overline{d}) & \xrightarrow{\quad\bullet\quad} & \pi_n(\wedge Z, \overline{d}) \end{array} \qquad (4.17)$$

commutes.

## Chapter 5

# The model of the fibre is the fibre of the model

Recall from §1.4, p. 17, that for any topological space, $X$, $H^*(X;\Bbbk) = H(A_{PL}(X))$ is denoted by $H(X)$.

### 5.1 The main theorem

Suppose $Y \stackrel{p}{\leftarrow} X \stackrel{j}{\leftarrow} F$ satisfies the fibration condition (4.1):

$$\left.\begin{array}{l} \text{- } Y \text{ is path connected and pointed by } * \in Y, \\ \text{- } X \text{ is path connected,} \\ \text{- } j \text{ is the inclusion of } p^{-1}(*), \text{ and} \\ \text{- either } \xi \text{ is a fibration, or else it is a Serre fibration in} \\ \quad \text{which } F \text{ is a CW complex.} \end{array}\right\} \tag{5.1}$$

Then (3.8) in §3.7 provides a commutative cdga diagram

$$\begin{array}{ccccc} (\wedge V, d) & \xrightarrow{\lambda} & (\wedge V \otimes \wedge Z, d) & \xrightarrow{\rho} & (\wedge Z, \overline{d}) \\ {\scriptstyle m_Y}\downarrow{\scriptstyle \simeq} & & {\scriptstyle m_X}\downarrow{\scriptstyle \simeq} & & {\scriptstyle m_F}\downarrow \\ A_{PL}(Y) & \xrightarrow{A_{PL}(p)} & A_{PL}(X) & \xrightarrow{A_{PL}(j)} & A_{PL}(F) \end{array} \tag{5.2}$$

in which the top row is a minimal $\Lambda$-extension, $m_X$ and $m_Y$ are quasi-isomorphisms and $(\wedge V, d)$ is a minimal Sullivan algebra.

Our purpose here is to give a simpler proof of Theorem 20.3 in [28].

**Theorem 5.1.** *With the hypotheses and notation above suppose that $F$ is path connected, that $\pi_1(Y,*)$ acts locally nilpotently on $H(F;\Bbbk)$, and that either $H(F)$ or $H(Y)$ has finite type. Then $m_F : (\wedge Z, \overline{d}) \to A_{PL}(F)$ is a quasi-isomorphism.*

**Remark.** (i) The requirement that the holonomy action be locally nilpotent automatically implies that it is nilpotent (Lemma 4.10, p. 137) since it is the dual of the holonomy representation in homology. This hypothesis is necessary, as pointed out in the Corollary to Theorem 4.2. Moreover some finiteness hypothesis is necessary as the theorem fails for the product fibration of an infinite wedge of three spheres with itself.

(ii) Because $F$ is path connected it follows that $H^1(p)$ is injective since $\pi_1(p)$ and therefore $H_1(p)$ must be surjective. Thus in the diagram above $Z = Z^{\geq 1}$ and the top row is a minimal Sullivan extension.

**Corollary 5.1.** *Suppose $F \xrightarrow{j} X \xrightarrow{p} Y$ satisfies the hypotheses of Theorem 5.1, and that*

$$\begin{array}{ccccc} (\wedge V, d) & \longrightarrow & (\wedge V \otimes \wedge Z, d') & \longrightarrow & (\wedge Z, \overline{d}) \\ \simeq \downarrow \psi & & \downarrow \varphi & & \simeq \downarrow \chi \\ A_{PL}(Y) & \xrightarrow[A_{PL}(p)]{} & A_{PL}(X) & \xrightarrow[A_{PL}(j)]{} & A_{PL}(F) \end{array}$$

*is a commutative diagram of cdga morphisms in which the top row is a minimal Sullivan extension and $\psi$ and $\chi$ are minimal Sullivan models. Then $\varphi$ is a quasi-isomorphism.*

*proof:* Let $(\wedge V, d) \to (\wedge V \otimes \wedge Z, d) \to (\wedge Z, \overline{d})$ be the Sullivan extension of formula (5.2). Then it follows from Proposition 1.11 (p. 23) that there are automorphisms $\alpha$ of $(\wedge V, d)$ and $\beta$ of $(\wedge Z, \overline{d})$ such that $\psi \circ \alpha \sim m_Y$ and $\chi \circ \beta \sim m_F$. Define a Sullivan extension $(\wedge V \otimes \wedge Z, d'')$ by requiring that $(\alpha \otimes \beta) \circ d'' = d' \circ (\alpha \otimes \beta)$. It is sufficient to prove that $\varphi \circ (\alpha \otimes \beta)$ is a quasi-isomorphism and so without loss of generality we may suppose that $\psi \sim m_Y$ and $\chi \sim m_F$.

Let $\Phi : (\wedge V, d) \to \wedge(t, dt) \otimes A_{PL}(Y)$ be a homotopy from $\psi$ to $m_Y$. Then apply Corollary 3.1 (p. 99) to extend $(id \otimes A_{PL}(p)) \circ \Phi$ to a homotopy $\Psi$ : $(\wedge V \otimes \wedge Z, d') \to \wedge(t, dt) \otimes A_{PL}(X)$ from $\varphi$ to a morphism $\varphi'$ extending $m_Y$. By construction this factor to give a homotopy $\chi \sim \chi' : (\wedge Z, \overline{d}) \to A_{PL}(F)$. Thus without loss of generality we may assume $\psi = m_Y$.

Finally, apply Corollary 3.2 (p. 99) to obtain a morphism $\gamma : (\wedge V \otimes \wedge Z, d') \to (\wedge V \otimes \wedge Z, d)$ extending $id_{\wedge V}$ and such that $m_X \circ \gamma \sim \varphi$ rel $(\wedge V, d)$, and such that the quotient $\overline{\gamma} : (\wedge Z, \overline{d'}) \to (\wedge Z, \overline{d})$ satisfies $m_F \circ \overline{\gamma} \sim \chi$. Thus $\overline{\gamma}$ is a quasi-isomorphism of minimal Sullivan algebras and so it is an automorphism by Corollary 1.5 (p. 23). This implies that $\gamma$ itself is an isomorphism and so $\varphi$ is a quasi-isomorphism. $\square$

*proof of Theorem 5.1:* We fix a minimal Sullivan model $\tau_0 : (\wedge W, d_0) \to A_{PL}(F)$ and then adopt the following strategy:

(1) We reduce to the case $Y$ is a CW complex with a single 0-cell (its base point) and with cells attached by based maps.
(2) The *circle construction*: We consider the case $Y = S^1$.
(3) The *extension step* : We consider the case of a CW pair $(Y, T)$ in two scenarios:

*Scenario A:* Either $H(F)$ has finite type or $(Y, T)$ has finitely many cells of some dimension $r \geq 1$.
*Scenario B:* $H(Y, T; \mathbb{Q}) = 0$.

In either scenario we extend a suitable quasi-isomorphism $(A_{PL}(T) \otimes \wedge W, d) \xrightarrow{\simeq} A_{PL}(X_T)$ to a suitable quasi-isomorphism $(A_{PL}(Y) \otimes \wedge W, d) \xrightarrow{\simeq} A_{PL}(X)$.

(4) We complete the proof with an appropriate limit argument.

*Step one: Reduction to the case $Y$ is a CW complex.*
Let $S \to Y$ be a weak homotopy equivalence from a CW complex, and let $F \to X_S \to S$ be the pull-back of the fibration to $S$. The map $X_S \to X$ is a weak homotopy equivalence and so we obtain a diagram of the form (5.2) in which $m_S : (\wedge V, d) \to A_{PL}(S)$ and $m_{X_S} : (\wedge V \otimes \wedge Z, d) \to A_{PL}(X_S)$ are quasi-isomorphisms. Thus we lose no generality in assuming $S = Y$; i.e. that $(Y, *)$ is a CW complex with a single 0-cell, $*$, and with cells attached by based maps from their bounding spheres.

*Step Two: The circle construction.*
Here we consider a fibration

$$F \xrightarrow{j} X \xrightarrow{p} S^1$$

in which $F = p^{-1}(*)$ is path connected and the holonomy action of $[id_{S^1}]$ on $F$ is locally nilpotent. As in §4.1, let

$$\chi : (I, \partial I) \times F \to (X, F)$$

be a homotopy covering $id_{S^1}$ starting at $j$, ending at $h : F \to F$. Then $h$ represents $hol[id_{S^1}]$. Since $Hhol([id_{S^1}])$ is locally nilpotent, by Proposition 4.6 (p. 133) a Sullivan representative of $A_{PL}(h)$ has the form $\varphi = \exp \theta$ where $\theta$ is a locally nilpotent derivation in $(\wedge W, d_0)$ of degree zero.

The derivation, $\theta$, determines the Sullivan extension

$$(\wedge u, 0) \to (\wedge u \otimes \wedge W, d_\theta) \to (\wedge W, d_0)$$

in which $\deg u = 1$, and $d_\theta(1 \otimes \Phi) = 1 \otimes d_0\Phi - u \otimes \theta\Phi$: to be called the *circle construction.*

We shall construct a commutative diagram

$$\begin{array}{ccccc} (\wedge u, 0) & \longrightarrow & (\wedge u \otimes \wedge W, d_\theta) & \longrightarrow & (\wedge W, d_0) \\ \downarrow \simeq & & \simeq \downarrow \tau & & \simeq \downarrow \tau_0 \\ A_{PL}(S^1) & \xrightarrow{A_{PL}(p)} & A_{PL}(X) & \xrightarrow{A_{PL}(j)} & A_{PL}(F) \end{array} \tag{5.3}$$

in which $\tau$ is a quasi-isomorphism.

First recall from §1.4 that we have a commutative diagram

$$\begin{array}{ccc} A_{PL}(I) & \xrightarrow[\varepsilon_{c_1}]{\cong} & A_{PL}(\Delta^1) = \wedge(t, dt) \\ A_{PL}(\pi) \uparrow & & \uparrow \\ A_{PL}(S^1) & \xrightarrow[\varepsilon_{\overline{c}_1}]{} & A_{PL}(\Delta^1/\partial\Delta^1) = (\wedge u, 0) \end{array}$$

in which : the evaluations $\varepsilon_{c_1}$ at $c_1$ and $\varepsilon_{\overline{c}_1}$ at $\overline{c}_1$ are surjective quasi-isomorphisms, $\pi$ is the map $I \to I/\partial I = S^1$, $u$ represents the fundamental cohomology class of $\Delta^1/\partial\Delta^1$, and $u$ is mapped to $-dt$. Thus a representative $u_S$ of the fundamental class of $S^1$ is mapped to $u$ and a unique $t_I \in A_{PL}(I)$ is determined by the conditions

$$dt_I = -A_{PL}(\pi)u_S \quad \text{and } t_{I|\{0\}} = 0\,.$$

It follows that $t_I$ is mapped to $t$ in $A_{PL}(\Delta^1)$.

Now the map $t \mapsto t_I$ defines a morphism $\sigma_I : \wedge(t, dt) \to A_{PL}(I)$. We define also a morphism $\sigma_{S^1} : (\wedge u, 0) \to A_{PL}(S^1)$ by $u \mapsto u_S$. Then $\varepsilon_{c_1} \circ \sigma_I = id$ and $\varepsilon_{\overline{c}_1} \circ \sigma_{S^1} = id$. Thus we have the commutative diagram

$$\begin{array}{ccccc} (\wedge(t, dt) & \xrightarrow{\sigma_I} & A_{PL}(I) & \xrightarrow{\varepsilon_{c_1}} & \wedge(t, dt) \\ \uparrow & & A_{PL}(\pi) \uparrow & & \uparrow \\ (\wedge u, 0) & \xrightarrow{\sigma_{S^1}} & A_{PL}(S^1) & \xrightarrow{\varepsilon_{\overline{c}_1}} & (\wedge u, 0) \end{array}$$

Next, denote by $j_0, j_1 : F \to I \times F$ and by $\pi_I, \pi_F : I \times F \to I, F$ the inclusions opposite $0, 1$ and the projections onto $I$ and $F$. Denote $A_{PL}(\pi_I)(t_I)$ also by $t_I$. Then let $A$ be the sub cdga of $A_{PL}(I \times F)$ of elements $\omega$ satisfying

$$A_{PL}(h) \circ A_{PL}(j_0)\omega = A_{PL}(j_1)\omega\,.$$

Since $\chi \circ j_0 = j$ and $\chi \circ j_1 = h$, $A_{PL}(\chi)$ factors through $(A, d)$. Moreover since for $k = 0, 1$, the composition

$$A_{PL}((I, \partial I) \times F) \stackrel{i}{\hookrightarrow} A_{PL}(I \times F) \xrightarrow{A_{PL}(j_k)} A_{PL}(F)$$

is zero, the inclusion $i$ factors through $(A, d)$. Thus we obtain the commutative diagram.

$$\begin{array}{ccccccccc} 0 & \longrightarrow & A_{PL}(X,F) & \longrightarrow & A_{PL}(X) & \xrightarrow{A_{PL}(j)} & A_{PL}(F) & \longrightarrow & 0 \\ & & \downarrow{\scriptstyle A_{PL}(\chi)} & & \downarrow{\scriptstyle A_{PL}(\chi)} & & \downarrow{\scriptstyle id} & & \\ 0 & \longrightarrow & A_{PL}((I,\partial I)\times F) & \longrightarrow & (A,d) & \xrightarrow{A_{PL}(j_0)} & A_{PL}(F) & \longrightarrow & 0\,, \end{array} \tag{5.4}$$

which we show is row exact.

The top row in the diagram is exact as observed in §1.4. Next we check that $A_{PL}(j_0) : A \to A_{PL}(F)$ is surjective. But if $\omega \in A_{PL}(F)$ then

$$\Omega = A_{PL}(\pi_F)\omega + A_{PL}(\pi_I)(t_I) \cdot (A_{PL}(\pi_F)A_{PL}(h) - A_{PL}(\pi_F))\omega$$

is an element of $A$ and $A_{PL}(j_0)\Omega = \omega$. If $\omega \in A \cap \operatorname{Ker} A_{PL}(j_0)$, then $A_{PL}(j_1)(\omega) = 0$ and $\omega \in A_{PL}((I, \partial I) \times F)$. It follows that the bottom row in (5.4) is exact.

Now, observe that $A_{PL}(X, F) \to A_{PL}((I, \partial I) \times F)$ is a quasi-isomorphism. To see this let $0 < \delta < \frac{1}{2}$ and set $I_\delta = [0, \delta] \cup [1 - \delta, 1] \subset I$. Then $\pi$ maps $I_\delta$ to a closed interval $J \subset S^1$ with $* \in J$. Set $X_J = p^{-1}(J)$. Then the inclusions $F \to X_J$ and $\partial I \times F \to I_\delta \times F$ are homotopy equivalences and so it is sufficient to show that

$$A_{PL}(X, X_J) \xrightarrow{\simeq} A_{PL}((I, I_\delta) \times F)\,. \tag{5.5}$$

Deleting $[0, \delta) \cup (1 - \delta, 1]$ from $I$ yields a closed interval which is mapped homeomorphically by $p$ to the complement of the interior of $J$. Denoting these two intervals by $K$ we have that $\chi$ restricts to a homotopy equivalence $K \times F \xrightarrow{\simeq} X_K$. By excision, $H(X, X_J) \xrightarrow{\cong} H(X_K, X_{\partial K})$ and $H((I, I_\delta) \times F) \xrightarrow{\cong} H((K, \partial K) \times F)$. It follows that (5.5) is a quasi-isomorphism, and so all three vertical morphisms in (5.4) are quasi-isomorphisms.

Next, recall that $\tau_0 : (\wedge W, d_0) \to A_{PL}(F)$ denotes a minimal Sullivan model for $F$. Because $\exp\theta$ is a Sullivan representative of $A_{PL}(h)$ there is a homotopy $\Phi : (\wedge W, d_0) \to \wedge(t, dt) \otimes A_{PL}(F)$ from $\tau_0 \circ \exp\theta$ to $A_{PL}(h) \circ \tau_0$. Composition of this homotopy with

$$\wedge(t, dt) \otimes A_{PL}(F) \xrightarrow{\sigma_I \otimes id} A_{PL}(I) \otimes A_{PL}(F) \xrightarrow{A_{PL}(\pi_I) \wedge A_{PL}(\pi_F)} A_{PL}(I \times F)$$

defines a morphism $\Gamma : (\wedge W, d_0) \to A_{PL}(I \times F)$. Extend $\Gamma$ to the morphism

$$\Gamma : \wedge(t, dt) \otimes (\wedge W, d_0) \to A_{PL}(I \times F)$$

by setting $\Gamma(t) = A_{PL}(\pi_I)(t_I)$.

Now, let $(C, d) \subset (\wedge(t, dt) \otimes (\wedge W, d_0)$ be the subalgebra of elements $\xi$ satisfying $(\varepsilon_1 \otimes id)(\xi) = \exp\theta \circ (\varepsilon_0 \otimes id)(\xi)$. By construction, $A_{PL}(j_0) \circ \Gamma = \tau_0 \circ \exp\theta \circ (\varepsilon_0 \otimes id)$ and $A_{PL}(j_1) \circ \Gamma = (\varepsilon_1 \otimes id) \circ \Phi = A_{PL}(h) \circ \tau_0 \circ (\varepsilon_1 \otimes id)$. It follows that $\Gamma$ restricts to a morphism $\zeta : (C, d) \to (A, d)$. But $(\ker(\varepsilon_0 \otimes id)) \cap C = (\ker\varepsilon_0 \cap \ker\varepsilon_1) \otimes (\wedge W, d_0)$ and $(\varepsilon_0 \otimes id) : (C, d) \to (\wedge W, d_0)$ is surjective, because for $\xi \in \wedge W$, $\xi' = \xi + t(\exp\theta(\xi) - \xi) \in C$ and $(\varepsilon_0 \otimes id)(\xi') = \xi$. Thus we have the row exact commutative diagram

$$\begin{array}{ccccc}
0 \to A_{PL}((I, \partial I) \times F) & \longrightarrow & (A, d) & \xrightarrow{A_{PL}(j_0)} & A_{PL}(F) \to 0 \\
\uparrow {\scriptstyle (A_{PL}(\pi_I)\circ\sigma_I)\wedge(A_{PL}(\pi_F)\circ\tau_0)} & & \uparrow {\scriptstyle \zeta} & & \uparrow {\scriptstyle \tau_0} \\
0 \to (\operatorname{Ker}\varepsilon_0 \cap \operatorname{Ker}\varphi_1) \otimes (\wedge W, d_0) & \longrightarrow & (C, d) & \xrightarrow{\varepsilon_0 \otimes id} & (\wedge W, d_0) \to 0 .
\end{array}$$

Since $\operatorname{Ker}\varepsilon_0 \cap \operatorname{Ker}\varepsilon_1 \to A_{PL}(I, \partial I)$ is a quasi-isomorphism, it follows that $\zeta$ is a quasi-isomorphism.

Moreover, as in (4.8) (p. 128), a commutative diagram

$$\begin{array}{ccccc}
(\wedge u, 0) & \longrightarrow & (\wedge u \otimes \wedge W, d_\theta) & \longrightarrow & (\wedge W, d_0) \\
\downarrow & & \downarrow {\scriptstyle \psi} & & \| \\
(\wedge dt, 0) & \longrightarrow & (C, d) & \xrightarrow[\varepsilon_0 \otimes id]{} & (\wedge W, d_0)
\end{array}$$

is defined by $\psi : u \mapsto -dt$ and $\psi w = \sum t^k/k! \otimes \theta^k w$, $w \in W$. Now $\operatorname{Ker}\varepsilon_0 \cap C = (\operatorname{Ker}\varepsilon_0 \cap \operatorname{Ker}\varepsilon_1) \otimes (\wedge W, d_0)$. On the other hand $\psi$ restricts to a map $u \otimes (\wedge W, d_0) \to (\wedge u \otimes \wedge W, d_0)$ and, since it is straightforward to check that $(\Bbbk u, 0) \xrightarrow{\cong} (\operatorname{Ker}\varepsilon_0 \cap \operatorname{Ker}\varepsilon_1)$, it follows that $\psi$ restricts to a quasi-isomorphism

$$u \otimes (\wedge W, d_0) \xrightarrow{\cong} (\operatorname{Ker}\varepsilon_0 \cap \operatorname{Ker}\varepsilon_1) \otimes (\wedge W, d_0) .$$

Thus $\psi$ itself is a quasi-isomorphism.

By construction,

$$A_{PL}(\chi) \circ A_{PL}(p)(u_S) = (\zeta \circ \psi)(u \otimes 1) .$$

Thus it follows from Proposition 3.6 (p. 98) that there is a morphism $\tau' : (\wedge u \otimes \wedge W, d_\theta) \to A_{PL}(X)$ such that in the diagram,

$$\begin{array}{ccc}
(\wedge u, 0) & \xrightarrow{A_{PL}(p)\circ\sigma_{S^1}} & A_{PL}(X) \\
\downarrow & \nearrow^{\tau'} & \simeq \downarrow {\scriptstyle A_{PL}(\chi)} \\
(\wedge u \otimes \wedge W, d_\theta) & \xrightarrow{\zeta\circ\psi} & (A, d) ,
\end{array}$$

the upper triangle commutes while $A_{PL}(\chi)\circ\tau' \sim \zeta\circ\psi$ rel $\wedge u$.

Since $A_{PL}(j_0)\circ A_{PL}(\chi) = A_{PL}(j)$ we obtain a homotopy

$$A_{PL}(j)\circ\tau' \sim A_{PL}(j_0)\circ\zeta\circ\psi\,, \quad \mathrm{rel}\,(\wedge u, 0)\,.$$

But $A_{PL}(j_0)\circ\zeta = \tau_0\circ(\varepsilon_0\otimes id)$ and $(\varepsilon_0\otimes id)\circ\psi = \varepsilon_0\otimes id$. Thus we obtain a homotopy

$$A_{PL}(j)\circ\tau' \sim \tau_0\circ(\varepsilon_0\otimes id)\,, \qquad \mathrm{rel}\,(\wedge u, 0)\,.$$

Now Proposition 3.7 (p. 100) asserts that there is a homotopy $\tau'\sim\tau$ rel $(\wedge u, 0)$: $(\wedge u\otimes\wedge W, d_\theta)\to A_{PL}(X)$ such that $A_{PL}(j)\circ\tau = \tau_0\circ(\varepsilon_0\otimes id)$. In particular, $A_{PL}(\chi)\circ\tau \sim A_{PL}(\chi)\circ\tau' \sim \zeta\circ\psi$; and so $\tau$ is a quasi-isomorphism and, since $A_{PL}(\chi)$, $\zeta$ and $\psi$ are quasi-isomorphisms, $\tau$ is a quasi-isomorphism. This completes the construction of (5.3).

*Step Three: Extension of a model.*

Here we consider the case of a CW-pair $(Y,T)$ and denote the restriction of the fibration to $T$ by $\pi : X_T\to T$. We denote by

$$\sigma^Y : A_{PL}(T)\leftarrow A_{PL}(Y) \qquad \text{and} \quad \sigma^X : A_{PL}(X_T)\leftarrow A_{PL}(X)$$

the cdga morphisms corresponding to the inclusions $T\to Y$ and $X_T\to X$. Then we fix a minimal Sullivan model

$$\tau_0 : (\wedge W, d_0)\xrightarrow{\simeq} A_{PL}(F)$$

and assume given a commutative cdga diagram

$$\begin{array}{ccc}(\wedge W, d_0) & \xleftarrow{\varepsilon\otimes id} & (A_{PL}(T)\otimes\wedge W, d_T)\\ {\scriptstyle\simeq}\downarrow{\scriptstyle\tau_0} & & {\scriptstyle\simeq}\downarrow{\scriptstyle\tau_T}\\ A_{PL}(F) & \xleftarrow[A_{PL}(\rho)]{} & A_{PL}(X_T)\end{array}$$

in which:

- $(A_{PL}(T)\otimes\wedge W, d_T)$ is a relative Sullivan algebra,
- $\varepsilon$ is the augmentation induced by the inclusion $*\to T$,
- $\tau_T$ is a quasi-isomorphism, and
- $d_T - id\otimes d_0 : W\to A^+_{PL}(T)\otimes\wedge W$.

(5.6)

Given this we shall, in either of the two scenarios, construct a commutative cdga diagram,

$$\begin{array}{ccc}(A_{PL}(T)\otimes\wedge W, d_T) & \xleftarrow{\sigma_Y\otimes id} & (A_{PL}(Y)\otimes\wedge W, d)\\ {\scriptstyle\simeq}\downarrow{\scriptstyle\tau_T} & & {\scriptstyle\simeq}\downarrow{\scriptstyle\tau}\\ A_{PL}(X_T) & \xleftarrow[\sigma_X]{} & A_{PL}(X)\,,\end{array} \qquad (5.7)$$

in which $(A_{PL}(Y)\otimes\wedge W,d)$ is a relative Sullivan algebra, $\tau$ is a quasi-isomorphism, and $d-id\otimes d_0 : W\to A_{PL}^+(Y)\otimes\wedge W$.

*Scenario A.* Here we assume that $(Y,T)$ has the form

$$Y=T\cup_g\vee_\alpha D_\alpha^{k+1}$$

for some $k\geq 0$, where either $H^*(F;\Bbbk)$ has finite type or else the number of $(k+1)$-cells is finite.

In the process of constructing (5.7) we shall encounter families of surjections $\rho_\alpha : U_\alpha\to U$ in which case we shall write

$$\prod_U U_\alpha$$

for the subset of elements $\prod u_\alpha\in\prod U_\alpha$ satisfying $\rho_\alpha u_\alpha=\rho_\beta u_\beta$ for all $\alpha,\beta$.

We first consider the case $k=0$ and $T=*$. Write $Y=\vee_\alpha(S_\alpha^1,*)$ and let $X_\alpha\to S_\alpha^1$ be the restriction of the fibration to $S_\alpha^1$. The inclusions $(S_\alpha^1,*)\to(Y,*)$ define elements $g_\alpha$ in $\pi_1(Y,*)$ whose holonomy actions in $H(F)$ are represented in $(\wedge W,d_0)$ by automorphisms $\varphi_\alpha=\exp\theta_\alpha$, where the $\theta_\alpha$ are locally nilpotent derivations. In Step Two we constructed quasi-isomorphisms

$$\begin{array}{ccccc}
(\wedge u_\alpha,0) & \longrightarrow & (\wedge u_\alpha\otimes\wedge W,d_\alpha) & \longrightarrow & (\wedge W,d_0)\\
\downarrow\simeq & & \tau_\alpha\downarrow\simeq & & \tau_0\downarrow\\
A_{PL}(S_\alpha^1) & \longrightarrow & A_{PL}(X_\alpha) & \longrightarrow & A_{PL}(F)
\end{array}$$

from the circle construction for $\theta_\alpha$ to the polynomial forms on the fibration. These morphisms combine into quasi-isomorphisms

$$\begin{array}{ccccc}
\prod_{\Bbbk}(\wedge u_\alpha,0) & \longrightarrow & \prod_{\wedge W}(\wedge u_\alpha\otimes\wedge W,d_\alpha) & \longrightarrow & (\wedge W,d_0)\\
\downarrow & & \downarrow & & \downarrow\\
A_{PL}(Y) & \longrightarrow & \prod_{A_{PL}(F)} A_{PL}(X_\alpha) & \longrightarrow & A_{PL}(F)
\end{array}\quad .$$

Now, consider the inclusion

$$\left[\prod_{\Bbbk}(\wedge u_\alpha)\right]\otimes\wedge W\longrightarrow\prod_{\wedge W}(\wedge u_\alpha\otimes\wedge W,d_\alpha)\,. \tag{5.8}$$

If the number of 1-cells in $Y$ is finite (5.8) is an isomorphism.

On the other hand, if $H(\wedge W, d_0)$ has finite type then it follows from Proposition 4.7 that we may choose the $\theta_\alpha$ so that for any $\Phi \in \wedge W$, the $\theta_\alpha(\Phi)$ span a finite dimensional subspace of $\wedge W$. It follows that $(\prod_{\Bbbk} \wedge u_\alpha) \otimes \wedge W$ is preserved by the differential in $\prod\limits_{\wedge W}(\wedge u_\alpha \otimes \wedge W, d_\alpha)$; the inclusion (5.8) is then trivially a quasi-isomorphism. In either case we may replace $\prod\limits_{\wedge W}(\wedge u_\alpha \otimes \wedge W, d_\alpha)$ in the diagram above by $(\prod\limits_{\Bbbk} \wedge u_\alpha \otimes \wedge W, d)$. Since $\prod\limits_{\Bbbk}(\wedge u_\alpha) \otimes \wedge W$ is $\prod\limits_{\Bbbk}(\wedge u_\alpha)$-semifree (Proposition 3.4) we may apply Proposition 1.1 to obtain a quasi-isomorphism.

$$(A_{PL}(Y) \otimes \wedge W, d) \xrightarrow{\simeq} \prod_{A_{PL}(F)} A_{PL}(X_\alpha)\,. \tag{5.9}$$

Finally, it is a standard and easy fact that the singular simplices in $X$ with image in some $X_\alpha$ form a sub simplicial set $L \subset Sing\, X$ and that this inclusion induces an isomorphism of homology. But $A_{PL}(L) = \prod_{A_{PL}(F)} A_{PL}(X_\alpha)$ and so the restriction $A_{PL}(X) \to \prod_{A_{PL}(F)} A_{PL}(X_\alpha)$ is a surjective quasi-isomorphism. Lift the quasi-isomorphism (5.9) through this quasi-isomorphism, and as in the circle construction, use Proposition 3.7 (p. 100) to obtain the commutative diagram

$$\begin{array}{ccc} (\wedge W, d_0) & \longleftarrow & (A_{PL}(Y) \otimes \wedge W, d) \\ \simeq \big\downarrow \tau_0 & & \simeq \big\downarrow \tau \\ A_{PL}(F) & \longleftarrow & A_{PL}(X)\,. \end{array}$$

This completes the construction of (5.7), when $T = *$ and $k = 0$. Note that in this construction $d_1 - 1 \otimes d_0 : W \to A_{PL}^{\geq 1}(Y)$.

It remains to construct (5.7) for $k \geq 1$. Since $(A_{PL}(T) \otimes \wedge W, d_T)$ is a relative Sullivan algebra each $W^p$ has an increasing filtration $0 = F_0^p \subset \cdots \subset F_r^p \subset \cdots$, such that $W^p = \cup_r F_r^p$ and for all $p, r$

$$d_T : F_r^p \to A_{PL}(T) \otimes \wedge(W^{<p} \oplus F_{r-1}^p)\,. \tag{5.10}$$

To construct $\tau$ and $d$ we may suppose they are already constructed in some $W^{\leq n} \oplus F_{q-1}^{n+1}$, and that they make (5.7) commute and satisfy

$$d : F_j^i \to A_{PL}(Y) \otimes \wedge(W^{i-1} \oplus F_{j-1}^i) \qquad \text{and } \operatorname{Im}(d - id \otimes d_0) \subset A_{PL}^+(Y) \otimes \wedge W\,.$$

It is then sufficient to extend $d$ and $\tau$ to $F_q^{n+1}$ so that (5.7) commutes and these relations continue to hold.

For simplicity, we denote $W^{\leq n} \oplus F_{q-1}^{n+1}$ by $V$. Then we have a commutative diagram

$$\begin{array}{ccc}
(A_{PL}(Y,T) \otimes \wedge V, d) & \xrightarrow{\;\overline{\tau}\;} & A_{PL}(X, X_T) \\
\downarrow & & \downarrow \\
(A_{PL}(Y) \otimes \wedge V, d) & \xrightarrow{\;\tau\;} & A_{PL}(X) \\
{\scriptstyle \sigma^Y \otimes id}\downarrow & & \downarrow{\scriptstyle \sigma^X} \\
(A_{PL}(T) \otimes \wedge V, d_T) & \xrightarrow{\;\tau_T\;} & A_{PL}(X_T)\,.
\end{array}$$

The key observation is the following

**Lemma 5.1.** *With the hypotheses and notation above, $H^i(\overline{\tau})$ is an isomorphism for $i \leq n+2$ and injective for $i \leq n+3$.*

*proof:* Recall that $Y = T \cup \left(\vee_\alpha D_\alpha^{k+1}\right)$. The map $\vee_\alpha D_\alpha^{k+1} \to Y$ lifts to a map $\left(\vee_\alpha D_\alpha^{k+1}, \vee_\alpha S_\alpha^k\right) \times F \to (X, X_T)$ restricting to $j$ in $* \times F$. Just as in the proof of Step Two, the standard excision argument shows that this map is a cohomology isomorphism

$$H(X, X_T) \xrightarrow{\cong} \prod_\alpha (\,\Bbbk e_\alpha \otimes H(F)\,)$$

where $e_\alpha$ is the fundamental class in $H^{k+1}(D_\alpha^{k+1}, S_\alpha^k)$. Since either there are finitely many $(k+1)$-cells or else $H(F)$ has finite type, this is an isomorphism $H(X, X_T) \xrightarrow{\cong} (\prod_\alpha \Bbbk e_\alpha) \otimes H(F)$. Note as well that if $X_D$ denotes $p^{-1}(\vee_\alpha D_\alpha^{k+1})$ then the inclusion of $F$ in $X_D$ is a homotopy equivalence, which identifies $H(X_D) = H(F)$. With this identification the isomorphism $H(X, X_T) \cong (\prod_\alpha \Bbbk e_\alpha) \otimes H(F)$ is given by the cup product

$$e_\alpha \otimes \beta \mapsto H^*(p) e_\alpha \cup \beta\,. \tag{5.11}$$

On the other hand, choose a subspace $(U, d) \subset A_{PL}(Y,T)$ such that $A_{PL}^{\geq k+2}(Y,T) \subset U \subset A_{PL}^{\geq k+1}(Y,T)$, and $H(U,d) = 0$. Since $(d - id \otimes d_0) : V \to A_{PL}^+(Y) \otimes \wedge V$ it follows that $U \otimes \wedge V$ is preserved by $d$. The filtration of $V$ gives rise to a filtration of $U \otimes \wedge V$ whose quotients have the form $(U \otimes -, d_U \otimes id)$. It follows that $H(U \otimes \wedge V, d) = 0$. In particular, the quotient map

$$(\,A_{PL}(Y,T) \otimes \wedge V, d\,) \to (\,A_{PL}(Y,T)/U \otimes \wedge V, d\,)$$

is a quasi-isomorphism. But an identical argument shows that the inclusion $((\prod_\alpha \Bbbk e_\alpha)\otimes \wedge V, d_0) \to (A_{PL}(Y,T)/U \otimes \wedge V, d)$ is also a quasi-isomorphism. This combined with the previous one gives an isomorphism

$$\left(\prod_\alpha \Bbbk e_\alpha\right) \otimes H(\wedge V, d_0) \xrightarrow{\cong} H\left(A_{PL}(Y,T)\otimes \wedge V, d\right).$$

It is straightforward via (5.11) to check that these isomorphisms identify $H(\overline{\tau})$ with $id \otimes H(\tau_0) : (\prod_\alpha \Bbbk e_\alpha)\otimes H(\wedge V, d_0) \to (\prod_\alpha \Bbbk e_\alpha)\otimes H(F)$. Since the inclusion $\lambda : (\wedge V, d_0) \to (\wedge W, d_0)$ satisfies $H^i(\lambda)$ is an isomorphism for $i \leq n$, and is injective for $i \leq n+1$, and since $k+1 \geq 2$, the Lemma follows.

□

We can now extend $d$ and $\tau$ to $F_q^{n+1}$. Let $w \in F_q^{n+1}$. By (5.10), $d_T(\sigma^Y \otimes id)w$ is a cycle in $A^+_{PL}(Y_T)\otimes \wedge V + 1 \otimes d_0(\wedge V)$. Lift it to $\Phi \in A^+_{PL}(Y)\otimes \wedge V + 1\otimes d_0(\wedge V)$. Then $d\Phi$ is a cycle in $A^+_{PL}(Y,T)\otimes \wedge V$.

On the other hand, $\tau_T(\sigma^Y \otimes id)w$ lifts to $\Psi \in A_{PL}(X)$ and $\tau\Phi - d\Psi \in A_{PL}(X, X_T)$. Thus the cycle $d\tau\Phi$ in $A_{PL}(X,X_T)$ is a boundary there and so $[d\tau\Phi] = 0$ in $H(X,X_T)$. But then $H(\overline{\tau})[d\Phi] = [d\tau\Phi] = 0$, and since $\deg(d\Phi) = n+3$, it follows from Lemma 5.1 that $[d\Phi] = 0$.

We show now that $d\Phi = d\Omega$ for some $\Omega \in A^+_{PL}(Y,T)\otimes \wedge V$. In fact, since $d^2 = 0$ it follows that

$$d\Phi \in \left(A^{\geq 2}_{PL}(Y,T)\otimes \wedge V\right) \oplus \left(A^1_{PL}(Y,T)\otimes d_0(\wedge V)\right).$$

Thus for some $\Phi_1 \in A^1_{PL}(Y,T)\otimes \wedge V$, $d\Phi - d\Phi_1 \in A^{\geq 2}_{PL}(Y,T)\otimes \wedge V$. Moreover, if $J \subset A^{\geq 1}_{PL}(Y,T)$ satisfies $J \oplus d(A^0_{PL}(Y,T)) = A^{\geq 1}_{PL}(Y,T)$ then the inclusion $(J\otimes \wedge V, d) \to (A_{PL}(Y,T)\otimes \wedge V, d)$ is a quasi-isomorphism (same proof as in Lemma 5.1). Since $d(\Phi - \Phi_1) \in J \otimes \wedge V$ and $[d(\Phi-\Phi_1)] = [d\Phi] = 0$ it follows that $d(\Phi - \Phi_1) = d\Phi_2$ for some $\Phi_2 \in J\otimes \wedge V$.

Now set $d'w = \Phi - \Phi_1 - \Phi_2$. Then $\tau d'w$ is a cycle in $A_{PL}(X)$ and $\sigma^X \tau d'w = d\tau_T(\sigma^Y \otimes 1)w = d\sigma^X \Psi$. Thus $\tau d'w - d\Psi$ is a cycle in $A_{PL}(X,X_T)$ and therefore, by the proof of Lemma 5.1, we have

$$\tau d'w - d\Psi = \overline{\tau}\Gamma + d\Xi$$

with $\Gamma$ a cycle in $A^{\geq k+1}_{PL}(Y,T)\otimes \wedge W^{\leq n}$ and $\Xi \in A_{PL}(X,X_T)$. Set $dw = d'w - \Gamma$ and $\tau w = \Psi + \Xi$. Applying this construction to a basis of $F_q^{n+1}$ completes the construction of (5.7).

It remains to show that $\tau$ is a quasi-isomorphism. But with the construction of $d$ and $\tau$ we may replace $V$ by $W$ in the diagram immediately preceding Lemma 5.1. Then it follows from Lemma 5.1 that $\overline{\tau}$ is a quasi-isomorphism, while $\tau_T$ is a quasi-isomorphism by hypothesis. Thus $\tau$ is also a quasi-isomorphism and Step Two is complete for Scenario A.

*Scenario B.* Here we assume that $H^*(Y,T;\mathbb{k}) = 0$ and recall from Lemma 4.10 (p. 137) that $\pi_1(Y,*)$ acts nilpotently in each $H^k(F;\mathbb{k})$. This gives a finite filtration $0 = K_0 \subset \cdots \subset K_r = H^k(F)$ such that $\pi_1(Y,*)$ acts trivially in each $K_i/K_{i-1}$. Thus $H^*(Y,T;K_i/K_{i-1}) = 0$ and, since the filtration is finite the cohomology with local coefficients $H^*(Y,T;\mathcal{H}^k(F;\mathbb{k}))$ is also zero. Now a Serre spectral argument ([60], Theorem 4.9*) shows that $H(\sigma^X)$ is an isomorphism.

On the other hand, let $(F^p_q)$ be an increasing filtration of $W^p$, $p \geq 1$ so that $d_T : 1 \otimes F^p_q \to A_{PL}(T) \otimes \wedge(W^{<p} \oplus F^p_{q-1})$. Fix some $p,q$ so that $d$ has already been defined in $V = W^{<p} \oplus F^p_{q-1}$ so that $(A_{PL}(\sigma^Y) \otimes 1) \circ d = d_\tau \circ (A_{PL}(\sigma^Y) \otimes 1)$ and $\mathrm{Im}\,(d - 1 \otimes d_0) \subset A^+_{PL}(Y) \otimes \wedge W$.

Now fix $w \in F^p_q$ and lift $d_T(1 \otimes w)$ to an element $\Phi \in A_{PL}(Y) \otimes \wedge V$ of the form $\Phi = \Phi_1 + 1 \otimes d_0 w$ with $\Phi_1 \in A^+_{PL}(Y) \otimes \wedge V$. As in Scenario A it follows that

$$d\Phi \in A^{\geq 2}_{PL}(Y,T) \otimes \wedge V + A^{\geq 1}_{PL}(Y,T) \otimes d_0(\wedge W)\,.$$

But this implies that for some $\Psi \in A^{\geq 1}_{PL}(Y,T) \otimes \wedge V$,

$$d(\Phi - \Psi) \in A^{\geq 2}_{PL}(Y,T) \otimes \wedge V\,.$$

Now it follows exactly as in Scenario A that since $H(A_{PL}(Y,T)) = 0$ we have $\Phi - \Psi = d\Omega$ for some $\Omega \in A^+_{PL}(Y,T) \otimes \wedge V$. Thus we may define

$$d(1 \otimes w) = \Phi - \Psi - \Omega$$

and applying this construction to a basis of $F^p_q$ constructs the desired extension of $d$ to $1 \otimes F^p_q$. Proceeding this way constructs $d$ in $A_{PL}(Y) \otimes \wedge W$. Since $\sigma^Y$ is a quasi-isomorphism, so is $\sigma^Y \otimes id$.

Finally, since $\sigma^X$ is a surjective quasi-isomorphism we may lift $\tau_T \circ (\sigma^Y \otimes id)$ through $\sigma^X$ to construct the quasi-isomorphism $\tau$. This completes the construction of (5.7) for scenario B.

*Step Four. Completion of the proof of Theorem 5.1.*

We now consider a filtration $* = Y_0 \subset Y_1 \subset \cdots \subset Y_k \subset \cdots$ of $Y$ by subcomplexes with the following properties. If $H(F)$ has finite type we choose $Y_k$ to be the $k$-skeleton. If $H(Y)$ has finite type we choose the filtration so that either $H(Y_k, Y_{k-1}) = 0$ or else $Y_k$ is obtained by attaching finitely many cells of a single dimension to $Y_{k-1}$.

Let $X_k \xrightarrow{\pi} Y_k$ be the restriction of the fibration to $Y_k$.

To finish the proof of the theorem, apply $A_{PL}(Y) \otimes_{\wedge V} -$ to diagram (5.2) to obtain the commutative diagram

$$\begin{array}{ccccc} A_{PL}(Y) & \longrightarrow & (A_{PL}(Y) \otimes \wedge Z, d) & \longrightarrow & (\wedge Z, \overline{d}) \\ {\scriptstyle =}\downarrow & & {\scriptstyle m}\downarrow & & {\scriptstyle m_F}\downarrow \\ A_{PL}(Y) & \longrightarrow & A_{PL}(X) & \longrightarrow & A_{PL}(F) \end{array}$$

It follows from (3.6), p. 95, that the Sullivan extension $(A_{PL}(Y) \otimes \wedge Z, d)$ is $A_{PL}(Y)$-semifree. Thus by Proposition 1.1 (p. 7), $m$ is a quasi-isomorphism. Then for any $k$ applying $A_{PL}(Y_k) \otimes_{A_{PL}(Y)} -$ to this diagram yields the row exact diagram

$$\begin{array}{ccccc} 0 \leftarrow (A(Y_k) \otimes \wedge Z, d) & \longleftarrow & (A(Y) \otimes \wedge Z, d) & \longleftarrow & (A(Y, Y_k) \otimes \wedge Z, d) \leftarrow 0 \\ {\scriptstyle m(k)}\downarrow & & {\scriptstyle m}\downarrow & & \downarrow \\ 0 \leftarrow A(X_k) & \longleftarrow & A(X) & \longleftarrow & A(X, X_k) \leftarrow 0\,. \end{array}$$

(Here $A(-)$ means $A_{PL}(-)$.) Since the homology both of $A_{PL}(Y, Y_k) \otimes \wedge Z, d)$ and of $A_{PL}(X, X_k)$ vanishes in degrees $\leq k$ it follows that $H^i(m(k))$ is an isomorphism for $i < k$.

Next we may apply Proposition 3.6 (p. 98) to lift $m(k)$ up to homotopy rel $A_{PL}(Y_k)$ through the quasi-isomorphism $\tau_k$ to obtain

$$n(k) : (A_{PL}(Y_k) \otimes \wedge Z, d) \to (A_{PL}(Y_k) \otimes \wedge W, d_k)\,,$$

with $\tau_k \circ n(k) \sim m(k)$ rel $A_{PL}(Y_k)$. Thus, since $H(\tau_k)$ is an isomorphism, $H^i(n(k))$ is an isomorphism for $i < k$. Now the homotopy $\tau_k \circ n(k) \sim m(k)$ rel $A_{PL}(Y_k)$ factors to give a homotopy $\tau_0 \circ \overline{n}(k) \sim m(0)$ where

$$\overline{n}(k) : \Bbbk \otimes_{A_{PL}(Y_k)} n(k) : (\wedge Z, \overline{d}) \to (\wedge W, d_0)\,.$$

It follows from Proposition 3.11 (p. 109) that $H^i(\overline{n}(k))$ is an isomorphism for $i < k - 1$. Thus $H(m(0))$ is an isomorphism for $i < k - 1$, and since this is true for all $k$, $H(m(0))$ is an isomorphism. But $\tau_0 \circ m(0) \sim m_F$ and so $m_F$ is a quasi-isomorphism. □

**Example.** Let $X$ be a path connected space with minimal Sullivan model $(\wedge V, d)$ and suppose $f : X \to X$ is a homeomorphism. The *mapping torus* $E$ is the quotient space $X \times I/\sim$ where $\sim$ is the equivalence relation generated by $(x, 1) \sim (f(x), 0)$. The projection

$$p : E \to S^1 = \mathbb{R}/\mathbb{Z}\,, \qquad [x, t] \mapsto \bar{t}$$

is a fibre bundle, and it follows directly from the definition (see p. 119) that

$$hol([\mathrm{id}_{S^1}]) = H^*(f) : H^*(X;\mathbb{Q}) \to H^*(X;\mathbb{Q}) .$$

Now suppose that $H^*(f)$ is locally nilpotent. Then $f$ admits a Sullivan representative of the form $e^\theta$ for some derivation $\theta$ of the minimal Sullivan model $(\wedge V, d)$ for $X$, and the circle construction of Theorem 5.1 (p. 145) provides a commutative diagram

$$\begin{array}{ccccc}
(\wedge u, 0) & \xrightarrow{\iota} & (\wedge u \otimes \wedge V, d_\theta) & \xrightarrow{\varepsilon \otimes id} & (\wedge V, d) \\
\downarrow \simeq & & \downarrow \simeq & & \downarrow \simeq \\
A_{PL}(S^1) & \xrightarrow{A_{PL}(p)} & A_{PL}(E) & \longrightarrow & A_{PL}(X) .
\end{array}$$

□

**Example.** Consider the case of the homotopy fibration associated to the projection on the first factor $\pi : S^1 \vee S^2 \to S^1$. The homotopy fibre $F$ is the universal cover $\widetilde{S^1 \vee S^2}$ of $S^1 \vee S^2$. The space $F$ is the union of the real line $\mathbb{R}$ with spheres $S^2_n$ attached at each integer $n$ in $\mathbb{R}$. In this case $X = S^1 \vee S^2$, $Y = S^1$ and the holonomy representation of a generator $\alpha$ of $\pi_1(S^1)$ is induced by the translation in $F$ on the right by one unit. Therefore $H_2(F;\mathbb{Q}) \cong \mathbb{Q}[\alpha, \alpha^{-1}][S^2_0]$, with the action given by the multiplication by $\alpha$.

Denote by $\Omega$ a cocycle in $A^2_{PL}(S^1 \vee S^2)$ whose class generates $H^2(S^1 \vee S^2)$. Recall that each element of $H^2(F;\mathbb{Q})$ can be seen as a linear function $H_2(F;\mathbb{Q}) \to \mathbb{Q}$. Since each sphere $S^2_n$ maps to the sphere $S^2$ by the covering projection $j : F \to S^1 \vee S^2$, it follows that $H^2(j)[\Omega]$ takes the value 1 on each class $[S^2_n]$

On the other hand, a model of $S^1 \vee S^2$ has the form $(\wedge u \otimes \wedge Z, d)$ with $\deg u = 1$, $Z = Z^{\geq 2}$, and $Z^2$ is generated by the elements $y_0, y_1, \cdots, y_n, \cdots$ satisfying $dy_0 = 0$ and $dy_i = uy_{i-1}$ for $i > 0$. The holonomy representation for the $\Lambda$-extension $(\wedge u \otimes \wedge Z, d)$ is locally nilpotent and is given by

$$Hhol(\alpha)[y_n] = [y_n] + [y_{n-1}] + \frac{[y_{n-2}]}{2!} + \cdots + \frac{[y_0]}{n!} .$$

The minimal models $m_Y : (\wedge u, 0) \to A_{PL}(S^1)$ and $m_X : (\wedge u \otimes \wedge Z, d) \to A_{PL}(S^1 \vee S^2)$ can be chosen so that $\pi_1(m_Y) : \pi_1(Y, *) \to G_L$ maps $\alpha$ to $\exp_{L_0} x$ with $\langle u, x \rangle = -1$, and $\varphi_X(y_0) = \Omega$. In particular, $H^2(\varphi_F)[y_0] = H^2(j)[\Omega]$.

Then Theorem 4.2 gives a commutative diagram

$$\begin{array}{ccc} H^2(\wedge Z, \overline{d}) & \xrightarrow{H\,hol(\exp_{L_0} x)^{-1}} & H^2(\wedge Z, \overline{d}) \\ \Big\downarrow{\scriptstyle H^2(m_F)} & & \Big\downarrow{\scriptstyle H^2(m_F)} \\ H^2(F;\mathbb{Q}) & \xrightarrow{H\,hol(\alpha)} & H^2(F;\mathbb{Q}) . \end{array}$$

Here $H hol(\exp_L x)$ is locally nilpotent (but not nilpotent) and $H hol(\alpha)$ is not. A straightforward computation shows that $H^2(m_F)$ is injective and that its image is the maximal nilpotent subspace of $H^2(F;\mathbb{Q})$. In particular, $m_F$ is not an isomorphism. □

Again, consider a fibration

$$Y \xleftarrow{p} X \xleftarrow{j} F$$

satisfying (5.1), and let $\varphi : (\wedge W, d) \to (\wedge V, d)$ be a Sullivan representative for $A_{PL}(p)$. Then if $(\wedge W \otimes \wedge U, d)$ is the acyclic closure of $(\wedge W, d)$ we may form the $\Lambda$-algebra

$$(\wedge V \otimes \wedge U, d) := (\wedge V, d) \otimes_{\wedge W} (\wedge W \otimes \wedge U, d) .$$

**Proposition 5.1.** *Suppose the fibration $p : X \to Y$ satisfies the hypotheses of Theorem 5.1. Then there is a quasi-isomorphism*

$$(\wedge V \otimes \wedge U, d) \xrightarrow{\simeq} A_{PL}(F) .$$

*In particular, a minimal Sullivan model for $(\wedge V \otimes \wedge U, d)$ is a minimal Sullivan model for $F$.*

*proof:* Theorem 5.1 yields the commutative diagram

$$\begin{array}{ccccc} (\wedge W, d) & \xrightarrow{\lambda} & (\wedge W \otimes \wedge Z, d) & \xrightarrow{\rho} & (\wedge Z, \overline{d}) \\ {\scriptstyle m_Y}\Big\downarrow{\scriptstyle \simeq} & & {\scriptstyle m_X}\Big\downarrow{\scriptstyle \simeq} & & {\scriptstyle m}\Big\downarrow{\scriptstyle \simeq} \\ A_{PL}(Y) & \xrightarrow{A_{PL}(p)} & A_{PL}(X) & \xrightarrow{A_{PL}(j)} & A_{PL}(F) \end{array}$$

in which all three vertical arrows are quasi-isomorphism. This yields a quasi-isomorphism

$$(\wedge W \otimes \wedge Z, d) \otimes_{\wedge W} (\wedge W \otimes \wedge U, d) \xrightarrow{\simeq} (\wedge W \otimes \wedge Z, d) \otimes_{\wedge W} \mathbb{k} = (\wedge Z, \overline{d}) .$$

On the other hand, we may choose $m_Y$ so that there is a homotopy commutative diagram

$$\begin{array}{ccc} (\wedge W, d) & \xrightarrow{\varphi} & (\wedge V, d) \\ {\scriptstyle m_Y}\downarrow \simeq & & {\scriptstyle m}\downarrow \simeq \\ A_{PL}(Y) & \xrightarrow{A_{PL}(p)} & A_{PL}(X). \end{array}$$

Now lift $m$ up to homotopy through the quasi-isomorphism $m_X$ to obtain $\psi : (\wedge V, d) \xrightarrow{\simeq} (\wedge W \otimes \wedge U, d)$. Then

$$m_X \circ \lambda = A_{PL}(p) \circ m_Y \sim m \circ \varphi \sim m_X \circ \psi \circ \lambda .$$

Since $m_X$ is a quasi-isomorphism, $\varphi \sim \psi \circ \lambda$. It follows that the $\Lambda$-algebra $(\wedge V \otimes \wedge U, d)$ constructed via $\varphi$ is quasi-isomorphic to that constructed via $\psi \circ m$, and $\psi$ maps this quasi-isomorphically to $(\wedge W \otimes \wedge Z, d) \otimes_{\wedge W} (\wedge W \otimes \wedge U, d)$. □

## 5.2 The holonomy action of $\pi_1(Y, *)$ on $\pi_*(F)$

Fix a fibration

$$(Y, *) \overset{p}{\leftarrow} (X, *) \overset{j}{\leftarrow} (F, *)$$

in which $Y$ is path connected and $F$ is simply connected. Then cf. (3.8) (p. 115) there is a commutative diagram

$$\begin{array}{ccccc} (\wedge V, d) & \xrightarrow{\lambda} & (\wedge V \otimes \wedge Z, d) & \xrightarrow{\rho} & (\wedge Z, \overline{d}) \\ {\scriptstyle m_Y}\downarrow \simeq & & {\scriptstyle m_X}\downarrow \simeq & & {\scriptstyle \overline{m}}\downarrow \\ A_{PL}(Y) & \xrightarrow{A_{PL}(p)} & A_{PL}(X) & \xrightarrow{A_{PL}(j)} & A_{PL}(F) . \end{array}$$

Moreover, as described in §4.6 (p. 142), the holonomy action of $\pi_1(Y, *)$ induces a (right) $\pi_*$-holonomy representation of $\pi_1(Y, *)$ in $\pi_*(F) \otimes \Bbbk$. Recall also from Theorem 1.6 (p. 43) that $H(F)$ has finite type if and only if $\pi_*(F) \otimes \Bbbk$ has finite type.

**Theorem 5.2.** *With the hypotheses and notation above, suppose $H(F)$ has finite type. Then the following are equivalent:*

*(i) $\overline{m}$ is a quasi-isomorphism.*
*(ii) The representation $H(hol)$ of $\pi_1(Y, *)$ in $H(F)$ is nilpotent.*
*(iii) The $\pi_*$-representation of $\pi_1(Y, *)$ in $\pi_*(F) \otimes \Bbbk$ is nilpotent.*

*proof:* Theorem 5.1 states that (ii) $\Rightarrow$ (i). On the other hand, if (i) holds then it follows from Theorem 4.2 (p. 137) that $H(\overline{m})$ identifies the holonomy representation $H(hol)$ with the restriction of $\pi_1(\overline{m})(\pi_1(Y,*))$ to the representation of $G_L = \pi_1(\wedge Z, \overline{d})$ in $H(\wedge Z, \overline{d})$. Since $H(\wedge Z, \overline{d}) \cong H(F)$ has finite type, the latter is nilpotent. Hence so is $H(hol)$. Moreover, Theorem 1.6 states that

$$\pi_1(\overline{m}) : \pi_*(F) \otimes I\!k \xrightarrow{\cong} \pi_*(\wedge Z, \overline{d}) \cong Z^{\#} .$$

As observed in (4.16) this identifies the $\pi_*$-representation of $\pi_1(Y,*)$ with the restriction to $\pi_1(Y,*)$ of the representation in $Z^{\#}$ of $\pi_1(\wedge Z, \overline{d})$, which is also nilpotent. Then the $\pi_*$-representation is nilpotent as well. In summary, we have

$$\text{(ii)} \Longleftrightarrow \text{(i)} \Longrightarrow \text{(iii)} .$$

It remains to show that (iii) $\Rightarrow$ (i).

In the following, for a continuous map $f : S \to T$ we shall denote $\pi_i(f) \otimes I\!k$, $i \geq 2$, simply by $\pi_i(f)$. In particular, we say that our fibration satisfies *condition (r)* if

$$\pi_i(p) : \pi_i(X) \otimes I\!k \xrightarrow{\cong} \pi_i(Y) \otimes I\!k , \qquad i \geq r+1 .$$

We shall complete the proof by first reducing to the case of fibrations satisfying condition $(r)$, and then using induction on $r$. Note that if condition $(r)$ is satisfied then it follows from the long exact homotopy sequence that

$$\begin{cases} \pi_i(F) \otimes I\!k = 0 , & i > r, \text{ and} \\ \pi_r(j) : \pi_r(F) \otimes I\!k \to \pi_r(X) \otimes I\!k \text{ is injective.} \end{cases} \tag{5.12}$$

For the reduction choose any $r$ and extend $p$ to a map

$$p' : X \vee \vee_\alpha S^{r+1}_\alpha \to Y$$

so that $\pi_{r+1}(p')$ is surjective. Then, as with the construction in §3.2 of Postnikov decompositions, successively adjoin cells of dimension $\geq r+2$ to construct a space $X(r)$ and an extension $p(r)$ of $p'$ so that

$$\pi_i(p(r)) : \pi_i(X(r)) \otimes I\!k \xrightarrow{\cong} \pi_i(Y) \otimes I\!k , \quad i \geq r+1 .$$

This yields the following map of fibrations, where we adopt the notation of §3.1:

$$\begin{array}{ccccc} & & X & \xleftarrow{\quad j \quad} & F \\ & \swarrow^{p} & \Big\downarrow{\scriptstyle \ell(r)} & & \Big\downarrow{\scriptstyle f(r)} \\ Y & & & & \\ & \nwarrow_{q(r)} & & & \\ & & X(r) \times_Y MY & \xleftarrow{\quad j(r) \quad} & X(r) \times_Y P'Y\,. \end{array} \tag{5.13}$$

It follows from the construction of $X(r)$ that $\pi_i(\ell(r))$ is an isomorphism for $i \leq r$. Thus a long exact homotopy sequence argument using (5.12) shows that $\pi_i(f(r)) : \pi_i(F) \otimes \Bbbk \to \pi_i(X(r) \times_Y P'Y) \otimes \Bbbk$ is an isomorphism for $i < r$ and surjective for $i = r$. It follows that $\pi_r(\ell(r))$ is also surjective. In this way we also obtain that $\pi_i(X(r) \times_Y P'Y) = 0$ for $i \geq r + 1$. In particular, $\pi_*(f(r))$ is a surjection of $\pi_1(Y, *)$-modules.

Both $F$ and $F(r)$ are simply connected, and so $H(F)$ has finite type. Thus by Theorem 1.6 (p. 43), $\pi_*(F) \otimes \Bbbk$ has finite type. Hence $\pi_*(F(r))$ has finite type and again by Theorem 1.6, $H(F(r))$ has finite type. Thus the fibration $q(r) : X(r) \times_Y MY \to Y$ satisfies the hypotheses of the Theorem, and condition (iii).

Moreover, it follows by a simple Sullivan model argument that $H_i(f(r)) : H_i(F; \Bbbk) \to H_i(X(r) \times_Y P'Y; \Bbbk)$ is an isomorphism for $i < r$. Thus $H^i(f(r))$ is an isomorphism of $\pi_1(Y, *)$-modules for $i < r$. Since the fibration $q(r) : X \times_Y P'Y \to Y$ satisfies condition $(r)$, if the Theorem holds for such fibrations, $H^i(F)$ is a nilpotent $\pi_1(Y, *)$-module for $i < r$. Since $r$ was arbitrarily large this would imply that the fibration $X \to Y$ satisfies (ii), and hence (i).

To prove the Theorem for fibrations satisfying (iii) and condition $(r)$ we first establish it in the special case that for some $n \geq 2$,

$$\pi_i(F) \otimes \Bbbk = 0\,, \qquad i \neq n, n+1\,. \tag{5.14}$$

But in this case $H(F)$ is the free graded commutative algebra $\wedge Z$ on the finite dimensional space $Z = Z^n \oplus Z^{n+1}$ (as follows, for example, easily from Theorem 1.6), and the Hurewicz homomorphism is an isomorphism

$$\pi_*(F) \otimes \Bbbk \xrightarrow{\cong} Z^{\#}\,.$$

Thus according to (4.16) (p. 142) the induced representation of $\pi_1(Y, *)$ in $Z^{\#}$ is dual to the restriction to $Z$ of the cohomology holonomy representation $H(hol)$. Since $H(hol)$ acts by algebra automorphisms it follows that $H(hol)$ is nilpotent, which establishes (ii), and hence (iii) in this case.

It remains to carry out the induction. Note that if our fibration satisfies condition (1) that $\pi_k(F)\otimes \Bbbk = 0$ and so (i) and (ii) are trivially true. Suppose now that the fibration $Y \leftarrow X \leftarrow F$ satisfies both (iii), and condition $(r+1)$, and that the Theorem is proved for fibrations satisfying (iii) and condition $(r)$. Then in diagram (5.13) $\pi_i(\ell(r))$ will be an isomorphism for $i \neq r+1$.

Now for simplicity, denote $j(r) : X(r)\times_Y P'Y \to X(r)\times_Y MY$ simply by $j(r) : F(r) \to Y(r)$. Then convert $\ell(r)$ to a fibration $\ell : X \times_{Y(r)} MY(r) \to Y(r)$ to obtain the commutative diagram

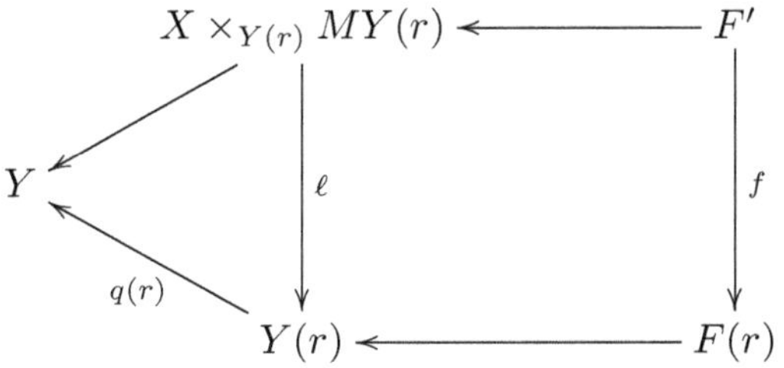


in which $f : F' \to F(r)$ is the restriction of the fibration $\ell$ to $F(r)$. Thus $F'$ is also the fibre of the fibration $q(r)\circ \ell : X \times_{Y(r)} MY(r) \to Y$, which restricts to the fibration $p : X \to Y$. It follows that the inclusion $F \to F'$ is a weak homotopy equivalence.

Next observe that since $p : X \to Y$ satisfies condition $(r+1)$ so does $q(r)\circ\ell : X \times_{Y(r)} MY(r) \to Y$. Thus for $i \geq r+2$, both $\pi_i(q(r)\circ\ell)$ and $\pi_i(q(r))$ are isomorphisms, and so $\pi_i(\ell)$ is an isomorphism for $i \geq r+2$. But since $\ell$ restricts to $\ell(r) : X \to Y(r)$ it follows as observed above that $\pi_i(\ell)$ is an isomorphism for $i \leq r$.

Denote the fibre of $\ell$ by $E$. Then it follows from the long exact homotopy sequence for the fibration $\ell$ that $\pi_i(E)\otimes\Bbbk = 0$, $i \neq r, r+1$ and $\pi_{r+1}(E)\otimes \Bbbk \xrightarrow{\cong} \operatorname{Ker}\pi_{r+1}(\ell)$, while $\pi_r(E)\otimes\Bbbk \cong \operatorname{Coker}\pi_{r+1}(\ell)$. But since $f$ is the restriction of $\ell$, $E$ is also the fibre of $f$. Since by (5.12) $\pi_i(F(r))\otimes\Bbbk = 0$, $i \geq r+1$ it follows that $\pi_{r+1}(E)\otimes\Bbbk \xrightarrow{\cong} \pi_{r+1}(F')\otimes\Bbbk$. Moreover, since $f$ restricts to $f(r)$ in $F$ it follows that $\pi_r(f)$ is surjective and $\pi_r(E)\otimes\Bbbk \xrightarrow{\cong} \operatorname{Ker}\pi_r(f)$.

Now the fibration $\ell : X \times_{Y(r)} MY(r) \to Y(r)$ determines a right representation of $\pi_1(Y,*) = \pi_1(Y(r),*)$ in $\pi_*(E)\otimes\Bbbk$, while the fibration $q(r)\circ\ell : X\times_{Y(r)} MY(r) \to Y$ determines a right representation of $\pi_1(Y,*)$ in $\pi_*(F')\otimes\Bbbk$. It is a straightforward consequence of the definition of holonomy that the map $\pi_*(E)\otimes\Bbbk \xrightarrow{\cong} \pi_*(F')\otimes\Bbbk$ is an isomorphism of

$\pi_1(Y,*)$-modules. In particular, the fibration $\ell : X \times_{Y(r)} MY(r) \to Y(r)$ satisfies the hypotheses of the Theorem and condition (iii).

Let $m_Y : (\wedge W, d) \xrightarrow{\simeq} A_{PL}(Y)$ be a minimal Sullivan model and extend this to a Sullivan model $m(r) : (\wedge W \otimes \wedge Z(r)) \xrightarrow{\simeq} A_{PL}(Y(r))$ restricting to $A_{PL}(q(r)) \circ m_Y$ in $\wedge W$. Then extend $m(r)$ to a Sullivan model

$$n : (\wedge W \otimes \wedge Z(r) \otimes \wedge Z, d) \xrightarrow{\simeq} A_{PL}(X \times_{Y(r)} MY(r))$$

restricting to $A_{PL}(\ell) \circ m(r)$ in $\wedge W \otimes \wedge Z(r)$. Applying $\Bbbk \otimes_{\wedge W} -$ then yields the commutative diagram

$$\begin{array}{ccccc}
(\wedge Z(r), \bar{d}) & \longrightarrow & (\wedge Z(r) \otimes \wedge Z, \bar{d}) & \longrightarrow & (\wedge Z, \bar{d}) \\
\downarrow{\scriptstyle \overline{m}(r)} & & \downarrow{\scriptstyle \overline{n}} & & \downarrow{\scriptstyle n'} \\
A_{PL}(F(r)) & \longrightarrow & A_{PL}(F') & \longrightarrow & A_{PL}(E)
\end{array}$$

Our induction hypothesis gives that $\overline{m(r)}$ is a quasi-isomorphism, and thus, since $F(r)$ is simply connected, $\pi_*(\overline{m(r)})$ is an isomorphism. In particular, $Z(r)^i = 0$, $i > r$. On the other hand, since $\pi_i(E) \otimes \Bbbk = 0$ for $i \neq r, r+1$ we showed above that $n'$ must also be a quasi-isomorphism, and so $\pi_*(n')$ is an isomorphism and $Z = Z^r \oplus Z^{r+1}$. It follows that $(\wedge Z(r) \otimes \wedge Z, \bar{d})$ is a minimal Sullivan algebra and that $\pi_*(\overline{n})$ is an isomorphism. Thus $\overline{n}$ is a quasi-isomorphism.

Finally, the homotopy equivalence $X \xrightarrow{\cong} X \times_{Y(r)} MY(r)$ gives the commutative diagram

$$\begin{array}{ccccc}
(\wedge W, d) & \longrightarrow & (\wedge W \otimes \wedge Z(r) \otimes \wedge Z, d) & \longrightarrow & (\wedge Z(r) \otimes \wedge Z, \bar{d}) \\
\simeq \downarrow{\scriptstyle m_Y} & & \simeq \downarrow{\scriptstyle m_X} & & \downarrow{\scriptstyle \overline{m}} \\
A_{PL}(Y) & \xrightarrow{A_{PL}(p)} & A_{PL}(X) & \xrightarrow{A_{PL}(j)} & A_{PL}(F)
\end{array}$$

in which $\overline{m}$ corresponds to $\overline{n}$ under the quasi-isomorphism $A_{PL}(F) \simeq A_{PL}(F')$. Thus $\overline{m}$ is a quasi-isomorphism and so (i), and therefore (ii) hold for the fibration $p : X \to Y$.

This closes the induction and completes the proof of the Theorem.

$\square$

## 5.3 The Sullivan model of a universal covering space

Suppose $(X,*)$ is a connected CW complex, and let $(B,*)$ be a classifying space for $\pi_1(X,*)$, so that (cf. §3.2) there is a map $h : (X,*) \to (B,*)$ for which $\pi_1(h)$ is an isomorphism. In particular, $h$ has a Sullivan representative

$$\varphi : (\wedge W, d) \to (\wedge V, d)$$

where $(\wedge W, d)$ and $(\wedge V, d)$ are respectively Sullivan minimal models for $B$ and for $X$. Thus if $(\wedge W \otimes \wedge U, d)$ is an acyclic closure for $(\wedge W, d)$ then as in Proposition 5.1 we may form the $\Lambda$-algebra

$$(\wedge V \otimes \wedge U, d) := (\wedge V, d) \otimes_{\wedge W} (\wedge W \otimes \wedge U, d) .$$

On the other hand, in §2.9 (p. 84) we recalled how the Whitehead product defines a natural right action of $\pi_1(X, *)$ on $\pi_k(X, *)$, $k \geq 2$, and we have the following

**Theorem 5.3.** *With the hypotheses and notation above suppose that $\pi_{\geq 2}(X) \otimes \Bbbk$ is a graded vector space of finite type in which $\pi_1(X, *)$ acts nilpotently. Then there is a quasi-isomorphism*

$$(\wedge V \otimes \wedge U, d) \xrightarrow{\simeq} A_{PL}(\widetilde{X}) ,$$

*$\widetilde{X}$ denoting the universal covering space of $X$.*

*proof:* Let

$$B \xleftarrow{\quad q \quad} X \times_B MB \xleftarrow{\quad j \quad} F$$

be the classifying fibration constructed from $h$ in §3.2 and recall from §3.1 the homotopy equivalence $\ell : X \to X \times_B MB$. Then we have isomorphisms

$$\pi_n(X) \otimes \Bbbk \xrightarrow[\cong]{\pi_n(\ell)} \pi_n(X \times_B MB) \otimes \Bbbk \xleftarrow[\cong]{\pi_n(j)} \pi_n(F) \otimes \Bbbk , \qquad n > 2.$$

It is straightforward to check from the definitions that the corresponding isomorphisms $\pi_n(X) \otimes \Bbbk \cong \pi_n(F) \otimes \Bbbk$ identify the Whitehead action of $\pi_1(X, *)$ with the holonomy $\pi_*$-representation in $\pi_*(F) \otimes \Bbbk$.

The latter representation is therefore nilpotent, and clearly $\pi_*(F) \otimes \Bbbk$ has finite type. Now it follows from Theorem 1.6 (p. 43) that $H_*(F)$ has finite type and from Theorem 5.2 (p. 160) that $\pi_1(X, *)$ acts nilpotently in $H(F)$. Thus the hypotheses of Theorem 5.1 are satisfied by the classifying fibration and Proposition 5.1 gives a quasi-isomorphism

$$(\wedge V \otimes \wedge U, d) \xrightarrow{\simeq} A_{PL}(F) .$$

But according to Lemma 4.3 (p. 124) there is a weak homotopy equivalence $\widetilde{X} \simeq F$, and so Theorem 5.3 follows.

□

## 5.4 The Sullivan model of a spatial realization

In this section we suppose our ground field is the field of rational numbers, $\mathbb{Q}$.

Recall from formula (1.11), p. 28, that a minimal Sullivan algebra $(\wedge V, d)$ determines a homotopy class of morphisms

$$m_{|\wedge V,d|} : (\wedge V, d) \to A_{PL}(|\wedge V, d|)\,,$$

where $|\wedge V, d|$ denotes the spatial realization of $(\wedge V, d)$. These morphisms need not be quasi-isomorphisms: if $V = V^3$ is countably infinite dimensional then $H^3(\wedge V, 0) = V^3$ cannot be the cohomology of a topological space. However we have

**Theorem 5.4.** *Let $(\wedge V, d)$ be a minimal Sullivan algebra defined over $\mathbb{Q}$.*

*(i) If $\dim V^1 < \infty$ then*

$$(\wedge V^1, d) \to A_{PL}(|\wedge V^1, d|)$$

*is a quasi-isomorphism.*

*(ii) If $(\wedge V^1, d) \to A_{PL}(|\wedge V^1, d|)$ is a quasi-isomorphism and if $\dim V^i < \infty$, $i \geq 2$, then*

$$m_{|\wedge V,d|} : (\wedge V, d) \to A_{PL}(|\wedge V, d|)$$

*is a quasi-isomorphism.*

*proof:* Recall the commutative diagram (3.9), p. 115,

$$\begin{array}{ccccc}
(\wedge V, d) & \longrightarrow & (\wedge V \otimes \wedge Z, d) & \longrightarrow & (\wedge Z, \overline{d}) \\
\Big\downarrow {\scriptstyle m_{|\wedge V,d|}} & & \Big\downarrow {\scriptstyle m_{|\wedge V \otimes \wedge Z,d|}} & & \Big\downarrow {\scriptstyle m_{|\wedge Z,\overline{d}|}} \\
A_{PL}(|\wedge V, d|) & \longrightarrow & A_{PL}(|\wedge V \otimes \wedge Z, d|) & \longrightarrow & A_{PL}(|\wedge Z, \overline{d}|)\,.
\end{array}$$

For (i) consider first the case that $V^1 = \mathbb{Q}v$. It follows from Theorems 1.3 (p. 37) and 2.4 (p. 75), that there is a group isomorphism $\pi_1(|\wedge v, 0|) \cong \mathbb{Q}$ and that $\pi_i(|\wedge v, 0|) = 0$, $i \geq 2$. Moreover, the proof of Theorem 1.3 shows that $1 \in \pi_1(|\wedge v, 0|)$ is represented by a continuous map $\varphi : S^1 \to |\wedge v, 0|$ such that $H^*(\varphi)[v]$ is the fundamental cohomology class of $S^1$, where $[v] \in H^1(|\wedge v, 0|)$ is the class represented by $m_{|\wedge v,0|}(v)$.

Now let $S^1_{\mathbb{Q}}$ be Sullivan's rational telescope as described for instance in §9 of [18]. Because $\pi_1(|\wedge v, 0|) = \mathbb{Q}$, the map $\varphi$ extends to a map

$\varphi_{\mathbb{Q}} : S^1_{\mathbb{Q}} \to |\wedge v, 0|$ which induces an isomorphism of homotopy groups and hence of cohomology. A direct computation shows that $H^1(S^1_{\mathbb{Q}}) = \wedge w$ where $w = H^*(\varphi_{\mathbb{Q}})[v]$ and thus $H^*(|\wedge v, 0|, \mathbb{Q}) = \wedge[v]$ and $m_{|\wedge v,0|}$ is a quasi-isomorphism.

The rest of (i) follows by induction. Suppose $\dim V^1 = k+1$ and (i) is true when $\dim V^1 = k$. Choose a basis element $v \in V^1$ with $dv = 0$. Then there is a Sullivan extension

$$(\wedge v, 0) \to (\wedge V^1, d) \to (\wedge Z, \overline{d})$$

where $Z = Z^1$. As noted above, this yields a commutative diagram

$$\begin{array}{ccccc}
(\wedge v, 0) & \longrightarrow & (\wedge V^1, d) & \longrightarrow & (\wedge Z, \overline{d}) \\
\downarrow{\scriptstyle m_{|\wedge v,0|}} & & \downarrow{\scriptstyle m_{|\wedge V^1,d|}} & & \downarrow{\scriptstyle m_{|\wedge Z,\overline{d}|}} \\
A_{PL}|\wedge v, 0| & \longrightarrow & A_{PL}|\wedge V^1, 0| & \longrightarrow & A_{PL}(|\wedge Z, \overline{d}|)\,.
\end{array}$$

We have shown that $m_{|\wedge v,0|}$ is a quasi-isomorphism; by induction $m_{|\wedge Z,\overline{d}|}$ is also a quasi-isomorphism.

But since $\dim Z < \infty$ and $Z = Z^1$ it follows that $\dim H(\wedge Z, \overline{d}) \leq \dim \wedge Z < \infty$. Moreover, $|\wedge Z, \overline{d}| \to |\wedge Y, d| \to |\wedge v, 0|$ is a fibre bundle of CW complexes ([18], Proposition 17.9) and hence satisfies the fibration condition (5.1). Finally, it follows from Corollary 4.3 (p. 138) that the isomorphism $H(m_{|\wedge Z,\overline{d}|})$ identifies the holonomy representation of $\pi_1(|\wedge v, 0|)$ in $H(|\wedge Z, \overline{d}|)$ with that in $H(\wedge Z, \overline{d})$ and the latter is nilpotent. Thus the hypotheses of Theorem 5.1 are satisfied and it follows from Corollary 5.1 (p. 146) that $m_{|\wedge V^1,d|}$ is a quasi-isomorphism. Assertion (i) follows by induction.

(ii) As in (i) we obtain from (3.9), a commutative diagram

$$\begin{array}{ccccc}
(\wedge V^1, d) & \longrightarrow & (\wedge V, d) & \longrightarrow & (\wedge V^{\geq 2}, \overline{d}) \\
\downarrow{\scriptstyle m_{|\wedge V^1,d|}} & & \downarrow{\scriptstyle m_{|\wedge V,d|}} & & \downarrow{\scriptstyle m_{|\wedge V^{\geq 2},\overline{d}|}} \\
A_{PL}(|\wedge V^1, d|) & \longrightarrow & A_{PL}(|\wedge V, d|) & \longrightarrow & A_{PL}(|\wedge V^{\geq 2}, \overline{d}|)\,.
\end{array}$$

Since each $V^i$, $i \geq 2$ is finite dimensional, $H(\wedge V^{\geq 2}, \overline{d})$ has finite type. Moreover, Theorem 17.12 (ii) of [18] asserts that $m_{|\wedge V^{\geq 2},\overline{d}|}$ is a quasi-isomorphism. Now it follows easily as in the proof of (i) that the holonomy representation of $\pi_1(|\wedge V^1, d|)$ in $H(|\wedge V^{\geq 2}, \overline{d}|; \mathbb{Q})$ is locally nilpotent. Finally, by hypothesis, $m_{|\wedge V^1,d|}$ is a quasi-isomorphism.

Thus the sequence $|\wedge V^{\geq 2}, \overline{d}| \to |\wedge V, d| \to |\wedge V^1, d|$ satisfies the hypotheses of Theorem 5.1 and so Corollary 5.1 (p. 146) asserts that $m_{|\wedge V,d|}$ is a quasi-isomorphism. □

# Chapter 6

# Loop spaces and loop space actions

Recall from §1.4 that for any topological space, $X$, $H^*(X;\Bbbk) = H(A_{PL}(X))$ is denoted by $H(X)$.

## 6.1 The loop cohomology coalgebra of $(\wedge V, d)$

*Throughout this section we consider a fixed minimal Sullivan algebra* $(\wedge V, d)$. We denote multiplication and the (unique) augmentation in $\wedge V$ respectively by

$$\mu_{\wedge V} : (\wedge V, d) \otimes (\wedge V, d) \to (\wedge V, d) \qquad \text{and } \varepsilon_{\wedge V} : (\wedge V, d) \to \Bbbk\,.$$

Then by Proposition 3.9 (p. 103) we may extend $\mu_{\wedge V}$ to a quasi-isomorphism

$$\mu : (\wedge V \otimes \wedge V \otimes \wedge U, d) \xrightarrow{\cong} (\wedge V, d) \tag{6.1}$$

from a minimal $\Lambda$-extension of $(\wedge V \otimes \wedge V, d)$, in which $\mu$ restricts to the augmentation

$$\varepsilon_{\wedge U} : U \to 0$$

and for $u \in U$,

$$d(1 \otimes 1 \otimes u) = (1 \otimes \alpha(u) - \alpha(u) \otimes 1) \otimes 1 + \Phi(u)\,, \tag{6.2}$$

where $\Phi(u) \in \wedge^+(V \oplus V)\wedge \wedge^+ (V \oplus V \oplus U)$, and

$$\alpha : U \xrightarrow{\cong} V$$

is a linear isomorphism of degree 1. In particular, the quotient differential in $\wedge U$ is zero.

Now observe that if we divide $\wedge V \otimes \wedge V \otimes \wedge U$ respectively by the ideals $\wedge^+ V \otimes \wedge V \otimes \wedge U$ and $\wedge V \otimes \wedge^+ V \otimes \wedge U$ then $\mu$ factors to give quasi-isomorphisms

$$\varepsilon = \varepsilon_{\wedge V} \otimes \varepsilon_{\wedge U} : (\wedge V \otimes \wedge U, d) \to \Bbbk \quad \text{and} \quad \varepsilon' = \varepsilon_{\wedge V} \otimes \varepsilon_{\wedge U} : (\wedge V \otimes \wedge U, d') \to \Bbbk\,,$$

both of which are acyclic closures of $\wedge V, d)$. While they may not be identical, they are related by a commutative diagram

$$\begin{array}{ccc} (\wedge V, d) & \longrightarrow & (\wedge V \otimes \wedge U, d) \\ \big\downarrow & \nearrow^{\chi} & \varepsilon \big\downarrow \simeq \\ (\wedge V \otimes \wedge U, d') & \xrightarrow[\varepsilon']{\simeq} & \mathbb{k} \end{array}$$

which we obtain from the interchange involution $\tau_{\wedge V} : \Phi \otimes \Psi \mapsto (-1)^{\deg \Phi \cdot \deg \Psi} \Psi \otimes \Phi$ of $\wedge V \otimes \wedge V$, as follows. Since $\mu_{\wedge V} \otimes \tau_{\wedge V} = \mu_{\wedge V}$, $\tau_{\wedge V}$ extends to a quasi-isomorphism $\tau : (\wedge V \otimes \wedge V \otimes \wedge U, d) \xrightarrow{\cong} (\wedge V \otimes \wedge V \otimes \wedge U, d)$ of augmented cochain algebras. Divide by $\wedge^+ V \otimes \wedge V \otimes \wedge U$ on the left, and by $\wedge V \otimes \wedge^+ V \otimes \wedge U$ on the right. Since $\tau$ preserves these ideals it factors to give a morphism $\chi$ making the diagram commute.

We now, and for the rest of this Chapter, *fix*

$$\varepsilon : (\wedge V \otimes \wedge U, d) \xrightarrow{\simeq} \mathbb{k} \tag{6.3}$$

as *the acyclic closure of* $(\wedge V, d)$. It follows from (6.2) that in $\wedge V \otimes \wedge U$,

$$d(1 \otimes u) - \alpha(u) \otimes 1 \in \wedge^+ V \wedge \wedge^+(V \oplus U) \,. \tag{6.4}$$

Now write $(\wedge V \otimes \wedge U, d) \otimes_{\wedge V} (\wedge V \otimes \wedge U, d) = (\wedge V \otimes \wedge U \otimes \wedge U, d)$. Then by Proposition 1.1 the morphisms

$$(\wedge U, 0) \xleftarrow[\varepsilon_{\wedge V} \otimes id \otimes \varepsilon_{\wedge U}]{\simeq} (\wedge V \otimes \wedge U \otimes \wedge U, d) \xrightarrow[\varepsilon_{\wedge V} \otimes \varepsilon_{\wedge U} \otimes id]{\simeq} (\wedge U, 0)$$

are surjective quasi-isomorphisms. We denote them respectively by

$$\varepsilon_L = \varepsilon_{\wedge V} \otimes id \otimes \varepsilon_{\wedge U} \quad \text{ and } \varepsilon_R = \varepsilon_{\wedge V} \otimes \varepsilon_{\wedge U} \otimes id \,.$$

**Definition.** The composite

$$\omega = H(\varepsilon_R) \circ H(\varepsilon_L)^{-1} : \wedge U \xrightarrow{\cong} \wedge U$$

is the *canonical conjugation* of $\wedge U$.

**Lemma 6.1.**

*(i) The canonical conjugation* $\omega$ *is an involution:*

$$\omega^2 = id\,.$$

*(ii) If* $\varphi : (\wedge V \otimes \wedge U, d) \to (\wedge V \otimes \wedge U', d')$ *is an isomorphism between acyclic closures extending* $id_{\wedge V}$ *then the isomorphism* $\Bbbk \otimes_{\wedge V} \varphi$ *is independent of the choice of* $\varphi$.

*(iii) The isomorphism* $\Bbbk \otimes_{\wedge V} \varphi$ *satisfies*

$$(\Bbbk \otimes_{\wedge V} \varphi) \circ \omega = \omega' \circ (\Bbbk \otimes_{\wedge V} \varphi)\,.$$

*proof:* (i) Denote by $\tau$ the interchange automorphism (§1.1, p. 2) of $(\wedge V \otimes \wedge U) \otimes_{\wedge V} (\wedge V \otimes \wedge U) = \wedge V \otimes \wedge U \otimes \wedge U$. It is a cdga morphism which interchanges the two tensorands $\wedge U$ and restricts to the identity in $\wedge V$. Since $\tau^2 = id$ and $\varepsilon_R \circ \tau = \varepsilon_L$ it follows that $\omega^2 = id$.

(ii) This follows from the diagram

$$\begin{array}{ccccc}
(\wedge U, 0) & \xleftarrow[\simeq]{\varepsilon_R} & (\wedge V \otimes \wedge U \otimes \wedge U, d) & \xrightarrow[\simeq]{\varepsilon_L} & (\wedge U, 0) \\
\Big\downarrow{\scriptstyle id} & & \Big\downarrow{\scriptstyle id \otimes_{\wedge V} \varphi} & & \Big\downarrow{\scriptstyle \Bbbk \otimes_{\wedge V} \varphi} \\
(\wedge U, 0) & \xleftarrow[\varepsilon'_R]{\simeq} & (\wedge V \otimes \wedge U \otimes \wedge U', d') & \xrightarrow[\varepsilon'_L]{\simeq} & (\wedge U', 0)\,,
\end{array}$$

since the horizontal rows are independent of $\varphi$.

(iii) This follows by composing the diagram of (ii) with the analogous diagram connecting $\wedge V \otimes \wedge U \otimes \wedge U'$ to $\wedge V \otimes \wedge U' \otimes \wedge U'$.

□

**Proposition 6.1.** *For the minimal Sullivan algebra* $(\wedge V, d)$, *and with the notation above, write* $\overline{\chi} := \Bbbk \otimes_{\wedge V} \chi : \wedge U \to \wedge U$. *Then*

$$\overline{\chi} = \omega : \wedge U \to \wedge U\,.$$

*proof:* It will be convenient to use identical copies $(\wedge V_i, d)$ of $(\wedge V, d)$. Then consider the cochain algebra

$$(E, d) = (\wedge V_1 \otimes \wedge U, d) \otimes_{\wedge V_1} (\wedge V_1 \otimes \wedge V_2 \otimes \wedge U, d) \otimes_{\wedge V_2} (\wedge V_2 \otimes \wedge U, d)\,.$$

Set

$$\rho_L = id \otimes_{\wedge V_2} \varepsilon_{\wedge V_2 \otimes \wedge U} : (E, d) \to (\wedge V \otimes \wedge U, d) \otimes_{\wedge V} (\wedge V \otimes \wedge U, d')$$

$$\rho_R = \varepsilon_{\wedge V_1 \otimes \wedge U} \otimes_{\wedge V_1} id : (E, d) \to (\wedge V \otimes \wedge U, d) \otimes_{\wedge V} (\wedge V \otimes \wedge U, d)$$

and define

$$\wedge U \xleftarrow{\varepsilon'_L} (\wedge V \otimes \wedge U, d) \otimes_{\wedge V} (\wedge V \otimes \wedge U, d') \xrightarrow{\varepsilon'_R} \wedge U$$

by setting $\varepsilon'_L = id \otimes_{\wedge V} \varepsilon'$ and $\varepsilon'_R = \varepsilon \otimes_{\wedge V} id$.

It is then easy to check that the diagram below is commutative:

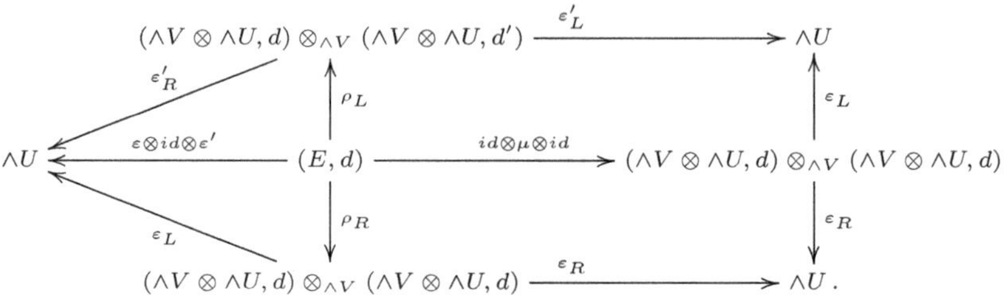

In view of Lemma 6.1, $\omega^2 = 1$ and $\omega \circ H(\varepsilon_L) = H(\varepsilon_R)$. Thus a simple diagram chase gives

$$H(\varepsilon'_R \circ \rho_L) = H(\varepsilon'_L \circ \rho_L)\,.$$

Finally, use the commutative diagram

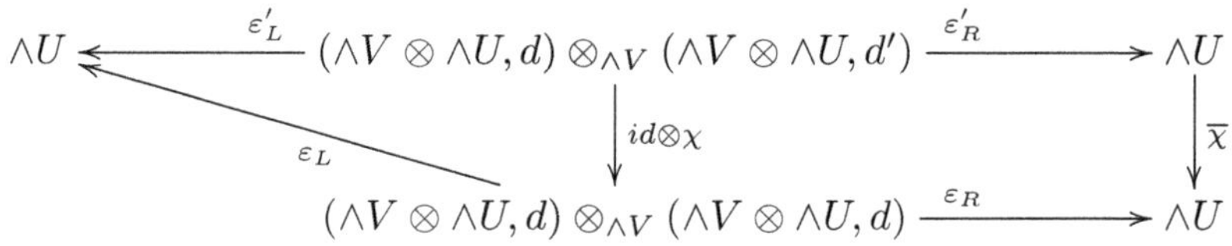

to deduce that $\overline{\chi} \circ H(\varepsilon'_R) \circ H(\varepsilon_L)^{-1} = \omega$. Since $\varepsilon'_L \circ \rho_L$ is a quasi-isomorphism, the Proposition follows.

□

Now consider a $\Lambda$-extension $(\wedge V \otimes \wedge Z, d)$ of the minimal Sullivan algebra $(\wedge V, d)$, with fibre $(\wedge Z, \overline{d})$. As above, set $(\wedge V \otimes \wedge Z \otimes \wedge U, d) = (\wedge V \otimes \wedge Z, d) \otimes_{\wedge V} (\wedge V \otimes \wedge U, d)$. Then (Proposition 1.1, p. 7)

$$id \otimes_{\wedge V} \varepsilon : (\wedge V \otimes \wedge Z \otimes \wedge U, d) \to (\wedge Z, \overline{d})$$

is a surjective quasi-isomorphism. Let $\sigma_{\wedge Z} : (\wedge Z, \overline{d}) \xrightarrow{\simeq} (\wedge V \otimes \wedge Z \otimes \wedge U, d)$ be a right inverse for $id \otimes_{\wedge V} \varepsilon$, and set

$$\Delta_{\wedge Z} = (\varepsilon_{\wedge V} \otimes id \otimes \omega) \circ \sigma_{\wedge Z} : (\wedge Z, \overline{d}) \to (\wedge Z, \overline{d}) \otimes (\wedge U, 0)\,.$$

Since $\sigma_{\wedge Z}$ is determined up to homotopy it follows that so is $\Delta_{\wedge Z}$. In particular, $H(\Delta_{\wedge Z})$ is independent of the choice of $\sigma_{\wedge Z}$ and we make the

**Definition.** (i) The morphism

$$H(\Delta_{\wedge Z}) : H(\wedge Z, \overline{d}) \to H(\wedge Z, \overline{d}) \otimes \wedge U$$

is the *holonomy coaction* of $\wedge U$ on $H(\wedge Z, \overline{d})$.

(ii) When the $\Lambda$-extension $(\wedge V \otimes \wedge Z, d)$ is the acyclic closure $(\wedge V \otimes \wedge U, d)$, the morphism

$$\Delta_{\wedge U} : \wedge U \to \wedge U \otimes \wedge U$$

is denoted simply by $\Delta$, and is called the *comultiplication* of $\wedge U$.

In particular, in the case of the acyclic closure we fix augmentation preserving right inverses $j_L$ and $j_R$ for $\varepsilon_L$ and $\varepsilon_R$, and then may take $\sigma_{\wedge U} = j_L$. Moreover, because the differential in the fibre, $\wedge U$, of an acyclic closure is zero, $\Delta = H(\Delta)$, and so is independent of the choice of $\sigma_{\wedge U}$.

On the other hand, recall from the start of this section the surjective quasi-isomorphism $\varepsilon' : (\wedge V \otimes \wedge U, d') \to \Bbbk$ and the quasi-isomorphism $\chi : (\wedge V \otimes \wedge U, d') \xrightarrow{\cong} (\wedge V \otimes \wedge U, d)$. Write

$$(\wedge V \otimes \wedge Z \otimes \wedge U, d') = (\wedge V \otimes \wedge Z, d) \otimes_{\wedge V} (\wedge V \otimes \wedge U, d') \qquad (6.5)$$

and note as above that

$$id \otimes_{\wedge V} \varepsilon' : (\wedge V \otimes \wedge Z \otimes \wedge U, d') \to (\wedge Z, \overline{d})$$

is a surjective quasi-isomorphism.

**Lemma 6.2.** *With the notation above, if* $\sigma'_{\wedge Z} : (\wedge Z, \overline{d}) \to (\wedge V \otimes \wedge Z \otimes \wedge U, d')$ *is a right inverse for* $id \otimes_{\wedge V} \varepsilon'$ *then*

$$\Delta_{\wedge Z} \simeq (\varepsilon_{\wedge V} \otimes id) \circ \sigma'_{\wedge Z} .$$

*proof.* Since $\varepsilon \circ \chi = \varepsilon'$, it follows that $(id \otimes_{\wedge V} \chi) \circ \sigma'_{\wedge Z}$ is a right inverse for $id \otimes_{\wedge V} \varepsilon$, and so $\sigma_{\wedge Z} \sim (id \otimes_{\wedge V} \chi) \circ \sigma'_{\wedge Z}$. Apply $\varepsilon_{\wedge V}$ to obtain

$$\Delta_{\wedge Z} \sim (id \otimes \omega\overline{\chi}) \circ (\varepsilon_{\wedge V} \otimes id) \circ \sigma'_{\wedge Z}$$

and recall from Lemma 6.1 that since $\omega^2 = id_{\wedge V}$ then $\omega\overline{\chi} = id_{\wedge U}$.

□

Next, let $(\wedge V_i, d)$, $i = 1, 2, 3$ be copies of $\wedge V$ and let $(\wedge V_i \otimes \wedge V_j \otimes \wedge U, d)$ be the corresponding copy of $(\wedge V \otimes \wedge V \otimes \wedge U, d)$ with $V_i$ corresponding to $V \otimes 1$ and $V_j$ corresponding to $1 \otimes V$. Then again by the Lifting Lemma there is a morphism $\nu$ making the diagram

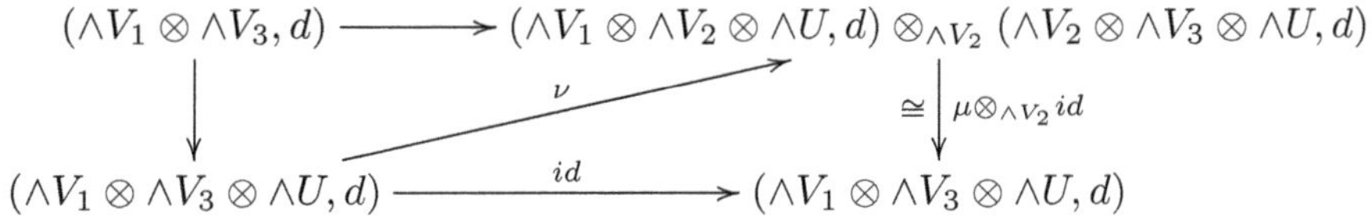

commutative, since $\mu \otimes_{\wedge V_2} id$ is a surjective quasi-isomorphism. Now set

$$\overline{\nu} = \Bbbk \otimes_{\wedge V_1 \otimes \wedge V_2 \otimes \wedge V_3} \nu : \wedge U \to \wedge U \otimes \wedge U\,.$$

**Lemma 6.3.** *With the above notation, $\overline{\nu} = \Delta$.*

*proof:* For simplicity we shall suppress the differentials from the notation where they are obvious. First observe that $(id \otimes_{\wedge V_2} \mu) \circ \nu$ and $id$ are all self maps of $(\wedge V_1 \otimes \wedge V_3 \otimes \wedge U, d)$, and that $\mu \circ (id \otimes_{\wedge V_2} \mu) \circ \nu = \mu \circ id$. Since $\mu$ is a quasi-isomorphism, it follows from the uniqueness assertion of Proposition 3.6 (p. 98) that

$$(id \otimes_{\wedge V_2} \mu) \circ \nu \sim id \qquad \text{rel } \wedge V_1 \otimes \wedge V_3\,.$$

Dividing by $(\wedge V_1 \otimes \wedge V_3)^+$ we obtain from $\nu$ a morphism

$$\widehat{\nu} : \wedge U \to (\wedge V_2 \otimes \wedge U, d) \otimes_{\wedge V_2} (\wedge V_2 \otimes \wedge U, d')$$

and a homotopy $(id \otimes_{\wedge V_2} \varepsilon') \circ \widehat{\nu} \sim id : \wedge U \to \wedge U$. Thus it follows from Lemma 6.2 that $\Delta = \overline{\nu}$.

□

Recall now that $\tau$ is the automorphism of $(\wedge V \otimes \wedge U \otimes \wedge U, d)$ interchanging the two tensorands, $\wedge U$. Thus $\overline{\tau} = \Bbbk \otimes_{\wedge V} \tau$ is the interchange automorphism of $\wedge U \otimes \wedge U$.

**Proposition 6.2.** *With the hypotheses and notation above*

*(i)* $(id \otimes \varepsilon_{\wedge U}) \circ \Delta_{\wedge Z} = id_{\wedge Z}$ *and* $(id \otimes \varepsilon_{\wedge U}) \circ \Delta = id_{\wedge U} = (\varepsilon_{\wedge U} \otimes id) \circ \Delta$.
*(ii)* $\overline{\tau} \circ (\omega \otimes \omega) \circ \Delta = \Delta \circ \omega$.
*(iii)* *If $\varphi : (\wedge V \otimes \wedge Z, d) \to (\wedge V \otimes \wedge Z', d)$ is a morphism of $\Lambda$-extensions extending $id_{\wedge V}$ then*

$$[(\Bbbk \otimes_{\wedge V} \varphi) \otimes id] \circ \Delta_{\wedge Z} \sim \Delta_{\wedge Z'} \circ (\Bbbk \otimes_{\wedge V} \varphi) : (\wedge Z, \overline{d}) \to (\wedge Z', \overline{d}) \otimes \wedge U\,.$$

*(iv)* $(id \otimes \Delta) \circ \Delta_{\wedge Z} \sim (\Delta_{\wedge Z} \otimes id) \circ \Delta_{\wedge Z}$.
*(v)* *In particular,* $(\Delta \otimes id) \circ \Delta = (id \otimes \Delta) \circ \Delta$.
*(vi)* $\mu_{\wedge U} \circ (id \otimes \omega) \circ \Delta = \varepsilon_{\wedge U} : \wedge U \to \Bbbk$, *where* $\mu_{\wedge U} : \wedge U \otimes \wedge U \to \wedge U$ *denotes multiplication.*

*proof:* (i) $(id \otimes \varepsilon_U) \circ \Delta_{\wedge Z} = (id \otimes_{\wedge V} \varepsilon) \circ \sigma_{\wedge Z} = id_{\wedge W}$. In particular, $(id \otimes \varepsilon_U) \circ \Delta = id_{\wedge U}$. Finally,

$$(\varepsilon_U \otimes id) \circ \Delta = \omega \circ (\varepsilon \otimes_{\wedge V} id) \circ j_L = \omega \circ \varepsilon_R \circ j_L = \omega^2 = id_{\wedge U}\,.$$

(ii) Recall that $j_L : (\wedge U, 0) \to (\wedge V \otimes \wedge U \otimes \wedge U, d)$ is a right inverse for $\varepsilon_L$. Then $j_R = \tau \circ j_L$ is a right inverse for $\varepsilon_R$. Thus by definition $H(j_R)^{-1} \circ H(j_L) = \omega$. Since $\omega^2 = id$,

$$H(j_L) \circ \omega = H(j_R)\,.$$

By definition, $\Delta = (id \otimes \omega) \circ H(\varepsilon_V \otimes id) \circ H(j_L)$ and so

$$\begin{aligned}
\Delta \circ \omega &= (id \otimes \omega) \circ H(\varepsilon_V \otimes id) \circ H(j_R) \\
&= (id \otimes \omega) \circ H(\varepsilon_V \otimes id) \circ H(\tau) \circ H(j_L) \\
&= \overline{\tau} \circ (\omega \otimes id) \circ H(\varepsilon_V \otimes id) \circ H(j_L) \\
&= \overline{\tau} \circ (\omega \otimes \omega) \circ (id \otimes \omega) \circ H(\varepsilon_V \otimes id) \circ H(j_L) \\
&= \overline{\tau} \circ (\omega \otimes \omega) \circ \Delta\,.
\end{aligned}$$

(iii) This is immediate from the definitions.

(iv) As in Lemma 6.3 let $(\wedge V_i, d)$ denote identical copies of $(\wedge V, d)$. Then let $(\wedge V_i \otimes \wedge V_j \otimes \wedge U, d)$ be the corresponding copy of $(\wedge V \otimes \wedge V \otimes \wedge U, d)$ and let $(\wedge V_i \otimes \wedge Z, d)$ be the corresponding copy of $(\wedge V \otimes \wedge Z, d)$.

Then by the Lifting Lemma 3.1 (p. 97) there are commutative diagrams of cochain algebras

$$\begin{array}{ccc}
(\wedge V_i, d) & \longrightarrow & (\wedge V_j \otimes \wedge Z, d) \otimes_{\wedge V_j} (\wedge V_i \otimes \wedge V_j \otimes \wedge U) \\
\downarrow & \nearrow^{\varphi_{i,j}} & \downarrow \\
(\wedge V_i \otimes \wedge Z, d) & \xrightarrow{=} & (\wedge V \otimes \wedge Z, d)\,.
\end{array}$$

Now, with the notation of (6.5), division by $\wedge^+ V_i$ turns $\varphi_{i,j}$ into a morphism $\widehat{\varphi}_{i,j} : (\wedge Z, \overline{d}) \to (\wedge V \otimes \wedge Z \otimes \wedge U, d')$ satisfying $(\varepsilon_{\wedge V} \otimes id_{\wedge Z} \otimes \varepsilon_{\wedge U}) \circ \widehat{\varphi}_{i,j} = id_{\wedge Z}$. It follows from Lemma 6.2 that $\Delta_{\wedge Z} \sim \overline{\varphi}_{i,j} := (\varepsilon_{\wedge V} \otimes id \otimes id) \circ \widehat{\varphi}_{i,j}$.

Next, as in the proof of Lemma 6.2 we suppress the differentials in the notation. Then we construct the commutative diagram,

$$\begin{array}{ccc}
\wedge V_3 & \longrightarrow & (\wedge(V_1 \oplus Z)) \otimes_{\wedge V_1} (\wedge(V_1 \oplus V_2 \oplus U)) \otimes_{\wedge V_2} (\wedge(V_2 \oplus V_3 \oplus U)) \\
\downarrow & \nearrow^{\Phi_1, \Psi_2} & \simeq \downarrow id \otimes_{\wedge V_1} \mu \otimes_{\wedge V_2} \mu \\
\wedge V_3 \otimes \wedge Z & \xrightarrow{\simeq} & \wedge V \otimes \wedge Z\,,
\end{array}$$

in which $\Psi_1$ and $\Psi_2$ are defined by :

$$\Psi_1 = (id_{\wedge V_1 \otimes \wedge Z} \otimes_{\wedge V_1} \nu) \circ \varphi_{3,1}$$

and

$$\Psi_2 = \varphi_{21} \otimes_{\wedge V_2} (id_{\wedge V_2 \otimes \wedge V_3 \otimes \wedge U}) \circ \varphi_{3,2}\,.$$

Since $(id\otimes\mu\otimes\mu)\circ\Psi_1=(id\otimes\mu\otimes\mu)\circ\Psi_2$, the uniqueness assertion of Proposition 3.6 (p. 98) implies that $\Psi_1\sim\Psi_2$ rel $\wedge V_3$. Dividing by $\wedge^+V_3$ then turns $\Psi_1$ and $\Psi_2$ into homotopic morphisms from $(\wedge Z,\overline{d})$, and so composing with $\varepsilon_{\wedge V_1}\otimes\varepsilon_{\wedge V_2}$ we obtain homotopic morphisms

$$\overline{\Psi}_1\sim\overline{\Psi}_2:(\wedge Z,\overline{d})\to(\wedge Z,\overline{d})\otimes\wedge U\otimes\wedge U\,.$$

By construction, and in view of Lemma 6.3, $\overline{\Psi}_1=(id_{\wedge Z}\otimes\overline{\nu})\circ\overline{\varphi}_{31}\sim(id\otimes\Delta)\circ\Delta_{\wedge Z}$, while $\overline{\Psi}_2=(\overline{\varphi}_{21}\otimes id)\circ\overline{\varphi}_{32}\sim(\Delta_{\wedge Z}\otimes id)\circ\Delta_{\wedge Z}$.

(vi) Since $\Delta=(\varepsilon_V\otimes id\otimes\omega)\circ j_L$, $(1\otimes\omega)\circ\Delta=(\varepsilon_V\otimes id\otimes id)\circ j_L$. Now note that the map $id\otimes\mu_{\wedge U}:(\wedge V\otimes\wedge U\otimes\wedge U,d)\to(\wedge V\otimes\wedge U,d)$ is just the multiplication morphism $(\wedge V\otimes\wedge U,d)\otimes_{\wedge V}(\wedge V\otimes\wedge U,d)\to(\wedge V\otimes\wedge U,d)$. Thus we have the commutative diagram

$$\begin{array}{ccc} (\wedge U,0) & & \\ \Big\downarrow{\scriptstyle j_L} & \searrow^{(1\otimes\omega)\circ\Delta} & \\ (\wedge V\otimes\wedge U\otimes\wedge U,d) & \xrightarrow{\varepsilon_V\otimes id\otimes id} & (\wedge U\otimes\wedge U,0)\\ \Big\downarrow{\scriptstyle id\otimes\mu_{\wedge U}} & & \Big\downarrow{\scriptstyle \mu_{\wedge U}}\\ (\wedge V\otimes\wedge U,d) & \xrightarrow{\varepsilon_V\otimes id} & (\wedge U,0)\,. \end{array}$$

Since the differential is zero in $\wedge U$ and $H(\wedge V\otimes\wedge U)\cong k\!\!k$, each cycle in $\ker\varepsilon\subset(\wedge V\otimes\wedge U,d)$ is in the ideal $\wedge^+V\otimes\wedge U$. This is in particular the case for $(1\otimes\mu_U)\circ j_L(\wedge^+U)$. Therefore $(\varepsilon_V\otimes id)\circ(id\otimes\mu_{\wedge U})\circ j_L(\wedge^+U)=0$. The result follows then from the commutativity of the diagram above.

□

By definition the comultiplication $\Delta:\wedge U\to\wedge U\otimes\wedge U$ is a morphism of commutative graded algebras. Thus it follows from Proposition 6.2 that it makes $\wedge U$ into a commutative graded Hopf algebra with counit $\varepsilon_U$.

**Definition.** $(\wedge U,\Delta)$ is the *loop cohomology Hopf algebra* for $(\wedge V,d)$.

## 6.2 The transformation map $\eta_L$

*Throughout this section we fix a minimal Sullivan algebra $(\wedge V,d)$ with homotopy Lie algebra, $L$, and an acyclic closure $(\wedge V\otimes\wedge U,d)$ with augmentation $\varepsilon_{\wedge U}:\wedge U\to k\!\!k$ given by $\varepsilon_{\wedge U}(U)=0$. We continue to use the other notation of §6.1.*

*Recall from §1.5 that $(\wedge V,d)$ is 1-finite if $\dim H^1(\wedge V,d)<\infty$ and that in this case its fundamental group $G_L=\exp_{L_0}(L_0)$ is defined (cf. §2.7).*

*Finally, recall from §2.4 that $I_L \subset UL$ is the ideal in the universal enveloping algebra generated by $L$, and that $\widehat{UL} = \varprojlim_n UL/I_L^n$.*

The comultiplication $\Delta : \wedge U \to \wedge U \otimes \wedge U$ defined in §6.1 dualizes to a product

$$\Delta^{\#} : (\wedge U)^{\#} \otimes (\wedge U)^{\#} \to (\wedge U)^{\#} .$$

It follows from Proposition 6.2(i) and (iv) (p. 174) that $\Delta^{\#}$ makes $(\wedge U)^{\#}$ into a graded algebra with identity $\varepsilon_{\wedge U}$. Moreover, because $\Delta$ is a morphism of graded algebras the multiplication in $\wedge U$ dualizes to a linear map,

$$\mu_{\wedge U}^{\#} : (\wedge U)^{\#} \to (\wedge U \otimes \wedge U)^{\#} ,$$

which is a morphism of graded algebras with respect to the products $\Delta^{\#}$ and $(\Delta \otimes \Delta)^{\#}$.

On the other hand, $\widehat{UL}$ is a graded algebra and the diagonal

$$\widehat{U\Delta} : \widehat{UL} \to \widehat{UL}\widehat{\otimes}\widehat{UL}$$

defined in §2.4 is a morphism of graded algebras.

Our first principal objective is to construct a morphism

$$\eta_L : \widehat{UL} \to (\wedge U)^{\#}$$

of graded algebras for which $\eta_L \otimes \eta_L$ extends (uniquely) to a morphism $\eta_L \widehat{\otimes} \eta_L : \widehat{UL}\widehat{\otimes}\widehat{UL} \to (\wedge U \otimes \wedge U)^{\#}$ and satisfies

$$\mu_{\wedge U}^{\#} \circ \eta_L = (\eta_L \widehat{\otimes} \eta_L) \circ \widehat{U\Delta} .$$

Then we prove that if $H^1(\wedge V, d)$ and each $V^i$, $i \geq 2$ are finite dimensional, then $\eta_L$ is an isomorphism.

To construct $\eta_L$ we first construct the classical anti-automorphism $\omega_L$ of $\widehat{UL}$ for any graded Lie algebra, $L$. It will correspond under $\eta_L$ to the dual, $\omega^{\#}$, of the canonical involution of $\wedge U$ defined in §6.1. Let $r : L \to \mathrm{End}(UL)$ be the linear map defined by

$$r(x)(a) = -(-1)^{\deg x \cdot \deg a} ax , \qquad a \in UL, x \in L .$$

It is straightforward to check that $r([x,y]) = [r(x), r(y)]$, and so $r$ extends to a homomorphism $Ur : UL \to \mathrm{End}(UL)$. We set

$$\omega_L(a) = r(a)(1) , \qquad a \in UL .$$

Then

$$\begin{aligned} \omega_L(xa) &= r(x) \circ r(a)(1) = r(x)(\omega_L(a)) = -(-1)^{\deg a \cdot \deg x} \omega_L(a) \cdot x \\ &= (-1)^{\deg a \cdot \deg x} \omega_L(a) \omega_L(x) . \end{aligned}$$

It follows that $\omega_L$ is an anti-automorphism of $UL$ satisfying $\omega_L^2 = id$ and $\omega_L(x) = -x$, $x \in L$. In particular, $\omega_L$ extends to an involutary anti-automorphism $\omega_L$ of $\widehat{UL}$, called the *canonical conjugation* of $\widehat{UL}$.

Now suppose $(\wedge V, d)$ is 1-finite, so that $G_L = \exp(L_0)$ is defined. Then for $a \in G_L$ we have $a = \exp x$, for some $x \in L_0$. Since $\omega_L(x) = -x$, $x \in L_0$, it follows that

$$\omega_L(a) = a^{-1}, \qquad a \in G_L. \tag{6.6}$$

To define $\eta_L$ we recall that, since the differential in $\wedge U$ is zero (Proposition 3.12, p. 111), the holonomy representation for the acyclic closure $(\wedge V \otimes \wedge U, d)$ is a Lie algebra homomorphism $\overline{\theta} : L \to \mathrm{Der}(\wedge U)$ (Lemma 4.8, p. 131). This extends to a homomorphism $U\overline{\theta} : UL \to \mathrm{End}(\wedge U)$. Because $L$ acts locally nilpotently in $\wedge U$, $U\overline{\theta}$ extends to a *holonomy representation* $\widehat{UL} \otimes \wedge U \to \wedge U$, which is denoted $a \otimes \Phi \mapsto a \cdot \Phi$.

This dualizes to the linear map

$$(\wedge U)^{\#} \otimes \widehat{UL} \to (\wedge U)^{\#}, \qquad f \otimes a \mapsto f \cdot a,$$

given by $(f \cdot a)(\Phi) = f(a \cdot \Phi)$, $\quad f \in (\wedge U)^{\#}$, $a \in \widehat{UL}$, $\Phi \in \wedge U$. Clearly

$$(f \cdot a) \cdot b = f \cdot (ab), \qquad f \in (\wedge U)^{\#},\ \ a, b \in \widehat{UL}. \tag{6.7}$$

**Definition.** The *transformation map,*

$$\eta_L : \widehat{UL} \to (\wedge U)^{\#}$$

is the linear map given by

$$\eta_L(a)(\Phi) = \varepsilon_{\wedge U}(a \cdot \Phi), \qquad a \in \widehat{UL},\ \ \Phi \in \wedge U. \tag{6.8}$$

When $a \in UL$, $\eta_L(a)(\Phi) = \varepsilon_{\wedge U}(U\overline{\theta}(a)(\Phi))$, $\qquad \Phi \in \wedge U$.

**Proposition 6.3.** *The transformation map, $\eta_L$, has the following properties*

*(i)* $\eta_L(1) = \varepsilon_{\wedge U}$, *and* $\eta_L(ab) = \eta_L(a) \cdot b$, $\quad a, b \in \widehat{UL}$.
*(ii)* $\eta_L$ *is a morphism of graded algebras, where* $(\wedge U)^{\#}$ *is equipped with the multiplication* $\Delta^{\#}$ *dual to* $\Delta$.
*(iii)* $\eta_L \circ \omega_L = \omega^{\#} \circ \eta_L$, *where* $\omega$ *is the canonical conjugation of §6.1.*
*(iv) The linear map* $\eta_L \widehat{\otimes} \eta_L : \widehat{UL} \widehat{\otimes} \widehat{UL} \to (\wedge U \otimes \wedge U)^{\#}$ *satisfies*

$$\mu_{\wedge U}^{\#} \circ \eta_L = \eta_L \widehat{\otimes} \eta_L \circ \widehat{\Delta_{UL}},$$

*where* $\mu_{\wedge U} : \wedge U \otimes \wedge U \to \wedge U$ *is the multiplication.*

*proof:* (i) This is immediate from (6.8).

(ii) Recall from p. 173 that $j_L : (\wedge U, 0) \to (\wedge V \otimes \wedge U \otimes \wedge U, d)$ denotes a right inverse for $\varepsilon_L$ where $\varepsilon_L$ is the projection on the left tensorand $\wedge U$. Let $\widehat{\theta} : L \to \mathrm{Der}(\wedge U \otimes \wedge U)$ denote the holonomy representation for the $\Lambda$-extension $(\wedge V \otimes \wedge U \otimes \wedge U, d)$ of $(\wedge V, d)$. It follows from Lemma 4.7 (p. 130) that for $x \in L$ and $\Phi \in \wedge U$,

$$\widehat{\theta}(x) \circ ((\varepsilon_V \otimes id) \circ j_L)(\Phi) = 0 .$$

For simplicity, we write $\varphi = (\varepsilon_{\wedge V} \otimes id) \circ j_L$. Since $(\wedge V \otimes \wedge U \otimes \wedge U, d) = (\wedge V \otimes \wedge U, d) \otimes_{\wedge V} (\wedge V \otimes \wedge U, d)$, it follows that $\widehat{\theta}(x) = \overline{\theta}(x) \otimes id + id \otimes \overline{\theta}(x)$. In particular, for $x \in L$,

$$\left(\overline{\theta}(x) \otimes id\right) \circ \varphi = -\left(id \otimes \overline{\theta}(x)\right) \circ \varphi .$$

Now for $\Phi \in \wedge^+ U$ we may write

$$\varphi\Phi = \Phi \otimes 1 + \Omega + 1 \otimes \Psi$$

with $\Omega \in \wedge^+ U \otimes \wedge^+ U$. Moreover, by the definition of $\omega$, $\Psi = \omega\Phi$. Thus we may rewrite the equation above as

$$\varphi\Phi = \Phi \otimes 1 + \sum \Phi_i \otimes \Psi_i + 1 \otimes \omega\Phi ,$$

with $\Phi_i, \Psi_i \in \wedge^+ U$. Then for $x \in L$,

$$\left.\begin{aligned} 0 &= \widehat{\theta}(x)\varphi\Phi \\ &= \overline{\theta}(x)\Phi \otimes 1 + \textstyle\sum \overline{\theta}(x)\Phi_i \otimes \Psi_i + (-1)^{\deg x \cdot \deg \Phi_i} \Phi_i \otimes \overline{\theta}(x)\Psi_i \\ &\qquad + (1 \otimes \overline{\theta}(x))\omega\Phi . \end{aligned}\right\} \tag{6.9}$$

Applying $\varepsilon_U \otimes \varepsilon_U$ gives

$$\eta_L(x)\Phi = -\eta_L(x) \cdot \omega\Phi = \eta_L(\omega_L(x))\omega\Phi . \tag{6.10}$$

Next, for $a \in I_L$ apply $U\overline{\theta}(a) \otimes id$ to (6.9). This gives

$$U\overline{\theta}(ax)\Phi = -\sum U\overline{\theta}(ax)\Phi_i \otimes \Psi_i - (-1)^{\deg x \cdot \deg \Phi_i} U\overline{\theta}(a)\Phi_i \otimes \overline{\theta}(x)\Psi_i .$$

Compose with $\varepsilon_U \otimes \varepsilon_U$ to obtain from (6.10) that

$$\begin{aligned} \eta_L(ax)\Phi &= (\eta_L(a) \otimes \eta_L(\omega_L x))\left(\textstyle\sum \Phi_i \otimes \Psi_i\right) \\ &= (\eta_L(a) \otimes \eta_L(x))\left(\textstyle\sum \Phi_i \otimes \omega\Psi_i\right) \\ &= (\eta_L(a) \otimes \eta_L(x))(\Delta\Phi) \\ &= \eta_L(a) \cdot \eta_L(x)(\Phi) . \end{aligned}$$

This shows that $\eta_L(ax) = \eta_L(a)\eta_L(x)$, and it follows that the restriction of $\eta_L$ to $UL$ is a homomorphism. Since for any $\Phi \in \wedge U$, some $I^n_{UL} \cdot \Phi = 0$ it follows that $\eta_L$ is a homomorphism.

(iii) It follows from Lemma 6.1(i) (p. 171), and Proposition 6.2(ii) (p. 174), that $\omega^{\#}$ is an involutory antiautomorphism. Now (6.10) applied to $\omega\Phi$ gives for $x \in L$ that

$$\eta_L(\omega_L x) = \eta_L(x) \circ \omega$$

and so $\eta_L \circ \omega_L$ and $\omega^{\#} \circ \eta_L$ agree in $L$. Since both maps are antiautomorphisms they agree in $UL$ and thus, just as in (ii), they agree in $\widehat{UL}$.

(iv) Note first that if $x \in L$ then $\Delta_{UL}(x) = x \otimes 1 + 1 \otimes x$, while $\overline{\theta}(x)$ is a derivation in $\wedge U$. It follows after a simple calculation from the definitions that

$$\begin{aligned} \mu^{\#}_{\wedge U}(\eta_L(x)) &= \eta_L(x) \otimes \varepsilon_U + \varepsilon_U \otimes \eta_L(x) \\ &\in (\wedge U)^{\#} \otimes (\wedge U)^{\#} \subset (\wedge U \otimes \wedge U)^{\#} \,. \end{aligned}$$

Since $\mu^{\#}_{\wedge U}$ and $\eta_L$ are algebra morphisms and $(\wedge U)^{\#} \otimes (\wedge U)^{\#}$ is a subalgebra of $(\wedge U \otimes \wedge U)^{\#}$ this implies that $\mu^{\#} \circ \eta_L$ maps $UL$ into $(\wedge U)^{\#} \otimes (\wedge U)^{\#}$. A second easy calculation shows that for $a \in UL$,

$$\mu^{\#}_{\wedge U} \circ \eta_L(a) = (\,\eta_L \otimes \eta_L\,) \circ \Delta_{UL}(a)\,.$$

Finally, again because of the local nilpotence of the holonomy representation, $\eta_L \otimes \eta_L$ extends uniquely to $\eta_L \widehat{\otimes} \eta_L : \widehat{UL} \widehat{\otimes} \widehat{UL} \to (\wedge U \otimes \wedge U)^{\#}$ and $\mu^{\#}_{\wedge U} \circ \eta_L = (\eta_L \widehat{\otimes} \eta_L) \circ \widehat{\Delta_{UL}}$ in $\widehat{UL}$.

□

**Theorem 6.1.** *Suppose $H^1(\wedge V, d)$ and each $V^i$, $i \geq 2$ is finite dimensional. Then $\eta_L : \widehat{UL} \xrightarrow{\cong} (\wedge U)^{\#}$ is an isomorphism.*

*proof:* Recall from Theorem 2.3 (p. 71) the linear isomorphism $\gamma_L : (\wedge U)^{\#} \xrightarrow{\cong} \widehat{UL}$. If $\dim V < \infty$ then, as established there, the linear maps $\wedge^k L \to UL$,

$$\gamma : x_1 \wedge \cdots \wedge x_k \mapsto \frac{1}{k!} \sum_{\sigma \in \Sigma_k} \varepsilon_\sigma x_{\sigma(1)} \cdot \cdots \cdot x_{\sigma(k)}$$

induce an isomorphism

$$\prod_k \wedge^k L \xrightarrow{\cong} \widehat{UL}\,. \tag{6.11}$$

As described in §2.1, the pairing $\langle , \rangle : L \times U \to {\rm l\!k}$ extends to pairings $\langle\; ,\; \rangle : \wedge^k L \times \wedge^k U \to {\rm l\!k}$, and when $L$ has finite type these identify

$$\wedge^k L = (\wedge^k U)^{\#}\,, \qquad\qquad k \geq 1\,,$$

and hence identify $\prod_k \wedge^k L = (\wedge U)^{\#}$. With this identification $\gamma_L$ is the isomorphism (6.11).

On the other hand, let $a \in \wedge^j L$. Then it follows from what was just observed that $\eta_L\gamma_L(a)$ vanishes on $\wedge^{>j} U$ and that

$$\eta_L\gamma_L(a)(\Phi) = \langle a, \Phi\rangle \qquad\qquad \text{if } \Phi \in \wedge^j U\,.$$

Since $\gamma_L$ is an isomorphism this implies that $\eta_L$ is one too.

In the general case as in §2.6 write $(\wedge V, d)$ as the increasing union of sub Sullivan algebras $(\wedge V_n, d)$ with finite dimensional homotopy Lie algebras $E(n+1)$. According to Lemma 2.17 (p. 73), $\widehat{UL} = \varprojlim_n U\widehat{E(n+1)}$. This identifies $\eta_L = \varprojlim_n \eta_{E(n+1)}$, and so $\eta_L$ is an isomorphism. $\square$

Next observe that a product is defined in the set $\pi_0(\wedge U)$ of augmentations in $\wedge U$ by

$$(f \star g)(\Phi) = (f \otimes g)(\Delta\Phi)\,, \qquad\qquad \Phi \in \wedge U\,,$$

and recall from (6.2) (p. 169) the linear isomorphism $\alpha : U \xrightarrow{\cong} V$ of degree 1. Recall also from Proposition 3.13 (p. 112) that the acyclic closure $(\wedge V \otimes \wedge U, d)$ decomposes as a $\Lambda$-extension of acyclic closures

$$(\wedge V^1 \otimes \wedge U^0, d) \to (\wedge V^1 \otimes \wedge U^0 \otimes \wedge V^{\geq 2} \otimes \wedge U^{\geq 1}, d) \xrightarrow{\varepsilon_0 \otimes id} (\wedge V^{\geq 2} \otimes \wedge U^{\geq 1}, \overline{d})$$

where $\varepsilon_0$ vanishes in $V^1$ and $U^0$. It follows at once from the definition of homotopy Lie algebras in §2.1 that $L_{\geq 1}$ is the homotopy Lie algebra of $(\wedge V^{\geq 2}, \overline{d})$.

**Proposition 6.4.** *Suppose $(\wedge V, d_1)$ is the associated quadratic Sullivan algebra of a Sullivan Lie algebra $E$. Then the diagonal $\Delta : \wedge U \to \wedge U \otimes \wedge U$ satisfies*

$$\Delta(a \cdot \Phi) = (id \otimes U\overline{\theta}(a))(\Delta\Phi)\,, \qquad\qquad a \in UE\,, \Phi \in \wedge U\,.$$

*proof:* Since $E$ is a Sullivan Lie algebra, the homotopy Lie algebra $L$ of $(\wedge V, d_1)$ is the completion $\widehat{E} = \varprojlim_n E/E^n$ of $E$. It follows that

$$\dim H^1(\wedge V, d_1) = \dim L_0/[L_0, L_0] = \dim E_0/[E_0, E_0] < \infty$$

and

$$\dim V^{i+1} = \dim L_i = \dim E_i < \infty\,, \qquad i \geq 1\,.$$

Thus by Theorem 6.1,

$$\eta_L : \widehat{UL} \xrightarrow{\cong} (\wedge U)^{\#}\,.$$

We now apply Proposition 6.3 (i) and (ii) to obtain for $b, c \in \widehat{UL}$ and $a \in UE$ that

$$\langle \eta_L(bca), \Phi\rangle = \langle \eta_L(bc), a\cdot\Phi\rangle = \langle \eta_L(b)\otimes\eta_L(c), \Delta(a\cdot\Phi)\rangle\,.$$

Similarly, writing $\Delta\Phi = \sum \Phi_i \otimes \Psi_i$, we have

$$\begin{aligned}\langle \eta_L(bca), \Phi\rangle &= \langle \eta_L(b)\otimes\eta_L(ca), \Delta\Phi)\\ &= \textstyle\sum_i (-1)^{\deg ca\cdot \deg\Phi_i}\langle\eta_L(b),\Phi_i\rangle\,\langle\eta_L(c), U\overline{\theta}(a)\Psi_i\rangle\\ &= \textstyle\sum_i (-1)^{\deg a\cdot\deg\Phi_i}\langle\eta_L(b)\otimes\eta_L(c), \Phi_i\otimes U\overline{\theta}(a)\Psi_i\rangle\\ &= \langle\eta_L(b)\otimes\eta_L(c), (id\otimes U\overline{\theta}(a))\Delta\Phi\rangle\,.\end{aligned}$$

Since only $0 \in \wedge U\otimes\wedge U$ is annihilated by $(\wedge U)^{\#}\otimes(\wedge U)^{\#}$, the Proposition follows.

□

**Proposition 6.5.** *Suppose $\dim H^1(\wedge V, d) < \infty$. Then for $x \in L$,*

*(i)* $\eta_L(x)(u) = -\langle\alpha(u), x\rangle\,, u \in U\,,$ *and* $\eta_L(x)(\wedge^{\geq 2}U) = 0$. *In particular,* $\eta_L : L \xrightarrow{\cong} U^{\#}$.

*(ii)* $\eta_L$ *restricts to a product preserving bijection* $G_L \xrightarrow{\cong} \pi_0(\wedge U)$. *In particular,* $\pi_0(\wedge U)$ *is a group and this is an isomorphism of groups.*

*(iii) The restriction of* $\eta_L$ *to* $UL_{\geq 1}$ *satisfies*

$$\eta_L = (\varepsilon_0\otimes id)^{\#}\circ\eta_{L_{\geq 1}} : UL_{\geq 1} \to (\wedge U)^{\#}\,.$$

*proof.* (i) Suppose $x \in L$ and $u \in U$. Then it follows from (6.2) (p. 169) that

$$d(1\otimes u) = \alpha(u)\otimes 1 + \sum v_i\otimes\Phi_i + \Omega$$

with $\Phi_i \in \wedge^+ U$ and $\Omega \in \wedge^{\geq 2}V\otimes\wedge U$. Thus by (4.6)

$$\overline{\theta}(x)(u) = -\langle\alpha(u), x\rangle - \sum\langle v_i, x\rangle\Phi_i$$

and in particular $\varepsilon_{\wedge U}\overline{\theta}(x) = -\langle \alpha(u), x\rangle$ . Moreover, since $\overline{\theta}(x)$ is a derivation it maps $\wedge^{\geq 2}U$ into $\wedge^+ U$ and so $\eta_L(x)(\wedge^{\geq 2}U) = 0$.

(ii) Elements $a \in G_L$ are characterized by $a = 1 + b$, $b \in I_L$ and $\Delta_L(a) = a \otimes a$. On the other hand, for $f_1, f_2, f_3 \in [\mathrm{Hom}(\wedge U, \Bbbk)]^0$,

$$\mu^{\#}_{\wedge U}(f_1)(a\otimes b) = f_1(ab) \qquad \text{and} \qquad (f_2\widehat{\otimes} f_3)(a\otimes b) = f_2(a)f_3(b)\,.$$

Thus the augmentations $f \in \mathrm{Hom}_0(\wedge U, \Bbbk)$ are characterized by $f(1) = 1$ and $\mu^{\#} f = f \otimes f$. Since by Proposition 6.3, $\eta_L$ preserves diagonals, (ii) follows.

(iii) If $a \in UL_{\geq 1}$, then the holonomy action of $a$ in $\wedge U$ commutes with multiplication by $U^0$ and so if $\Phi \in \wedge^+ U^0 \otimes \wedge U^+$ then $\eta_L(a)(\Phi) = \varepsilon_{\wedge U}(a\cdot\Phi) = 0$. This gives (iii).

□

Now recall from Lemma 2.20 (p. 82) that if $\dim H^1(\wedge V, d) < \infty$ and $\dim V^i < \infty$, $i \geq 2$ then multiplication defines an isomorphism of Hopf algebras

$$\widehat{UL_0} \otimes UL_{\geq 1} \xrightarrow{\cong} \widehat{UL}\,,$$

which restricts to an inclusion

$$\Bbbk[G_L]\otimes UL_{\geq 1} \to \widehat{UL}\,.$$

By Proposition 6.5 this gives the commutative diagram

$$\begin{array}{ccc} \Bbbk[G_L]\otimes UL_{\geq 1} & \xrightarrow{\text{mult}} & \widehat{UL} \\ \Big\downarrow{\scriptstyle \eta_{L_0}\otimes\eta_{L_{\geq 1}}} & & \cong\Big\downarrow{\scriptstyle \eta_L} \\ (\wedge U^0)^{\#}\otimes(\wedge U^{\geq 1})^{\#} & \xrightarrow[\cong]{\text{mult}} & (\wedge U)^{\#}\,. \end{array} \tag{6.12}$$

## 6.3 The graded Hopf algebra, $H_*(|\wedge U|;\mathbb{Q})$

*In this section,* $\Bbbk = \mathbb{Q}$.

Fix a minimal Sullivan algebra $(\wedge V, d)$ and denote by $(\wedge V\otimes\wedge U, d)$ its acyclic closure, and by $|\wedge U|$ the spatial realization of $(\wedge U, 0)$ - cf. §1.6. By Propositions 1.3 (p. 11) and 1.12 (p. 26) the spatial realization $|\wedge U\otimes\wedge U|$ of $\wedge U\otimes\wedge U$ is the product $|\wedge U|\times|\wedge U|$. Thus the diagonal $\Delta : \wedge U\to\wedge U\otimes\wedge U$ and the canonical involution $\omega : \wedge U\to\wedge U$ constructed in §6.1, together with the augmentation $\varepsilon_{\wedge U} : \wedge U\to\mathbb{Q}$ and the multiplication $\mu_{\wedge U}$, realize to continuous maps

$$|\Delta| : |\wedge U|\times|\wedge U|\to|\wedge U|\,,\quad |\omega| : |\wedge U|\to|\wedge U|\,,\quad |\varepsilon_{\wedge U}| : pt\to|\wedge U|\,,$$

$$\text{and } |\mu_{\wedge U}| : |\wedge U| \to |\wedge U| \times |\wedge U| .$$

Thus $|\mu_{\wedge U}|(z) = (z,z)$, $z \in |\wedge U|$ and it follows from Proposition 6.2 (p. 174) that $|\Delta|$ makes $|\wedge U|$ into a topological group with inversion map $|\omega|$ and identity $|\varepsilon_{\wedge U}|$. In particular these maps make $H_*(|\wedge U|;\mathbb{Q})$ into a graded Hopf algebra. We denote this simply by $H_*(|\wedge U|)$.

Observe now that the natural morphism

$$m_{|\wedge U|} : (\wedge U, 0) \to A_{PL}(|\wedge U|)$$

(§1.6, p. 26) induces a homomorphism $\wedge U \to H^*(|\wedge U|)$ which then dualizes to a linear map

$$H_*(m_{|\wedge U|}) : H_*(|\wedge U|) \to (\wedge U)^\# .$$

Because the multiplication in $|\wedge U|$ and in $(\wedge U)^\#$ are defined in terms of $\Delta$ it follows at once that $H_*(m_{|\wedge U|})$ is a morphism of graded algebras.

Recall now that if $\dim H^1(\wedge V, d) < \infty$ and $\dim V^i < \infty$, $i \geq 2$ then the transformation map $\eta_L : \widehat{UL} \xrightarrow{\cong} (\wedge U)^\#$ restricts to an inclusion

$$\widehat{\eta_L} : \mathbb{Q}[G_L] \otimes UL_{\geq 1} \hookrightarrow (\wedge U)^\#$$

of graded Hopf algebras, where $L$ is the homotopy Lie algebra of $(\wedge V, d)$. Our first objective is to prove

**Theorem 6.2.** *If $H^1(\wedge V, d)$ and $V^i$, $i \geq 2$ are finite dimensional, then there is a unique isomorphism of graded Hopf algebras,*

$$\rho_{\wedge U} : \mathbb{Q}[G_L] \otimes UL_{\geq 1} \xrightarrow{\cong} H_*(|\wedge U|) ,$$

*satisfying $H_*(|m_{\wedge U}|) \circ \rho_{\wedge U} = \widehat{\eta_L}$.*

*proof:* First observe that the 0-simplices in $|\wedge U|$ are the augmentations $\varepsilon : \wedge U \to \mathbb{Q}$ and that $H_*(m_{|\wedge U|})([\varepsilon]) = \varepsilon$. Moreover, since 1 is the only cycle of degree 0 in $(A_{PL})_n$ - cf §1.4, it follows that in $|\wedge U^0|$ the only simplices are the 0-simplices and so $|\wedge U^0|$ is the discrete set of its 0-simplices. Moreover, since $\wedge U = \wedge U^0 \otimes \wedge U^{\geq 1}$ it follows that

$$|\wedge U| = |\wedge U^0| \times |\wedge U^{\geq 1}| = \coprod_{\varepsilon \in |\wedge U^0|} \{\varepsilon\} \times |\wedge U^{\geq 1}| .$$

In particular, $\pi_0(|\wedge U|)$ is mapped bijectively to the set $\pi_0(\wedge U)$ of augmentations of $\wedge U$ and so Proposition 6.5 provides isomorphisms

$$\mathbb{Q}[G_L] \xrightarrow[\eta_L]{\cong} \mathbb{Q}[\pi_0(\wedge U)] \xleftarrow[H_*(m_{|\wedge U|})]{\cong} \mathbb{Q}[\pi_0(|\wedge U|)] . \qquad (6.13)$$

On the other hand, because $U^{\geq 1}$ has finite type, Theorem 1.5 (p. 42) asserts that $(\wedge U^{\geq 1}, 0) \xrightarrow{\simeq} A_{PL}(|\wedge U^{\geq 1}|)$. Together with Proposition 6.5 and Theorem 6.1, this implies that the morphisms

$$UL_{\geq 1} \xrightarrow[\eta_{L_{\geq 1}}]{\cong} (\wedge U^{\geq 1})^{\#} \xleftarrow[H(m_{|\wedge U^{\geq 1}|})]{\cong} H_*(|\wedge U^{\geq 1}|)$$

are also isomorphisms. Since $|\wedge U|$ is a topological group, multiplication is an isomorphism

$$\mathbb{Q}[\pi_0(|\wedge U|)] \otimes H_*(|\wedge U^{\geq 1}|; \mathbb{Q}) \xrightarrow{\cong} H_*(|\wedge U|)\,.$$

Now the Theorem follows from (6.12).

□

Now suppose $(\wedge V \otimes \wedge Z, d)$ is a $\Lambda$-extension of the minimal Sullivan algebra $(\wedge V, d)$ and recall the homomorphism $H(\Delta_{\wedge Z})$ of §6.1 (p. 173). It is convenient to use the interchange isomorphism to regard $H(\Delta_{\wedge Z})$ as a homomorphism

$$H(\Delta_{\wedge Z}) : H(\wedge Z, \overline{d}) \to \wedge U \otimes H(\wedge Z, \overline{d})\,.$$

Thus we may define

$$\xi : H_*(|\wedge U|) \otimes H(\wedge Z, \overline{d}) \to H(\wedge Z, \overline{d})$$

as the composite

$$\xi : H_*(|\wedge U| \otimes H(\wedge Z, \overline{d}) \xrightarrow{H_*(m_{|\wedge U|}) \otimes H(\Delta_{\wedge Z})} (\wedge U)^{\#} \otimes \wedge U \otimes H(\wedge Z, \overline{d})$$

$$\xrightarrow{\langle\, ,\, \rangle \otimes id} H(\wedge Z, \overline{d})\,.$$

Now recall from §4.3 that associated with a $\Lambda$-extension $(\wedge V, d) \to (\wedge V \otimes \wedge Z, d)$ is the holonomy representation of $\mathbb{Q}[G_L] \otimes UL_{\geq 1}$ in $H(\wedge Z, \overline{d})$.

**Theorem 6.3.** *Suppose $H^1(\wedge V, d)$ and $V^i$, $i \geq 2$, are finite dimensional. Then the isomorphism $\rho_{\wedge U} : \mathbb{Q}[G_L] \otimes UL_{\geq 1} \xrightarrow{\cong} H_*(|\wedge U|)$ identifies $\xi$ with the holonomy representation of $\mathbb{Q}[G_L] \otimes UL_{\geq 1}$ in $H(\wedge Z)$.*

*proof:* Let $\xi' : \widehat{UL} \otimes H(\wedge Z, \overline{d}) \to H(\wedge Z, \overline{d})$ be the composite

$$\xi' : \widehat{UL} \otimes H(\wedge Z, \overline{d}) \xrightarrow{\eta_L \otimes H(\Delta_{\wedge Z})} (\wedge U)^{\#} \otimes \wedge U \otimes H(\wedge Z, \overline{d})$$

$$\xrightarrow{\langle\, ,\, \rangle \otimes id} H(\wedge Z, \overline{d})\,.$$

Then it is immediate from the definitions that $\rho_{\wedge U}$ identifies $\xi$ with the restriction of $\xi'$ to $(\mathbb{Q}[G_L]\otimes UL_{\geq 1})\otimes H(\wedge Z,\overline{d})$. Thus we have only to show that $\xi'$ is the holonomy representation, and this follows at once, if we establish.

$$(\eta_L(a)\otimes id)\circ H(\Delta_{\wedge Z}) = \left(\varepsilon_U\otimes \widehat{U\overline{\theta}_{\wedge Z}(a)}\right)\circ H(\Delta_{\wedge Z}),\quad a\in\widehat{UL}\,. \tag{6.14}$$

As always it is sufficient to show this for $a\in UL$.

Since by Lemma 4.8 (p. 131), $U\overline{\theta}$ is a representation of $UL$ in $H(\wedge Z,\overline{d})$, (6.14) will be proved if we can show that for $b\in UL$, $x\in L$ we have

$$\left[\,(\eta_L(xb)\otimes id) - (id\otimes\overline{\theta}_{\wedge Z}(x))\circ(\eta_L(b)\otimes id)\,\right]\circ H(\Delta_{\wedge Z}) = 0\,. \tag{6.15}$$

For this, as in §6.1, let $\sigma_{\wedge Z} : (\wedge Z,\overline{d})\to(\wedge V\otimes\wedge U\otimes\wedge Z,d)$ be a right inverse for the quasi-isomorphism $\varepsilon\otimes id$ and set

$$\psi = (\varepsilon_V\otimes id)\circ\sigma_{\wedge V} : (\wedge Z,\overline{d})\to(\wedge U,0)\otimes(\wedge Z,\overline{d})\,.$$

Then denote by $\overline{\theta}$, $\overline{\theta}_{\wedge Z}$, and $\widehat{\theta}$ the holonomy representations of $L$ in $\wedge U$, $H(\wedge Z,\overline{d})$ and $\wedge U\otimes H(\wedge Z,\overline{d})$. Thus,

$$\widehat{\theta}(x) = \overline{\theta}(x)\otimes id + id\otimes\overline{\theta}_{\wedge Z}(x)\,,\qquad x\in L\,,$$

and by Lemma 4.7 (p. 130),

$$\widehat{\theta}(x)H(\psi)\beta = 0\,,\qquad \beta\in H(\wedge Z,\overline{d})\,.$$

Now fix $\beta\in H(\wedge Z,\overline{d})$ and write $H(\psi)\beta = \sum\Phi_i\otimes\beta_i$. Thus

$$\sum_i\overline{\theta}(x)\Phi_i\otimes\beta_i = -(id\otimes\overline{\theta}_{\wedge Z}(x))(\sum\Phi_i\otimes\beta_i)\,.$$

For $b\in UL$ apply $U\overline{\theta}(b)$ to both sides to obtain

$$\sum_i U\overline{\theta}(bx)\Phi_i\otimes\beta_i = (-1)^{\deg x\cdot\deg b}(id\otimes\overline{\theta}_{\wedge Z}(x))(\sum U\overline{\theta}(b)\Phi_i\otimes\beta_i)\,.$$

Then since $H(\Delta_{\wedge Z}) = (\omega\otimes id)\circ\psi$ and since $\eta_L\circ\omega_L = \omega^{\#}\circ\eta_L$ it follows that

$$\begin{aligned}(\eta_L(xb)\otimes id)(H(\Delta_{\wedge Z})\beta) &= \textstyle\sum_i(-1)^{\deg b\cdot\deg x}\eta_L(\omega_L b\,\omega_L x)(\Phi_i)\otimes\beta_i\\ &= \textstyle\sum_i(-1)^{\deg b\cdot\deg x}\varepsilon_{\wedge U}\big((U\overline{\theta}(\omega_L b)\overline{\theta}(\omega_L x))(\Phi_i)\otimes\beta_i\big)\\ &= (1\otimes\overline{\theta}_{\wedge Z}x)(\varepsilon_{\wedge U}\overline{\theta}(b)\otimes 1)(\omega\Phi_i\otimes\beta_i)\\ &= (id\otimes\overline{\theta}_{\wedge Z}(x))\circ(\eta_L(b)\otimes id)(H(\Delta_{\wedge Z}\beta))\,.\end{aligned}$$

This completes the proof of (6.15), and with it of Theorem 6.3.

□

## 6.4 Connecting Sullivan algebras with topological spaces

*Throughout this section we consider a path connected CW complex $(Y,*)$ and a fixed Sullivan model $m_Y : (\wedge V, d) \to A_{PL}(Y)$ from a minimal Sullivan algebra with acyclic closure $(\wedge V \otimes \wedge U, d)$ as defined in §6.1, and we fix a continuous map $h : (X,*) \to (Y,*)$ from a path connected space $X$.*

Recall from §3.1 that $MY$ denotes the space of pairs $(f,s)$ in which $f : \mathbb{R} \to Y$ is a continuous map satisfying $f(t) = f(0)$, $t \leq 0$ and $f(t) = f(s)$, $t \geq s$. Denote by $p_L$ and $p_R : MY \to Y$ the maps $(f,s) \mapsto f(0)$ and $(f,s) \mapsto f(s)$. We use $p_L$ and $p_R$ respectively to define $- \times_Y MY$ and $MY \times_Y -$.

In §3.1 we associated with the continuous map $h$ the commutative diagram

$$\begin{array}{ccccc} & & X & & \\ & \swarrow^{h} & \downarrow{\scriptstyle \ell} & & \\ Y & \xleftarrow[q]{} & X \times_Y MY & \xleftarrow[j]{} & X \times_Y P'Y\,, \end{array} \tag{6.16}$$

in which $q(x,(f,s)) = f(s)$, and $P'Y$ is the subspace of $MY$ of pairs $(f,s)$ with $f(s) = *$. The row in the diagram is a fibration and $\ell$ is the homotopy equivalence given by $\ell(x) = (x, c_{h(x)})$, $c_{h(x)}$ denoting the path of length 0 at $h(x)$. We say we have *converted $h$ to this fibration.* When $h$ is the inclusion $* \to Y$ this is the *path space fibration,*

$$Y \leftarrow PY \leftarrow \Omega Y\,,$$

in which $PY$ is the subspace of $MY$ of paths $(f,s)$ with $f(0) = *$ and $\Omega Y$ is the subspace of paths satisfying $f(0) = * = f(s)$.

As in §3.1 if $(f,s) \in MY$ we write $|f| = s$ and denote $(f,s)$ simply by $f$ with the understanding that $|f|$ is defined. Thus a continuous map $MY \times_Y MY \to MY : (f,g) \mapsto f \bullet g$, is defined by

$$(f \bullet g)(t) = \begin{cases} f(t)\,, & t \leq |f| \\ g(t - |t|)\,, & t \geq |f|\,, \end{cases}$$

with $|f \bullet g| = |f| + |g|$.

In particular, as described in §4.1, a continuous map

$$\Phi_X : X \times_Y MY \times_Y P'Y \longrightarrow X \times_Y P'Y \tag{6.17}$$

is given by $((x,f),g) \mapsto (x, f \bullet g)$, and $\Phi_X$ restricts to the *holonomy action,*

$$\mu_X : (X \times_Y P'Y) \times \Omega Y \to X \times_Y P'Y\,,$$

which is a right action of the topological monoid on the fibre $X \times_Y P'Y$ of (6.16).

Our principal objective is to show that the algebraic constructions in §6.1 correspond directly to the topological constructions above. For this, as in §3.7, we extend $m_Y$ to the commutative diagrams below,

$$\begin{array}{ccccc} (\wedge V, d) & \longrightarrow & (\wedge V \otimes \wedge Z, d) & \longrightarrow & (\wedge Z, \overline{d}) \\ \simeq \big\downarrow m_Y & & \simeq \big\downarrow m_X & & \big\downarrow m_F \\ A_{PL}(Y) & \xrightarrow{A_{PL}(q)} & A_{PL}(X \times_Y MY) & \xrightarrow{A_{PL}(j)} & A_{PL}(X \times_Y P'Y) \end{array}$$

and

$$\begin{array}{ccccc} (\wedge V, d) & \longrightarrow & (\wedge V \otimes \wedge U, d) & \longrightarrow & (\wedge U, 0) \\ \simeq \big\downarrow m_Y & & \simeq \big\downarrow m_P & & \big\downarrow m_\Omega \\ A_{PL}(Y) & \longrightarrow & A_{PL}(PY) & \longrightarrow & A_{PL}(\Omega Y)\,, \end{array}$$

noting that $m_F$ and $m_\Omega$ may not be quasi-isomorphisms. Our goal is

**Theorem 6.4.** *With the hypotheses and notation above and from §6.1 the diagram*

$$\begin{array}{ccc} (\wedge Z, \overline{d}) & \xrightarrow{\Delta_{\wedge Z}} & (\wedge Z, \overline{d}) \otimes \wedge U \\ \big\downarrow m_F & & \big\downarrow m_F \otimes m_\Omega \\ A_{PL}(X \times_Y P'Y) & \xrightarrow{A_{PL}(\mu_X)} & A_{PL}((X \times_Y P'Y) \times \Omega Y) \end{array}$$

*is homotopy commutative.*

**Corollary 6.1.** *If $Y$ is simply connected and $H(Y)$ has finite type then $\Delta_{\wedge Z}$ is a Sullivan representative for $\mu_X$.*

*proof:* In this case it follows from Theorem 5.1 (p. 145) that $m_F$ and $m_\Omega$ are quasi-isomorphisms and that $\wedge U$ has finite type. It follows that $m_F \otimes m_\Omega$ is a quasi-isomorphism as well.

□

Before proceeding to the proof of Theorem 6.4 we need to identify the topological analogue of the canonical involution $\omega$ of $\wedge U$, defined in §6.1. Now, as described in §4.1, an involution $\omega_M$ of $MY$ is given by

$$(\omega_M f)(t) = f(|t| - t) \quad \text{and } |\omega_M f| = |f|\,.$$

Then $\omega_M$ restricts to homeomorphisms $PY \xrightarrow{\cong} P'Y$ and $P'Y \xrightarrow{\cong} PY$, and also restricts to the involution $\omega_Y$ of $\Omega Y$ denoted by $f \mapsto f^{-1}$ in §4.1. We establish

**Proposition 6.6.** *With the notation and hypotheses above,*

$$m_\Omega \circ \omega \sim A_{PL}(\omega_Y) \circ m_\Omega \,.$$

*proof:* An argument similar to the proof that $q$ is a fibration shows that the map $p = (p_L, p_R) : MY \to Y \times Y$ sending $(f,s)$ to $(f(0), f(s))$ is a fibration; evidently the fibre at $(*,*)$ is the loop space $\Omega Y$. Moreover, the homotopy equivalence $\ell_Y : Y \to MY$ which maps $y$ to the path of length 0 at $y$ satisfies $p \circ \ell_Y = \Delta_Y : y \mapsto (y,y)$. By definition, the diagram

$$\begin{array}{ccc} \Omega Y & \xrightarrow{\omega_Y} & \Omega Y \\ \downarrow & & \downarrow \\ MY & \xrightarrow{\omega_M} & MY \\ \downarrow & & \downarrow \\ Y \times Y & \xrightarrow{T} & Y \times Y \end{array}$$

commutes, where $T(y_1, y_2) = (y_2, y_1)$.

Now recall from (6.1) (p. 169) that we fixed a Sullivan model

$$\mu : (\wedge V \otimes \wedge V \otimes \wedge U, d) \xrightarrow{\simeq} (\wedge V, d)$$

for the multiplication homomorphism $\mu_{\wedge V} : \wedge V \otimes \wedge V \to \wedge V$. Thus we may lift $m_Y \circ \mu$ through the quasi-isomorphism $A_{PL}(\ell_Y)$ to obtain the commutative diagram

$$\begin{array}{ccccc} (\wedge V \otimes \wedge V, d) & \longrightarrow & (\wedge V \otimes \wedge V \otimes \wedge U, d) & \longrightarrow & (\wedge U, 0) \\ \downarrow{\simeq} & & \downarrow{m} & & \downarrow{\overline{m}} \\ A_{PL}(Y \times Y) & \xrightarrow{A_{PL}(p)} & A_{PL}(MY) & \longrightarrow & A_{PL}(\Omega Y)\,. \end{array} \qquad (6.18)$$

On the other hand, recall from §6.1 that the interchange automorphism $\tau_{\wedge V}$ of $\wedge V \otimes \wedge V$ extends to a quasi-isomorphism $\tau : (\wedge V \otimes \wedge V \otimes \wedge U, d) \xrightarrow{\cong} (\wedge V \otimes \wedge V \otimes \wedge U, d)$ such that $\mu \circ \tau = \mu$. Thus we have

$$A_{PL}(\ell_Y) \circ m \circ \tau = m_Y \circ \mu \circ \tau = m_Y \circ \mu = A_{PL}(\ell_Y) \circ m = A_{PL}(\ell_Y) \circ A_{PL}(\omega_M) \circ m \,.$$

Let $\lambda : \wedge V \otimes \wedge V \to \wedge V \otimes \wedge V \otimes \wedge U$ be the inclusion. Then

$$m \circ \tau \circ \lambda = m \circ \lambda \circ \tau_{\wedge V} = A_{PL}(p) \circ (m_Y \otimes m_Y) \circ \tau_{\wedge V}$$

$$= A_{PL}(p) \circ A_{PL}(T) \circ (m_Y \otimes m_Y) = A_{PL}(\omega_M) \circ m \circ \lambda .$$

Since $A_{PL}(\ell_Y)$ is a quasi-isomorphism, it follows from the uniqueness assertion in Proposition 3.6 (p. 98) that

$$m \circ \tau \sim A_{PL}(\omega_M) \circ m \qquad \text{rel } \wedge V \otimes \wedge V .$$

Now by construction, after division by $(\wedge V \otimes \wedge U)^+$, $\tau$ induces the morphism $\overline{\chi}$ of Proposition 6.1 (p. 171). By that Proposition, $\overline{\chi} = \omega$. Thus dividing by $(\wedge V \otimes \wedge V)^+$ and projecting onto $A_{PL}(\Omega Y)$ we obtain

$$\overline{m} \circ \omega \sim \omega_Y \circ \overline{m} .$$

It remains to show that $\overline{m} \sim m_\Omega$. Restricting the fibration $p : MY \to Y \times Y$ to $* \times Y$ gives the path fibration $PY \to Y$. Thus applying $\varepsilon_{\wedge V} \otimes id \otimes id$ to (6.18) we obtain the commutative diagram,

$$\begin{array}{ccccc} (\wedge V, d) & \longrightarrow & (\wedge V \otimes \wedge U, d) & \longrightarrow & (\wedge U, 0) \\ {\scriptstyle m_Y} \downarrow \simeq & & {\scriptstyle \widehat{m}} \downarrow \simeq & & {\scriptstyle \overline{m}} \downarrow \\ A_{PL}(* \times Y) & \longrightarrow & A_{PL}(PY) & \longrightarrow & A_{PL}(\Omega Y) , \end{array}$$

in which necessarily $\widehat{m} \sim m_P$ rel $\wedge V$. Thus $\overline{m} \sim m_\Omega$.

□

*proof of Theorem 6.4:* Pull the product fibration $(X \times_Y MY) \times PY \to Y \times Y$ to the fibration

$$Y \leftarrow (X \times_Y MY) \times_Y PY \xleftarrow{\ i\ } (X \times_Y P'Y) \times \Omega Y$$

via the diagonal map $\Delta_Y : y \mapsto (y, y)$. Then define homotopy inverse maps

$$\sigma_X : ((X \times_Y MY) \times_Y PY) \to X \times_Y P'Y , \qquad \text{and}$$

$$\lambda_X : X \times_Y P'Y \to ((X \times_Y MY) \times_Y PY) ,$$

by setting

$$\sigma_X : ((x, f), g) \mapsto (x, f \bullet \omega_M g) \qquad \text{and } \lambda_X : (x, f) \mapsto ((x, f), c_*) ,$$

where $c_*$ is the path of length 0 at the base point, $*$. The fibration yields the commutative diagram

$$\begin{array}{ccccc} (\wedge V, d) & \longrightarrow & (\wedge V \otimes \wedge Z, d) \otimes_{\wedge V} (\wedge V \otimes \wedge U, d) & \xrightarrow{\varepsilon_{\wedge V} \otimes id \otimes id} & (\wedge Z, \overline{d}) \otimes \wedge U \\ \simeq \downarrow {\scriptstyle m_Y} & & \downarrow {\scriptstyle m_X \otimes m_P} & & \downarrow {\scriptstyle m_F \otimes m_\Omega} \\ A_{PL}(Y) & \longrightarrow & A_{PL}\left((X \times_Y MY) \times_Y PY\right) & \xrightarrow{A_{PL}(i)} & A_{PL}\left((X \times_Y P'Y) \times \Omega Y\right) . \end{array}$$

(*Note*: $m_X \otimes m_P$ is not always a quasi-isomorphism !)

Now, as in §6.1 write $(\wedge V\otimes\wedge Z, d)\otimes_{\wedge V}(\wedge V\otimes\wedge U, d) = (\wedge V\otimes\wedge Z\otimes\wedge U, d)$. Then it is immediate from the construction that

$$A_{PL}(\lambda_X)\circ(m_X\otimes m_P) = m_F\circ(\varepsilon_{\wedge V}\otimes id_{\wedge Z}\otimes\varepsilon_{\wedge U})\,.$$

Thus if $\sigma_{\wedge Z} : (\wedge Z, \overline{d}) \to (\wedge V\otimes\wedge Z\otimes\wedge U, d)$ is a right inverse for the quasi-isomorphism $\varepsilon_{\wedge V}\otimes id_{\wedge Z}\otimes\varepsilon_{\wedge U}$, it follows that

$$(m_X\otimes m_P)\circ\sigma_{\wedge Z}\sim A_{PL}(\sigma_X)\circ m_F\,.$$

This in turn implies that

$$(m_F\otimes m_\Omega)\circ(\varepsilon_{\wedge V}\otimes id\otimes id)\circ\sigma_{\wedge V}\sim A_{PL}(i)\circ A_{PL}(\sigma_X)\circ m_F\,.$$

But by the definition (6.17) of the holonomy action $\Phi_X$, $(id\times\omega_Y)\circ\Phi_X = \sigma_X\circ i$. Moreover, by the definition of $\Delta_{\wedge Z}$, $(id\otimes\omega)\circ\Delta_{\wedge Z} = (\varepsilon_{\wedge V}\otimes id\otimes id)\circ\sigma_{\wedge Z}$. Thus we obtain

$$(m_F\otimes(m_\Omega\circ\omega))\circ\Delta_{\wedge Z}\sim A_{PL}(id\times\omega_Y)\circ A_{PL}(\Phi_X)\circ m_F\,.$$

Since by Proposition 6.6, $m_\Omega\circ\omega\sim A_{PL}(\omega_Y)\circ m_\Omega$ and since $\omega$ is an isomorphism, this gives the Theorem.

□

**Corollary 6.2.** *The diagram*

$$\begin{array}{ccc}
(\wedge U, 0) & \xrightarrow{\quad\Delta\quad} & (\wedge U,0)\otimes(\wedge U,0)\\
\Big\downarrow{\scriptstyle m_\Omega} & & \Big\downarrow{\scriptstyle m_\Omega\otimes m_\Omega}\\
A_{PL}(\Omega Y) & \xrightarrow{A_{PL}(\mu_*)} & A_{PL}(\Omega Y\times\Omega Y)
\end{array}$$

*is homotopy commutative.*

In the proof of Theorem 6.4 we may interchange the factors in the fibre product $(X\times_Y MY)\times_Y PY$ and interchange the tensorands in the tensor product $(\wedge V\otimes\wedge Z)\otimes_{\wedge V}(\wedge V\otimes\wedge U)$, and thereby obtain

**Corollary 6.3.** *With the interchanges above the diagram*

$$\begin{array}{ccc}
(\wedge Z, \overline{d}) & \xrightarrow{\quad\Delta_{\wedge Z}\quad} & (\wedge U,0)\otimes(\wedge Z,\overline{d})\\
\Big\downarrow{\scriptstyle m_Z} & & \Big\downarrow{\scriptstyle m_Z\otimes m_\Omega}\\
A_{PL}(X\times_Y P'Y) & \xrightarrow{A_{PL}(\mu_X)} & A_{PL}(\Omega Y\times(X\times_Y P'Y))
\end{array}$$

*is homotopy commutative.*

In the rest of this section $\Bbbk = \mathbb{Q}$ and for any topological space we write $H_*(S;\mathbb{Q}) = H_*(S)$. In particular, we shall apply Theorem 6.4 to interpret the holonomy representation of $\mathbb{Q}[G_L]\otimes UL_{\geq 1}$ in $H(\wedge Z, \bar{d})$ (§4.3, p. 131) in terms of the linear map $H(\mu_X)$. *For this we shall assume that $H^1(\wedge V, d)$ and $V^i$, $i \geq 2$ are finite dimensional and denote by $L$ the homotopy Lie algebra of $(\wedge V, d)$.*

Observe first that, with the notation of Theorem 6.4, $H(m_\Omega) : \wedge U \to H(\Omega Y)$ dualizes to a linear map

$$H(m_\Omega)^\# : H_*(\Omega Y) \to (\wedge U)^\# .$$

**Proposition 6.7.** *With the hypotheses above and notation of Theorem 6.2, there is a morphism of graded Hopf algebras, $H_*(m_\Omega) : H_*(\Omega Y) \to \mathbb{Q}[G_L]\otimes UL_{\geq 1}$, such that $H(m_\Omega)^\#$ is the composite*

$$H(m_\Omega)^\# = \eta_L \circ H_*(m_\Omega) .$$

*proof:* Recall from Theorem 6.2 (p. 184) the canonical morphism $m_{|\wedge U|} : (\wedge U, 0) \to A_{PL}(|\wedge U|)$, and that $H(m_{|\wedge U|}) : (\wedge U, 0) \to H(|\wedge U|)$ dualizes to a morphism

$$H_*(m_{|\wedge U|}) = H(m_{|\wedge U|})^\# : H_*(|\wedge U|) \to (\wedge U)^\# ,$$

and that $H_*(m_{|\wedge U|})$ is an isomorphism of graded Hopf algebras from $H_*(|\wedge U|)$ to $\eta_L(\mathbb{Q}[G_L] \otimes H_*(UL_{\geq 1}))$.

On the other hand, $m_\Omega$ realizes to a map $|m_\Omega| : |Sing\,\Omega Y| \to |\wedge U|$. Use Corollary 6.2 to obtain a commutative diagram in cohomology, then dualize and restrict to $H_*(\Omega Y)\otimes H_*(\Omega Y) \subset H(\Omega Y \times \Omega Y)^\#$. The resulting diagram shows that $H_*(|m_\Omega|)$ is a morphism of graded Hopf algebras. But, according to Proposition 1.15 (p. 29), $A_{PL}(|m_\Omega|) \circ m_{|\wedge U|} \sim \lambda_{\Omega Y} \circ m_\Omega$, where $\lambda_{\Omega Y} : |Sing\,\Omega Y| \to \Omega Y$ is the canonical weak homotopy equivalence. Thus identifying $H(|Sing\,\Omega Y|) = H(\Omega Y)$, we obtain $H(m_\Omega) = H(|m_\Omega|) \circ H(m_{|\wedge U|})$. Now it follows from Corollary 1.7 (p. 30) that

$$H(m_\Omega)^\# = H(m_{|\wedge U|})^\# \circ H_*(|m_\Omega|) .$$

The Proposition follows with $H_*(m_\Omega) = H(m_{|\wedge U|})^\# \circ H_*(|m_\Omega|)$.

□

Now, given any two topological spaces $S$ and $T$ we denote by

$$a \otimes \beta \mapsto \langle a, \beta \rangle$$

the linear map $H_*(S) \otimes H(S\times T) \to H(T)$ given by

$$\langle b, \langle a, \beta\rangle\rangle = (-1)^{\deg b \cdot \deg a} \langle a \otimes b, \beta\rangle , \qquad b \in H_*(T), a \in H_*(S), \beta \in H(S\times T) .$$

Then, with the notation of Corollary 6.3, the *holonomy action* of $H_*(\Omega Y)$ on $H(X \times_Y P'Y)$ is given by

$$a \cdot \beta = \langle a, H(\mu_X)\beta \rangle \,, \qquad a \in H_*(\Omega Y), \beta \in H(X \times_Y P'Y) \,.$$

On the other hand, the *holonomy representation* of $\mathbb{Q}[G_L] \otimes UL_{\geq 1}$ in $H(\wedge Z, \overline{d})$, defined in §4.3 (p. 131), is given by

$$a \cdot \beta = (\widehat{U\overline{\theta}})(a)\beta \,, \qquad a \in \mathbb{Q}[G_L] \otimes UL_{\geq 1}, \beta \in H(\wedge Z, \overline{d}) \,.$$

**Theorem 6.5.** *Suppose $(\wedge V, d)$ is a minimal Sullivan algebra and that $\dim H^1(\wedge V, d)$ and $\dim V^i$, $i \geq 2$ are finite. Then, with the notation and hypotheses established at the start of this section, the holonomy representation of $\mathbb{Q}[G_L] \otimes UL_{\geq 1}$ is connected with the holonomy action of $H_*(\Omega Y)$ on $H(X \times_Y P'Y)$ by the relation*

$$H(m_F)(H_*(m_\Omega)(a) \cdot \beta) = a \cdot H(m_F)\beta \,, \qquad a \in H_*(\Omega Y), \beta \in H(\wedge Z, \overline{d}) \,.$$

*proof:* In analogy with the map $\langle \, , \rangle : H_*(S) \otimes H(S \times T) \to H(T)$ we denote by

$$\langle \, , \rangle : (\wedge U)^\# \otimes (\wedge U \otimes H(\wedge Z, \overline{d})) \to H(\wedge Z, \overline{d})$$

the map $\gamma \otimes \Phi \otimes \beta \mapsto \langle \gamma, \Phi \rangle \beta$. Then by Proposition 6.7 and Theorem 6.4,

$$(H_*(m_\Omega)a) \cdot \beta = \langle H(m_\Omega)^\# a, H(\Delta_{\wedge Z})\beta \rangle \,, \qquad a \in H_*(\Omega Y), \beta \in H(\wedge Z, \overline{d}) \,.$$

Now it follows from Corollary 6.3 that

$$H(m_F)\left((H_*(m_\Omega)a) \cdot \beta\right) = \langle a, H(\mu_X)H(m_F)\beta \rangle = a \cdot H(m_F)\beta \,.$$

□

**Remark.** Theorem 6.5 generalizes the second assertion of Theorem 4.2 (p. 137) from $\mathbb{Q}[G_L]$ to $\mathbb{Q}[G_L] \otimes UL_{\geq 1}$.

# Chapter 7

# Sullivan spaces

*In this chapter, all groups are discrete. Recall from §1.4 that for any topological space, $X$, $H^*(X;\Bbbk) = H(A_{PL}(X))$ is denoted by $H(X)$.*

## 7.1 Sullivan spaces

Suppose $\varphi : (\wedge V, d) \to A_{PL}(X)$ is a morphism from a minimal Sullivan algebra to the forms on a path connected based space $(X, *)$. In §1.8 we constructed linear maps

$$\pi_k(\varphi) : \pi_k(X) \otimes_{\mathbb{Z}} \Bbbk \to \pi_k(\wedge V, d)\,, \qquad k \geq 2\,.$$

If $X$ is simply connected and if $H(X)$ is a graded space of finite type then Theorem 1.6 (p. 43) asserts that if $\varphi$ is a quasi-isomorphism then the linear maps $\pi_k(\varphi)$, $k \geq 2$, are all isomorphisms.

This may no longer be true if $X$ is not simply connected. When it is true, however, then the Sullivan model of $X$ does reflect the homotopy invariants of $X$, and so we make the

**Definition.** A *Sullivan space* is a path connected based space $(X, *)$ with universal covering space $(\widetilde{X}, *)$ such that

(i) $\dim H^1(X) < \infty$ and $\dim H^k(\widetilde{X}) < \infty$, $k \geq 2$;
(ii) A minimal model $m_X : (\wedge V, d) \xrightarrow{\simeq} A_{PL}(X)$ satisfies

$$\pi_k(m_X) : \pi_k(X) \otimes \mathbb{Q} \xrightarrow{\cong} \pi_k(\wedge V, d)\,, \qquad k \geq 2\,.$$

Note that (i) and (ii) are independent of the choice of $\Bbbk$.

The key additional ingredient in the study of non-simply connected spaces $(X, *)$ is the fundamental group $\pi_1(X)$, which may be any discrete group $G$. Recall from §3.2 (p. 93) that a classifying space for $G$ is a based CW complex $(B,*)$ such that $\pi_i(B,*)=0$, $i\geq 2$, together with an isomorphism $G\xrightarrow{\cong}\pi_1(B,*)$. Then we have

**Theorem 7.1.** *A connected CW complex $(X,*)$ is a Sullivan space if and only if*

*(i) $\dim H^1(X)<\infty$ and $\dim H^k(\widetilde{X})<\infty$, $k\geq 2$;*
*(ii) The minimal Sullivan model for the classifying space $B$ of $\pi_1(X,*)$ is a Sullivan 1-algebra $(\wedge V^1,d_1)$;*
*(iii) The action by covering transformations of $\pi_1(X,*)$ in $H(\widetilde{X})$ is locally nilpotent.*

Now suppose $(X,*)$ is a connected CW complex and that $(B,*)$ is a classifying space for $\pi_1(X,*)$. Then, as described in §3.2, there is a classifying space fibration

$$(B,*)\xleftarrow{q}(X\times_B MB,*)\xleftarrow{j}(F,*)\tag{7.1}$$

and (Proposition 3.2, p. 93), $x\mapsto(x,c_*)$ defines a based homotopy equivalence

$$\ell:(X,*)\to(X\times_B MB,*)\tag{7.2}$$

such that $\pi_1(q)$ is an isomorphism. Thus $F$ is simply connected, and $\pi_k(j)$ is an isomorphism for $k\geq 2$. Moreover, by Lemma 4.3 (p. 124) $F$ and $\widetilde{X}$ have the same weak homotopy type.

Now let $m_B:(\wedge V^1\otimes\wedge W^{\geq 2},d)\xrightarrow{\simeq}A_{PL}(B)$ be a Sullivan minimal model, so that the restriction $m:(\wedge V^1,d)\to A_{PL}(B)$ satisfies $H^1(m)$ is an isomorphism and $H^2(m)$ is injective. On the other hand, since $F$ is simply connected a classical Serre spectral sequence argument shows that $H^1(q)$ is an isomorphism and $H^2(q)$ is injective. Thus we may extend $m_B$ to the commutative diagram

$$\begin{array}{ccccc}(\wedge V^1\otimes\wedge W^{\geq 2},d) & \longrightarrow & (\wedge V^1\otimes\wedge W^{\geq 2}\otimes\wedge Z^{\geq 2},d) & \longrightarrow & (\wedge Z^{\geq 2},\overline{d})\\ \simeq\big\downarrow m_B & & \simeq\big\downarrow\varphi & & \big\downarrow\overline{\varphi}\\ A_{PL}(B) & \longrightarrow & A_{PL}(X\times_B MB) & \longrightarrow & A_{PL}(F)\,.\end{array}\tag{7.3}$$

In particular we may extend $m$ to the commutative diagram

$$\begin{array}{ccccc} (\wedge V^1,d) & \longrightarrow & (\wedge V,d) & \xrightarrow{\quad\zeta\quad} & (\wedge V^{\geq 2},\bar{d}) \\ \Big\downarrow m & & \simeq\Big\downarrow m_X & & \Big\downarrow \overline{m} \\ A_{PL}(B) & \xrightarrow{A_{PL}(q)} & A_{PL}(X\times_B MB) & \xrightarrow{A_{PL}(j)} & A_{PL}(F)\,, \end{array} \tag{7.4}$$

in which $A_{PL}(\ell)\circ m_X$ is a minimal Sullivan model for $X$. Theorem 7.1 is then contained in

**Theorem 7.2.** *With the notation above, if $(X,*)$ is a Sullivan CW complex then*

*(i) $m$ is a quasi-isomorphism and $H(B)$ is a graded vector space of finite type.*
*(ii) $\overline{m}$ is a quasi-isomorphism*
*(iii) $\dim H^1(X)<\infty$ and $H(\widetilde{X})$ is a graded vector space of finite type.*
*(iv) $H(X)$ is a graded vector space of finite type*
*(v) $G=\pi_1(X,*)$ acts nilpotently via covering transformation in each $H^k(\widetilde{X})$.*

*Conversely, if $(X,*)$ is a path connected, based CW complex for which $m$ is a quasi-isomorphism. If (v), and one of (iii) or (iv) holds, or if (ii) and (iii) hold, then $(X,*)$ is a Sullivan space.*

**Corollary 7.1.** *Let $(\wedge V,d)$ be the minimal Sullivan model of a Sullivan space. Then*

*(i) $\dim H^1(\wedge V,d)<\infty$ and $\dim V^i<\infty\,,\quad i\geq 2\,.$*
*(ii) The homotopy Lie algebra, $L$, of $(\wedge V,d)$ is a Sullivan Lie algebra.*
*(iii) The morphism*

$$\eta_L:\widehat{UL}\longrightarrow(\wedge U)^\#$$

*of §6.2 is an isomorphism, where $(\wedge V\otimes\wedge U,d)$ is the acyclic closure of $(\wedge V,d)$.*

*proof:* (i) and (ii) are immediate from Theorem 7.2 and (iii) follows from Theorem 6.1 (p. 180). □

*proof of Theorem 7.2*: We abuse notation and write $m_X=A_{PL}(\ell)\circ m_X:(\wedge V,d)\xrightarrow{\cong}A_{PL}(X)$; this is then a minimal Sullivan model for $X$. Thus the isomorphisms $\pi_k(j)\otimes lk$ and $\pi_k(\zeta)$, $k\geq 2$, identify $\pi_k(m_X)$ with $\pi_k(\overline{m})$.

On the other hand, (7.4) gives an isomorphism $H^1(B) \cong H^1(X)$ while Lemma 4.3 (p. 124) gives an isomorphism $H(F) \cong H(\widetilde{X})$, which identifies the action by covering transformations of $\pi_1(X,*)$ in $H(\widetilde{X})$ with the holonomy representation in $H(F)$. Thus $(X,*)$ is a Sullivan space if and only if

$$\left.\begin{aligned} &\dim H^1(B) < \infty\,,\\ &H(F) \text{ is a graded vector space of finite type, and}\\ &\pi_k(\overline{m}) \text{ is an isomorphism, } k \geq 2\,. \end{aligned}\right\} \qquad (7.5)$$

Now suppose $(X,*)$ is a Sullivan CW complex, and let $n : (\wedge Z, d) \xrightarrow{\simeq} A_{PL}(F)$ be a minimal Sullivan model. Then the Lifting Lemma 1.2 (p. 19) gives a homotopy commutative diagram

$$\begin{array}{ccc} (\wedge Z, d) & \xrightarrow[n]{\simeq} & A_{PL}(F) \\ \psi \Big\uparrow & \nearrow_{\overline{m}} & \\ (\wedge V^{\geq 2}, \overline{d})\,. & & \end{array}$$

Since $H(F)$ has finite type, Theorem 1.6 (p. 43) asserts that each $\pi_k(n)$ is an isomorphism. By (7.5) each $\pi_k(\overline{m})$ is an isomorphism. It follows that $\psi$ is an isomorphism and $\overline{m}$ is a quasi-isomorphism. This proves (ii) and (iii).

On the other hand, it follows from Proposition 4.4 (p. 132) that the fundamental group $G_L$ for the Sullivan algebra $(\wedge V, d)$ acts nilpotently in each $H^k(\wedge V^{\geq 2}, d)$. Thus, since $\overline{m}$ is a quasi-isomorphism, Theorem 4.2 (p. 137) asserts that $G$ acts nilpotently in each $H^k(F)$, thereby establishing (v).

Next, in the diagram (7.3) note that because $H(F)$ has finite type and $G$ acts nilpotently in each $H^k(F)$, Theorem 5.1 (p. 145) asserts that $\overline{\varphi}$ is a quasi-isomorphism.

Let $\rho : (\wedge W^{\geq 2} \otimes \wedge Z^{\geq 2}, \overline{d}) \to (\wedge Z^{\geq 2}, \overline{d})$ be the surjection $k\!\!k \otimes_{\wedge W} -$. Now, Corollary 3.2 (p. 99) extends $id_{\wedge V^1}$ to a morphism

$$\chi : (\wedge V^1 \otimes \wedge W^{\geq 2} \otimes \wedge Z^{\geq 2}, d) \to (\wedge V^1 \otimes \wedge V^{\geq 2}, d)$$

such that $m_X \circ \chi \sim \varphi$ and the induced morphism

$$\overline{\chi} : (\wedge W^{\geq 2} \otimes \wedge Z^{\geq 2}, \overline{d}) \to (\wedge V^{\geq 2}, \overline{d})$$

satisfies $\overline{m} \circ \overline{\chi} \sim \overline{\varphi} \circ \rho$. Since $m_X$ and $\varphi$ are quasi-isomorphisms so is $\chi$. Thus it follows from Proposition 3.10 (p. 107) that $\overline{\chi}$ is a quasi-isomorphism. Since $\overline{m}$ and $\overline{\varphi}$ are also quasi-isomorphisms, so is $\rho$. Now Proposition 3.3 (p. 96) gives $W = 0$, and therefore $m = m_B$ is a quasi-isomorphism.

Moreover, by (7.5) $\dim H^1(B) < \infty$. It follows from the construction of minimal Sullivan models that $V^1$ is at most countably infinite dimensional. This is then true for $H(\wedge V^1, d) \cong H(B) = H_*(B; \Bbbk)^\#$. Since the dual of a vector space can never be countably infinite dimensional, $H(B)$ must have finite type, which completes the proof of (i).

Finally, since $H(F)$ and $H(B)$ have finite type, their minimal Sullivan models $(\wedge V^1, d)$ and $(\wedge V^{\geq 2}, d)$ are at most countably infinite dimensional. It follows that for each $k$, $H^k(\wedge V, d) \cong H^k(X)$ is at most countably infinite dimensional and thus finite dimensional, which proves (iv).

Conversely, suppose $m$ is a quasi-isomorphism and that (v), as well as one of (iii) or (iv) is satisfied. Then since $H^1(X) \cong H^1(B)$ it follows that $\dim H^1(B) < \infty$. Thus $V^1$ is at most countably infinite dimensional, and by the argument just above, $H(B)$ is a graded vector space of finite type. In view of (v) we may apply Theorem 5.1 (p. 145) to conclude that $\overline{m}$ is a quasi-isomorphism. Moreover, since one of $H(F)$ and $H(X)$ has finite type, it follows as above that each $(\wedge V)^k$ is at most countably infinite dimensional. Since $H(\wedge V^{\geq 2}, \overline{d}) \cong H(F)$, it follows that $H(F)$ has finite type. Thus by Theorem 1.6 (p. 43) each $\pi_k(\overline{m})$ is an isomorphism. In view of (7.5) it follows that $X$ is a Sullivan space.

Finally if (ii) and (iii) hold then it follows from Theorem 1.6 (p. 43) that $\pi_k(\overline{m})$, $k \geq 2$ is an isomorphism. Since $H^1(X) \cong H^1(B)$, (7.4) is satisfied and $X$ is a Sullivan space.

□

## 7.2 The classifying space $BG$

Let $G$ be an arbitrary group. The functorial construction of a classifying space for a group $G$ as the spatial realization $|SG|$ of a simplicial set $SG$ is defined as follows:

$$(SG)_n = \begin{cases} * & n = 0 \\ \underbrace{G \times \cdots \times G}_{n}, & n \geq 1 \end{cases}$$

with

$$\partial_i(g_1, \cdots, g_n) = \begin{cases} (g_2, \cdots, g_n), & i = 0 \\ (g_1, \cdots, g_i g_{i+1}, \cdots, g_n), & 1 \leq i < n \\ (g_1, \cdots, g_{n-1}), & i = n \end{cases}$$

and

$$s_j(g_1, \cdots, g_n) = \begin{cases} (e_G, g_1, \cdots, g_n), & j = 0 \\ (g_1, \cdots, g_{j-1}, e_G, g_j, \cdots, g_n), & 1 \le j < n \\ (g_1, \cdots, g_n, e_G), & j = n. \end{cases}$$

Thus a homomorphism $\varphi : H \to G$ determines the simplicial map $\langle\varphi\rangle : SH \to SG$ defined by $\langle\varphi\rangle(x_1, \cdots, x_n) = (\varphi x_1, \cdots, \varphi x_n)$.

**Notation.** The spatial realization of $SG$ will be denoted by $BG$. If $\varphi : H \to G$ is a homomorphism of groups the spatial realization of $\langle\varphi\rangle$ will be denoted $B\varphi : BH \to BG$.

By construction $BG$ is a CW complex pointed by the single 0-cell corresponding to the inclusion $e_G \to G$. Then one simplices, $g \in G = (SG)_1$ define simplicial maps $\langle g\rangle : \Delta[1] \to SG$ and we denote by $|g| : (I, 0, 1) \to BG$, the realization of $\langle g\rangle$.

**Proposition 7.1.** *For any group $G$:*

*(i) $SG$ is a Kan simplicial set.*
*(ii) $\pi_i(BG) = 0$, $i \ge 2$, and a natural isomorphism $G \xrightarrow{\cong} \pi_1(BG)$ is given by $g \mapsto [|g|]^{-1}$.*

*In particular, $BG$ is a classifying space for $G$.*

*proof:* (i) is ([65], Exercise 8.2.4). (ii) is ([62], Example 8.3.3).

□

Now suppose $K$ is a normal subgroup of $G$. The right adjoint action of $G$ on $K$ is the anti-homomorphism $G \to Aut\,K$ defined by $a \cdot g = g^{-1}ag$, $a \in K$, $g \in G$. It extends to the action of $G$ on $SK$ defined by

$$(\mathrm{Ad}\, g) \cdot (a_1, \cdots, a_n) = (g^{-1}a_1 g, \cdots, g^{-1}a_n g), \qquad g \in G, a_i \in K.$$

This defines a right action of $G$ on $BK$.

When $K = G$, a simplicial homotopy between the right action of $G$ on $SG$ and the identity is given by the map $\Delta[1] \times SG \to SG$,

$$\begin{aligned} &((\underbrace{0, \cdots, 0}_{r}, \underbrace{1, \cdots, 1}_{n-r+1}), (a_1, \cdots, a_n)) \\ &\mapsto (a_1, \cdots, a_{r-1}, a_r g, g^{-1}a_{r+1}g, \cdots, g^{-1}a_n g), \end{aligned} \tag{7.6}$$

since

$$\begin{cases} ((0, \cdots, 0), (a_1, \cdots, a_n)) \mapsto (a_1, \cdots, a_n), \\ ((1, \cdots, 1), (a_1, \cdots, a_n)) \mapsto (g^{-1}a_1 g, \cdots, g^{-1}a_n g) \end{cases}.$$

In particular, for $g \in K$, $Ad\, g \sim id_{SK} : SK \to SK$. It follows that the induced anti-homomorphism $G \to \mathcal{G}_{BK}$ factors to yield an anti-homomorphism $G/K \to \mathcal{G}_{BK}$, where (p. 119) $\mathcal{G}_{BK}$ denotes the group of homotopy classes of homotopy equivalences of $BK$.

**Proposition 7.2.** *Denote by* $K \xrightarrow{i} G \xrightarrow{\rho} G/K$ *the short exact sequence corresponding to the inclusion of a normal subgroup* $K$ *in a group* $G$*. Then*

*(i)* $B\rho : BG \to B(G/K)$ *is a fibre bundle with fibre at* $*$ *given by the map* $Bi : BK \to BG$.
*(ii) The holonomy action of* $G/K$ *on* $BK$ *corresponding to this fibre bundle is the anti-homomorphism* $G/K \to \mathcal{G}_{BK}$ *obtained from the right adjoint action of* $G$ *on* $K$.

*proof:* (i) As in ([18], p. 245) it is sufficient to construct simplicial isomorphisms

$$\begin{array}{ccc} \Delta[n] \times_{S(G/K)} SG & \xrightarrow[\varphi_\sigma]{\cong} & \Delta[n] \times SK \\ & \searrow \quad \swarrow & \\ & \Delta[n] & \end{array}$$

for each $\sigma : \Delta[n] \to S(G/K)$. If $\sigma = (y_1, \cdots, y_n)$ fix $g_j \in G$ with $\rho g_j = y_j$, ensuring that if $y_j = e_{G/K}$ then $g_j = e_G$. This defines a simplicial map $\tau : \Delta[n] \to SG$ such that $S\rho \circ \tau = \sigma$.

Now recall from the Example (p. 8) that a $k$-simplex in $\Delta[n]$ is a sequence of integers $(i_0, \cdots, i_k)$ with $0 \le i_0 \le \cdots \le i_k \le n$, with face and degeneracy maps defined by

$$\partial_j(i_0, \cdots, i_k) = (i_0, \cdots \widehat{i_j}, \cdots, i_k), \qquad 0 \le j \le k$$

$$s_j(i_0, \cdots, i_k) = (i_0, \cdots, i_j, i_j, \cdots, i_k), \qquad 0 \le j \le k.$$

In particular $\sigma(i_0, \cdots, i_k) = (v_1, \cdots, v_k)$ with

$$v_j = \begin{cases} e & \text{if } i_j = i_{j+1} \\ y_{i_j+1} \cdots y_{i_{j+1}} & \text{if } i_j < i_{j+1}. \end{cases}$$

Therefore if $((i_0, \cdots, i_k), (x_1, \cdots, x_k)) \in \Delta[n] \times_{S(G/K)} SG$, then

$$\rho x_\ell = \begin{cases} e_G & \text{if } i_\ell = i_{\ell+1} \\ y_{i_\ell+1} \cdots y_{i_{\ell+1}} & \text{if } i_\ell < i_{\ell+1}. \end{cases}$$

Now set

$$\varphi_\sigma((i_0, \cdots, i_k), (x_1, \ldots, x_k)) = ((i_0, \ldots, i_k), (z_1, \ldots, z_k)),$$

with $z_\ell = g_{i_1}\cdots g_{i_\ell+1}x_\ell g_{i_{\ell+1}}^{-1}\cdots g_1^{-1}$. It follows from the relations above that $z_\ell \in K$ and that $\varphi_\sigma$ is a morphism of simplicial sets. Finally, the formula for $z_\ell$ can be inverted to express $x_\ell$ in terms of $z_\ell$, and this defines an inverse to $\varphi_\sigma$, which is therefore an isomorphism.

(ii) Fix $g \in G$ and write $\rho g = \overline{g}$. The homotopy (7.6) restricts to a simplicial homotopy $\Delta[1]\times SK \to SG$, and this is precisely the isomorphism

$$\varphi_{\langle g\rangle}^{-1} : \Delta[1] \times_{S(G/K)} SG \xleftarrow{\cong} \Delta[1] \times SK\,,$$

constructed in (i). But

$$|\varphi_{\langle g\rangle}^{-1}| : I \times BK \to BG$$

is a homotopy covering $|g|$ from the inclusion of $BK$ to the right adjoint action of $g$. By definition the restriction of $|\varphi_{\langle g\rangle}^{-1}|$ to $\{1\} \times BK$ is the holonomy action, and this is exactly the restriction of $|Ad\, g|$ to $BK$.

□

Next, recall from §1.4 that associated with the simplicial set $SG$ of a group $G$ is the chain complex $(C_*(SG), \partial)$ in which $C_n(SG)$ is the free abelian group with basis the non-degenerate simplices of $(SG)_n$. Thus $C_0(SG) = \mathbb{Z}$, $C_1(SC) = \mathbb{Z}[G]/\mathbb{Z}[e_G]$, $\mathbb{Z}[G]$ denoting the group ring of $G$, and $C_*(SG) = TC_1(SG)$. We denote $H(C_*(SG))$ by $H_*(BG;\mathbb{Z})$.

Further, as defined in §1.4, $C^*(SG)$ is the cochain algebra $\mathrm{Hom}_{\mathbb{Z}}(C_*(SG), \Bbbk) = (C_*(SG) \otimes_{\mathbb{Z}} \Bbbk)^{\#}$. Thus, as observed after Corollary 1.3 (p. 16), there is a natural commutative square of isomorphisms

$$\begin{array}{ccc} H(BG) & \xrightarrow{\cong} & H(C^*(SG)) \\ \cong\uparrow & & \uparrow\cong \\ H(A_{PL}(BG)) & \xrightarrow{\cong} & H(A_{PL}(SG)) \end{array}$$

As stated there, we identify these four algebras, denoting all four simply by $H(BG)$.

Finally, recall from §2.5 that if $G$ is a discrete group we write $[a,b] = aba^{-1}b^{-1}$ for $a, b \in G$ and denote by $[H,K]$ the subgroup generated by the elements $[a,b]$ with $a \in H$ and $b \in K$. In particular, the *lower central series* for $G$ is the sequence $G = G^1 \supset \cdots \supset G^n \supset \cdots$ defined inductively by $G^{n+1} = [G, G^n]$.

Now suppose $K \subset G$ is a normal subgroup. Then the projection $\rho : G \to G/K$ factors as $G \xrightarrow{\rho'} G/[G,K] \xrightarrow{\rho''} G/K$. The next Lemma is a classical fact, but because of its importance in what follows we provide an elementary proof for the convenience of the reader.

**Lemma 7.1.** *For every group $G$ and every normal subgroup $K \subset G$,*

*(i) The Hurewicz homomorphism $G = \pi_1(BG) \xrightarrow{hur} H_1(BG;\mathbb{Z})$ factors to give an isomorphism*

$$G/[G,G] \xrightarrow{\cong} H_1(BG;\mathbb{Z})\,.$$

*(ii) $Im\,H_2(B\rho'') = Im\,H_2(B\rho)$.*

*proof:* (i) The isomorphism $G \xrightarrow{\cong} \pi_1(BG)$ of Proposition 7.1 is given by $g \mapsto [\|g^{-1}\|]$. Thus for $g \in G$ the simplicial map $\langle g^{-1}\rangle : \Delta[1] \to SG$ represents $g$ as the corresponding element of $\pi_1(BG)$. The Hurewicz homomorphism maps $\langle g^{-1}\rangle$ to the homology class, $H(g^{-1})$, of the cycle $g^{-1} \in C_1(SG)$.

Now by definition, for $a, b \in G$, $\partial(a \otimes b) = b - ab + a$. It follows that $H(ab) = H(a) + H(b)$. In particular, $H(a^k) = kH(a)$, $k \in \mathbb{Z}$, and $H(ab) = H(ba)$. Thus $H : a \mapsto H(a)$ induces a surjection $h : G/[G,G] \to H_1(BG;\mathbb{Z})$ of abelian groups. Since $C_1(SG)$ is a free abelian group the surjective homomorphism $C_1(SG) \to H_1(BG;\mathbb{Z})$ lifts through $h$ to define a surjective homomorphism $\varphi : C_1(SG) \to G/[G,G]$. It is immediate that $\varphi \circ \partial = 0$ and so $\varphi$ factors to provide an inverse to $h$.

(ii) First recall that for any homomorphism $\varphi : H \to G$, $B\varphi$ is the spatial realization of the simplicial map $\langle\varphi\rangle$ sending $(b_1, \cdots, b_n) \to (\varphi b_1, \cdots, \varphi b_n)$. Thus the identification $H_*(BG;\mathbb{Z}) = H(C_*(SG))$ identifies $H_*(B\varphi)$ with the homomorphism $H(\langle\varphi\rangle)$.

Now fix an injection of sets $\sigma : G/[G,K] \to G$ satisfying $\rho'\sigma = id$, and suppose $z = \sum_i n_i\, a_i \otimes b_i$ is a cycle in $C_2(S(G/[G,K]))$ with $a_i, b_i \in G/[G,K]$ and $n_i \in \mathbb{Z}$. Set

$$w = \sum_i n_i\, \sigma(a_i) \otimes \sigma(b_i)\,.$$

Then $\partial w$ is a cycle in $C_1(SG)$ and $\rho'\partial w = 0$. Write $\partial w = \sum_j \sum_\ell m_{\ell j}\sigma(\widehat{g}_j)b_{\ell j}$ with $b_{\ell j} \in [G,K]$ and the $\widehat{g}_j$ distinct elements of $G/[G,K]$. Then

$$0 = \rho'\partial w = \sum_j (\sum_\ell m_{\ell j})\widehat{g}_j$$

and so for each $j$, $\sum_\ell m_{\ell j} = 0$. It follows that

$$\partial w + \partial\left(\sum_{\ell,j} m_{\ell j}\,\sigma(\widehat{g}_j) \otimes b_{\ell j}\right) = \sum_{\ell,j} m_{\ell j} b_{\ell j}\,.$$

On the other hand, every $b \in [G,K]$ can be written as $b = \partial u$ with $u \in C_2(SG)$ and $(\rho \otimes \rho)u = e_G \otimes e_G$. Indeed if $b = a_1a_2$ with $a_1, a_2 \in [G,K]$ then $b + \partial(a_1 \otimes a_2) = a_1 + a_2$, which reduces the argument to the case $b = [g,c]$ with $g \in G$ and $c \in K$. But in this case $b = \partial u$ with

$$u = -gc \otimes g^{-1}c^{-1} - g \otimes c - g^{-1} \otimes c^{-1} + g \otimes g^{-1} + c \otimes c^{-1} + (g + g^{-1}) \otimes e_G$$

and clearly $(\rho \otimes \rho)u = e_{G/K} \otimes e_{G/K}$.

Thus writing $b_{\ell j} = \partial u_{\ell j}$ with $(\rho \otimes \rho)(u_{\ell j}) = e_{G/K} \otimes e_{G/K}$ we have

$$x = w + \left( \sum m_{\ell j}\, \sigma(\widehat{g_j}) \otimes b_{\ell j} - \sum m_{\ell j} u_{\ell j} \right)$$

is a cycle in $C_2(SG)$ such that $(\rho \otimes \rho)(x) = (\rho'' \otimes \rho'')(z)$.

□

**Lemma 7.2.** *If $H \to G$ is a homomorphism of abelian groups such that $H \otimes \mathbb{Q} \xrightarrow{\cong} G \otimes \mathbb{Q}$ then $H_*(BH; \Bbbk) \xrightarrow{\cong} H_*(BG; \Bbbk)$.*

*proof:* It is sufficient to prove this for $\Bbbk = \mathbb{Q}$. But if $A$ and $B$ are rational vector spaces then the natural map $A \otimes_{\mathbb{Z}} B \to A \otimes_{\mathbb{Q}} B$ is an isomorphism. It follows that the natural map $C_*(SG) \otimes_{\mathbb{Z}} \mathbb{Q} \to C_*(S(G \otimes \mathbb{Q})) \otimes_{\mathbb{Z}} \mathbb{Q}$ is an isomorphism.

□

Much of the rest of this Chapter applies the material of Chapters 1 and 2 to the classifying spaces $BG$. For the convenience of the reader we provide the following brief summary:

Suppose $(\wedge V, d)$ is a minimal Sullivan algebra. Its homotopy Lie algebra $L$ is given by $L_p = (V^{p+1})^\#$ with the bracket determined by the differential, $d$ (§2.1). If $\dim H^1(\wedge V, d)$ and $\dim V^i$, $i \geq 2$ are finite then $L$ is pronilpotent (Proposition 2.1, p. 55). In this case an exponential group $G_L$ is defined together with inverse bijections, $\exp_{L_0} : L_0 \to G_L$ and $\log_{L_0} : G_L \to L_0$, (Lemma 2.9, p. 57, and Proposition 2.3, p. 59). If $d = 0$ then $[L,L] = 0$ and (Proposition 2.4, p. 65) $\exp_{L_0} : L_0 \to G_L$ is an isomorphism of abelian groups.

On the other hand, the spatial realization of the simplicial set $\langle \wedge V, d \rangle$ (cf §1.6) is denoted by $|\wedge V, d|$, and bijections

$$\iota_n : \pi_n(|\wedge V, d|) \to (V^n)^\#, \qquad n \geq 1$$

are established in Theorem 1.3 (p. 37). In particular, $\iota_n$, $n \geq 2$, and $\exp_{L_0} \circ \iota_1 : \pi_1(|\wedge V, d|) \to G_L$ are isomorphisms of groups (Theorem 1.4(i), p. 37, and Theorem 2.4, p. 75). Thus we set

$$\pi_1(\wedge V, d) = G_L \quad \text{and } \pi_n(\wedge V, d) = (V^n)^\#, \quad n \geq 2.$$

In particular, if $V = V^1$ then $\pi_1(| \wedge V, d|) = G_L$ and $\pi_n(| \wedge V, d|) = 0$, $n \geq 2$. Thus, since $| \wedge V, d|$ is a CW complex, $| \wedge V, d|$ is a classifying space for $G_L$, and so there is a based homotopy equivalence $| \wedge V, d| \simeq B_{G_L}$.

## 7.3 The Sullivan 1-model of $BG$

Recall that the Sullivan 1-model of a commutative cochain algebra $(C, d)$ with $H^0(C, d) = \Bbbk$ is a morphism $\varphi : (\wedge W, d) \to (C, d)$ such that $W = W^1$, and $H^1(\varphi)$ is an isomorphism and $H^2(\varphi)$ is injective. By Corollary 3.5 (p. 109) these exist and are unique.

Our principal objective here, Theorem 7.4 (p. 208), provides an explicit construction of the Sullivan 1-model $(\wedge W, d) \to A_{PL}(BG)$ of the classifying space of any group $G$ for which $\dim G/[G,G] \otimes \Bbbk < \infty$. In particular (Theorem 7.3), we exhibit it as the minimal Sullivan model of the commutative cochain algebra $\varinjlim A_{PL}(B\,G/G^{n+1})$. We first consider the case $G$ is abelian, then proceed by induction to the groups $G/G^{n+1}$, and finally pass to the limit with the aid of Lemma 7.2 (p. 204).

Recall that the inverse suspension $s^{-1}M$ of a graded vector space $M$ is defined by $(s^{-1}M)^i = M^{i-1}$.

**Proposition 7.3.** *Suppose $G$ is an abelian group and $G \otimes_{\mathbb{Z}} \Bbbk$ is finite dimensional. Then*

*(i) The Hurewicz homomorphism tensored with $\Bbbk$ is an isomorphism of degree 1,*

$$G \otimes_{\mathbb{Z}} \Bbbk \xrightarrow{\cong} H_1(BG; \Bbbk)\,.$$

*(ii) The inverse dual, $s^{-1}Hom_{\mathbb{Z}}(G, \Bbbk) \xrightarrow{\cong} H^1(BG)$, extends uniquely to an isomorphism of graded algebras*

$$\wedge s^{-1}\, Hom_{\mathbb{Z}}(G, \Bbbk) \xrightarrow{\cong} H(BG; \Bbbk)\,.$$

*(iii) If $z_i \in A^1_{PL}(BG)$ are cycles representing the cohomology classes corresponding to a basis $x_i$ of $Hom_{\mathbb{Z}}(G, \Bbbk)$ then the correspondence $x_i \mapsto z_i$ extends to a minimal Sullivan model for $BFG$,*

$$(\wedge s^{-1} Hom_{\mathbb{Z}}(G, \Bbbk), 0) \xrightarrow{\simeq} A_{PL}(BG)\,.$$

*(iv) If $m : (\wedge V, 0) \to A_{PL}(BG)$ is any minimal Sullivan model then $\pi_1(m) : G \to \pi_1(\wedge V, 0)$ extends to an isomorphism*

$$G \otimes_{\mathbb{Z}} \Bbbk \xrightarrow{\cong} \pi_1(\wedge V, 0)\,.$$

*proof:* The first assertion follows from Lemma 7.1(i) (p. 203). For the second, choose elements $u_1, \cdots, u_r \in G$ such that the $u_i \otimes 1$ form a basis of $G \otimes_{\mathbb{Z}} \Bbbk$. It follows from Lemma 7.2 (p. 204) that it is sufficient to prove (ii) for the subgroup $\mathbb{Z}u_1 \times \cdots \times \mathbb{Z}u_r \subset G$. But it is immediate from the definition that the functor $G \rightsquigarrow SG$ preserves products, and so we only need to prove (ii) for $G = \mathbb{Z}u$, in which case $BG = S^1$ and (ii) is obvious. Evidently (iii) is an immediate consequence of (ii).

Finally, to prove (iv) observe from (iii) that $V = V^1$ and recall the notation and remarks at the end of §7.2. Since $d = 0$ in $\wedge V$, $\exp_{L_0} : (V^1)^\# \to G_L$ is an isomorphism of groups, endowing $G_L$ with the structure of a $\Bbbk$-vector space. Thus $\pi_1(m)$ extends to a linear map $G \otimes_{\mathbb{Z}} \Bbbk \to G_L = \pi_1(\wedge V, d)$. Now apply Proposition 1.19 (p. 42) with $X = BG$ and $\varphi = m$ to obtain the commutative diagram

$$\begin{array}{ccc} G \otimes_{\mathbb{Z}} \Bbbk & \xrightarrow[\cong]{hur} & H_1(BG; \Bbbk) \\ {\scriptstyle \log_{L_0} \circ \pi_1(m)} \downarrow & & \downarrow {\scriptstyle (H^1(m))^\#} \\ \mathrm{Hom}(V^1, \Bbbk) & \xrightarrow{\cong} & \mathrm{Hom}(H^1(\wedge V, 0), \Bbbk) \, . \end{array}$$

Since $H^1(m) : H^1(\wedge V, 0) \xrightarrow{\cong} H^1(BG)$ is an isomorphism, it follows that $\log_{L_0} \circ \pi_1(m)$ is a bijection. Thus $\pi_1(m)$ is an isomorphism.

□

Now we return to the consideration of the lower central series $G^{n+1} = [G, G^n]$ of a general group $G$ for which $\dim G/[G,G] \otimes \Bbbk < \infty$. In particular, for each $n$ the short exact sequence $G^n/G^{n+1} \to G/G^{n+1} \to G/G^n$ translates (Proposition 7.2, p. 205) to a fibre bundle

$$B\, G/G^n \leftarrow B\, G/G^{n+1} \leftarrow B\, G^n/G^{n+1}$$

in which the holonomy action of $G/G^n$ in $B\, G^n/G^{n+1}$ is trivial (because the right adjoint action in $G^n/G^{n+1}$ is trivial).

Moreover we have

**Lemma 7.3.** *For any group $G$,*

*(i) The maps $G/[G,G] \otimes_{\mathbb{Z}} G^n/G^{n+1} \to G^{n+1}/G^{n+2}$, $(a,b) \mapsto [a,b]$ are surjective homomorphisms.*

*(ii) In particular, if $\dim G/[G,G] \otimes \Bbbk < \infty$ then $\dim G^{n+1}/G^{n+2} \otimes \Bbbk < \infty$ for all $n \geq 0$.*

*proof:* As observed in §2.7 (p. 81), the first assertion follows from a classical identity of Hall ( [45], Theorem 5.3). The second assertion follows by induction on $n$.

□

Again consider a group $G$ for which $\dim G/[G,G] \otimes k\!\!k < \infty$. Denote by

$$\rho_n : G/G^{n+1} \to G/G^n \quad \text{and} \quad \zeta_{n+1} : G^n/G^{n+1} \to G/G^{n+1}$$

the standard homomorphisms. Then $G^{n+1}/G^{n+2}$ satisfies the hypotheses of Proposition 7.3 (p. 205). Moreover the adjoint action of $G/G^{n+1}$ on $G^{n+1}/G^{n+2}$ is trivial, and so by Proposition 7.2(ii) (p. 201) we may apply Corollary 5.1 (p. 146). This provides a Sullivan model for $A_{PL}(B\,G/G^{n+2})$ of the form

$$\begin{array}{ccccc} (\wedge W_{n-1}, d) & \longrightarrow & (\wedge W_{n-1} \otimes \wedge M_n, d) & \longrightarrow & (\wedge M_n, 0) \\ \downarrow \simeq & & \simeq \downarrow & & \simeq \downarrow \overline{m}_n \\ A_{PL}(B\,G/G^{n+1}) & \xrightarrow[A_{PL}(B\,\rho_{n+1})]{} & A_{PL}(B\,G/G^{n+2}) & \xrightarrow[A_{PL}(B\,\zeta_{n+2})]{} & A_{PL}(B\,G^{n+1}/G^{n+2}) \end{array} \qquad (7.7)$$

in which $M_n = s^{-1}\mathrm{Hom}_{\mathbb{Z}}(G^{n+1}/G^{n+2}, k\!\!k)$ and $\overline{m}_n$ is the quasi-isomorphism of Proposition 7.3.

Moreover, since the quotient differential in $\wedge M_n$ is zero it follows that the component of $d(1 \otimes M_n)$ in $W_{n-1} \otimes M_n$ defines the holonomy representation in $\wedge M_n$ for the $\Lambda$-extension. Since (Proposition 7.2(ii), p. 201) the holonomy action of $G/G^{n+1}$ in $H(BG^{n+1}/G^{n+2})$ is trivial, it follows that

$$d : M_n \to \wedge^2 W_{n-1} . \qquad (7.8)$$

Now define an increasing sequence of Sullivan 1-algebras, $(\wedge W_n, d)$, by setting

$$(\wedge W_n, d) = (\wedge W_{n-1} \otimes \wedge M_n, d) ,$$

and note that the direct decomposition $W_n = \oplus_{k \le n} M_k$ gives an isomorphism

$$W_n = \oplus_{k \le n} \mathrm{Hom}_{\mathbb{Z}}(G^{k+1}/G^{k+2}, k\!\!k) .$$

An inductive assembly of the left hand squares of (7.7) then establishes

**Theorem 7.3.** *Let $G$ be a group for which $\dim G/[G,G] \otimes \Bbbk < \infty$.*

*(i) For $n \geq 1$ the minimal Sullivan model of each $A_{PL}(BG/G^{n+1})$ has the form*

$$m_{n-1} : (\wedge W_{n-1}, d) \xrightarrow{\simeq} A_{PL}(BG/G^{n+1})$$

*in which $W_{n-1}$ is concentrated in degree 1.*

*(ii) These morphisms may be constructed to combine into a commutative diagram*

$$\begin{array}{ccccc} (\wedge W_0, 0) & \xrightarrow{\lambda_0} \cdots & (\wedge W_{n-1}, d) & \xrightarrow{\lambda_{n-1}} & (\wedge W_n, d) \to \cdots \\ {\scriptstyle m_0}\downarrow \simeq & & {\scriptstyle m_{n-1}}\downarrow \simeq & & {\scriptstyle m_n}\downarrow \simeq \\ A_{PL}(BG/[G,G]) & \longrightarrow \cdots & A_{PL}(BG/G^{n+1}) & \longrightarrow & A_{PL}(BG/G^{n+2}) \to \cdots \end{array} \tag{7.9}$$

*where each square is the left hand square of (7.7) for the appropriate $n$.*

*(iii) Taking direct limits provides a quasi-isomorphism*

$$(\wedge W, d) = \varinjlim_n (\wedge W_n, d) \xrightarrow{\simeq} \varinjlim A_{PL}(BG/G^{n+2}),$$

*thereby exhibiting $(\wedge W, d)$ as the minimal Sullivan model of $\varinjlim_n A_{PL}(BG/G^{n+2})$.*

Now observe that the surjections $G \to G/G^{n+2}$ induce morphisms $A_{PL}(BG/G^{n+2}) \to A_{PL}(BG)$ and hence a morphism

$$\varinjlim_n A_{PL}(BG/G^{n+2}) \to A_{PL}(BG). \tag{7.10}$$

Precomposing this with the quasi-isomorphism of Theorem 7.3 gives a morphism $m : (\wedge W, d) \to A_{PL}(BG)$.

Now recall the definition in §7.1 of a Sullivan 1-model. In particular, if $(\wedge V, d)$ is a minimal Sullivan algebra then the morphism $(\wedge V^1, d) \to (\wedge V, d)$ is its minimal Sullivan 1-model.

**Theorem 7.4.** *Let $G$ be a group for which $\dim G/[G,G] \otimes \Bbbk < \infty$. The morphism*

$$m : (\wedge W, d) \to A_{PL}(BG)$$

*is the Sullivan 1-model of $BG$.*

**Corollary 7.2.** *If $G = \pi_1(X, *)$ and $\dim H^1(X) < \infty$, then a minimal Sullivan model for $X$ has the form $(\wedge W \otimes \wedge V^{\geq 2}, d)$.*

*proof:* This is immediate from (7.4) (p. 197), Theorem 7.4, and the uniqueness of Sullivan 1-models (Corollary 3.5, p. 109).

□

The main step in the proof of Theorem 7.4 is

**Lemma 7.4.** *With the hypotheses and notation of Theorem 7.3, if $\alpha \in H^2(\wedge W_{n-1}, d)$ and $H^2(m)\alpha = 0$ then $H^2(\lambda_{n-1})(\alpha) = 0$, and so the image of $\alpha$ in $H(\wedge W, d)$ is zero.*

*proof:* Suppose this is not the case and write $\rho : G \to G/G^{n+1}$ as $G \overset{\rho'}{\to} G/G^{n+2} \overset{\rho''}{\to} G/G^{n+1}$. By hypothesis there is a class $[z'] \in H_2(B\,G/G^{n+2}; \mathbb{Z})$ such that $\langle \alpha, H_2(B\rho'')[z']\rangle \neq 0$. But then by Lemma 7.1 (p. 203) for some $[z] \in H_2(BG; \mathbb{Z})$ we have $H(B\rho)[z] = H(B\rho'')[z']$ and so $\langle H(m)\alpha, [z]\rangle = \langle \alpha, H(B\,\rho)[z]\rangle = \langle \alpha, H(B\rho'')[z]\rangle \neq 0$, contrary to our hypothesis that $H(m)\alpha = 0$.

□

*proof of Theorem 7.4:* We have to show that $H^1(m)$ is an isomorphism and $H^2(m)$ is injective. But it follows immediately from Lemma 7.1(i) (p. 203) that the linear maps

$$H^1(B\,G/[G,G]) \to H^1(B\,G/G^n) \quad \text{and} \quad H^1(B\,G/[G,G]) \to H^1(BG)$$

are isomorphisms. This implies that (in the notation of (7.8)) the linear maps

$$H(d) : M_n \to H^2(\wedge W_{n-1}, d)$$

are injective and that $H^1(m) : H^1(\wedge W, d) \to H^1(BG)$ is an isomorphism.

On the other hand, if $\alpha \in H^2(\wedge W, d)$ and $H^2(m)\alpha = 0$ then since $\alpha$ is in the image of some $H^2(\wedge W_n, d)$, Lemma 7.4 implies that $\alpha = 0$.

□

Recall from §1.5 that a continuous map $\varphi : X \to Y$ is called a rational homology equivalence if $H^*(\varphi; \mathbb{Q})$ is an isomorphism. Then we have the following analogue of the Serre-Whitehead theorem for Sullivan spaces:

**Proposition 7.4.** *A map of Sullivan spaces, $\varphi : (X,*) \to (Y,*)$, with fundamental groups $G_X$ and $G_Y$ is a rational homology equivalence if and only if:*

*(i)* $\pi_1(\varphi) : G_X^n/G_X^{n+1} \otimes \mathbb{Q} \to G_Y^n/G_Y^{n+1} \otimes \mathbb{Q}$ *is an isomorphism for* $n \geq 1$, *and*

*(ii)* $\pi_k(\varphi) : \pi_k(X) \otimes \mathbb{Q} \to \pi_k(Y) \otimes \mathbb{Q}$ *is an isomorphism for* $k \geq 2$.

*proof:* Let $m_\varphi : (\wedge V, d) \to (\wedge W, d)$ be a Sullivan representative for $A_{PL}(\varphi) : A_{PL}(Y) \to A_{PL}(X)$. Then since $X$ and $Y$ are Sullivan spaces the restriction of $m_\varphi$ to the morphism $m_\varphi(1) : (\wedge V^1, d) \to (\wedge W^1, d)$ is a Sullivan representative for $B(\pi_1(\varphi))$.

Thus if $H(\varphi)$ is an isomorphism, so is $m_\varphi$. Hence so are each $\pi_k(m_\varphi)$ and $m_\varphi(1)$. Now $V^1$ and $W^1$ have the canonical filtrations $(V_p^1)$ and $(W_n^1)$ defined in §2.1, and since $m_\varphi$ is an isomorphism it restricts to isomorphisms $V_n^1 \xrightarrow{\cong} W_n^1$. In this case (i) follows from (7.7) (p. 207). Moreover, since $X$ and $Y$ are Sullivan spaces, Theorem 7.2 (p. 197) identifies $V^k$ and $W^k$ respectively with the dual spaces of $\pi_k(Y) \otimes_{\mathbb{Z}} \Bbbk$ and $\pi_k(X) \otimes_{\mathbb{Z}} \Bbbk$. Thus if $H(\varphi)$ is an isomorphism so are the linear maps $\pi_k(\varphi) : \pi_k(X) \otimes_{\mathbb{Z}} \Bbbk \to \pi_k(Y) \otimes_{\mathbb{Z}} \Bbbk$, $k \geq 2$.

Conversely, if (i) holds then, again by (7.7), $m_\varphi$ restricts to an isomorphism $V^1 \xrightarrow{\cong} W^1$. Moreover, for $k \geq 2$, by definition, the models $(\wedge V, d) \to A_{PL}(X)$ and $(\wedge W, d) \to A_{PL}(Y)$ induce isomorphisms

$$\pi_k(X) \otimes_{\mathbb{Z}} \Bbbk \xrightarrow{\cong} \pi_k(\wedge V, d) \quad \text{and} \quad \pi_k(Y) \otimes_{\mathbb{Z}} \Bbbk \xrightarrow{\cong} \pi_k(\wedge W, d)$$

because $X$ and $Y$ are Sullivan spaces. Thus if (ii) holds, $\pi_k(m_\varphi)$ is an isomorphism for $k \geq 2$, and hence $m_\varphi$ is an isomorphism. It follows that $H(\varphi) = H(m_\varphi)$ is an isomorphism.

□

For any group $G$ with minimal model $(\wedge V, d)$ for $BG$, Lemma 7.1 (p. 203) gives $H^1(\wedge V, d) = H^1(BG) \cong [G/[G,G] \otimes_{\mathbb{Z}} \Bbbk)^\#$. On the other hand, if $L$ is the homotopy Lie algebra of $(\wedge V, d)$ then Corollary 2.1 (p. 51) states that $\dim L_0/[L_0, L_0] < \infty$ if and only if $\dim H^1(\wedge V, d) < \infty$. It follows that

$$\dim G/[G,G] \otimes_{\mathbb{Z}} \Bbbk < \infty \iff \dim L_0/[L_0, L_0] < \infty . \tag{7.11}$$

Finally, by Theorem 2.1 (p. 50), $L_0$ is pronilpotent. Thus this equivalence can be restated as

**Lemma 7.5.** *A group $G$ satisfies $\dim G/[G,G] \otimes_{\mathbb{Z}} \Bbbk < \infty$ if and only if $L_0$ is a pronilpotent Sullivan Lie algebra.*

Now recall from Theorem 7.3 (p. 208) the Sullivan 1-algebra $(\wedge W, d) = \varinjlim_n (\wedge W_n, d)$ associated with a group $G$ for which $\dim G/[G,G] \otimes_{\mathbb{Z}} \Bbbk < \infty$.

**Proposition 7.5.** *If* $\dim G/[G,G] \otimes_{\mathbb{Z}} \Bbbk < \infty$, *then*

$$W_{n+1} = W \cap d^{-1}(\wedge^2 W_n)\,.$$

*Thus the filtration* $(W_n)$ *coincides with that constructed in §2.2 (p. 49) directly from the Sullivan 1-algebra* $(\wedge W, d)$.

*proof:* First observe that since $H^1(B\,G/[G,G]) \xrightarrow{\cong} H^1(BG)$ it follows from (7.4) that $W_0 = W \cap \ker d$. Next note that it follows from (7.8) that $d : W_{n+1} \to \wedge^2 W_n$. On the other hand, if for some $w \in W$, $dw \in \wedge^2 W_n$ then $[dw] \in H^2(\wedge W_n)$ is in the kernel of the morphism $H(\wedge W_n, d) \to H(\wedge W, d)$ induced by the inclusion $\wedge W_n \hookrightarrow \wedge W$. Thus it is in the kernel of $H^2(n)$. By Lemma 7.4 (p. 209) $H(\lambda_n)[dw] = 0$ and so $dw = dw'$ for some $w' \in W_{n+1}$. Thus $w - w' \in W_0$ and $w \in W_{n+1}$.

□

Next recall from §2.1 the Cartan-Chevalley-Eilenberg construction $C^*(\,)$ and, as in §2.1 (p. 46), let $L_0 = L_0^1 \supset L_0^2 \supset \cdots$ denote the central descending series for the fundamental Lie algebra $L_0$ of $G$. According to Theorem 2.1 (p. 50) the duality $L_0 = W^\#$ induces isomorphisms $L_0/L_0^{n+2} \xrightarrow{\cong} (W_n)^\#$, $n \geq 0$. Moreover, Theorem 2.1 combined with Theorem 7.3 (p. 208) gives the commutative diagram

$$\begin{array}{ccccccc}
\cdots \longrightarrow & C^*(L_0/L_0^{n+1}) & \longrightarrow & C^*(L_0/L_0^{n+2}) & \longrightarrow \cdots & \\
 & \cong\uparrow & & \cong\uparrow & & \\
\cdots \longrightarrow & (\wedge W_{n-1}, d) & \longrightarrow & (\wedge W_n, d) & \longrightarrow \cdots & \qquad (7.12)\\
 & \simeq\downarrow & & \simeq\downarrow & & \\
\longrightarrow & A_{PL}(B\,G/G^{n+1})) & \longrightarrow & A_{PL}(B\,G/G^{n+2}) & \longrightarrow \cdots &
\end{array}$$

In particular, the morphisms

$$A_{PL}(BG) \xleftarrow{m} (\wedge W, d) \xrightarrow{\cong} \varinjlim C^*(L_0/L_0^{n+2})\,,$$

identify the Sullivan 1-model $(\wedge W, d)$ of $BG$ as the associated quadratic Sullivan algebra of the Sullivan Lie algebra $L_0$.

In particular, if $G$ is the fundamental group of a Sullivan space $X$ then Corollary 7.2 shows that the minimal Sullivan model of $X$ has the form

$$\Big( (\varinjlim C^*(L_0/L_0^{n+1})) \otimes \wedge V^{\geq 2}, d \Big)$$

where $L_0$ is the fundamental Lie algebra of $X$.

Again, consider a group $G$ for which $\dim G/[G,G] \otimes k < \infty$ and let $L$ denote the homotopy Lie algebra of the minimal Sullivan model $(\wedge V, d)$ for $BG$. Since $L_0$ is a pronilpotent Sullivan Lie algebra the corresponding group $G_L \subset \widehat{UL_0}$ is defined (§2.4, p. 59), and we have

**Proposition 7.6.** *With the hypotheses and notation above, the associated quadratic Sullivan algebra for $L_0$, $\varinjlim_n C^*(L_0/L_0^n))$ is the Sullivan 1-model for $BG$, $BG_L$ and $C^*(L_0)$. Moreover, if $G = \pi_1(X,*)$ this is also the Sullivan 1-model for $X$.*

*proof:* It follows from diagram (7.3) (p. 196) that $X$ and $BG$ have the same Sullivan 1-model if $G = \pi_1(X,*)$. On the other hand, since $\dim G/[G,G] \otimes_{\mathbb{Z}} k = \dim H^1(BG) < \infty$, the Sullivan 1-model of $BG$ is the morphism $m : (\wedge W, d) \to A_{PL}(BG)$ of Theorem 7.4, and diagram (7.12) identifies this as $\varinjlim_n C^*(L_0/L_0^n)$.

Now the spatial realization $|\wedge W, d|$ is a classifying space for $BG_L$, as follows directly from Theorem 1.4 (p. 37) and Theorem 2.4 (p. 75). Moreover, according to Corollary 2.4(ii) (p. 69),

$$G_L/[G_L, G_L] \cong L_0/[L_0, L_0]\,.$$

It follows with respect to the filtration $W_n$ of §2.1 that the inclusion $W_0 \hookrightarrow \wedge W \to A_{PL}(|\wedge W, d|)$ induces an isomorphism $H^1(\wedge W, d) \xrightarrow{\cong} H^1(|\wedge W, d|) \cong H^1(BG_L)$. Since $m_{|\wedge W,d|} : (\wedge W, d) \to A_{PL}(|\wedge W, d|)$ induces an injection in cohomology (Proposition 1.13, p. 27) it follows that $(\wedge W, d)$ is a Sullivan 1-model for $BG_L$.

Finally, Proposition 2.1 (p. 55) defines a morphism of cochain algebras $\mu : (\wedge W, d_1) \to C^*(L_0)$. Now $H^1(C^*(L_0)) = (L_0/[L_0, L_0])^\# = W_0$, since $\dim H^1(\wedge W, d) < \infty$ by hypothesis. Thus $H^1(\mu)$ is an isomorphism. Suppose $\alpha \in H^2(\wedge W, d_1)$ satisfies $H(\mu)\alpha = 0$. Then $\alpha$ is represented by a cycle $z \in \wedge^2 W_n$, some $n$, and $\mu z = d\omega$ for some $\omega \in (\wedge^2 sL_0)^\#$. But the restriction of $\omega$ to $\wedge^2 s\,L/L^{n+1}$ is an element $\Phi$ in $\wedge^2 W_n$ whose boundary is $z$; thus $z = d\Phi$ and $\alpha = 0$.

Altogether, $H^1(\mu)$ is an isomorphism and $H^2(\mu)$ is injective, and it follows that $(\wedge W, d)$ is a Sullivan 1-model for $C^*(L_0)$.

□

Finally, suppose that $\sigma : K \to G$ is any group homomorphism. Then $B\sigma : BK \to BG$ preserves base points, and so a Sullivan representative

$$\varphi_\sigma : (\wedge V, d) \to (\wedge U, d)$$

for $B\sigma$ is determined up to based homotopy. In particular, $\varphi_\sigma$ restricts to a morphism

$$\psi_\sigma : (\wedge V^1, d) \to (\wedge U^1, d),$$

which preserves the filtrations of $U^1$ and $W^1$, and the diagrams

$$\begin{array}{ccc} (\wedge V^1_n, d) & \xrightarrow{\psi_\sigma} & (\wedge U^1_n, d) \\ \simeq \big\downarrow & & \simeq \big\downarrow \\ A_{PL}(B\,G/G^{n+1}) & \xrightarrow[A_{PL}(B\sigma)]{} & A_{PL}(B\,K/K^{n+1}) \end{array}$$

are based homotopy commutative.

**Lemma 7.6.** *The morphism $\psi_\sigma$ is independent of the choice of $\varphi_\sigma$.*

*proof:* Given a second representative $\varphi'_\sigma$ we have that $\varphi_\sigma$ and $\varphi'_\sigma$ are based homotopic and so by Lemma 1.4(ii) (p. 22) they induce the same map from $\wedge^+ V^1/\wedge^{\geq 2} V^1 \to \wedge^+ U^1/\wedge^{\geq 2} U^1$. But this coincides with their restrictions to $V^1$.

□

In particular, suppose

$$\varphi : \Gamma \to \operatorname{Aut} G$$

is a homomorphism from a group $\Gamma$ to the group of automorphisms of a group $G$. For $\sigma \in \Gamma$, let $\psi_\sigma$ be the restriction to $(\wedge V^1, d)$ of a Sullivan representative of $B(\varphi(\sigma))$, where $(\wedge V, d)$ is the minimal model of $BG$. As an immediate consequence of Lemma 7.6 we have

**Proposition 7.7.** *With the hypotheses and notation above the map $\sigma \mapsto \psi_\sigma$ is a representation of $\Gamma$ in $(\wedge V^1, d)$.*

## 7.4 Malcev completions

The beginnings of rational homotopy can in part be traced back to the concept of Malcev completion ([46], [42], [58], [35]).

**Definition.** A group $\Gamma$ is *Malcev $\Bbbk$-complete* if each $\Gamma^n/\Gamma^{n+1}$ is a $\Bbbk$-vector space and

$$\Gamma = \varprojlim_n \Gamma/\Gamma^n .$$

A *Malcev* $k\!\!k$*-completion* of group $G$ is a homomorphism $\varphi : G \longrightarrow \Gamma$ in which $\Gamma$ is Malcev $k\!\!k$-complete and $\varphi$ induces isomorphisms

$$G^n/G^{n+1} \otimes_{\mathbb{Z}} k\!\!k \xrightarrow{\cong} \Gamma^n/\Gamma^{n+1}\,.$$

Our first objective here is to show how Sullivan models provide the construction of Malcev completions. For this, suppose given a group $G$ for which $\dim G/[G,G] \otimes k\!\!k < \infty$, and let

$$m : (\wedge W, d) \longrightarrow A_{PL}(BG)\,,$$

be the Sullivan 1-model of Theorem 7.4. For simplicity we denote the homotopy Lie algebra of $(\wedge W, d)$ by $L$. Thus $L = L_0 = W^{\#}$.

Now recall from the end of §7.2 (see §1.8 for more details) that the Sullivan 1-model determines the homomorphism

$$\pi_1(m) : \pi_1(BG) \to \pi_1(\wedge W, d)\,.$$

Since $\pi_1(BG) = G$ and $\pi_1(\wedge W, d)$ is the exponential group, $G_L$, of $L$, this is just a homomorphism

$$\pi_1(m) : G \to G_L\,.$$

The purpose of this section is to establish:

**Theorem 7.5.** *If* $m : (\wedge W, d) \to A_{PL}(BG)$ *is the Sullivan* 1*-model for a group* $G$*, and if* $(G/[G,G]) \otimes_{\mathbb{Z}} k\!\!k$ *is finite dimensional, then* $\pi_1(m) : G \to G_L$ *is a Malcev* $k\!\!k$*-completion. In particular* $G_L$ *is Malcev* $k\!\!k$*-complete.*

**Corollary 7.3.** *Suppose* $k\!\!k = \mathbb{Q}$*. If* $G$ *is a Malcev* $\mathbb{Q}$*-complete group and* $\dim_{\mathbb{Q}} G/[G,G] < \infty$ *then* $\pi_1(m) : G \cong G_L$*.*

*proof:* By Theorem 7.5, $\pi_1(m)$ induces isomorphisms $G^n/G^{n+1} \otimes_{\mathbb{Z}} \mathbb{Q} \xrightarrow{\cong} G_L^n/G_L^{n+1}$. Since $G$ is Malcev $\mathbb{Q}$-complete, $G^n/G^{n+1}$ is a $\mathbb{Q}$-vector space, and so these are isomorphisms $G^n/G^{n+1} \xrightarrow{\cong} G_L^n/G_L^{n+1}$. It follows by induction on $n$ that $\pi_1(m)$ induces isomorphisms $G/G^n \xrightarrow{\cong} G_L/G_L^n$. But by hypotheses $G = \varprojlim G/G^n$, while by Theorem 2.2 (p. 64) $G_L = \varprojlim G_L/G_L^n$. It follows that $\pi_1(m)$ is an isomorphism.

□

**Corollary 7.4.** *Suppose* $X$ *is a path connected space and* $\dim H^1(X) < \infty$*. If* $n : (\wedge V, d) \xrightarrow{\simeq} A_{PL}(X)$ *is a minimal Sullivan model, then*

$$\pi_1(n) : \pi_1(X) \to \pi_1(\wedge V, d)$$

*is a Malcev* $k\!\!k$*-completion.*

*proof:* Without loss of generality we may assume $X$ is a CW complex. Denote $\pi_1(X)$ by $G$. Then there is a continuous map $p : X \to BG$ such that $\pi_1(p) = id_G$. Now let $m : (\wedge W, d) \to A_{PL}(BG)$ be the Sullivan 1-model of Theorem 7.4 (p. 208). Then, it follows from diagram (7.3) (p. 196) that $A_{PL}(p) \circ m$ is a Sullivan 1-model for $X$. Since Sullivan 1-models are unique (Corollary 3.5, p. 109) there is an isomorphism

$$\varphi : (\wedge W, d) \overset{\cong}{\to} (\wedge W^1, d)$$

such that $n \circ \varphi \sim_* A_{PL}(p) \circ m : (\wedge V^1, d) \to A_{PL}(X)$. Since $\pi_1(p)$ and $\pi_1(\varphi)$ are isomorphisms, this identifies $\pi_1(n)$ with $\pi_1(m) : G \to \pi_1(\wedge W, d)$. Thus the Corollary follows from Theorem 7.5.

□

*proof of Theorem 7.5:* First note that it follows from Corollary 2.4 (p. 69) that each $G_L^n/G_L^{n+1}$ is a $\Bbbk$-vector space. Since by Theorem 2.2 (p. 64) $G_L = \varprojlim G_L/G_L^n$, $G_L$ is Malcev $\Bbbk$-complete.

Next, according to Proposition 7.5 (p. 211), the filtration of $W$ defined in (7.8) coincides with that defined "abstractly" in §2.2, p. 49. Let $L_0 = L_0^1 \supset \cdots \supset L_0^k \supset \cdots$ be the lower central series of $L$ defined by $L_0^{k+1} = [L_0, L_0^k]$. Then since $\dim L_0/[L_0, L_0] < \infty$, Theorem 2.1 (p. 50) provides isomorphisms

$$W_n \overset{\cong}{\longrightarrow} (L_0/L_0^{n+2})^\#$$

and so Proposition 2.5(iv) (p. 68) identifies

$$\pi_1(\wedge W_n, d) = G_L/G_L^{n+2} \quad \text{and} \quad \pi_1(\wedge(W_n/W_{n-1}), 0) = G_L^{n+1}/G_L^{n+2}\,.$$

On the other hand, the morphisms $m_n$ and $\overline{m}_n$ of (7.7, p. 207) induce morphisms

$$G/G^{n+2} \overset{\pi_1(m_n)}{\longrightarrow} \pi_1(\wedge W_n, d) \quad \text{and} \quad G^{n+1}/G^{n+2} \overset{\pi_1(\overline{m}_n)}{\longrightarrow} \pi_1(\wedge W_n/W_{n-1})$$

as described in §1.8. Thus from Theorem 7.3 we obtain the commutative diagram

$$\begin{array}{ccccc}
\cdots \leftarrow G/G^{n+1} & \longleftarrow & G/G^{n+2} & \longleftarrow & \cdots \\
\downarrow{\scriptstyle \pi_1(m_{n-1})} & & \downarrow{\scriptstyle \pi_1(m_n)} & & \\
\cdots \leftarrow G_L/G_L^{n+1} & \longleftarrow & G_L/G_L^{n+2} & \longleftarrow & \cdots
\end{array} \tag{7.13}$$

and from the construction of the model for $B\,G^{n+1}/G^{n+2}$ we obtain the commutative diagram of short exact sequences

$$\begin{array}{ccccccccc} 1 & \longleftarrow & G/G^{n+1} & \longleftarrow & G/G^{n+2} & \longleftarrow & G^{n+1}/G^{n+2} & \longleftarrow & 1 \\ & & \big\downarrow & & \big\downarrow & & \big\downarrow{\scriptstyle \pi_1(\overline{m}_n)} & & \\ 1 & \longleftarrow & G_L/G_L^{n+1} & \longleftarrow & G_L/G_L^{n+2} & \longleftarrow & G_L^{n+1}/G_L^{n+2} & \longleftarrow & 1\,. \end{array} \tag{7.14}$$

By Proposition 7.3(iv), $\pi_1(\overline{m}_n)$ extends to an isomorphism $G^{n+1}/G^{n+2} \otimes_{\mathbb{Z}} \Bbbk \overset{\cong}{\rightarrow} G_L^{n+1}/G_L^{n+2}$.

Finally, since $G_L = \varprojlim G_L/G_L^n$ (Theorem 2.2, p. 64), it follows from (7.14) that $\pi_1(m)$ is a Malcev $\Bbbk$-completion.

$\square$

Theorem 7.5 shows that if $G$ is any group for which $G/[G,G] \otimes_{\mathbb{Z}} \Bbbk$ is finite dimensional, then the fundamental group of the Sullivan 1-model of $BG$ is the Malcev $\Bbbk$-completion of $G$. Conversely, we begin with a Sullivan 1-algebra and establish a result in the reverse direction, restricting now to the case $\Bbbk = \mathbb{Q}$.

**Theorem 7.6.** *When $\Bbbk = \mathbb{Q}$ the correspondence*

$$(\wedge W, d) \rightsquigarrow \pi_1(\wedge W, d)$$

*is a bijection between isomorphism classes of Sullivan 1-algebras $(\wedge W, d)$ satisfying $\dim H^1(\wedge W, d) < \infty$ and Malcev $\mathbb{Q}$-complete groups $G$ satisfying $\dim G/[G,G] < \infty$.*

*proof:* Let $L$ denote the fundamental Lie algebra of a Sullivan 1-algebra $(\wedge W, d)$ for which $\dim H^1(\wedge W, d) < \infty$. Then $\pi_1(\wedge W, d) = G_L$. By Lemma 7.1 (p. 203) $G_L/[G_L, G_L] \otimes_{\mathbb{Z}} \Bbbk \cong H^1(B_{G_L}) \cong H^1(\wedge W, d) < \infty$. Thus Theorem 7.5 asserts that $G_L$ is Malcev $\mathbb{Q}$-complete.

On the other hand, suppose $G$ is a Malcev $\mathbb{Q}$-complete group satisfying $\dim G/[G,G] < \infty$. Then Theorem 7.5 exhibits a Malcev completion of $G$ as the fundamental group of a Sullivan 1-algebra $(\wedge W, d)$ satisfying $\dim H^1(\wedge W, d) < \infty$. But a Malcev $\mathbb{Q}$-completion of a Malcev $\mathbb{Q}$-complete group is just an isomorphism of groups, and so $G$ is isomorphic to the fundamental group of $(\wedge W, d)$.

It follows that the correspondence of Theorem 7.6 is surjective. On the other hand, two isomorphic Sullivan 1-algebras have isomorphic fundamental groups while isomorphic groups have homotopy equivalent classifying spaces and these therefore have isomorphic Sullivan 1-models. Thus the correspondence is a bijection.

□

Finally, we make the

**Definition.** A *weighted Lie algebra* is a graded Lie algebra $L$ together with a decomposition $L = \oplus_p L(p)$ such that $[L(p), L(q)] \subset L(p+q)$.

Now suppose $L$ is a graded Lie algebra and set $\mathcal{L}(L)(p) = L^p/L^{p+1}$. As observed in Corollary 2.5 (p. 70) the Lie bracket in $L$ induces linear maps

$$L^p/L^{p+1} \otimes L^q/L^{q+1} \xrightarrow{[\ ,\ ]} L^{p+q}/L^{p+q+1},$$

and it follows from the Jacobi identity in $L$ that these brackets make $\mathcal{L}(L) = \oplus_p L^p/L^{p+1}$ into a weighted Lie algebra.

On the other hand, suppose $G$ is a group. As observed in Corollary 2.5, it follows from a classical identity of Hall ( [45], Theorem 5.3) that conjugation $[\ ,\ ] : a, b \mapsto aba^{-1}b^{-1}$ induces additive maps

$$\left(G^p/G^{p+1}\right) \otimes \left(G^q/G^{q+1}\right) \longrightarrow G^{p+q}/G^{p+q+1}$$

which make the direct sum $\mathcal{L}(G) = \oplus_p \left(G^p/G^{p+1}\right)$ into a weighted Lie algebra concentrated in degree zero, and defined over $\mathbb{Z}$.

**Theorem 7.7.** *If* $\dim G/[G,G] \otimes \Bbbk < \infty$ *and* $L$ *is the homotopy Lie algebra of a minimal Sullivan model for* $BG$*, then the weighted algebras* $\mathcal{L}(G) \otimes \Bbbk$ *and* $\mathcal{L}(L_0)$ *are isomorphic.*

**Corollary 7.5.** *If* $G$ *is the fundamental group of a path connected space* $(X, *)$ *for which* $\dim H^1(X) < \infty$*, and if* $E$ *is the homotopy Lie algebra of a minimal Sullivan model for* $X$ *then* $\mathcal{L}(G) \otimes \Bbbk \cong \mathcal{L}(E_0)$.

*proof:* Since $\dim G/[G,G] \otimes_{\mathbb{Z}} \Bbbk < \infty$, it follows from (7.11) that $\dim L_0/[L_0, L_0] < \infty$ and so, in particular the group $G_L$ is defined (§2.4). Now Corollary 2.5 states that $\mathcal{L}(L_0) \cong \mathcal{L}(G_L)$.

On the other hand, by Theorem 7.5 the natural morphism $G \to G_L$ induces isomorphisms $G^p/G^{p+1} \otimes \Bbbk \xrightarrow{\cong} G_L^p/G_L^{p+1}$. These clearly define an isomorphism $\mathcal{L}(G) \otimes \Bbbk \xrightarrow{\cong} \mathcal{L}(G_L)$.

□

**Remark 1.** In the literature there are a number of constructions of the Malcev completion $\widehat{G}$ of a group $G$, leading always to the same group. These include the original constructions of Malcev ([46]) of torsion-free nilpotent groups and the homological construction of Hilton-Mislin-Roitberg ([35]). For general groups, we refer the reader to the Quillen construction ([58], [55]) and the Bousfield-Kan construction ([7]).

**Remark 2.** The Lie algebra $\mathcal{L}(G)$ is also a well-known invariant of $G$ and appears frequently in the literature. Theorem 7.7 and its Corollary show that if $\dim G/[G,G]\otimes \Bbbk < \infty$ and if $\dim H^1(X) < \infty$ then $\mathcal{L}(G)$ and $\mathcal{L}(\pi_1(X,*))\otimes \Bbbk$ arise as "associated weighted Lie algebras" of $L_0$, where $L$ is respectively the homotopy Lie algebra of $BG$ and of $X$.

## 7.5 The morphism $m_{|\wedge V,d|} : (\wedge V, d) \to A_{PL}(|\wedge V, d|)$

*In this section* $\Bbbk = \mathbb{Q}$.

Recall from §1.6 (p. 28) that associated with a minimal Sullivan algebra $(\wedge V, d)$ is a morphism $m_{|\wedge V,d|} : (\wedge V, d) \to A_{PL}(|\wedge V, d|)$. In Theorem 1.5 (p. 42) we showed that if $V^1 = 0$ then $m_{|\wedge V,d|}$ was a quasi-isomorphism if and only if $V$ was a graded vector space of finite type. Here we establish

**Theorem 7.8.** *For any minimal Sullivan algebra $(\wedge V, d)$ the following conditions are equivalent:*

*(i) $H^1(\wedge V, d)$ and each $V^i$, $i \geq 2$ are finite dimensional, and $m_{|\wedge V,d|}$ is a quasi-isomorphism.*
*(ii) $|\wedge V, d|$ is a Sullivan space.*

We first establish

**Proposition 7.8.** *Suppose $(\wedge W^1, d)$ is a Sullivan 1-algebra. Then $\dim H^1(\wedge W^1, d) < \infty$ if and only if $\dim H^1(|\wedge W^1, d|) < \infty$. In this case $m_{|\wedge W^1,d|}$ extends to a minimal Sullivan model of the form*

$$\psi : (\wedge W^1 \otimes \wedge Z^{\geq 2}, d) \xrightarrow{\simeq} A_{PL}(|\wedge W^1, d|)$$

*and $\dim H^1(|\wedge W^1, d|) = \dim H^1(\wedge W, d)$.*

*proof:* Extend $m_{|\wedge W^1,d|}$ to a quasi-isomorphism from a minimal Sullivan model $(\wedge W^1 \otimes \wedge Z, d)$, and let $\lambda : (\wedge W^1, d) \to (\wedge W^1 \otimes \wedge Z, d)$ denote the inclusion. Then, as observed in §1.8, $m_{|\wedge W^1,d)|}$ and $\psi$ determine the maps

$$\pi_1(m_{|\wedge W^1,d|}) : \pi_1(|\wedge W^1|) \to (W^1)^\#\,, \quad \text{and}$$

$$\pi_1(\psi) : \pi_1(|\wedge W^1|) \to (W^1)^\# \oplus (Z^1)^\# .$$

Moreover, by Theorem 1.3 (p. 37), $\pi_1(m_{|\wedge W^1,d|})$ is a bijection, while $\lambda^\# \circ \pi_1(\psi) = \pi_1(m_{|\wedge W_1,d|})$. Moreover, the same proof as in Proposition 1.18 (p. 32) shows that the diagram

$$\begin{array}{ccccc}
\pi_1(|\wedge W^1,d|) & \xrightarrow{\pi_1(\psi)} & (W^1)^\# \oplus (Z^1)^\# & \xrightarrow{\lambda^\#} & (W^1)^\# \\
\Big\downarrow hur & & \Big\downarrow \xi^\# & & \Big\downarrow \xi^\# \\
H_1(|\wedge W,d|) & \xrightarrow{H^1(\psi)^\#} & H^1(\wedge W \otimes \wedge Z,d)^\# & \xrightarrow{H^1(\lambda)^\#} & H^1(\wedge W^1,d)^\#
\end{array}$$

commutes, where $\xi : H^+(\wedge W \otimes \wedge Z, d) \to W \oplus Z$ is the map induced by the surjection $\wedge^{\geq 1}((W \oplus Z) \to W \oplus Z$ with kernel $\wedge^{\geq 2}(W \oplus Z)$.

Since $H^1(\wedge W^1, d) = W^1 \cap \ker d$, $\xi : H^1(\wedge W, d) \to W^1$ is injective and so $\xi^\#$ is surjective. It follows that $H^1(\lambda)^\# \circ H^1(\psi)^\#$ is surjective and so, if $\dim H^1(|\wedge W^1, d|) < \infty$, we must also have $\dim H^1(\wedge W^1, d) < \infty$.

In the other direction suppose $\dim H^1(\wedge W^1, d) < \infty$. Then according to Theorems 1.3 and 1.4, $\pi_1(m_{|\wedge W^1,d|}) : \pi_1(|\wedge W^1, d|) \to \pi_1(\wedge W^1, d)$ is an isomorphism of groups. Moreover $\pi_1(|\wedge W^1, d|)$ is the exponential group $G_L$ of the homotopy Lie algebra $L$ of $(\wedge W^1, d)$. But as observed in Lemma 7.1 together with Theorem 2.2,

$$H_1(|\wedge W^1, d|) = G_L/[G_L, G_L] \otimes \mathbb{Q} \cong L_0/[L_0, L_0] \cong H^1(\wedge W^1, d)^\# .$$

Thus $\dim H_1(|\wedge W^1, d|) < \infty$ and

$$\dim H^1(|\wedge W^1, d|) = \dim H^1(\wedge W^1, d) .$$

Finally, if $\dim H^1(\wedge W^1, d) < \infty$, then by Theorem 2.2 (p. 64),

$$G_L \xrightarrow{\cong} \varprojlim_n G_L/G_L^n$$

is Malcev $\mathbb{Q}$-complete. Thus it follows from Corollary 7.3 (p. 214) that

$$\pi_1(\psi) : \pi_1(|\wedge W^1, d|) \to \pi_1(\wedge W^1 \otimes \wedge Z^1, d)$$

is an isomorphism, and hence $\pi_1(\lambda) : W^1 \to W^1 \oplus Z^1$ is an isomorphism. It follows that $Z^1 = 0$, which completes the proof.

□

*proof of Theorem 7.8.* Consider the $\Lambda$-extension $(\wedge V^1,d) \xrightarrow{i} (\wedge V,d) \xrightarrow{p} (\wedge V^{\geq 2},\overline{d})$. If either (i) or (ii) holds, by Proposition 7.8, $H^1(\wedge V,d)$ is finite dimensional and it follows from Theorem 1.4 (p. 37) that

$$|\wedge V^1,d| \xleftarrow{|i|} |\wedge V,d| \xrightarrow{|p|} |\wedge V^{\geq 2},\overline{d}|$$

is a classifying space fibration for $|\wedge V,d|$. Moreover, again by Proposition 7.8, $m_{|\wedge V^1,d|}$ extends to a minimal Sullivan model of $|\wedge V^1,d|$ of the form

$$\psi : (\wedge V^1 \otimes \wedge Z^{\geq 2},d) \xrightarrow{\simeq} A_{PL}(|\wedge V^1,d|)\,.$$

Thus we obtain a commutative diagram

$$\begin{array}{ccccc}
(\wedge V^1 \otimes \wedge Z^{\geq 2},d) & \longrightarrow & (\wedge V^1 \otimes \wedge Z^{\geq 2} \otimes \wedge U^{\geq 2},d) & \longrightarrow & (\wedge U^{\geq 2},\overline{d}) \\
\simeq \big\downarrow \psi & & \simeq \big\downarrow \varphi & & \big\downarrow \overline{\varphi} \\
A_{PL}(|\wedge V^1,d|) & \xrightarrow{A_{PL}(|i|)} & A_{PL}(|\wedge V,d|) & \xrightarrow{A_{PL}(|p|)} & A_{PL}(|\wedge V^{\geq 2},\overline{d}|)\,.
\end{array} \tag{7.15}$$

Then lift $m_{|\wedge V,d|}$ through $\varphi$ to obtain the commutative diagram.

$$\begin{array}{ccccc}
(\wedge V^1,d) & \longrightarrow & (\wedge V,d) & \longrightarrow & (\wedge V^{\geq 2},\overline{d}) \\
\big\downarrow \lambda & & \big\downarrow \beta & & \big\downarrow \gamma \\
(\wedge V^1 \otimes \wedge Z^{\geq 2},d) & \longrightarrow & (\wedge V^1 \otimes \wedge Z^{\geq 2} \otimes \wedge U^{\geq 2},d) & \longrightarrow & (\wedge Z^{\geq 2} \otimes \wedge U^{\geq 2},\overline{d})\,,
\end{array}$$

in which $\lambda$ is the inclusion.

By hypothesis

$$m_{|\wedge V^1,d|} = \psi \circ \lambda \quad \text{and } m_{|\wedge V,d|} \sim \varphi \circ \beta \quad \text{rel}\,(\wedge V^1,d)\,.$$

By naturality we have $A_{PL}(|p|) \circ m_{|\wedge V,d|} = m_{|\wedge V^{\geq 2},\overline{d}|} \circ p$. Let $\rho : (\wedge Z^{\geq 2} \otimes \wedge U^{\geq 2},\overline{d}) \to (\wedge U^{\geq 2},\overline{d})$ be the surjection. Then it follows that

$$\overline{\varphi} \circ \rho \circ \gamma \sim m_{|\wedge V^{\geq 2},\overline{d}|}\,.$$

Now suppose (i) holds. Since $V^{\geq 2}$ has finite type, Theorem 1.5 (p. 42) asserts that $m_{|\wedge V^{\geq 2},\overline{d}|}$ is a quasi-isomorphism. Since $\pi_1(\wedge V,d)$ acts locally nilpotently in $H(\wedge V^{\geq 2},\overline{d})$ and since $\pi_1(m_{|\wedge V^1,d|})$ is an isomorphism it follows from Theorem 4.2 (p. 137) that $\pi_1(|\wedge V^1,d|)$ acts locally nilpotently in $H(|\wedge^{\geq 2},\overline{d}|)$. Thus Theorem 1.5, applied to diagram (7.15) implies that $\overline{\varphi}$ is a quasi-isomorphism. Hence $\rho\circ\gamma$ is a quasi-isomorphism. Moreover, $m_{|\wedge V,d|}$ is a quasi-isomorphism by hypothesis, and so $\beta$ is also a quasi-isomorphism.

In particular we obtain the commutative diagram of $\Lambda$-extensions

$$\begin{array}{ccccc}
(\wedge V^1, d) & \longrightarrow & (\wedge V, d) & \longrightarrow & (\wedge V^{\geq 2}, \overline{d}) \\
\downarrow \lambda & & \downarrow \beta & & \downarrow \rho\circ\gamma \\
(\wedge V^1 \otimes \wedge Z^{\geq 2}, d) & \longrightarrow & (\wedge V^1 \otimes \wedge Z^{\geq 2} \otimes \wedge U^{\geq 2}, d) & \longrightarrow & (\wedge U^{\geq 2}, \overline{d})
\end{array}$$

and it follows from Proposition 3.10 (p. 107) that $\lambda$ is a quasi-isomorphism and $Z = 0$. Now Theorem 7.2 (p. 197) implies that $| \wedge V, d|$ is a Sullivan space.

Conversely, suppose $| \wedge V, d|$ is a Sullivan space. Then $H(| \wedge V^{\geq 2}, \overline{d}|)$ has finite type and Theorem 5.1 (p. 145) asserts that $\wedge V^{\geq 2}$ has finite type and that $m_{|\wedge V^{\geq 2}, \overline{d}|}$ is a quasi-isomorphism. Moreover, by Proposition 7.8 (p. 218), the minimal Sullivan model for $| \wedge V^1, d|$ has the form $(\wedge V^1 \otimes \wedge Z^{\geq 2}, d)$, and since $| \wedge V, d|$ is a Sullivan space, $Z = 0$. Thus we have the commutative diagram

$$\begin{array}{ccccc}
(\wedge V^1, d^1) & \longrightarrow & (\wedge V^1 \otimes \wedge V^{\geq 2}, d) & \longrightarrow & (\wedge V^{\geq 2}, \overline{d}) \\
\simeq \downarrow m_{|\wedge V^1, d^1|} & & \downarrow m_{|\wedge V, d|} & & \simeq \downarrow m_{|\wedge V^{\geq 2}, \overline{d}|} \\
A_{PL}(| \wedge V^1, d^1|) & \longrightarrow & A_{PL}(| \wedge V, d|) & \longrightarrow & A_{PL}(| \wedge V^{\geq 2}, \overline{d}|) \,,
\end{array}$$

and by hypothesis $\pi_1(| \wedge V, d|)$ acts locally nilpotently on $H(| \wedge V^{\geq 2}, \overline{d}|)$. Now Corollary 5.1 (p. 146) states that $m_{|\wedge V, d|}$ is a quasi-isomorphism.

□

**Corollary 7.6.** *Let $(\wedge V, d) \to (\wedge V \otimes \wedge Z, d) \to (\wedge Z, \overline{d})$ be a Sullivan extension in which $| \wedge Z, \overline{d}|$ is a Sullivan space. Then in the fibration*

$$| \wedge V, d| \leftarrow | \wedge V \otimes \wedge Z, d| \leftarrow | \wedge Z, \overline{d}|$$

*$\pi_1(| \wedge V, d|)$ acts locally nilpotently in each $H^k(\wedge Z, \overline{d})$.*

*proof:* The morphism $\pi_1(| \wedge V, d|) \to \pi_1(\wedge V, d)$ makes $H(\wedge Z, d)$ into a locally nilpotent $\pi_1(| \wedge V, d|)$-module. Moreover, by Theorem 4.2 (p. 137) $m_{(\wedge Z, \overline{d})} : H(\wedge Z, \overline{d}) \xrightarrow{\cong} H(| \wedge Z, \overline{d}|)$ is a morphism of $\pi_1(| \wedge V, d|)$-modules, and by Theorem 7.8, $m_{|\wedge Z, \overline{d}|}$ is a quasi-isomorphism.

□

## 7.6 When $BG$ is a Sullivan space

**Proposition 7.9.** *Let $G$ be a group and let $(\wedge V, d)$ be the minimal Sullivan model of $BG$.*

*(i) $BG$ is a Sullivan space if and only if $H^1(BG)$ is finite dimensional and $(\wedge V, d)$ is a Sullivan 1-algebra.*
*(ii) $BG$ is a Sullivan space if and only if $H^1(BG)$ is finite dimensional and $\varinjlim H(B\,G/G^{n+1}) \xrightarrow{\cong} H(BG)$.*
*(iii) If $BG$ is a Sullivan space then $H(BG)$ is a graded vector space of finite type.*

*proof:* (i) is immediate from (7.5) (p. 198). Also (i) and (ii) are equivalent because of Theorem 7.4 (p. 208) and (iii) follows from Theorem 7.2 (p. 197).

**Theorem 7.9.** *Suppose $K \to G \to G/K$ is a short exact sequence of groups in which the adjoint representation of $G/K$ in $H^1(BK)$ is nilpotent. If $BK$ and $B(G/K)$ are Sullivan spaces then so is $BG$.*

*proof:* We first show that the adjoint representation of $G$ in $H(BK)$ is nilpotent. Since $BK$ is a Sullivan space the minimal Sullivan model of $BK$ has the form $(\wedge W, d) \xrightarrow{\cong} A_{PL}(BK)$, with $W = W^1$. Thus it follows from Proposition 7.7 (p. 213) that the adjoint representation of $G$ in $A_{PL}(BK)$ lifts to a representation in $(\wedge W, d)$.

Next note that there are induced representations of $G$ in each $K^n/K^{n+1} \otimes_{\mathbb{Z}} \Bbbk$. Moreover, the surjective linear maps

$$[\,,\,] : (K/[K,K] \otimes_{\mathbb{Z}} \Bbbk) \otimes_{\Bbbk} \left(K^n/K^{n+1} \otimes_{\mathbb{Z}} \Bbbk\right) \to K^{n+1}/K^{n+2} \otimes_{\mathbb{Z}} \Bbbk\,,$$

of Lemma 7.3 (p. 206) are clearly maps of $G$-modules, where $G$ acts on the left by the tensor product of the adjoint representations. It follows by induction on $n$ that the adjoint representation of $G$ in each $K^n/K^{n+1} \otimes_{\mathbb{Z}} \Bbbk$ is nilpotent.

Now an inductive argument applied to diagram (7.7) (p. 207) shows that the representation of $G$ in each $(\wedge W_n, d)$ is nilpotent, and hence the representation in $(\wedge W, d)$ is locally nilpotent. This in turn implies that the adjoint representation of $G$ in $H(BK)$ is locally nilpotent. Since $H(BK)$ has finite type (Proposition 7.7(ii), p. 213) it follows that this representation is nilpotent.

On the other hand, by Proposition 7.2(ii) (p. 201) the adjoint representation of $G/K$ in $H(BK)$ coincides with the holonomy representation for the fibre bundle

$$BK \to BG \to BG/K\,.$$

Since $BG/K$ is a CW complex and $H(BK)$ has finite type it follows from Theorem 5.1 (p. 145) that $BG$ has a Sullivan model of the form $(\wedge V \otimes \wedge Z, d)$ in which $(\wedge V, d)$ and $(\wedge Z, \overline{d})$ are respectively minimal Sullivan models for $A_{PL}(G/K)$ and $A_{PL}(K)$.

By Proposition 7.9 (p. 222) $V$ and $Z$ are concentrated in degree 1, and so $(\wedge V \otimes \wedge Z, d)$ is a Sullivan 1-algebra. Since $BK$ and $B(G/K)$ are Sullivan spaces, by definition both $H^1(\wedge V, d)$ and $H^1(\wedge Z, \overline{d})$ are finite dimensional. It follows that $\dim H^1(\wedge V \otimes \wedge Z, d) < \infty$, and so $BG$ is a Sullivan space.

□

Recall next that the free product $G_1 \# G_2$ of two groups $G_1$ and $G_2$ is their coproduct in the category of groups ( [40], §33).

**Theorem 7.10.** *If $BG_1$ and $BG_2$ are Sullivan spaces, then so is $B(G_1 \# G_2)$.*

For the proof of the Theorem we require two Lemmas.

**Lemma 7.7.** *Let $(A, d)$ be a commutative cochain algebra satisfying $H(A, d) = H^1(A, d) \oplus \Bbbk$. Then the minimal Sullivan model of $(A, d)$ is a Sullivan 1-algebra.*

*proof:* This follows by the same argument as in Example 7, §12(d) of [18] for the 1-connected case.

□

Now recall that if $(A, d)$ and $(B, d)$ are augmented cochain algebras defined over $\Bbbk$ then their wedge, $(A, d) \vee (B, d)$ is defined by

$$(A, d) \vee (B, d) = (A, d) \times_{\Bbbk} (B, d)\,.$$

**Lemma 7.8.** *If $(\wedge V, d)$ and $(\wedge W, d)$ are Sullivan 1-algebras, then the canonical morphism*

$$\varphi : (\wedge V, d) \otimes (\wedge W, d) \to (\wedge V, d) \vee (\wedge W, d)$$

*has a minimal Sullivan model of the form*

$$(\wedge V, d) \otimes (\wedge W, d) \xrightarrow{i} (\wedge V \otimes \wedge W \otimes \wedge U, d) \xrightarrow[\simeq]{\psi} (\wedge V, d) \vee (\wedge W, d)\,,$$

*in which the fibre $(\wedge U, \overline{d})$ is a Sullivan 1-algebra and $H(\wedge U, \overline{d}) = H^1(\wedge U, \overline{d}) \oplus \Bbbk$.*

*proof:* First observe that $H^1(\varphi)$ is an isomorphism and so $\varphi$ does have a minimal Sullivan model (Theorem 3.1(ii), p. 102) whose fibre is a minimal Sullivan algebra. Denote by $(CV,d)$ and $(CW,d)$ acyclic closures $(\wedge V \otimes \wedge sV, d)$ and $(\wedge W \otimes \wedge sW, d)$. Tensoring over $(\wedge V, d) \otimes (\wedge W, d)$ then yields an inclusion

$$(CV,d) \otimes (CW,d) \to (CV \otimes CW \otimes \wedge U, d)$$

of $(CV,d) \otimes (CW,d)$ semifree modules. Thus it follows from Proposition 1.1 (p. 7) that $\Bbbk \otimes_{CV \otimes CW}$ defines a quasi-isomorphism

$$(CV \otimes CW \otimes \wedge U, d) \xrightarrow{\simeq} (\wedge U, \bar{d}) .$$

On the other hand we may exhibit $(CV \otimes CW \otimes \wedge U, d)$ as the $(\wedge V \otimes \wedge W \otimes \wedge U, d)$-semifree module

$$(\wedge V \otimes \wedge W \otimes \wedge U \otimes \wedge sV \otimes \wedge sW, d) .$$

Applying the morphism $\psi \otimes id_{\wedge sV \otimes \wedge sW}$ yields (Proposition 1.1, p. 7) a quasi-isomorphism

$$(\wedge V \otimes \wedge W \otimes \wedge U \otimes \wedge sV \otimes \wedge sW, d) \xrightarrow{\simeq} ((\wedge V \vee \wedge W) \otimes \wedge sV \otimes \wedge sW, d) .$$

Together with the quasi-isomorphism above this yields an isomorphism

$$H(\wedge U, \bar{d}) \cong H\left((\wedge V \vee \wedge W) \otimes \wedge sV \otimes \wedge sW, d\right) .$$

Next observe that the short exact sequence

$$0 \to (\wedge^+ V \otimes \wedge sV, d) \to (\wedge V \otimes \wedge sV, d) \to (\wedge sV, 0) \to 0$$

has a connecting homomorphism

$$\delta_V : \wedge^+ sV \xrightarrow{\cong} H(\wedge^+ V \otimes \wedge sV, d) ,$$

which is a linear isomorphism of degree one. In particular, $H(\wedge^+ V \otimes \wedge sV, d)$ is concentrated in degree one. The identical observation applies, of course to $H(\wedge^+ W \otimes \wedge sW, d)$.

On the other hand, apply $- \otimes_{\wedge V \otimes \wedge W} CV \otimes CW$ to the short exact sequence

$$0 \to \wedge^+ V \otimes \wedge^+ W \to \wedge V \otimes \wedge W \to \wedge V \vee \wedge W \to 0$$

and then pass to homology; the resulting connecting homomorphism, $\delta$, restricts to an isomorphism

$$\delta : H^+\left((\wedge V \vee \wedge W) \otimes \wedge sV \otimes \wedge sW, d\right) \xrightarrow{\cong} H(\wedge^+ V \otimes \wedge sV) \otimes H(\wedge^+ W \otimes \wedge sW) .$$

Again, $\delta$ has degree one and since

$$H(\wedge U,\overline{d}) \cong H\left((\wedge V \vee \wedge W)\otimes \wedge sV \otimes \wedge sW, d\right) ,$$

it follows that $H(\wedge U,\overline{d}) = \Bbbk \oplus H^1(\wedge U,\overline{d})$. Now Lemma 7.7 implies that $U = U^1$.

$\square$

*proof of Theorem 7.10:* Suppose $BG_1$ and $BG_2$ are Sullivan spaces. It follows from the Van Kampen theorem ( [32], Theorem 1.20) that $\pi_1(BG_1 \vee BG_2) = G_1\#G_2$. On the other hand, let $S_i$ be the mapping cylinder for the map $B\{e\} \to BG_i$. Then $S = S_1 \cup_{B\{e\}} S_2$ has the homotopy type of $BG_1 \vee BG_2$. Thus it follows from ( [32], Theorem 1B.11) that $BG_1 \vee BG_2$ is a classifying space for $G_1\#G_2$. In particular, as observed at the start of §7.1, there is a homotopy equivalence $BG_1 \vee BG_2 \xrightarrow{\simeq} B(G_1\#G_2)$.

On the other hand, according to Example 1, §12(c) of [18] the canonical morphism

$$A_{PL}(BG_1 \vee BG_2) \to A_{PL}(BG_1) \vee A_{PL}(BG_2)$$

is a quasi-isomorphism. Thus if $(\wedge V, d) \xrightarrow{\simeq} A_{PL}(BG_1)$ and $(\wedge W, d) \xrightarrow{\simeq} A_{PL}(BG_2)$ are minimal Sullivan models then a minimal Sullivan model for $(\wedge V, d) \vee (\wedge W, d)$ is a minimal Sullivan model for $B(G_1\#G_2)$.

But since $BG_1$ and $BG_2$ are Sullivan spaces we know from Proposition 7.2 (p. 201) that $V$ and $W$ are concentrated in degree 1. But then Lemma 7.8 (p. 223) asserts that $(\wedge V, d) \vee (\wedge W, d)$ has a Sullivan model of the form $(\wedge V \otimes \wedge W \otimes \wedge U, d)$ with $U = U^1$. Thus it is indeed a Sullivan 1-algebra, and so necessarily minimal.

Finally, almost by definition, if $G = G_1\#G_2$ then $G/[G,G] = G_1/[G_1,G_1] \oplus G_2/[G_2,G_2]$ and so, because $G_1/[G_1,G_1] \otimes_{\mathbb{Z}} \Bbbk$ and $G_2/[G_2,G_2] \otimes_{\mathbb{Z}} \Bbbk$ are finite dimensional, so is $G/[G,G] \otimes_{\mathbb{Z}} \Bbbk$.

$\square$

## Chapter 8

# Examples

### 8.1 Nilpotent and rationally nilpotent groups

A group $G$ is *nilpotent* if some term in the lower central series is the identity: $G^{n+2} = \{e\}$. We say that $G$ is *rationally nilpotent* if for some $n$, $H_*(BG^{n+2};\mathbb{Q}) = \mathbb{Q}$.

Suppose $G$ is rationally nilpotent. Then it follows from Theorem 5.1 (p. 145) (or by a Serre spectral sequence argument) that the morphism $A_{PL}(B\,G/G^{n+2}) \to A_{PL}(BG)$ is a quasi-isomorphism. Thus if $G/[G,G]\otimes \Bbbk$ is finite dimensional, Theorem 7.3 (p. 208) asserts that the minimal Sullivan model of $BG$ has the form

$$(\wedge W_n, d) \xrightarrow{\simeq} A_{PL}(BG)\,.$$

In particular if $G$ is a rational nilpotent group with $\dim G/[G,G]\otimes \Bbbk < \infty$, then $BG$ is a Sullivan space.

Finally, it follows from the quasi-isomorphism $(\wedge W_n, d) \xrightarrow{\simeq} A_{PL}(BG)$ that the fundamental Lie algebra, $L_0$, of a rationally nilpotent group $G$ is a nilpotent Lie algebra and satisfies $L_0^{n+2} = 0$. Moreover, as observed in (7.12) (p. 211), in this case the minimal Sullivan model of $BG$ is the Cartan-Eilenberg-Chevalley cochain complex for $L_0$.

### 8.2 Nilpotent and rationally nilpotent spaces

A connected CW complex, $Y$, is *nilpotent* if $\pi_1(Y,*)$ is a nilpotent group and acts nilpotently (via covering transformations) on each $\pi_n(\widetilde{Y})$, $n \geq 2$, $\widetilde{Y}$ denoting the universal cover of $Y$. The CW complex $Y$ is *rationally nilpotent* if $\pi_1(Y,*)$ is a rationally nilpotent group acting nilpotently in each $\pi_n(\widetilde{Y})\otimes\mathbb{Q}$. A path connected topological space is *rationally nilpotent* if it has the weak homotopy type of a rationally nilpotent CW complex.

**Proposition 8.1.** *A connected CW complex $(Y,*)$ pointed by a 0-cell is rationally nilpotent if and only if $\pi_1(Y,*)$ is rationally nilpotent and the representation of $\pi_1(Y,*)$ induced by covering transformations in each $H_k(\widetilde{Y};\mathbb{Q})$ is nilpotent.*

*proof:* We show by induction on $k$ that the representation in each $H_i(\widetilde{Y};\mathbb{Q})$, $i \leq k$ is nilpotent if and only if the representation in each $\pi_i(\widetilde{Y})\otimes\mathbb{Q}$, $i \leq k$, is nilpotent. This is trivial for $k=1$. Suppose it is established for some $k$. The standard Postnikov construction provides a fibration

$$F(k+2) \to \widetilde{Y} \xrightarrow{\rho_{k+1}} B(k+1)$$

in which $F(k+2)$ is $k+1$-connected and both $\pi_i(\rho_{k+1})$ and $H_i(\rho_{k+1})$ are isomorphisms for $i \leq k+1$. Moreover the action of $G$ in $\widetilde{Y}$ descends to an action "up to homotopy" in $B(k+1)$.

In the same way there is a fibration

$$F_{k+1} \to B(k+1) \to B(k)$$

inducing the obvious isomorphisms $\pi_i(B(k+1)) \xrightarrow{\cong} \pi_i(B(k))$, $i \leq k$. Moreover the action "up to homotopy" of $G$ on $B(k+1)$ and $B(k)$ induce an action up to homotopy on $F_{k+1}$. The representation of $G$ in $\pi_{k+1}(\widetilde{Y})\otimes\mathbb{Q}$ coincides with the representation of $G$ in $\pi_{k+1}(F_{k+1})\otimes\mathbb{Q}$, and this is identified via the Hurewicz homomorphism with the representation in $H_{k+1}(F_{k+1};\mathbb{Q})$. Now the inductive step follows from a standard Serre spectral sequence argument.

□

Given a connected CW complex $(Y,*)$ pointed by a 0-cell and with $\pi_1(Y,*)$ denoted by $G$, we may as described in Lemma 4.3 (p. 124) construct a fibration $F \xrightarrow{j} X \xrightarrow{p} BG$ and a commutative diagram

$$\begin{array}{ccc} (Y,*) & \xleftarrow{\quad\rho\quad} & (\widetilde{Y},*) \\ {\scriptstyle\pi_Y}\uparrow & & \uparrow{\scriptstyle\sigma} \\ (X,*) & \xleftarrow{\quad j\quad} & (F,*) \end{array}$$

in which the vertical arrows are weak homotopy equivalences and $\rho$ is the projection of the universal cover. Thus from Lemma 4.3 (p. 124) we deduce the following corollary to Proposition 8.1.

**Corollary 8.1.** *A connected CW complex $(Y,*)$ is rationally nilpotent if and only if $\pi_1(Y,*)$ is rationally nilpotent and the holonomy representation of $\pi_1(Y,*)$ in each $H_k(F;\mathbb{Q})$ is nilpotent.*

In particular, from Theorem 7.2 (p. 197) we obtain

**Corollary 8.2.** *If $(Y,*)$ is a rationally nilpotent CW complex with rational cohomology of finite type then $Y$ is a Sullivan space.*

**Proposition 8.2.** *Suppose $\varphi : (\wedge V, d) \to A_{PL}(X)$ is the rational minimal Sullivan model of a rationally nilpotent CW complex, $X$. Then*

$$|\varphi| : X \to |\wedge V, d|$$

*is a rational homology equivalence.*

*proof:* Since $\pi_1(X,*)$ is rationally nilpotent it follows that the minimal Sullivan model of $B\pi_1(X,*)$ has the form $(\wedge W^1, d)$ with $\dim W^1 < \infty$. Since $X$ is a Sullivan space its minimal Sullivan model has the form $(\wedge V, d)$ with $(\wedge V^1, d) = (\wedge W^1, d)$ and $\dim V^k < \infty$, $k \geq 2$. Now the Proposition simply restates Theorem 5.4 (p. 166).

□

**Corollary 8.3.** *Two rationally nilpotent spaces with rational cohomology of finite type have the same rational homology type if and only if their minimal Sullivan models are isomorphic.*

## 8.3 The groups $\mathbb{Z}\#\cdots\#\mathbb{Z}$

The Van Kampen theorem ([32], Theorem 1.20) provides homotopy equivalences

$$\varphi : (S^1,*) \vee \cdots \vee (S^1,*) \xrightarrow{\simeq} B(\mathbb{Z}\#\cdots\#\mathbb{Z}),$$

in which the restriction of $\varphi$ to the $k^{th}$ circle induces the inclusion of the $k^{th}$ copy of $\mathbb{Z}$. Theorem 7.10 (p. 223) asserts that $B(\mathbb{Z}\#\cdots\#\mathbb{Z})$ is a Sullivan space, but in this case we can determine the Sullivan model of $BG = (S^1,*) \vee \cdots \vee (S^1,*)$ explicitly.

For this let $r$ be the number of circles in $S^1 \vee \cdots \vee S^1$ and denote $H(S^1 \vee \cdots \vee S^1;\mathbb{Q})$ simply by $H$. Thus

$$H = k\!\!k \oplus H^1,$$

$\dim H^1 = r$, and $H^1 \cdot H^1 = 0$. It is straightforward to check that if $(B, d)$ is any commutative cochain algebra for which $H(B) \cong H$, then $(H, 0)$ is connected to $(B, d)$ by quasi-isomorphisms. In particular, a Sullivan minimal model for $(H, 0)$ is also a Sullivan minimal model for $S^1 \vee \cdots \vee S^1$.

Now "rescale" $H$ to a graded commutative algebra $A$ by setting $A = \Bbbk \oplus A^3$ with $A^3 = H^1$. According to Example 2 in §23 of [18], $(A, 0)$ has a minimal Sullivan model of the form

$$(\wedge U, d) \xrightarrow{\simeq} (A, 0)$$

in which $(\wedge U, d)$ is the Cartan-Chevalley-Eilenberg cochain algebra on a free graded Lie algebra $\mathbb{L}_F$. Here $F$ is a graded $r$-dimensional space concentrated in degree 2, $U^3$ is the dual of $sF$, $d = 0$ in $U^3$, and $(\Bbbk \oplus U^3, 0) \to (\wedge U, d)$ is a quasi-isomorphism. Composed with the quasi-isomorphism above this gives an isomorphism $U^3 \xrightarrow{\cong} A^3$.

More generally, $U = \{U^{2k+1}\}_{k \geq 1}$ and each $U^{2k+1}$ is dual to the suspension of the vector space $[F, \cdots, F]$ spanned by the Lie brackets of length $k$ in $F$. Thus for $k \geq 1$,

$$d : U^{2k+1} \to \bigoplus_{1 \leq i \leq k-i \leq k-1} U^{2i+1} \wedge U^{2(k-i)+1}$$

is injective.

Next, "rescale" $U$ to a vector space $W$ concentrated in degree 1 by setting $W = \oplus_{k \geq 0} V_k$ and $V_k = U^{2(k+1)+1}$. The isomorphism $W \xrightarrow{\cong} U$ extends to an isomorphism $\wedge W \xrightarrow{\cong} \wedge U$ of (non-graded) algebras because $W$ and $U$ are concentrated in odd degrees. This isomorphism transports $d$ to a differential, $d$, in $\wedge W$ mapping $W \to \wedge^2 W$. Thus $d$ is homogeneous of degree 1 and $(\wedge W, d)$ is a cdga. We also have

$$d : V_k \to \bigoplus_{0 \leq i \leq k-i-1 \leq k-1} V_i \wedge V_{k-i-1} ,$$

which exhibits $(\wedge W, d)$ as a Sullivan 1-algebra: in the notation of §2.2,

$$W_n = \bigoplus_{i \leq n} V_i \, .$$

Finally, note that with this rescaling the quasi-isomorphism $(\wedge U, d) \xrightarrow{\simeq} (A, 0)$ becomes a quasi-isomorphism

$$(\wedge W, d) \xrightarrow{\simeq} (H, 0) ,$$

thereby exhibiting $(\wedge W, d)$ as the minimal Sullivan model of $S^1 \vee \cdots \vee S^1$. Moreover the graded Lie algebra $\mathbb{L}_F$ rescales to a Lie algebra $\mathbb{L}_E$ concentrated in degree 0, with $V_k$ dual to the suspension of the space $[E, \cdots, E]$ of iterated brackets of length $k+1$. Thus (Proposition 2.1, p. 55)

$$L = \varprojlim_n \mathbb{L}_E / \mathbb{L}_E^n$$

is the fundamental Lie algebra of $S^1 \vee \cdots \vee S^1$.

## 8.4 Semidirect products

If $G$ and $K$ are groups and $\varphi : G \to \operatorname{Aut} K$ is a homomorphism then the *semidirect product* is the group $K \ltimes G$ which coincides with $K \times G$ as a set but with multiplication given by

$$(k_1, g_1)(k_2, g_2) = (k_1\varphi(g_1)(k_2), g_1g_2)\,.$$

Thus $K$ is a normal subgroup of $K \ltimes G$ and projection on $G$ is an isomorphism $(K \ltimes G)/K \xrightarrow{\cong} G$. In particular, this identifies the adjoint action of $G$ on $K$ with the homomorphism $\varphi$.

Now as a special case of Theorem 7.9 (p. 222) we have

**Proposition 8.3.** *If $BG$ and $BK$ are Sullivan spaces and $\varphi : G \to Aut K$ induces a nilpotent representation of $G$ in $K/[K, K] \otimes \mathbb{Q}$ then $B(K \ltimes G)$ is a Sullivan space.*

For instance, suppose $K = K_1 \times K_2$ in which $K_1 = \mathbb{Z}_1 \# \cdots \# \mathbb{Z}_r$, each $\mathbb{Z}_i$ is a copy of $\mathbb{Z}$, and $K_2$ is a finitely generated nilpotent group.

For $1 \leq i \leq r$, denote the generator of $\mathbb{Z}_i$ by $a_i$ and let $b_i$ be a central element in $K_2$. Then a homomorphism $\varphi : \mathbb{Z}_1 \# \cdots \# \mathbb{Z}_r \to \operatorname{Aut} K$ is defined by

$$\varphi(a_i)(a_j) = a_j b_i \quad \text{ and } \quad \varphi(a_i)(b) = b\,, \qquad b \in K_2\,.$$

It is immediate from the definition that the induced representation $\overline{\varphi}$ of $\mathbb{Z}_1 \# \cdots \# \mathbb{Z}_r$ in $K/[K, K] \otimes_{\mathbb{Z}} \mathbb{Q}$ reduces to the identity in $K_2/[K_2, K_2] \otimes_{\mathbb{Z}} \mathbb{Q}$ and satisfies

$$(\overline{\varphi}(a) - id) : K_1/[K_1, K_1] \otimes_{\mathbb{Z}} \mathbb{Q} \to K_2/[K_2, K_2] \otimes_{\mathbb{Z}} \mathbb{Q}\,.$$

Thus this representation is nilpotent and so Proposition 8.3 asserts that $B\,(K \ltimes (\mathbb{Z}_1 \# \cdots \# \mathbb{Z}_r))$ is a Sullivan space.

As a second example consider the right adjoint action of a group $G$, on itself,

$$a \cdot g = g^{-1}ag\,, \qquad g, a \in G\,.$$

The induced right representation of $G$ in $G/[G, G] \otimes_{\mathbb{Z}} \mathbb{Q}$ is trivial and so it follows from Proposition 8.3 that if $BG$ is a Sullivan space then $B(G \ltimes G)$ is again a Sullivan space.

## 8.5 Orientable Riemann surfaces

Denote by $S_g$ the orientable Riemann surface of genus $g$; $S_g$ is the connected sum of $g$ copies of $S^1 \times S^1$. The cohomology algebra $H_g = H^*(S_g;\mathbb{Q})$ is then given explicitly by $H_g^k = 0$, $k \geq 3$ and

$$H_g^k = \begin{cases} \mathbb{Q}\,1\,, & k=0 \\ \oplus_{i=1}^g (\mathbb{Q}\alpha_i \oplus \mathbb{Q}\beta_i)\,, & k=1 \\ \mathbb{Q}\,\omega_g\,, & k=2 \end{cases}$$

where $\alpha_i$ and $\beta_i$ restrict to the dual of the cycles represented by the two circles in the $i^{th}$ torus. The multiplication in $H_g$ is given by

$$\alpha_i\alpha_j = 0 = \beta_i\beta_j \quad \text{and} \quad \alpha_i\beta_j = \delta_{ij}\omega_g\,;$$

$\delta_{ij}$ denoting the Kronecker delta.

In [13] the authors use Hodge theory and Galois descent to show that $S_g$ and $(H_g, 0)$ have the same minimal Sullivan model; early work of Sullivan then applies to construct that model explicitly. Here we prove

**Proposition 8.4.** *Any commutative cochain algebra $(A,d)$ with $H(A) \cong H_g$ has the same minimal Sullivan model as $(H_g,0)$.*

In the classical rational homotopy language, the result in [13] asserts that $S_g$ is *formal*; the extension here asserts that $(H_g, 0)$ is *intrinsically formal.*

*proof:* To begin, choose an arbitrary commutative cochain algebra $(A, d_A)$ for which $H(A, d_A) = H_g$ and let $(\wedge U, d) \xrightarrow{\simeq} (A, d_A)$ be a minimal Sullivan model. Choose a differential ideal $I \subset (\wedge U)^{\geq 2}$ to satisfy

$$I^{\geq 3} = (\wedge U)^{\geq 3} \quad \text{and} \quad I^2 \oplus (ker\, d)^2 = (\wedge U)^2\,.$$

Thus $\wedge U \to \wedge U/I$ is a quasi-isomorphism, and so we need only consider the case that $A^0 = \mathbb{Q}$, $A^{\geq 3} = 0$ and $A^2 = \mathbb{Q}\Phi \oplus d_A(A^1)$ where $\Phi$ is a cycle representing $\omega_g$. We assume henceforth that $(A, d_A)$ satisfies these conditions.

Next, let $H$ be the graded algebra $H_g/\omega_g$. As observed in §8.3, $(H,0)$ has a minimal Sullivan algebra of the form $(\wedge W, d)$ with

$$W = \underset{n\geq 0}{\oplus} V_n \quad \text{and} \quad d(V_0) = 0\,, \quad \text{and with } d: V_n \to \underset{i+j=n-1}{\oplus} V_i \wedge V_j$$

an injective map for $n \geq 1$. In particular $V_0$ has a basis $u_1, \cdots, u_g$, $v_1, \cdots, v_g$ representing the classes $\alpha_1, \cdots, \alpha_g, \beta_1, \cdots, \beta_g$ respectively. Then $V_1$ has a basis of the form $u_{ij}$ $(i < j)$, $v_{ij}$ $(i < j)$ and $w_{ij}$ (all $i, j$) with

$$du_{ij} = u_i \wedge u_j \quad dv_{ij} = v_i \wedge v_j \quad \text{and } dw_{ij} = u_i \wedge v_j \,. \tag{8.1}$$

On the other hand, consider the cdga $(B, d_B) = A \oplus \mathbb{Q}\,\Psi$ defined by $A^+\Psi = 0$ and $d\Psi = \Phi$. Define a morphism $\varphi : (\wedge V_0 \otimes \wedge V_1, d) \to (B, d)$ so that $\varphi u_i$, $\varphi v_j$ are cycles in $A^1$ representing $\alpha_i$, $\beta_j$, and so that $\varphi u_{ij}, \varphi v_{ij} \in A^1$ are chosen so that $d_A \varphi = \varphi d$, and

$$\varphi w_{ij} = a_{ij} + \delta_{ij} \Psi$$

where $a_{ij} \in A^1$ and $da_{ij} = u_i v_j - \delta_{ij}\Phi$. Since $H^2(B, d_B) = 0$, we may then extend $\varphi$ inductively to each $V_n$ to obtain a linear function $\mathcal{L} : W \to \mathbb{Q}$ and a quasi-isomorphism

$$\varphi : (\wedge W, d) \xrightarrow{\simeq} (B, d_B)$$

such that $\varphi w = \gamma(w) + \mathcal{L}(w)\Psi$ where $\gamma : W \to A^1$.

Thus the function $\mathcal{L}$ defines an element $x$ in the fundamental Lie algebra, $L$ of $\wedge W$,

$$\mathcal{L}(w) = \langle w, x \rangle \,,$$

and it satisfies

$$\langle V_0, x \rangle = 0 \,, \qquad \langle u_{ij}, x \rangle = \langle v_{ij}, x \rangle = 0 \qquad \text{and } \langle w_{ij}, x \rangle = \delta_{ij} \,. \tag{8.2}$$

Denote by $K(x) \subset W$ the codimension one space of $W$ defined by $K(x) = \{ w \,|\, \langle w, x \rangle = 0 \}$. Then it is easy to verify that $\varphi$ restricts to a quasi-isomorphism

$$(\mathbb{Q}\,1 \oplus K(x) \oplus \wedge^{\geq 2} W, d) \xrightarrow{\simeq} (A, d_A) \,.$$

The intrinsic formality of $H_g$ now follows from the following Lemma.

□

**Lemma 8.1.** *The cdga's $(A(x), d) := (\mathbb{Q}\,1 \oplus K(x) \oplus \wedge^{\geq 2} W, d)$ for which $x \in L$ satisfies (8.2) are all isomorphic.*

*proof:* Fix any $x \in L$ satisfying (8.2). As observed in §8.3,

$$L = \varprojlim_n \frac{\mathbb{L}(x_1, \cdots, x_g, y_1, \cdots, y_g)}{\mathbb{L}(x_1, \cdots, y_g)^n}$$

where $x_1, \cdots, y_g$ are dual to $u_1, \cdots, v_g$, and $x_i, y_j$ vanish on $V_{\geq 1}$. It follows from (8.2) that

$$x = \sum_{i=1}^{g} [x_i, y_i] + z$$

for some $z \in L^3$.

Our main step is to construct an automorphism $\sigma : L \xrightarrow{\cong} L$ such that

$$\sigma\left(\sum_{i=1}^{g} [x_i, y_i]\right) = \sum_{i=1}^{g} [x_i, y_i] + z\,.$$

Suppose then that for some $n$ we have constructed $x_i(k), y_i(k) \in L^{\geq 2}$, $k \leq n$, such that

$$\sum_{i=1}^{g} [x_i, y_i] + z = \sum_{i=1}^{g} [x_i + x_i(k), y_i + y_i(k)] + z(k)\,,$$

with $x_i(k+1) - x_i(k), y_i(k+1) - y_i(k) \in L^{\geq k+2}$, and with $z(k) \in L^{\geq k+3}$. Recall from the proof of Theorem 2.1 (p. 50) that $L^{n+3} = \sum_{i=1}^{g} [x_i, L^{n+2}] + \sum_{i=1}^{g} [y_i, L^{n+2}]$. Then we may write

$$z(n) = \sum_{i=1}^{k} [x_i + x_i(n), b] + [y_i + y_i(n), a]$$

with $a$, $b \in L^{\geq n+2}$. Then

$$\sum_{i=1}^{g} [x_i, y_i] + z - \sum_{i=1}^{g} [x_i + x_i(n) - a, y_i + y_i(n) + b] \in L^{\geq n+3}\,.$$

Set $x_i(n+1) = x_i(n) - a$ and $y_i(n+1) = y_i(n) + b$. This completes an inductive construction of elements $x_i(n)$ and $y_i(n)$.

Now set $x_i' = \lim x_i(n)$ and $y_i' = \lim y_i(n)$. The map $x_i \mapsto x_i'$, $y_i \mapsto y_i'$ extends to a homomorphism $\mathbb{L}(x_1, \cdots, x_g, y_1, \cdots, y_g) \to L$, which completes to a homomorphism $\sigma : L \to L$. Because $x_i - x_i', y_i - y_i' \in L^{\geq 2}$, $\sigma$ is an automorphism, and by construction, $\sigma(\sum_i [x_i, y_i]) = \sum_i [x_i, y_i] + z$.

Finally, recall from Theorem 2.1 (p. 50) that

$$(\wedge W, d) = \lim_{\to n} C^*(L/L^{n+2})\,.$$

It follows that $\sigma$ induces an automorphism $\tau$ of $(\wedge W, d)$, which restricts to an isomorphism between $\mathbb{Q}1 \oplus K(\overline{x}) \oplus \wedge^{\geq 2} W$ and $\mathbb{Q}1 \oplus K(x) \oplus \wedge^{\geq 2} W$, where $\overline{x} = \sum [x_i, y_i]$. In particular, these cdga's have the same minimal Sullivan model. □

**Theorem 8.1.** *(i) An orientable Riemann surface, $S_g$, $g \geq 1$, is a Sullivan space.*

*(ii) A minimal Sullivan model for $S_g$ has the form $(\wedge V, D)$ with $V = V^1 = \oplus_{n\geq 0} M_n$, where*

- *$M_0$ has a basis $u_i, v_i$ $1 \leq i \leq g$.*
- *$M_1$ has a basis $u_{ij}, v_{ij}$ and $w_{k,\ell}$ with $1 \leq i < j \leq g$ and $1 \leq k, \ell \leq g$ but with $k + \ell > 2$.*
- *$Du_{ij} = u_i u_j$, $Dv_{ij} = v_i v_j$, $Dw_{ij} = u_i v_j$ if $i \neq j$, and*

$$Dw_{ii} = u_i v_i - u_1 v_1 .$$

*Moreover, for $n \geq 2$, $D : M_n \to \oplus_{i+j=n-1} M_i \wedge M_j$.*

*proof:* First observe that a Riemann surface is a classifying space for its fundamental group; i.e. all its higher homotopy groups are zero. Denote by $Y$ the universal cover of $S_g$. Then $Y$ is a surface so $H_q(Y) = 0$ if $q > 2$. Moreover since $g \geq 1$, the fibre over a point is closed, discrete and infinite. Therefore, $Y$ is not compact and so $H_2(Y) = 0$. It follows then by the Hurewicz theorem that all the homotopy groups of $Y$ are zero, and so $N_g$ is aspherical. Thus we have only to show that a minimal Sullivan model for the Riemann surface is a Sullivan 1-algebra of the form stated in the Theorem.

Denote by $(\wedge V, d)$ the minimal Sullivan model for the space

$$X = (S^1 \times S^1) \vee (\vee_{i=3}^{2g} S_i^1) .$$

It follows from Theorems 7.9 (p. 222) and 7.10 (p. 223) that the minimal Sullivan model $(\wedge V, d)$ for $X$ is a Sullivan 1-algebra : $V = V^1$. Moreover, writing $Y = S^1 \times S^1$ and $Z = \vee_{i=3}^{2g} S_i^1$, we have quasi-isomorphisms

$$(\wedge V, d) \xrightarrow{\simeq} A_{PL}(X) \xrightarrow{\simeq} A_{PL}(Y) \oplus_{\mathbb{Q}} A_{PL}(Z) \xrightarrow{\simeq} (H(Y) \oplus_{\mathbb{Q}} H(Z), 0) .$$

Thus (cf. §2.10) $X$ is formal.

Now apply Proposition 2.7 (p. 88), to write $V = \oplus_{n\geq 0} M_n$, and let $\wedge V = \oplus_{n\geq 0} (\wedge V)_n$ denote the induced decomposition of $\wedge V$. We say $(\wedge V)_n$ has *lower degree* $n$. The decomposition of Proposition 2.7 has the following properties:

(i) $d$ decreases lower degree by 1

(ii) The corresponding lower gradation of $H(\wedge V, d)$ satisfies $H_k(\wedge V, d) = 0$, $k \geq 1$.

In particular, $d : M_1 \to \wedge^2 M_0$ may be described as follows. Assign $M_0$ the basis $u_i, v_i$, $1 \leq i \leq g$, where $[u_1]$ and $[v_1]$ correspond to the two circles in $Y$ and the $[u_i]$, $[v_i]$, $i \geq 2$ correspond to the circles in $Z$. Then $M_1$ may be assigned the basis : $u_{ij}$ $(1 \leq i < j \leq g)$, $v_{ij}$ $(1 \leq i < j \leq g)$ and $w_{ij}$ $(1 \leq i, j \leq g$ and $i + j > 2)$, with

$$du_{ij} = u_i \wedge u_j\,, \qquad dv_{ij} = v_i \wedge v_j \qquad \text{and } dw_{ij} = u_i \wedge v_j\,.$$

Next we introduce direct decompositions $M_n = \oplus_{r\geq 0} M_n(r)$ and denote the corresponding decompositions of $\wedge V$ by $\wedge V = \oplus_{n,r}(\wedge V)_n(r)$. We refer to this as the *central degree.* For this assign the basis elements $u_i, v_i, u_{ij}, v_{ij}$ and $w_{ij}$ central degree zero if $2 \leq i, j \leq g$. Then assign $u_1, v_1, u_{1j}, v_{1j}, w_{1j}$ central degree 1, and observe that $d : M_1 \to (\wedge^2 V)_0$ preserves the central degree. Now assume by induction that for some $n \geq 1$, the central degree has been defined in $M_k$, $k \leq n$, and that $d$ preserves the central degree. It follows from condition (ii) above that

$$d : M_{n+1} \xrightarrow{\cong} (\wedge^2 V)_n \cap \ker d\,.$$

Thus the central degree decomposition of $(\wedge^2 V)_n \cap \ker d$ defines a decomposition $M_{n+1} = \bigoplus_{r\geq 0} M_{n+1}(r)$ for which $d : M_{n+1}(r) \to (\wedge^2 V)_n(r)$. This completes the construction of the central degree.

Finally we perturb $d$ by a sequence of derivations $\delta_i$, $i \geq 2$ in $\wedge V$ satisfying : $\delta_i = 0$ in $M_0$ and

$$\delta_i : M_n(r) \to (\wedge^2 V)_{n-1}(r + i)\,, \qquad i \geq 2, n \geq 1\,.$$

Since each $M_n$ is finite dimensional, $\delta_i$ will vanish in $M_n$ for $i$ sufficiently large, and so $\delta = \sum_{i\geq 2} \delta_i$ will be a well-defined derivation. We shall construct the $\delta_i$ so that $(d + \delta)^2 = 0$ and $(\wedge V, d + \delta)$ is a Sullivan model for $(H_g, 0)$, thereby establishing the Proposition.

First, define $\delta_2$ in $M_1$ by setting $\delta_2(w_{ii}) = -u_1 \wedge v_1$ and requiring $\delta_2$ to vanish on the other basis elements. Then set $\delta_i(M_1) = 0$, $i \geq 3$. It is immediate that $(d+\delta)^2 = 0$ in $M_1$. Now suppose for some $n \geq 1$ that $\delta$ has been defined in $M_k$, $k \leq n$. Suppose further that for some $q$, $\delta_i$ has been defined in $M_{n+1}$ for $i < q$, and that for $r \geq 0$,

$$(d + \delta) \circ (d + \sum_{i<q} \delta_i) : M_{n+1}(r) \to \bigoplus_{j\geq q} (\wedge^3 V)_{n-1}(r + j)\,.$$

Since $\delta$ increases central degree and $(d + \delta)^2 = 0$ in $(\wedge V)_{\leq n}$ it follows that for $w \in M_{n+1}(r)$ the component $\Phi$ of $(d + \delta) \circ (d + \sum_{i<q} \delta_i)w$ in $(\wedge^3 V)_{n-1}(r + q)$ is a $d$-cycle. In view of condition (ii) above we may write $\Phi = d\Psi$ with $\Psi \in (\wedge^2 V)_n(r + q)$.

Now apply this observation to a basis $w_\ell$ for $M_{n+1}(r)$ to obtain elements $\Psi_\ell \in (\wedge^2 V)_n(r+q)$ and then define $\delta_q$ in $M_{n+1}(r)$, each $r$, by setting $\delta_q w_\ell = -\Psi_\ell$. It is immediate that

$$(d+\delta)\circ(d+\sum_{i\leq q}\delta_i) : M_{n+1}(r) \to \bigoplus_{j\geq q+1}(\wedge^3 V)_{n-1}(r+j)$$

and so the existence of $\delta$ follows by double induction, first on $q$ and then on $n$. It is immediate from the calculation that $(d+\delta)^2 = 0$.

It remains to show that $(\wedge V, d+\delta)$ is a Sullivan model for $(H_g, 0)$. Recall the description of $(H_g, 0)$ at the start of this section and define $\rho : \wedge V \to H_g$ by setting $\rho : u_i \mapsto \alpha_i$, $\rho : v_i \mapsto \beta_i$, and $\rho(M_n) = 0$, $n \geq 1$. By construction, $(d+\delta)(w_{ii}) = u_i \wedge v_i - u_1 \wedge v_1$. Thus $\rho \circ (d+\delta)(w_{ii}) = \alpha_i\beta_i - \alpha_1\beta_1 = \omega_g - \omega_g = 0$. Since $(d+\delta)(M_{\geq 2}) \subset (\wedge V)_{\geq 1}$ it follows that $\rho\circ(d+\delta) = 0$ and so $\rho$ is a morphism of cochain algebras. Moreover, assigning $\alpha_i, \beta_i$, $i \geq 2$ central degree 0, $\alpha_1, \beta_1$ central degree 1, and $\omega_g$ central degree 2 we see that $\rho$ preserves central degree. Since $\rho \circ d = 0$ and by construction, $\rho$ induces an isomorphism

$$H(\wedge V, d) \xrightarrow{\cong} H(Y) \oplus_{\mathbb{Q}} H(Z)$$

a spectral sequence argument gives that $H(\rho) : H(\wedge V, d+\delta) \xrightarrow{\cong} (H_g, 0)$. □

## 8.6 The classifying space of the pure braid group $P_n$ is a Sullivan space

The pure braid group with $n$ strands, $P_n$, can be defined as the fundamental group of the configuration space of $n$ points in $\mathbb{C}$,

$$F(\mathbb{C}, n) = \{(z_1, \cdots, z_n) \,|\, z_i \neq z_j \text{ for } i \neq j\}.$$

The symmetric group $\Sigma_n$ acts freely on $F(\mathbb{C}, n)$ and the quotient $C_n(\mathbb{C})$ is called the unordered configuration space of $n$ points in $\mathbb{C}$. Its fundamental group is the classical braid group $B_n$ on $n$ strands. Thus $F(\mathbb{C}, n)$ is a covering space of $C_n(\mathbb{C})$, and this gives a short exact sequence

$$1 \to P_n \to B_n \to \Sigma_n \to 1.$$

For $r < s$, denote by $A_{rs}$ the braid in $P_n$ where the strands different from $r$ and $s$ are vertical lines and the $r^{th}$ strand turns once around the $s^{th}$ one. The elements $A_{rs}$ generate $P_n$.

By forgetting the last strand we get a map $P_n \to P_{n-1}$ that fits into a short exact sequence ([5])

$$1 \to F_{n-1} \to P_n \to P_{n-1} \to 1$$

where $F_{n-1}$ is the free group on $n-1$ generators generated by the elements $A_{in}$, $i < n$. The action of the other generators $A_{rs}$ on $F_{n-1}$ is given by relations of the form

$$A_{rs}^{-1} A_{in} A_{rs} = C_{irs} A_{in} C_{irs}^{-1}$$

where $C_{irs}$ are commutators in the variables $A_{jn}$ ( [5]).

Since $P_2 = \mathbb{Z}$, and the action of $P_{n-1}$ on $F_{n-1}/[F_{n-1}, F_{n-1}]$ is trivial, the spaces $BP_n$ are all Sullivan spaces (Theorem 7.9, p. 222).

## 8.7 The Heisenberg group

The Heisenberg group $G$ is the group of upper triangular matrices of the form

$$\begin{pmatrix} 1 & a & c \\ 0 & 1 & b \\ 0 & 0 & 1 \end{pmatrix}$$

with coefficients $a, b, c \in \mathbb{Z}$. This group is generated by the matrices $A$, $B$ and $C$,

$$A = \begin{pmatrix} 1 & 1 & 0 \\ 0 & 1 & 0 \\ 0 & 0 & 1 \end{pmatrix}, \qquad B = \begin{pmatrix} 1 & 0 & 0 \\ 0 & 1 & 1 \\ 0 & 0 & 1 \end{pmatrix}, \qquad C = \begin{pmatrix} 1 & 0 & 1 \\ 0 & 1 & 0 \\ 0 & 0 & 1 \end{pmatrix}.$$

The group $G$ is 2-nilpotent. The subgroup $\mathbb{Z}$ generated by the matrix $C$ is central. This gives a short exact sequence $0 \to \mathbb{Z} \to G \to \mathbb{Z}^2 \to 0$. Denote by $G_{\mathbb{R}}$ and $G_{\mathbb{Q}}$ the groups of the analogous matrices with coefficients in $\mathbb{R}$ and $\mathbb{Q}$. Then $G_{\mathbb{R}}$ is a contractible space on which $G$ acts freely. This shows that the quotient $G_{\mathbb{R}}/G$ is a $K(G, 1)$.

Since the group $G$ is 2-nilpotent, the minimal model of $BG$, $(\wedge V, d)$, is a 1-minimal model and is given by the group structure: $(\wedge V, d) \cong (\wedge(x, y, z), d)$ with $dx = dy = 0$ and $dz = xy$.

On the other hand, $G_{\mathbb{Q}}$ is a divisible group and each of its elements has a power in $G$. Therefore the group $G_{\mathbb{Q}}$ is isomorphic to the Malcev completion of $G$.

## 8.8 Seifert manifolds

Let $\alpha$ be a non-zero rational number. Then rotation by $2\pi\alpha$ is a homeomorphism $\omega$ of the disk $D^2$ with finite period, and the orbit space $D^2/\omega$ is again a disk. A *standard fibred torus* is a space of the form $D^2 \times I/\sim$ where the equivalence relation identifies $(x,1)$ with $(\omega(x),0)$. The fibres of the projection $D^2\times I/\sim\to D^2/\omega$ are all circles and the fibre over 0 is called the *central fibre*; it is *ordinary* if $\alpha$ is an integer; otherwise it is *exceptional.*

A Seifert manifold $M$ is a closed 3-manifold together with a decomposition into a disjoint union of circles (called fibres) such that each fibre has a tubular neighborhood that is a union of circles forming a standard fibered torus. If $M$ is compact there are only finitely many exceptional fibres. Moreover [52] the orbit space $M/S^1$ is a two-dimensional compact manifold. We only consider the case when $M/S^1$ is oriented, and thus a Riemann surface $S_g$.

Choose an exceptional fibre and the associated fibered torus $T$ with boundary $\partial T$ and interior $T_0$. We denote by $T'$ the trivial bundle $D^2\times S^1$ and use the obvious homeomorphism $\partial T'\cong\partial T$ to replace $T'$ by $T$. Denote by $M'$ the new manifold. We will prove that $M$ and $M'$ have the same minimal Sullivan models. This will imply by induction on the number of exceptional fibres that $M$ has the minimal Sullivan model of a principal $S^1$-bundle over $S_g$.

Now recall from Proposition 13.5 of [18] that there is a quasi-isomorphism

$$A_{PL}(M)\xrightarrow{\simeq}A_{PL}(M\backslash T_0)\times_{A_{PL}(\partial T)}A_{PL}(T)\,.$$

Denote by $p,p':T,T'\to S^1$ the projections onto the central fibre. Since $p$ and $p'$ are homotopy equivalences it follows from the exact Mayer-Vietoris homology sequences that they give quasi-isomorphisms

$$\begin{array}{c}A_{PL}(M\backslash T_0)\times_{A_{PL}(\partial T)}A_{PL}(T)\\ {\scriptstyle id\times A_{PL}(p)}\big\uparrow{\scriptstyle\simeq}\\ A_{PL}(M\backslash T_0)\times_{A_{PL}(\partial T)}A_{PL}(S^1)\\ {\scriptstyle\simeq}\big\downarrow{\scriptstyle id\times A_{PL}(p')}\\ A_{PL}(M\backslash T_0)\times_{A_{PL}(\partial T)}A_{PL}(T')\,.\end{array}$$

This implies the result.

Now let $(\wedge W,d)$ be the minimal model of $N_g$.

**Proposition 8.5.** *A minimal Sullivan model for an orientable Seifert manifold, $M$, has the form $(\wedge W \otimes \wedge u, d)$ where $\deg u = 1$ and either $du = 0$ or $du$ represents the fundamental class of $S_g$. In either case $B(\pi_1(M))$ is a Sullivan space.*

*proof:* Denote by $(\wedge V, d)$ the minimal Sullivan model of $M$. In view of the discussion above we may suppose $M$ is a principal $S^1$-bundle over $S_g$. It follows from Proposition 15.17 of [18] that a Sullivan model for $S_g$ has the form $(\wedge x \otimes \wedge V, D)$ where $\deg x = 2$, $Dx = 0$ and $D(1 \otimes v) = 1 \otimes dv + \sum_{k\geq 1} x^k \otimes \theta_k(v)$. We have therefore a quasi-isomorphism

$$\varphi : (\wedge x \otimes \wedge V, D) \to (\wedge W, d)\,.$$

We consider now the Sullivan model $(\wedge x \otimes \wedge V \otimes \wedge u, D)$ with $Du = x$, whose minimal model is clearly $(\wedge V, d)$. This gives quasi-isomorphisms

$$(\wedge V, d) \xrightarrow{\simeq} (\wedge x \otimes \wedge V \otimes \wedge u, D) \xrightarrow[\simeq]{\varphi \otimes id} (\wedge W \otimes \wedge u, D')\,,$$

with $D'(u) = \varphi(x)$.

The last assertion follows because, by Proposition 8.1, $W = W^1$, while according to Orlik [52], $M$ is a classifying space for $\pi_1(M, *)$.

□

## 8.9 Arrangement of hyperplanes

An *arrangement of hyperplanes* is a finite collection $\{H_1, \cdots, H_n\}$ of codimension 1 linear subspaces of $\mathbb{C}^\ell$, some $\ell \geq 1$. To such an arrangement we associate the manifold $X = \mathbb{C}^\ell \setminus \cup_{i=1}^n H_i$. Write each hyperplane $H_i$ as the kernel of a linear form $\varphi_i$, and denote by $A$ the subalgebra of de Rham forms on $X$ generated by the closed forms

$$\omega_i = \frac{1}{2\pi i}\frac{d\varphi_i}{\varphi_i}\,.$$

A theorem of Brieskorn ([8]) states that the morphism that associates to $\omega_i$ its cohomology class $[\omega_i]$ induces an isomorphism $A \xrightarrow{\cong} (H^*(X;\mathbb{R}), 0)$. This implies that $X$ is a formal space.

A complete description of $A$ is then given by Orlik and Solomon ([53]): Let $C = \wedge V_1$ where $V_1$ is the vector space concentrated in degree 1 with basis $e_1, \cdots, e_n$ corresponding to the hyperplanes $H_1, \cdots, H_n$. For $J = \{i_1, \cdots, i_p\} \subset \{1, \cdots, n\}$ we write $[e_J] = \sum_k (-1)^{k-1} e_{i_1} \wedge \cdots \wedge \widehat{e_{i_k}} \cdots e_{i_p}$.

Denote by $\varphi : C \to A$ the morphism defined by $\varphi(e_i) = \omega_i$. Then the kernel of $\varphi$ is generated by the $[e_J]$ such that codim $H_{i_1} \cap \cdots \cap H_{i_p} < p$.

The arrangement is fiber-type if there is a tower of bundle maps

$$X = X_r \overset{p_r}{\to} X_{r-1} \to \cdots \to X_2 \overset{p_2}{\to} X_1 = \mathbb{C}^*,$$

where $X_k$ is the complement of an arrangement in $\mathbb{C}^k$, $p_k$ is a linear projection $\mathbb{C}^k \to \mathbb{C}^{k-1}$, and the fibre $F_k$ of $p_k$ is a copy of $\mathbb{C}$ with finitely many points removed.

**Proposition 8.6.** *The space $X$ associated to a fiber-type arrangement is a $K(G,1)$ and $BG$ is a Sullivan space.*

*proof:* Since $\mathbb{C}$ with finitely many points removed has the homotopy type of a wedge of circles, this is a Sullivan space. Now the proof works by induction on $r$ because as seen in [16], $\pi_1(X_{k-1})$ acts trivially on $H_*(F_k)$. □

## 8.10 Connected sum of real projective spaces

Let $X$ be the connected sum $X = \mathbb{R}P^{n+1}\#\mathbb{R}P^{n+1}$ with $n > 0$, $n$ even. Then we show that $B(\pi_1(X))$ is a Sullivan space, but $X$ is not a Sullivan space.

First let $S^n \times I$ be the collar in the connected sum and attach the disk, $D^{n+1}$, to $X$ by identifying its boundary with $S^n \times \{0\}$. This gives a space $Y$ of the same homotopy type as $\mathbb{R}P^{n+1} \vee \mathbb{R}P^{n+1}$. Since $X$ and $Y$ have the same 2-skeleton with respect to the standard cellular decomposition it follows from the cellular approximation theorem for continuous maps ([18], p. 9) and the van Kampen theorem that $\pi_1(X,*) = \pi_1(Y,*) = \mathbb{Z}_2\#\mathbb{Z}_2$.

On the other hand, $\mathbb{R}P^\infty \vee \mathbb{R}P^\infty$ is the classifying space for $\mathbb{Z}_2\#\mathbb{Z}_2$, and its rational homology is trivial. Thus $\mathbb{Z}_2\#\mathbb{Z}_2$ is rationally nilpotent and so $B(\mathbb{Z}_2\#\mathbb{Z}_2)$ a Sullivan space.

Then denote by $D$ the subspace of $S^{n+1}$ formed by the $n+2$-uples whose last component is $> 1/2$, and by $-D$ its antipodal image in the sphere. We denote by $Y$ the quotient of $S^{n+1}\backslash(D \cup -D)$ by the antipodal action of $\mathbb{Z}_2$. Thus $X$ is obtained by the union of two copies of $Y$ whose boundaries have been identified. Clearly $Y$ is homeomorphic to the quotient of $S^n \times [-\frac{1}{2}, \frac{1}{2}]$ by the relation identifying $(x,y)$ to $(-x,-y)$, or to $S^n \times [\frac{1}{2}, \frac{3}{2}]$ by the relation $(x,y) \sim (-x, 2-y)$.

Denote by $a$, $b$ the generators of $\mathbb{Z}_2\#\mathbb{Z}_2$ in $\mathbb{Z}_2\#\{e\}$ and $\{e\}\#\mathbb{Z}_2$. Then $G$ consists of the elements $c_1\cdots c_p$ $(p\geq 1)$ with $c_i = a$ or $b$ and $c_i \neq c_{i+1}$, together with the identity. Thus the elements of $G$ are given by the following: $\{(ab)^n, n\geq 0\}$, $\{(ba)^n, n>0\}$, $\{(ab)^n a, n\geq 0\}$ and $\{b(ab)^n, n\geq 0\}$. The group $G$ acts on $S^n\times\mathbb{R}$ by $a\cdot(u,t) = (-u,-t)$ and $b\cdot(u,t) = (-u, 2-t)$. Thus $a$ acts by the antipodal action on $S^n$ and symmetry around 0 on $\mathbb{R}$, and $b$ by the antipodal action on $S^n$ and symmetry around 1 on $\mathbb{R}$. Since

$$(ab)^n\cdot(u,t) = (u, t-2n)\,,\qquad (ba)^n\cdot(u,t) = (u,t+2n)\,,$$

$$(ab)^n a\cdot(u,t) = (-u,-t-2n)\quad\text{and}\quad b(ab)^n\cdot(u,t) = (-u,-t+2n+2)\,,$$

this action is free and division by the action is a covering projection. Moreover, it follows from the description above of $Y$ that the quotient is homeomorphic to $X$. This shows that the universal cover of $X$ is $S^n\times\mathbb{R}$. However, since $n$ is even, $X$ has the same rational cohomology as $S^{n+1}$, and its Sullivan minimal model is $(\wedge u, 0)$, with $u$ in degree $n+1$. Therefore $X$ is not a Sullivan space because $\pi_n(X)\otimes\mathbb{Q}\neq 0$.

## 8.11 A final example

Let $I_1$ and $I_2$ be two copies of the interval $I$ pointed by 0. We construct the projection $q : I_1\vee I_2 \to S^1_1\vee S^1_2$ by identifying the endpoints of the intervals. Let now $f_1$ and $f_2$ be the maps of the torus $T = S^1_1\times S^1_2$ given by $f_1(x,y) = (x,xy)$ and $f_2(x,y) = (xy,y)$. Then form

$$E = (I_1\vee I_2)\times T/\sim$$

where $(1_1, z)\sim(0, f_1(z))$ and $(1_2,z)\sim(0,f_2(z))$. We obtain in this way a fibration

$$T\to E\xrightarrow{\pi} S^1_1\vee S^1_2\,.$$

This is not a nilpotent fibration because the holonomy representations associated to the canonical generators of $\pi_1(S^1_1\vee S^1_2)$ on $H^1(T;\mathbb{Q})$ are given by the matrices

$$\begin{pmatrix}1&0\\1&1\end{pmatrix}\quad\text{and}\quad\begin{pmatrix}1&1\\0&1\end{pmatrix}$$

and there is no basis in which those matrices can be simultaneously triangulated.

However the pullback fibrations of $\pi$ along the two injections of $S^1$ into $S^1_1 \vee S^1_2$ are nilpotent fibrations. In particular, it follows from Theorem 5.1 (p. 145) that minimal Sullivan models for the restrictions $E_1$ and $E_2$ to $S^1_1$ and $S^1_2$ have the form

$$(\wedge u_1 \otimes \wedge(z_1, z_2), d) \quad \text{and} \quad (\wedge u_2 \otimes \wedge(z_1, z_2), d)$$

in which $u_1$ and $u_2$ are cycles representing the fundamental classes of $S^1_1$ and $S^1_2$.

Since $\pi : E \to S^1 \vee S^1$ is obtained by gluing these fibrations along a single fibre, $A_{PL}(E)$ is quasi-isomorphic to the cdga $(\wedge(u_1, u_2)/(u_1u_2) \otimes \wedge(z_1, z_2), d)$. Moreover it follows from Theorem 4.2 (p. 137) that the differential is given by $du_i = 0$, $dz_1 = u_2z_2$ and $dz_2 = u_1z_1$.

A straightforward computation shows that a Sullivan model for $A_{PL}(\pi)$ has the form $(\wedge u_1, u_2)/(u_1u_2) \otimes \wedge V, d)$ with $V = V^{\geq 2}$. The fibre, $(\wedge V, \overline{d})$, of the model is in this case very far from the model of the fibre $(\wedge(z_1, z_2), 0)$.

# Chapter 9

# Lusternik-Schnirelmann category

## 9.1 The LS category of topological spaces and commutative cochain algebras

The *LS (Lusternik-Schnirelmann) category of a topological space* $X$, denoted $\mathrm{cat}\, X$, is the least integer $m$ (or $\infty$) such that $X$ can be covered by $m+1$-open sets each contractible in $X$. This is an invariant of the homotopy type of $X$, and this and its elementary properties are presented in detail in §28 of [18]. For another presentation see [12] [36].

A clearly related concept is that of *$n$-cones*: there are based topological spaces $(P, *)$ of the form

$$(*) = P_0 \subset P_1 \subset \cdots \subset P_n = P$$

where each $P_{k+1}$ is a well-based space of the form $P_{k+1} = P_k \cup_{h_k} \overline{C}\Sigma^k Y_k$. (Here $\overline{C}\Sigma^k Y_k$ is the based cone on the $k^{th}$ suspension of a well-based space $Y_k$ and $h_k : \Sigma^k Y_k \to Y_k$ is a based continuous map.) The connection is provided by

**Theorem 9.1.** *([18], Th. 27.10; [23]) If $X$ is a normal space then $\mathrm{cat}\, X \leq n$ if and only if $X$ is retract of an $n$-cone. In particular, if $X$ is an $N$-dimensional CW complex with a single 0-cell then $\mathrm{cat}\, X \leq N$.*

The LS category of a commutative cochain algebra is defined in an analogous way, as we recall from [18] §29. First, let $(\wedge V, d)$ be a minimal Sullivan algebra and denote by

$$\lambda_m : (\wedge V, d) \to (\wedge V \otimes \wedge Z(m), d)$$

a minimal Sullivan model for the projection $\rho_m : (\wedge V, d) \to (\wedge V/\wedge^{>m} V, \overline{d})$:

$$\begin{array}{ccc} (\wedge V, d) & \xrightarrow{\lambda_m} & (\wedge V \otimes \wedge Z(m), d) \\ & \searrow^{\rho_m} & \simeq \downarrow \zeta_m \\ & & (\wedge V/\wedge^{>m} V, \overline{d}) \end{array} \tag{9.1}$$

**Definition.**

(i) The LS *category, cat($\wedge V, d$), of the minimal Sullivan algebra* $(\wedge V, d)$ is the least integer $m$ (or $\infty$) for which there is a cdga morphism,

$$\pi_m : (\wedge V \otimes \wedge Z(m), d) \to (\wedge V, d)\,,$$

such that $\pi_m \circ \lambda_m \sim id$.

(ii) The LS category, *cat*$(A, d)$, of a commutative cochain algebra $(A, d)$ with $H^0(A) = {I\!\!k}$ is the LS category of its minimal Sullivan model.

(iii) The *rational LS category*, $\mathrm{cat}_0(X)$, of a path connected topological space is the LS category of the cochain algebra $A_{PL}(X)$ defined over $\mathbb{Q}$.

Note that it follows from Corollary 3.1 (p. 99) that $\pi_m$ can be chosen so that $\pi_m \circ \lambda_m = id$.

**Proposition 9.1.** *(i) If $(A, d)$ is a commutative cochain algebra such that $A^0 = {I\!\!k}$ and $(A^+)^{m+1} = 0$ then cat$(A, d) \leq m$.*

*(ii) If $(A, d) \overset{\alpha}{\to} (B, d) \overset{\beta}{\to} (C, d)$ are morphisms of commutative cochain algebras satisfying $H^0(-) = {I\!\!k}$, and if $\beta \circ \alpha$ is a quasi-isomorphism then cat$(A, d) \leq$ cat$(B, d)$.*

*(iii) The category of a Sullivan 1-algebra $(\wedge V, d)$ is the maximum $m$ (or $\infty$) such that $H^m(\wedge V, d) \neq 0$.*

*proof:* (i) Let $\varphi : (\wedge V, d) \to (A, d)$ be a minimal Sullivan model. Then $\varphi$ factors as $(\wedge V, d) \overset{\rho_m}{\to} (\wedge V/\wedge^{>m} V, d) \overset{\overline{\varphi}}{\to} (A, d)$. Thus, with the notation of (9.1), $\overline{\varphi}\circ\zeta_m : (\wedge V \otimes \wedge Z(m), d) \to (A, d)$ satisfies $(\overline{\varphi}\circ\zeta_m)\circ\lambda_m = \varphi$. Since $\varphi$ is a quasi-isomorphism, by Proposition 1.10 (p. 21) $\overline{\varphi}\circ\zeta_m$ lifts up to homotopy through $\varphi$. This yields a morphism $\pi_m : (\wedge V \otimes \wedge Z(m), d) \to (\wedge V, d)$ such that $\varphi \circ \pi_m \sim \overline{\varphi} \circ \zeta_m$. In particular $\varphi \circ \pi_m \circ \lambda_m \sim \varphi$, and since $\varphi$ is a quasi-isomorphism, $\pi_m \circ \lambda_m \sim id$ ([18], Proposition 12.29).

(ii) Replacing $\alpha$ and $\beta$ by their Sullivan representatives reduces (ii) to the case that these cochain algebras are minimal Sullivan algebras, $(\wedge V, d)$, $(\wedge W, d)$ and $(\wedge U, d)$. In this case we obtain the commutative diagram

$$\begin{array}{ccccc} (\wedge V, d) & \xrightarrow{\alpha} & (\wedge W, d) & \xrightarrow{\beta} & (\wedge U, d) \\ \downarrow{\scriptstyle \rho_m} & & \downarrow{\scriptstyle \rho_m} & & \\ (\wedge V/\wedge^{>m} V, d) & \xrightarrow{\overline{\alpha}} & (\wedge W/\wedge^{>m} W, d)\,. & & \end{array}$$

Now consider the morphisms

$$(\wedge V, d) \xrightarrow{\lambda_m} (\wedge V \otimes \wedge Z, d) \underset{\simeq}{\xrightarrow{\zeta_m}} (\wedge V/\wedge^{>m} V, \overline{d})$$

defined in (9.1). If $\mathrm{cat}\,(\wedge W, d) \leq m$ a straightforward argument (Proposition 1.10 (p. 21)) produces a morphism

$$\gamma : (\wedge V \otimes \wedge Z, d) \to (\wedge U, d)$$

such that $\gamma \circ \lambda_m \sim \beta \circ \alpha$. Lift $\gamma$, up to homotopy, through the quasi-isomorphism $\beta \circ \alpha$ to obtain $\pi_m : (\wedge V \otimes \wedge Z, d) \to (\wedge V, d)$ such that $(\beta \circ \alpha) \circ \pi_m \sim \gamma$. Thus $(\beta \circ \alpha) \circ \pi_m \circ \lambda_m \sim \gamma \circ \lambda_m \sim \beta \circ \alpha$, and since $\beta \circ \alpha$ is a quasi-isomorphism it follows that $\pi_m \circ \lambda_m \sim id$; i.e. $\mathrm{cat}\,(\wedge V, d) \leq m$.

(iii) If $\mathrm{cat}\,(\wedge V, d) = m$, then the projection

$$\rho_m : (\wedge V, d) \to (\wedge V/\wedge^{>m} V, \overline{d}) = (\wedge V/(\wedge V)^{>m}, \overline{d})$$

is injective in cohomology, and so $H^{>m}(\wedge V, d) = 0$. Conversely, suppose $H^{>m}(\wedge V, d) = 0$. Form the acyclic differential ideal $I = \wedge^{>m} V \oplus S^m$ where $S^m \oplus (\ker d)^m = (\wedge V)^m$. Then the projection $p : (\wedge V, d) \to (\wedge V/I, \overline{d})$ is a quasi-isomorphism and it follows from (i) that $\mathrm{cat}(\wedge V, d) \leq m$. □

A connection between topological and rational LS category is provided by

**Theorem 9.2.** *Let $X$ be a path connected normal topological space. Then*

$$cat_0(X) \leq cat\, X\,.$$

*proof:* Suppose $\mathrm{cat}\, X = m$. Then $X$ is a retract of an $m$-cone $Y$. Let $(\wedge V, d)$ (respectively $(\wedge W, d)$) be the minimal Sullivan model of $X$ (respectively of $Y$).

The proof of (i) $\Rightarrow$ (ii) in Theorem 29.1 of [18] does not require simple connectivity, and therefore provides a quasi-isomorphism $(\wedge W, d) \xrightarrow{\simeq} (A, d)$

with $A^0 = \mathbb{k}$ and $(A^+)^{m+1} = 0$. It follows from Proposition 9.1(i) that $\mathrm{cat}(\wedge W, d) \leq m$. Since $X$ is a retract of $Y$, there are cdga morphisms

$$(\wedge V, d) \xrightarrow{\alpha} (\wedge W, d) \xrightarrow{\beta} (\wedge V, d)$$

such that $\beta \circ \alpha \sim id$. Thus the Proposition follows from Proposition 9.1(ii).

□

## 9.2 The mapping theorem

Here we extend Theorem 29.5 of [18] to the non-simply connected case.

**Theorem 9.3.** *Let $\varphi : (\wedge V, d) \to (\wedge W, d)$ be a morphism of minimal Sullivan algebras.*

*(i) If $\varphi$ is surjective, then $cat(\wedge V, d) \geq cat(\wedge W, d)$.*

*(ii) If there are $r$ elements of odd degrees $w_1, \cdots, w_r \in W$ such that $\wedge W = \wedge(w_1, \cdots, w_r)\, Im\, \varphi$, then*

$$cat(\wedge V, d) + r \geq cat(\wedge W, d)\,.$$

*proof:* For (i) we suppose $\mathrm{cat}(\wedge V, d) = m$. Thus in the notation of (9.1) we have morphisms $(\wedge V, d) \overset{\lambda_m}{\to} (\wedge V \otimes \wedge Z, d) \overset{\pi_m}{\to} (\wedge V, d)$ with $\pi_m \circ \lambda_m = id$.

On the other hand, since $\varphi$ is surjective we may choose $V$ so that $\varphi : V \to W$. Let

$$\chi : (\wedge V \otimes \wedge U, d) \overset{\simeq}{\to} (\wedge W, d)$$

be a $\Lambda$-model for $\varphi$ as constructed in Proposition 3.9 (p. 103). Then applying $\wedge V \otimes \wedge U \otimes_{\wedge V} -$ we obtain the diagram

$$\begin{array}{ccc} (\wedge V \otimes \wedge U, d) & \underset{id \otimes \lambda_m}{\overset{id \otimes \pi_m}{\rightleftarrows}} & (\wedge V \otimes \wedge U \otimes \wedge Z, d) \\ & {\scriptstyle \rho_m \otimes id} \searrow \qquad \swarrow {\scriptstyle id \cdot \zeta_m}\, \simeq & \\ & ((\wedge V / \wedge^{>m} V) \otimes \wedge U, d)\,. & \end{array}$$

On the other hand, dividing $\wedge V \otimes \wedge U$ by $\wedge^{\geq 2} V \otimes \wedge U$ we obtain the cdga $((V \oplus \mathbb{k}) \otimes \wedge U, \overline{d})$ with $\overline{d} : \wedge^+ U \to V \otimes \wedge U$ an injective map. Dividing by a direct summand of $\mathrm{Im}\, \overline{d}$ in $V \otimes \wedge U$ then provides a surjection $(\wedge V \otimes \wedge U, d) \to (\mathrm{Im}\, \overline{d} \oplus \wedge U, \overline{d})$ onto a cochain algebra whose cohomology is $\mathbb{k}$ in degree 0 and 0 in the other degrees. This factors through $\rho_m \otimes id$ to a surjection

$$\psi : (\wedge V / \wedge^{>m} V \otimes \wedge U, d) \to (\mathrm{Im}\, \overline{d} \oplus \wedge U, \overline{d})\,.$$

Set $A = \psi^{-1}(\Bbbk \cdot 1)$. Then the inclusion $(A,d) \hookrightarrow (\wedge V/\wedge^{>m} V \otimes \wedge U, d)$ is a quasi-isomorphism and $A \subset (\wedge^+ V/\wedge^{>m} V \otimes \wedge U) \oplus \Bbbk$. Thus $A^0 = \Bbbk$ and $(A^+)^{m+1} = 0$. It follows from Proposition 9.1(i) that $\mathrm{cat}\,(A,d) \leq m$. Since $(A,d)$ is connected by quasi-isomorphisms with $(\wedge V \otimes \wedge U \otimes \wedge Z, d)$, it follows that $\mathrm{cat}(\wedge V \otimes \wedge U \otimes \wedge Z, d) \leq m$. But $(\wedge V \otimes \wedge U, d)$ is a retract of that cochain algebra. Thus, since $(\wedge V \otimes \wedge U, d) \simeq (\wedge W, d)$, it follows from Proposition 9.1(ii) that $\mathrm{cat}\,(\wedge W, d) \leq m$.

For (ii) we suppose that $w_1 \in W^n$, $dw_1 \in \wedge W$ and $w_2, \cdots, w_r \in W^{\geq n}$. Choose the generating space $V$ so that $\varphi : V^{\leq n} \to W^{\leq n}$. Then the ideal $I$ generated by $V^{\leq n} \cap \ker \varphi$, is preserved by $d$, and the quotient $(\wedge V/I, d)$ is a minimal Sullivan algebra $(\wedge V', d)$. The map $\varphi$ factors as

$$(\wedge V, d) \xrightarrow{\psi} (\wedge V', d) \xrightarrow{\varphi'} (\wedge W, d).$$

Since $\psi$ is surjective, by (i) it is sufficient to prove that $\mathrm{cat}\,(\wedge V', d) + r \geq \mathrm{cat}\,(\wedge W, d)$.

First, note that $\wedge W = \wedge(w_1, \cdots, w_r)\,\mathrm{Im}\,\varphi'$ and that $\varphi'$ is injective in $(\wedge V')^{<n}$. Now we construct a new cdga $(\wedge V' \otimes \wedge u, d)$ with $du = (\varphi')^{-1}(d(w_1))$, and extend $\varphi'$, to

$$\varphi'' : (\wedge V' \otimes \wedge u, d) \to (\wedge W, d)\,, \qquad \text{with } \varphi''(u) = w_1\,.$$

By induction,

$$\mathrm{cat}\,(\wedge V' \otimes \wedge u, d) + (r-1) \geq \mathrm{cat}(\wedge W, d)\,.$$

Finally suppose $(\wedge V', d)$ has category $m$. Then $(\wedge V', d)$ is a retract of the Sullivan model of a cdga (B,d) in which $B = B^+ \oplus \Bbbk$ and $(B^+)^{n+1} = 0$. It follows that $(\wedge V' \otimes \wedge u, d)$ is a retract of a Sullivan model of the cdga $(B \otimes \wedge u)$ for which $[(B \otimes \wedge u)^+]^{m+2} = 0$. Thus $\mathrm{cat}\,(\wedge V', d) + 1 \geq \mathrm{cat}(\wedge V' \otimes \wedge u, d)$, which completes the proof.

□

## 9.3 Module category and the Toomer invariant

Fix a minimal Sullivan algebra $(\wedge V, d)$, and let $(P,d)$ be a semifree resolution of a $(\wedge V, d)$-module $(M,d)$ (cf. §1.2 (p. 7)). Note that we permit $M$ to be $\mathbb{Z}$-graded: $M = \{M_i\}_{i\in\mathbb{Z}}$. Consider the commutative diagram

$$\begin{array}{ccc} (P,d) & \xrightarrow{\lambda_k} & (P(k),d) \\ & \searrow^{\rho(k)} & \simeq \downarrow \zeta(k) \\ & & (P/\left(\wedge^{>k} V \cdot P\right), d)\,, \end{array}$$

in which $\zeta(k)$ is a semifree resolution (cf. §1.2) of $\left(P/\left(\wedge^{>k}V\cdot P\right),d\right)$.

Recall from Proposition 6.4 in [18], that any morphism from a semifree $(\wedge V,d)$-module lifts up to homotopy through a quasi-isomorphism. In particular, two semifree resolutions of $(M,d)$ are quasi-isomorphic. Thus the invariants defined next are independent of the choice of the semifree resolution $(P,d)$ of $(M,d)$.

**Definition.**

(1) The *module category* of $(M,d)$, denoted $\mathrm{mcat}(M,d)$, is the least integer $m$ (or $\infty$) for which there is a morphism $\eta:(P(m),d)\to(P,d)$ such that the $(\wedge V,d)$-module morphisms $\eta\circ\lambda_m$ and $id$ are homotopic (cf. §1.1): $\eta\circ\lambda(m)\sim id$.
(2) The *Toomer invariant* of $(M,d)$, denoted $e(M,d)$, is the least integer $m$ (or $\infty$) such that $H(\rho(m))$ is injective.

**Proposition 9.2.** *Let $(\wedge V,d)$ be a minimal Sullivan algebra.*

*(i) If $(M,d)$ is a $(\wedge V,d)$ module and $(\wedge^{\geq m+1}V)\cdot M=0$ then $mcat(M,d)\leq m$.*
*(ii) If $(M,d)\stackrel{\alpha}{\to}(N,d)\stackrel{\beta}{\to}(Q,d)$ are morphisms of $(\wedge V,d)$-modules and $\beta\circ\alpha$ is a quasi-isomorphism, then $mcat(M,d)\leq mcat(N,d)$.*

*proof:* This follows in exactly the same way as Proposition 9.1 is proved. □

**Proposition 9.3.** *Let $(\wedge V,d)$ be a minimal Sullivan algebra, and $(M,d)$ be any $(\wedge V,d)$-module. Then*

*(i) $e(M,d)\leq mcat(M,d)$.*
*(ii) $mcat(M,d)\leq mcat(\wedge V,d)\leq cat(\wedge V,d)$.*

*proof:* This is a restatement of Proposition 29.15 (i) and (ii) of [18], which place no further restrictions on $(\wedge V,d)$ or $(M,d)$. □

## 9.4 cat = mcat

The main result of this section is the

**Theorem 9.4.** (Hess, [33]; [18] Theorem 29.9) *Let $(\wedge V,d)$ be a minimal Sullivan algebra. Then $cat\,(\wedge V,d)=mcat(\wedge V,d)$.*

*proof:* This follows from Proposition 9.4 below by an argument identical with that of the proof of Theorem 28.9 in [18]. (Note that Proposition 9.4 simply extends Proposition 29.10 of [18] to the non-simply connected case.) □

**Proposition 9.4.** *The quotient morphism* $\rho : (\wedge V, d) \to (\wedge V/\wedge^{>m} V, d)$ *for a minimal Sullivan algebra extends to a minimal Sullivan model*

$$\chi : (\wedge V \otimes \wedge Z, d) \xrightarrow{\simeq} (\wedge V/\wedge^{>m} V, d)$$

*in which* $\chi(Z) = 0$, *and* $Z$ *can be written as a direct sum,* $Z = \oplus_{q\geq 0} Z_q$ *such that (setting* $Z_{-1} = 0$*) for each* $q \geq 0$,

$$d : 1 \otimes Z_q \to \wedge^{>m+q(m-1)} V \oplus \left[\bigoplus_{i=0}^{q} \wedge^{>i(m-1)} V \otimes Z_{q-i}\right] \oplus (1 \otimes (\wedge^2 Z)_{q-1}).$$

*proof:* We introduce a second gradation in $\wedge V$ (the *left grading*) by setting $V^p = V^{1,p}$. Then a minimal Sullivan algebra $(\wedge V, d_1)$ is defined by $d_1 : V \to \wedge^2 V$ and $(d - d_1) : V \to \wedge^{\geq 3} V$. The differential $d_1$ is called the *quadratic part* of $d$ and it increases the left degree by 1.

Exactly as in the construction of a Sullivan model in the singly graded case (see also [18], Example 6, p. 159), we may extend $\rho : (\wedge V, d_1) \to (\wedge V/\wedge^{>m} V, d_1)$ to a bigraded Sullivan model

$$\chi : (\wedge V \otimes \wedge Z, d_1) \xrightarrow{\cong} (\wedge V/\wedge^{>m} V, d_1)$$

in which each $Z^p = \oplus_{k\geq m} Z^{k,p}$, $d_1$ increases left degree by 1, and $\chi(Z) = 0$.

Because this is a Sullivan extension, $Z$ is the increasing union of bigraded subspaces

$$0 = Z(0) \subset \cdots \subset Z(r) \subset \cdots$$

such that

$$d_1 : 1 \otimes Z^p(r+1) \to \wedge V \otimes \wedge(Z^{<p} \oplus Z^p(r)).$$

For simplicity of notation we write

$$W(p, r) = Z^{<p} \oplus Z^p(r).$$

The condition above on $d_1$ implies that each $\wedge V \otimes \wedge W(p, r)$ is preserved by $d_1$.

The Sullivan fibre of the extension $(A, d_1) = (\wedge V \otimes \wedge Z, d_1)$ will be denoted $(\wedge Z, \overline{d_1})$; it is the quotient of $A$ by the ideal $\wedge^+ V \otimes \wedge Z$. However, the expression of $(A, d)$ as a tensor product depends on the choice of the subspace $1 \otimes Z$. This choice can be any linear inclusion $\ell$ of the bigraded space $Z$ such that the surjection $\pi : A \to \wedge Z$ satisfies $\pi \circ \ell = id$ and $\chi \circ \ell = 0$. We refer to this as the *choice of a fibre subspace.* Note that the condition on $d_1$ above is independent of the choice of fibre subspace.

The rest of the proof is a sequence of five steps.

Step One: *The inclusion*

$$(\Bbbk \oplus \ker \overline{d_1} \cap Z^{m,*}, 0) \to (\wedge Z, \overline{d_1})$$

*is a quasi-isomorphism.*

Step Two: *The generating space $Z \subset \wedge Z$ can be chosen so that $Z = \oplus_{q \geq 0} Z_q$ with $\overline{d_1} : Z \to \wedge^2 Z$, and*

$$Z_0 = \ker \overline{d_1} \cap Z^{m,*} \quad \text{and } \overline{d_1} : Z_q \xrightarrow{\cong} \ker \overline{d_1} \cap (\wedge^2 Z)_{q-1}\,, \quad q \geq 1\,.$$

*Moreover, if $m = 1$ then $Z = Z^{m,*}$ while if $m \geq 2$, then $Z_q = Z^{q(m-1)+m,*}$.*

Step Three: *The fibre subspace $1 \otimes Z$ may be chosen so that*

$$d_1 - (id \otimes \overline{d_1}) : 1 \otimes Z_q^p(r+1) \to \wedge V \oplus \left(\wedge^+ V \otimes W_{\leq q}(p,r)\right)\,.$$

*In particular, $\wedge V \otimes (\Bbbk \oplus Z_0)$ is stable under $d_1$. Moreover, $\chi$ restricts to a quasi-isomorphism*

$$(\wedge V \otimes (\Bbbk \oplus Z_0), d_1) \xrightarrow{\simeq} (\wedge V / \wedge^{>m} V, d_1)\,.$$

Step Four: *The differential $d_1$ in $(\wedge V \otimes \wedge Z)$ may be perturbed to a differential $d$ such that $(\wedge V \otimes \wedge Z, d)$ is a Sullivan extension of $(\wedge V, d)$ and $d - d_1$ increases left degrees by at least two, while for $q \geq 0$,*

$$(d - id \otimes \overline{d_1}) : 1 \otimes Z_q \to \wedge^{>m+q(m-1)} V \oplus \left[\bigoplus_{i=0}^{q} \wedge^{>i(m-1)} V \otimes Z_{q-i}\right]\,.$$

Step Five: *Conclusion of the proof.*

It remains to carry out this program.

*Step One. The inclusion*

$$(\Bbbk \oplus \ker \overline{d_1} \cap Z^{m,*}), 0) \to (\wedge Z, \overline{d_1})$$

*is a quasi-isomorphism.*

*proof:* In the acyclic closure $(\wedge V \otimes \wedge U, d_1)$ of $(\wedge V, d_1)$ as constructed in Proposition 3.9 (p. 103) we may suppose by Corollary 3.4 (p. 105) that $d_1$ :

$U \to V \otimes \wedge U$. Thus assigning $\wedge U$ the left degree zero makes $(\wedge V \otimes \wedge U, d_1)$ into a bigraded algebra in which $d_1$ increases left degrees by 1.

Now let $\varepsilon$ be the augmentation in $\wedge V \otimes \wedge U$ given by $\varepsilon(U) - \varepsilon(V) = 0$ and consider the morphisms

$$\begin{array}{ccc} ((\wedge V/\wedge^{>m} V) \otimes \wedge U, d_1) & \xleftarrow{\chi \otimes id} & (\wedge V \otimes \wedge Z, d_1) \otimes_{\wedge V} (\wedge V \otimes \wedge U, d_1) \\ & & \Big\downarrow {\scriptstyle id \otimes \varepsilon} \\ & & (\wedge Z, \overline{d_1})\,. \end{array}$$

Both $(\wedge V \otimes \wedge Z, d_1)$ and $(\wedge V \otimes \wedge U, d_1)$ are $(\wedge V, d_1)$-semifree and both $\chi$ and $\varepsilon$ are quasi-isomorphisms. It follows by Proposition 6.7 in §6 of [18] that $\chi \otimes id$ and $id \otimes \varepsilon$ are quasi-isomorphisms. But $\wedge V/\wedge^{>m} V \otimes \wedge U$ vanishes in left degrees $> m$ while $\wedge^+ Z$ vanishes in left degrees $< m$. It follows that $H(\wedge Z, \overline{d_1}) = I\!k \oplus H^{m,*}(\wedge Z, \overline{d_1})$.

Regard $\wedge^m V \otimes \wedge U$ as a subspace of $(\wedge V/\wedge^{>m} V) \otimes \wedge U$. It follows that there is a subspace $H = H^{m,*} \subset \ker d_1 \cap (\wedge^m V \otimes \wedge U)$ such that the inclusion $(I\!k \oplus H, 0) \to (\wedge V/\wedge^{>m} V \otimes \wedge U, d_1)$ is a quasi-isomorphism. Since this is automatically a cdga quasi-isomorphism there is then a cdga quasi-isomorphism

$$\sigma : (\wedge Z, \overline{d_1}) \xrightarrow{\simeq} (I\!k \oplus H, 0) \tag{9.2}$$

respecting the bidegrees. Since $H \wedge H = 0$, $\sigma$ restricts to an isomorphism

$$\ker \overline{d_1} \cap Z^{m,*} \xrightarrow{\cong} H\,.$$

Thus the inclusion

$$(I\!k \oplus (\ker \overline{d_1} \cap Z^{m,*}), 0) \to (\wedge Z, \overline{d_1})$$

is a quasi-isomorphism.

*Step Two: The generating space $Z \subset \wedge^+ Z$ can be chosen so that $Z = \oplus_{q \geq 0} Z_q$, with*

$$Z_0 = \ker \overline{d_1} \cap Z^{m,*} \quad \text{and } \overline{d_1} : Z_q \xrightarrow{\cong} (\ker \overline{d_1}) \cap (\wedge^2 Z)_{q-1}, \quad \text{if } q \geq 1\,.$$

*Moreover, if $m = 1$ then $Z = Z^{m,*}$ while if $m \geq 2$ then $Z_q = Z^{q(m-1)+m,*}$.*

*proof:* Set $Z_0 = \ker \overline{d_1} \cap Z^{m,*}$. The first assertion then follows from (9.2), as asserted in Example 7 of §12(d) in [18]. The second assertion follows by a straightforward induction on $q$.

*Step Three: The fibre subspace $1 \otimes Z$ may be chosen so that for $q \geq 0$,*

$$d_1 - (id \otimes \overline{d_1}) : 1 \otimes Z_q^p(r+1) \to \wedge V \oplus (\wedge^+ V \otimes W_{\leq q}(p,r))\,.$$

*In particular,* $\wedge V\otimes(\Bbbk\oplus Z_0)$ *is stable under* $d_1$. *Moreover,* $\chi$ *restricts to a quasi-isomorphism* $(\wedge V\otimes(\Bbbk\oplus Z_0),d_1)\xrightarrow{\simeq}(\wedge V/\wedge^{>m}V,d_1)$.

*proof:* Assume for some $p\geq 1$, $r\geq 0$ that the fibre subspace has been chosen so that the conclusion of Step Three holds in $1\otimes W(p,r)$. For simplicity, denote $W(p,r)$ simply by $W$. Fix some $q\geq 0$ and recall from Step Two that $Z_q\subset Z^{k,*}$ for some $k$. Then fix a basis for a direct summand of $Z_q^p(r)$ in $Z_q^p(r+1)$. Let $z$ be an element of that basis. Then

$$d_1(1\otimes z)-1\otimes\overline{d_1}(z)=\sum_{j=0}^{n}\Phi(j)$$

with $\Phi(j)\in\wedge^+V\otimes\wedge^jW$.

If $n\geq 2$ then our assumption implies that $(id\otimes\overline{d_1})\Phi(n)=0$. Thus it follows from Step Two that $\Phi(n)=(id\otimes\overline{d_1})\Psi(n)$ for some $\Psi(n)\in\wedge^+V\otimes(\wedge^{n-1}Z)^{<p}$. Since for any $s$, $\wedge V\otimes\wedge Z^{\leq s}$ is preserved by $d_1$, and since $Z^{<p}\subset W$, we obtain

$$d_1(1\otimes z)-1\otimes\overline{d_1}z-d_1\Psi(n)\in\wedge^+V\otimes(\bigoplus_{j=0}^{n-1}\wedge^jW).$$

Iterate this process to obtain $\Psi=\sum_{j\leq n}\Psi(j)\in\wedge^+V\otimes(\wedge^+W)^{<p}$ such that

$$d_1(1\otimes z-\Psi)-1\otimes\overline{d_1}z\in\wedge V\oplus(\wedge^+V\otimes W).$$

We may clearly choose $\Psi$ to have the same left degree as $z$.

Write

$$d_1(1\otimes z-\Psi)-1\otimes\overline{d_1}z=\Phi_0+\Phi_1$$

with $\Phi_0\in\wedge V$ and $\Phi_1\in\wedge^+V\otimes W$. It follows from our assumption and Step Two that

$$(1\otimes\overline{d_1})\Phi_1\in\wedge^+V\otimes\wedge^2W\qquad\text{and } d_1(\Phi_0+\Phi_1)-(1\otimes\overline{d_1})\Phi_1\in\wedge V\otimes(\Bbbk\oplus W).$$

On the other hand,

$$d_1(\Phi_0+\Phi_1)=-d_1(1\otimes\overline{d_1}z)\in\wedge V\otimes(\Bbbk\oplus W)\oplus\left(\wedge^+V\otimes(\wedge^2W)_{<q}\right).$$

These two relations imply that $(1\otimes\overline{d_1})\Phi_1\in\wedge^+V\otimes(\wedge^2W)_{<q}$. Now it follows from Step Two that $\Phi_1\in\wedge^+V\otimes W_{\leq q}$.

Thus we may replace $1\otimes z$ by $1\otimes z-\Psi$ for each element of the chosen basis to arrange that the conclusion of Step Three holds in $W(p,r+1)$. The first assertion of Step Three now follows by induction, first on $r$, then on $p$.

Since $\overline{d_1} = 0$ in $Z_0$, it follows from what we have shown that $d_1 : Z_0 \to \wedge V \otimes (\Bbbk \oplus Z_0)$. Filter both $\wedge V \otimes (\Bbbk \oplus Z_0)$ and $\wedge V \otimes \wedge Z$ by the subspaces $\wedge^{\geq k} V \otimes -$. The inclusion $i : \wedge V \otimes (\Bbbk \oplus Z_0) \to \wedge V \otimes \wedge Z$ induces a morphism of the resulting spectral sequences, beginning with the inclusion

$$(\wedge V \otimes (\Bbbk \oplus Z_0), 0) \to (\wedge V, 0) \otimes (\wedge Z, \overline{d_1}),$$

which is a quasi-isomorphism by Step One. Since $\chi$ is a quasi-isomorphism, so is its restriction to $(\wedge V \otimes (\Bbbk \oplus Z_0), d_1)$.

*Step Four: The differential $d_1$ in $\wedge V \otimes \wedge Z$ may be perturbed to a differential $d$ such that $(\wedge V \otimes \wedge Z, d)$ is a Sullivan extension of $(\wedge V, d)$ and $d - d_1$ increases left degrees by at least two, while*

$$d - (id \otimes \overline{d_1}) : 1 \otimes Z_q \to \wedge^{>m+q(m-1)} V \bigoplus \left[ \oplus_{i=0}^{q} \wedge^{>i(m-1)} V \otimes Z_{q-i} \right].$$

*proof:* Suppose for some $p \geq 1$ and $i \geq 0$ that $d$ has been constructed in $W(p,r)$ to satisfy these conditions, and again denote $W(p,r)$ simply by $W$. By Step Two any $Z^{k,*}$ is the direct sum of spaces $Z_q^{k,*}$ for which $Z_{<q} \subset Z^{\leq k,*}$. It follows from Step Three and our assumption that each $\wedge V \otimes \wedge W^{\leq k,*}$ is stable under $d - id \otimes \overline{d_1}$ and under $d_1 - id \otimes \overline{d_1}$.

Now fix $q \geq 0$ and let $z$ be an element of a basis of a direct summand of $Z_q^p(r)$ in $Z_q^p(r+1)$. By Step Two, $Z_q \subset Z^{k,*}$ for some $k$.

Since $d - d_1$ increases left degrees by at least two, and since $d - d_1 = (d - id \otimes \overline{d_1}) - (d_1 - id \otimes \overline{d_1})$ it follows that

$$d - d_1 : W_s^{j,*} \to \wedge^{\geq j+2} V \oplus (\wedge^{\geq 2} V \otimes W_{\leq s}^{\leq j}). \tag{9.3}$$

It follows from this together with Step Three and our assumption that

$$dd_1(1 \otimes z) = (d - d_1) d_1 (1 \otimes z) \in \wedge^{\geq 3} V \otimes (\Bbbk \oplus W) \oplus (\wedge^{\geq 2} V \otimes (\wedge^2 W)_{<q}).$$

In particular the component of $dd_1(1 \otimes z)$ in $\wedge^{\geq 2} V \otimes (\wedge^2 W)_{<q}$ is an $(id \otimes \overline{d_1})$-cycle and hence by Step Two of the form $(id \otimes \overline{d_1})\Omega$ for some $\Omega \in \wedge^{\geq 2} V \otimes Z_{\leq q}$.

Since $\deg z = p$ it follows that $\deg \Omega = p + 1$ and thus that

$$\Omega \in \wedge^{\geq 2} V \otimes Z_{\leq q}^{<p} \subset \wedge^{\geq 2} V \otimes W_{\leq q}^{<p}.$$

Thus $d\Omega$ is defined and by assumption

$$(d - id \otimes \overline{d_1})\Omega \in \wedge V \oplus \left[ \wedge^{\geq 3} V \otimes W_{\leq q}^{<p} \right].$$

It follows that

$$d(d_1(1 \otimes z) - \Omega) \in \wedge V \oplus (\wedge^{\geq 3} V \otimes W_{\leq q}^{<p}).$$

But then the component in $\wedge V \otimes W$ of this element is an $(id \otimes \overline{d_1})$-cycle and so by definition in $\wedge^{\geq 3} V \otimes Z_0^{<p}$. Thus

$$d(d_1(1 \otimes z) - \Omega) \in \wedge V \otimes (k\!\!k \oplus Z_0^{<p}) .$$

Next consider the filtration $F^j = (-)^{\geq j,*}$. Since $z \in Z^{k,*}$ it follows from our assumption that $dd_1(1 \otimes z) \in F^{k+3}$. Thus we may suppose that $\Omega \in F^{k+2}$, $d\Omega \in F^{k+3}$ and thus that

$$d(d_1(1 \otimes z) - \Omega) \in F^{k+3} .$$

In particular, the component of $d_1(1 \otimes z) - \Omega$ of left degree $k+2$ is a $d_1$-cycle. But by construction $k \geq m$ and according to Step Three, $H^{j,*}(\wedge V \otimes (k\!\!k \oplus Z_0), d_1) = 0$ if $j > m$. Thus the component of left degree $k+2$ has the form $d_1\Phi(0)$ for some $\Phi(0) \in [\wedge V \otimes (k\!\!k \oplus Z_0)]^{k+1,*}$. Since $Z_0 \subset Z^{m,*}$ and since

$$d(d_1(1 \otimes z) - \Omega) \in \wedge V \oplus (\wedge^{\geq 3} V \otimes Z_0^{<p})$$

it follows that $\Phi(0) \in \wedge V \oplus (\wedge^+ V \otimes Z_0^{<p})$. Thus

$$d(d_1(1 \otimes z) - \Omega - d_1\Phi(0)) \in F^{k+4} .$$

Iterating this process yields an element $\Phi \in (\wedge^{\geq 2} V \otimes Z_0^{<p}) \cap F^{k+2}$ such that $d(d_1(1 \otimes z) - \Omega - \Phi) = 0$. Extend $d$ to $1 \otimes Z_q^p(r+1)$ by setting $d = d_1(1 \otimes z) - \Omega - \Phi$. This, together with induction, completes the construction of $d$.

Finally, recall from Step Two that the left degree, $k$, of $z$ is $q(m-1)+m$. Since $d_1$ and $id \otimes \overline{d_1}$ increase left degrees by 1, it follows after a simple calculation that

$$(d_1 - id \otimes \overline{d_1})(1 \otimes z) \in \wedge^{k+1} V \oplus \left( \bigoplus_{i=0}^{q} \wedge^{>i(m-1)+k-m} V \otimes Z_{q-i} \right) .$$

In the same way,

$$\Omega \in \wedge^{\geq k+2} V \oplus \left( \bigoplus_{i=0}^{q} \wedge^{>i(m-1)+k-m+1} V \otimes Z_{q-i} \right)$$

and

$$\Phi \in \wedge^{\geq k+2} V \oplus \left( \wedge^{>k-m+1} V \otimes Z_0 \right) .$$

Thus $d$ satisfies all the conditions required by Step Four.

*Step Five: Completion of the proof of Proposition 9.4*

*proof:* In Step Four we constructed a commutative cochain algebra $(\wedge V \otimes \wedge Z, d)$ and it is immediate from Step Four that this is a Sullivan extension of $(\wedge V, d)$. Moreover, it follows from Steps Three and Four that $\ker\chi = \wedge^{\geq m+1}V \oplus (\wedge V \otimes \wedge^+ Z)$ is preserved by $d$, and so $\chi$ is a morphism

$$\chi : (\wedge V \otimes \wedge Z, d) \to (\wedge V/ \wedge^{>m} V, d)\,.$$

Filtering both cochain algebras by $F^k = (-)^{\geq k,*}$ produces a morphism of spectral sequences beginning with the quasi-isomorphism

$$\chi : (\wedge V \otimes \wedge Z, d_1) \to (\wedge V/ \wedge^{>m} V, d_1)$$

and it follows that $\chi : (\wedge V \otimes \wedge Z, d) \to (\wedge V/ \wedge^{>m} V, d)$ is also a quasi-isomorphism. The other conditions on $d$ are established in Step Four.

□

**Corollary 9.1.** *In the construction of Proposition 9.4, $\chi$ restricts to a $(\wedge V, d)$- semifree resolution*

$$\chi : (\wedge V \otimes (\Bbbk \oplus Z_0), d) \xrightarrow{\simeq} (\wedge V/ \wedge^{>m} V, d)\,.$$

*proof:* It follows from Step Four that $\wedge V \otimes (\Bbbk \oplus Z_0)$ is preserved by $d$. Filter both $(\wedge V, d)$-models by $F^k = (-)^{\geq k,*}$ and observe that the morphism of spectral sequences induced by $\chi$ begins with the quasi-isomorphism

$$\chi : (\wedge V \otimes (\Bbbk \oplus Z_0), d_1) \xrightarrow{\simeq} (\wedge V/ \wedge^{>m} V, d_1)\,.$$

□

If $(\wedge V, d)$ is a minimal Sullivan algebra and we set $(\wedge V)^{m,p} = (\wedge^m V)^p$ then the quadratic part of $d$, $d_1$, increases both bidegrees by 1. Thus, as in the proof of Theorem 9.4, the homology $H(\wedge V, d_1)$ inherits a bigrading $H(\wedge V, d_1) = \{\, H^{m,p}(\wedge V, d_1)\,\}$.

**Proposition 9.5.** *Suppose $(\wedge V, d)$ is a minimal Sullivan algebra and denote by $d_1$ the quadratic part of the differential.*

*(i) If $H^{m+1,*}(\wedge V, d_1) = 0$, then $H^{j,*}(\wedge V, d_1) = 0$ for $j \geq m+1$.*
*(ii) $cat(\wedge V, d_1) = e(\wedge V, d_1) \leq m$.*
*(iii) $cat(\wedge V, d) \leq e(\wedge V, d_1)$.*

*proof:* (i) Recall from Corollary 9.1 the $(\wedge V, d)$-semifree resolution $\chi : (\wedge V \otimes (\Bbbk \oplus Z_0), d) \xrightarrow{\simeq} (\wedge V/ \wedge^{>m} V, d)$. By construction, the differential in

$\wedge V \otimes (\Bbbk \oplus Z_0)$ preserves the decreasing filtration by left degree. Moreover, by Step Three in the proof of Proposition 9.4,

$$\chi : (\wedge V \otimes (\Bbbk \oplus Z_0), d_1) \xrightarrow{\simeq} (\wedge V/ \wedge^{>m} V, d_1)$$

is a $(\wedge V, d_1)$-semifree resolution.

Now if $H^{m+1,*}(\wedge V, d_1) = 0$, we shall construct a morphism $\psi : (\wedge V \otimes (\Bbbk \oplus Z_0), d_1) \to (\wedge V, d_1)$ of bigraded $(\wedge V, d_1)$-modules such that $\psi$ is the identity in $\wedge V \otimes 1$. For this recall the filtration $Z(r)$ introduced at the start of the proof of Proposition 9.4. If $\psi$ is already defined in some $Z_0(r)$ and if $z$ is an element of $Z_0(r+1)$ then $d_1(1 \otimes z) \in [\wedge V \otimes (\Bbbk \oplus Z_0(r))]^{m+1,*}$ and $\psi(d_1(1 \otimes z))$ is a $d_1$-cycle in $\wedge^{m+1} V$. By hypothesis, $\psi(d_1(1 \otimes z)) = d_1 \Phi$ for some $\Phi \in \wedge^m V$ and as in the proof of Proposition 9.4 we may extend $\psi$ to $Z_0(r+1)$ and then by induction on $r$ to all of $\wedge V \otimes (\Bbbk \oplus Z_0)$. It follows that $\mathrm{mcat}(\wedge V, d_1) \leq m$, and that $H^{j,*}(\wedge V, d_1) = 0$ for $j \geq m+1$.

(ii) Recall from Proposition 9.3 that $e(\wedge V, d_1) \leq \mathrm{mcat}\,(\wedge V, d_1)$. But if $e(\wedge V, d_1) = q$ then $H^{q+1,*}(\wedge V, d_1) = 0$ and so by the argument above, $\mathrm{mcat}\,(\wedge V, d_1) \leq q$. Thus $\mathrm{mcat}(\wedge V, d_1) = e(\wedge V, d_1)$ and by Theorem 9.4 $\mathrm{cat}(\wedge V, d_1) = e(\wedge V, d_1)$.

(iii) Suppose $e(\wedge V, d_1) = m$ and consider the retraction

$$\psi : (\wedge V \otimes (\Bbbk \oplus Z_0), d_1) \to (\wedge V, d_1)$$

constructed above. We perturb $\psi$ to a retraction

$$\varphi : (\wedge V \otimes (\Bbbk \oplus Z_0), d) \to (\wedge V, d)\,,$$

establishing thereby that $\mathrm{cat}(\wedge V, d) = \mathrm{mcat}(\wedge V, d) \leq m$.

For this suppose $\varphi$ has been constructed in some $Z_0(r)$ such that $\varphi(1 \otimes z) - \psi(1 \otimes z) \in \wedge^{\geq m+1} V$. If $z \in Z_0(r+1)$ then $d(1 \otimes z) \in \wedge V \otimes (\Bbbk \oplus Z_0(r))$ and $\varphi(d(1 \otimes z))$ is a $d$-cycle in $\wedge^{\geq m+1} V$, satisfying

$$\varphi d(1 \otimes z) - d_1 \varphi(1 \otimes z) \in \wedge^{\geq m+2} V\,.$$

Thus $d\psi(1 \otimes z) - \varphi d(1 \otimes z)$ is a $d$-cycle in $\wedge^{\geq m+2} V$. Since $H^{>m,*}(\wedge V, d_1) = 0$, $d\psi(1 \otimes z) - \varphi d(1 \otimes z) = d_1 \Psi(0) + \Phi(0)$ with $\Phi(0) \in \wedge^{\geq m+3} V$. If follows that $d(\psi(1 \otimes z) - \Psi(0))$ is a $d$-cycle in $\wedge^{\geq m+3} V$. Iterating this process yields an element $\Psi$ such that $\varphi d(1 \otimes z) = d\Psi$ and $\Psi - \psi(1 \otimes z) \in \wedge^{\geq m+2} V$. We may extend $\varphi$ to $1 \otimes z$ by setting $\varphi(1 \otimes z) = \Psi$. The existence of $\varphi$ then follows by a standard induction argument.

□

**Corollary 9.2.** *If $(\wedge W, d)$ is a Sullivan 1-algebra, then*

$$e(\wedge W, d) = cat(\wedge W, d) = \max\{\, i \mid H^i(\wedge W, d) \neq 0\,\}\,.$$

*Moreover, $H^i(\wedge W, d) \neq 0$ for $i \leq e(\wedge W, d)$.*

*proof:* This follows from Proposition 9.5 because for Sullivan 1-algebras, $d = d_1$, and the left degree coincides with the ordinary degree.

□

**Proposition 9.6.** *Suppose $(\wedge V, d)$ is a minimal Sullivan algebra for which $\dim V^i < \infty$ if $i \geq 2$. If $cat(\wedge V, d) = m < \infty$ then $cat(\wedge V^1, d) < \infty$ and for some $N$, $H^i(\wedge V^1, d) = 0, i \geq N$.*

*proof:* Since each $V^i$, $i \geq 2$, is finite dimensional there is a finite dimensional subspace $U \subset V^1$ such that

$$d(1 \otimes V^i) \subset \wedge U \otimes \wedge V^{\leq i}, \qquad 2 \leq i \leq m.$$

Without loss of generality we may suppose $d : U \to \wedge^2 U$, and thus obtain a Sullivan extension $(\wedge U \otimes \wedge V^{[2,m]}, d) \to (\wedge V, d) \to (\wedge W, \overline{d})$ in which $W^i = 0$ for $2 \leq i \leq m$. The Mapping Theorem 9.3 then asserts that $\text{cat}(\wedge W, \overline{d}) \leq m$.

Since the differential in $\wedge W^1$ is purely quadratic, Step Three in the proof of Proposition 9.4 gives a $(\wedge W^1, \overline{d})$-semifree resolution

$$\chi : (\wedge W^1 \otimes (\Bbbk \oplus Z_0), \overline{d}) \xrightarrow{\simeq} (\wedge W^1 / \wedge^{>m} W^1, \overline{d})$$

in which $Z_0$ is the increasing union of subspaces $0 = Z_0(0) \subset \cdots \subset Z_0(r) \subset \cdots$ with $\overline{d} : 1 \otimes Z_0(r) \to \wedge^{m+1} W^1 \oplus (W^1 \otimes Z_0(r-1))$. It follows by induction on $r$ that $Z_0$ is concentrated in degree $m$ (i.e. left degree coincides with the normal degree).

Now let $(\wedge W \otimes (\Bbbk \oplus M), \overline{d}) \xrightarrow{\simeq} (\wedge W / \wedge^{>m} W, \overline{d})$ be a semifree $(\wedge W, \overline{d})$ resolution restricting to the obvious surjection in $\wedge W \otimes 1$. Then $\chi$ lifts to a morphism of $(\wedge W^1, \overline{d})$-modules,

$$\varphi : (\wedge W^1 \otimes (\Bbbk \oplus Z_0), \overline{d}) \to (\wedge W \otimes (\Bbbk \oplus M), \overline{d}),$$

extending the inclusion of $\wedge W^1$ in $\wedge W$. But since by the Mapping Theorem 9.3, cat $(\wedge W, \overline{d}) \leq m$, there is a retraction $\psi : (\wedge W \otimes (\Bbbk \oplus M), \overline{d}) \to (\wedge W, \overline{d})$ of $(\wedge W, \overline{d})$-modules extending the identity in $(\wedge W, \overline{d})$.

Now consider the morphism $\psi \circ \varphi : (\wedge W^1 \otimes (\Bbbk \oplus Z_0), \overline{d}) \to (\wedge W, \overline{d})$ of $(\wedge W^1, \overline{d})$-modules. It extends the identity in $\wedge W^1$ and satisfies

$$(\psi \circ \varphi)(Z_0) \subset (\wedge W)^m = (\wedge W^1)^m.$$

Thus $\psi \circ \varphi$ is a retraction $(\wedge W^1 \otimes (\Bbbk \oplus Z_0), \overline{d}) \to (\wedge W^1, \overline{d})$. It follows from Hess' Theorem 9.4, that cat $(\wedge W^1, \overline{d}) \leq m$, and therefore by Proposition 9.5 that $H^i(\wedge W^1, \overline{d}) = 0$, for $i > m$.

Finally, we have a spectral sequence converging from $\wedge U^1 \otimes H(\wedge W^1, \overline{d})$ to $H(\wedge V^1, d)$ and it follows that $H^i(\wedge V^1, d) = 0$, $i > \dim U^1 + m$. In particular, cat $(\wedge V^1, d) \leq \dim U^1 + m$.

□

## 9.5 cat $= e(-)^\#$

The Toomer invariant was originally introduced in the hope that $e(\wedge V, d)$ would provide ( [63]) an algebraic description of rational category. While this proved not to be the case ( [18], §29(b), Example 2), the next Theorem shows that the idea was close to being correct. Moreover, Corollary 9.4 shows that this idea is correct if $H(\wedge V, d)$ is a Poincaré duality algebra, or if $d : V \to \wedge^2 V$ is purely quadratic.

Recall now that if $(M, d)$ is a $(\wedge V, d)$ module then $M^\#$ is the $(\wedge V, d)$-module defined by

$$(a \cdot f)(m) = (-1)^{\deg a \cdot \deg f} f(a \cdot m) \quad \text{and} \quad (df)(m) = -(-1)^{\deg f} f(dm)\,,$$

for $a \in \wedge V, m \in M$ and $f \in M^\#$.

**Theorem 9.5.** *Suppose $(\wedge V, d)$ is a minimal Sullivan algebra and $H(\wedge V, d)$ has finite type. Then*

$$cat\,(\wedge V, d) = e((\wedge V)^\#)\,.$$

*proof:* Given Theorem 9.4, this is identical with the proof of Theorem 19.16 in [18]. While that Theorem states as hypotheses that $(\wedge V, d)$ is simply connected and of finite type, the only use of these hypotheses is the fact that the natural morphism of $(\wedge V, d)$-modules

$$\sigma : (\wedge V, d) \to (((\wedge V)^\#)^\#, d)$$

is a quasi-isomorphism. But this is true if $H(\wedge V, d)$ has finite type. Now, if $H(\sigma)$ is an isomorphism and $\rho : (P, d) \xrightarrow{\simeq} ((\wedge V)^\#, d)$ is a semifree resolution, it follows that the composite

$$(\rho)^\# \circ \sigma : (\wedge V, d) \to (P^\#, d)$$

is a quasi-isomorphism. With this, the proof of Theorem 29.16 in [18] applies verbatim to prove Theorem 9.5. □

**Corollary 9.3.** *Suppose $X$ and $Y$ are path connected topological spaces, that $(\wedge V, d)$ and $(\wedge W, d)$ are minimal Sullivan algebras, and that $H(X), H(Y), H(\wedge V, d)$ and $H(\wedge W, d)$ all have finite type. Then*

$$cat_0(X \times Y) = cat_0 X + cat_0 Y$$

*and*

$$cat((\wedge V, d) \otimes (\wedge W, d)) = cat(\wedge V, d) + cat(\wedge W, d)\,.$$

*proof:* There is an obvious morphism

$$\mathrm{Hom}(\wedge V, \Bbbk) \otimes \mathrm{Hom}(\wedge W, \Bbbk) \to \mathrm{Hom}(\wedge V \otimes \wedge W, \Bbbk)$$

of $(\wedge V, d) \otimes (\wedge W, d)$-modules, and since $H(\wedge V, d)$ and $H(\wedge W, d)$ have finite type, this is a quasi-isomorphism. Now the Corollary follows from the Theorem by an argument identical to the proof of Theorem 30.2 in [18]. □

**Corollary 9.4.** *Suppose $(\wedge V, d)$ is a minimal Sullivan algebra and that $H(\wedge V, d)$ is a Poincaré duality algebra. Then*

$$cat\,(\wedge V, d) = e\,(\wedge V, d)\,.$$

*proof:* Let $z \in (\wedge V)^n$ be a cycle representing the fundamental class and let $f : \wedge V \to \Bbbk$ be a linear map of degree $-n$ such that $f(z) = 1$ and $f \circ d = 0$. Then it follows in a straightforward way from the definitions that a morphism of $(\wedge V, d)$-modules

$$\varphi : (\wedge V, d) \to (\wedge V)^{\#}$$

is defined by $\varphi(\Phi)(\Psi) = f(\Phi \wedge \Psi)$. The fact that $H(\wedge V, d)$ is a Poincaré duality algebra implies that $\varphi$ is a quasi-isomorphism, and the Corollary follows.

□

## 9.6 Jessup's Theorem

Suppose $M$ is a graded vector space and $f \in \mathrm{Hom}_k(M, M)$ for some $k \leq 0$. Then $f$ is *locally conilpotent* if $\cap_p f^p(M) = 0$, and $f$ is *conilpotent* if for all $i$ there is some $h(i)$ such that $f^{h(i)}(M) \cap M^i = 0$. Clearly conilpotence implies local conilpotence, and if $M$ has finite type then these two conditions are equivalent.

**Definition.** An *Engels derivation* is a locally nilpotent derivation $\theta$ in $(\wedge W, d)$ of degree $\leq 0$ for which either

(i) $W$ has finite type and $H(\theta)$ is conilpotent, or
(ii) $\theta$ is conilpotent.

Now suppose $\theta$ is a derivation of degree $\leq 0$ in a minimal Sullivan algebra $(\wedge W, d)$ and define a linear map $f : W \to W$ by requiring that $\theta - f : W \to \wedge^{\geq 2} W$.

**Lemma 9.1.** *If $f$ is conilpotent, then so is $\theta$.*

*proof:* Straightforward arguments show first that $f$ extends to a conilpotent derivation $\theta_f$ in $\wedge W$ and then, because $\theta - \theta_f : \wedge^k W \to \wedge^{\geq k+1} W$, that $\theta$ is conilpotent.

□

**Lemma 9.2.** *Suppose $\theta$ is an Engels derivation in a minimal Sullivan algebra $(\wedge W, d)$ and that $(x_n, z_n)_{n\geq 0}$ is an infinite sequence of pairs of elements in $\wedge W$ satisfying*

$$dz_n = 0 \qquad \text{and} \qquad \theta z_{n+1} = z_n + dx_n\,, \qquad n \geq 0\,.$$

*Then there is an infinite sequence of elements $y_n \in \wedge W$ such that*

$$dy_n = z_n \qquad \text{and} \qquad \theta y_{n+1} = y_n + x_n\,, \qquad n \geq 0\,.$$

*proof:* If $W$ has finite type and $H(\theta)$ is conilpotent the argument proving Lemma 31.14 in [18] applies verbatin to give the Lemma. Otherwise, $\theta$ itself is conilpotent. It follows that $H(\theta)$ is conilpotent and hence, if $\gamma_i$ is an infinite sequence of classes in $H(\wedge W, d)$ such that $H(\theta)\gamma_{i+1} = \gamma_i$ then each $\gamma_i = 0$.

In particular, in this case we may write $z_i = dw_i$. Now define, for each $s$, a sequence of elements $y_i(s)$, $i \leq s$ as follows:

$$y_s(s) = w_s \qquad \text{and } y_i(s) = \theta y_{i+1}(s) - x_i\,, \qquad i < s\,.$$

A quick calculation shows that $dy_i(s) = z_i$ and $\theta y_{i+1}(s) = y_i(s) + x_i$. In particular, $\theta(y_i(s+1) - y_i(s)) = y_{i-1}(s+1) - y_{i-1}(s)$.

Let $n(i)$ be the integer such that

$$\theta^{n(i)}(\wedge W) \cap (\wedge W)^{\deg y_i} = 0\,.$$

Then if $s - i \geq n(i)$

$$y_i(s+1) - y_i(s) = \theta^{s-i}(y_s(s+1) - y_s(s)) = 0\,.$$

Define the sequence $y_i$ by setting $y_i = y_i(s)$, any $s > i + n(i)$.

□

Now consider a Sullivan extension $(\wedge V \otimes \wedge Z, d)$ in which $(\wedge V \otimes \wedge Z, d)$ is itself a minimal Sullivan algebra. Then the inclusion $V \hookrightarrow V \oplus Z$ dualizes to a surjection $L \to E$ from the homotopy Lie algebra of $(\wedge V \otimes \wedge Z, d)$ to the homotopy Lie algebra of $(\wedge V, d)$. Recall also from §4.3 the holonomy representation $\overline{\theta}$ of $L$ in $H(\wedge Z, \overline{d})$.

We extend Jessup's Theorem 31.10 of [18] to

**Theorem 9.6.** *With the hypotheses above let $x_1, \cdots, x_r$ be linearly independent elements of $E_{even}$. Assume that for each $i$, either*

*(i) $V \oplus Z$ has finite type, $ad_E(x_i)$ is nilpotent, and $\overline{\theta}(x_i)$ is conilpotent, or*
*(ii) $x_i$ is the image of $y_i \in L$ and $ad_L y_i$ is nilpotent.*

*Then*

$$cat(\wedge V \otimes \wedge Z, d) \geq cat(\wedge Z, d) + r\,.$$

*proof:* In the case of hypothesis (i) the proof of Theorem 31.10 of [18] applies verbatim. In the case of hypothesis (ii) let the $x_i$ be ordered so that $\deg x_1 = \cdots = \deg x_s < \deg x_{s+1} \leq \cdots \leq \deg x_r$. Divide by $V^{\leq \deg x_1}$ and by a subspace of $V^{1+\deg x_1}$ to arrange that $V = \oplus_{i=1}^{s} \Bbbk v_i \oplus V^{>\deg x_1 + 1}$, where $\langle v_i, y_j \rangle = \delta_{ij}$. Then denote $U = \oplus_{i=2}^{s} \Bbbk v_i \oplus V^{>\deg x_1+1}$ and choose the direct summand, $Z$, of $V$ in $V \oplus Z$ so that $\langle Z, y_j \rangle = 0$, $1 \leq j \leq r$. By the Mapping Theorem 9.3 (p. 248), this does not increase the LS category of the Sullivan algebra.

Next in $\wedge v_1 \otimes \wedge U \otimes \wedge Z$ write $d(1 \otimes \Phi) = v_1 \otimes \theta\Phi + 1 \otimes \overline{d}\Phi$, $\Phi \in \wedge U \otimes \wedge Z$. Define $\chi : U \oplus Z \to U \oplus Z$ by requiring $\theta - \chi : U \oplus Z \to \wedge^{\geq 2}(U \oplus Z)$. Then a brief calculation (cf. §2.1) gives $\langle w, [y_1, y] \rangle = \langle \chi(w), y \rangle$ for $w \in U \oplus Z$ and $y \in L$. Since $ad_L\, y_1$ is nilpotent it follows that $\chi$ is conilpotent. By Lemma 9.1, $\theta$ is conilpotent and hence $\theta$ is an Engels derivation in $\wedge(U \oplus Z)$.

Now, in view of Lemma 9.2, the proof of Proposition 31.13 in [18] applies verbatim to show that

$$\text{cat}\,(\wedge V \otimes \wedge Z, d) \geq \text{cat}(\wedge U \otimes \wedge Z, \overline{d}) + 1\,.$$

Finally, by construction, $y_2, \cdots, y_r$ are in the image of the homotopy Lie algebra of $(\wedge(U \oplus Z), d)$. It follows by induction that $\text{cat}(\wedge U \otimes \wedge Z, \overline{d}) \geq \text{cat}(\wedge Z, \overline{d}) + r - 1$, which establishes the Theorem.

□

As a corollary we deduce the following application.

**Proposition 9.7.** *Let $X$ be a Sullivan CW complex with universal cover $\widetilde{X}$. Suppose $r$ linearly independent elements $x_i$ in the fundamental Lie algebra, $L_0$, of $X$ are such that each $ad_{L_0} x_i$ is nilpotent. Then*

$$cat_0(X) \geq r + cat_0(\widetilde{X})\,.$$

*In particular, if $L_0$ is finite dimensional then*

$$cat_0(X) \geq cat_0(\widetilde{X}) + dim\, L_0\,.$$

*proof:* Let $(\wedge U, d)$ be the minimal Sullivan model of $X$. We apply Theorem 9.6 to the Sullivan extension $\wedge U^1 \otimes \wedge U^{\geq 2}$, noting that (Theorem 7.2, p. 197) $(\wedge U^{\geq 2}, \overline{d})$ is the minimal Sullivan model of $\widetilde{X}$. □

Finally, we prove a variant of Theorem 9.6.

**Theorem 9.7.** *Let $L$ be the homotopy Lie algebra of a minimal Sullivan algebra $(\wedge V, d)$ such that $\dim V^1 = \infty$. If $cat(\wedge V, d) < \infty$ then there are elements $x, y \in L_0$ for which*

$$(ad\, x)^k y \neq 0\,, \qquad k \geq 0\,.$$

*proof:* By Proposition 9.6 (p. 259), cat $(\wedge V^1, d) < \infty$ and so we may assume without loss of generality that $V = V^1$. Choose a surjection $(\wedge V, d) \to (\wedge W, d)$ such that $\dim W = \infty$ and cat $(\wedge W, d)$ is minimized. Then the homotopy Lie algebra of $(\wedge W, d)$ is a sub Lie algebra of the homotopy Lie algebra of $(\wedge V, d)$, and it is sufficient to prove the Theorem for $(\wedge W, d)$. Denote the homotopy Lie algebra of $(\wedge W, d)$ by $L$.

Write $(\wedge W, d)$ as a Sullivan extension $(\wedge w \otimes \wedge Z, d)$ with $dw = 0$ and define $x \in L$ by $\langle w, x\rangle = 1$ and $\langle Z, x\rangle = 0$. Let $E \subset L$ be the homotopy Lie algebra of $(\wedge Z, \overline{d})$. Then (cf. §2.2) the differential in $\wedge W$ is given by

$$d(1 \otimes z) = 1 \otimes \overline{d}z - w \otimes \theta z,$$

where $\theta$ is the derivation of degree 0 in $(\wedge Z, \overline{d})$ given by

$$\langle \theta z, y\rangle = \langle z, [x, y]\rangle\,, \qquad y \in E\,.$$

It follows that $\ker ad\, x = [\, Z/\theta(Z)\,]^{\#}$.

Now if $ad\, x$ acts locally nilpotently in $E$ then $\ker ad\, x$ is infinite dimensional and so $Z/\theta(Z)$ is infinite dimensional. Moreover, the ideal $I$ generated by $\theta(Z)$ in $\wedge Z$ is clearly preserved by $\overline{d}$ and by $\theta$ and so division by $I$ gives a quotient Sullivan algebra of the form $(\wedge w \otimes \wedge(Z/\theta(Z)), id \otimes d')$. It follows from Theorem 9.3 (p. 248) and Proposition 9.1(iii) (p. 246) that

$$\text{cat}\,(\wedge W, d) \geq \text{cat}\,(\wedge w \otimes \wedge(Z/\theta(Z)), id \otimes d') = 1 + \text{cat}\,(\wedge(Z/\theta(Z)), d')\,,$$

contradicting our hypothesis on $(\wedge W, d)$. Thus $ad\, x$ is not locally nilpotent, which establishes the Theorem.

□

## 9.7 Example

Let $p : X \to S^3$ be the fibration associated with the projection of $S^3 \vee S^3$ on the first factor, and let $(A, d) \subset A_{PL}(X)$ be a sub cochain algebra such that $(A, d) \xrightarrow{\simeq} A_{PL}(X)$ and $A^{>3} = 0$. Then a quasi-isomorphism

$$(\wedge y \otimes \wedge(w_i, i \geq 1)/(w_i w_j), d) \xrightarrow{\simeq} (A, d)$$

is defined as follows: $dy = 0 = dw_1$, $y$ and $w_1$ are mapped respectively to representatives of the fundamental classes of the first and second factor $S^3$, and $dw_i = yw_{i-1}$, $i \geq 2$. Thus $\deg y = 3$ and $\deg w_i = 1 + 2i$. It follows that if $(\wedge y \otimes \wedge Z, d)$ is a minimal Sullivan model for $A_{PL}(p)$, then $(\wedge Z, d) \xrightarrow{\simeq} (\wedge(w_i/(w_i w_j), 0))$.

By Theorem 5.1 (p. 145) this implies that $\wedge w_i/(w_i w_j)$ is isomorphic to the cohomology algebra of the space $\vee_{i \geq 1} S^{2i+1}$. This algebra has $\Bbbk 1 \oplus (\oplus_{i \geq 1} \Bbbk w_i)$ as basis, and all products of elements in $H^+$ are zero. Thus the fibre, $F$, of $p$ has the rational homotopy type of $\vee_{i \geq 1} S^{2i+1}$.

Next note that the homotopy Lie algebra, $L$, of $(\wedge v, 0)$ is the abelian Lie algebra $\Bbbk a$ with degree $a = 2$. Thus Theorem 6.2 (p. 184) implies that $H_*(\Omega S^2 : \Bbbk)$ is the polynomial algebra $\wedge a$. Moreover, it follows from the quasi-isomorphism above that the holonomy representation of $L$ in $\oplus_{i \geq 1} \Bbbk w_i$ is given by $a \cdot w_{i+1} = w_i$, $i \geq 1$ and $a \cdot w_1 = 0$. Now applying Theorem 6.5 (p. 193) we find that the holonomy action $F \times \Omega S^3 \to \vee_{i \geq 1} S^{2i+1}$ induces the map on homology given by

$$\overline{w}_i \cdot a^k = \overline{w}_{i+k}\,, \qquad i \geq 1, k \geq 0\,;$$

where $\overline{w_i}$ is the homology class dual to $w_i$. Thus $H_+(F)$ is the free $H_*(\Omega S^3; \Bbbk)$ module generated by $\overline{w}_1$.

On the other hand, $\mathrm{cat}_0 S^3 = \mathrm{cat}_0 F = 1$. Here Theorem 9.6 does not apply, reflecting the fact that the holonomy representation of $L$ in $H_*(F; \Bbbk)$ is not nilpotent.

Consider now a continuous map $\varphi : B = S^1 \times S^1 \times S^1 \to S^3$ that induces an isomorphism on $H_3(-; \Bbbk)$, and let $q : E \to B$ be the pullback of $\varphi$ and $p$.

$$\begin{array}{ccc} E & \xrightarrow{\psi} & X \\ \downarrow{\scriptstyle q} & & \downarrow{\scriptstyle p} \\ S^1 \times S^1 \times S^1 & \xrightarrow{\varphi} & S^3\,. \end{array}$$

A model of $q$ is given by

$$(\wedge(v_1, v_2, v_3), 0) \to (\wedge(v_1, v_2, v_3) \otimes \wedge w_i/(w_i w_j), D)$$

with $Dw_1 = 0$, $Dw_i = v_1v_2v_3w_{i-1}$ for $i \geq 2$, and $\deg v_1 = \deg v_2 = \deg v_3 = 1$. Here the homotopy Lie algebra of $\wedge(v_1, v_2, v_3)$ is the abelian Lie algebra $\Bbbk x_1 \oplus \Bbbk x_2 \oplus \Bbbk x_3$, and the holonomy representation is trivial. Thus $E$ is a Sullivan space. Since $E$ has a model of nilpotency 4, $\mathrm{cat}_0(E) \leq 4$. Now by Proposition 9.7

$$4 \geq \mathrm{cat}_0(E) \geq \mathrm{cat}_0(S^1 \times S^1 \times S^1) + \mathrm{cat}_0(F) = 3 + 1 = 4\,.$$

Therefore, $\mathrm{cat}_0(E) = 4$.

## Chapter 10

# Depth of a Sullivan algebra and of a Sullivan Lie algebra

*In this Chapter, from §10.3 on, we shall be considering minimal Sullivan algebras* $(\wedge V, d)$ *and graded Lie algebras* $L$*, together with pairings* $\langle\ ,\ \rangle : V \times sL \to \Bbbk$*. Consistent with the conventions of Chapter 2 we shall often abuse notation and write these as pairings* $(v, x) \mapsto \langle v, x\rangle$ *where* $v \in V$ *and* $x \in L$*.*

## 10.1 Ext, Tor and the Hochschild-Serre spectral sequence

Let $A$ be a graded algebra. We recall the classical functors Ext and Tor in the graded context. Thus a free resolution $P_* \to M$ of a right $A$-module $M$ is an exact sequence of right $A$-modules of the form

$$\to P_{p+1} \to P_p \to \cdots \to P_0 \to M \to 0\,,$$

in which the $P_n$ are free $A$-modules. If $Q$ and $N$ are respectively a left and a right $A$-module then:

$$\to P_{p+1} \otimes_A Q \to P_p \otimes_A Q \to \cdots \to P_0 \otimes_A Q$$

and

$$\leftarrow \mathrm{Hom}_A(P_{p+1}, N) \leftarrow \mathrm{Hom}_A(P_p, N) \leftarrow \cdots \leftarrow \mathrm{Hom}_A(P_0, N)$$

are complexes of graded vector spaces, with corresponding homology denoted by

$$\mathrm{Tor}^A(M,Q) = \{\, \mathrm{Tor}^A_p(M,Q)\,\} \quad \text{ and } \quad \mathrm{Ext}_A(M,N) = \{\, \mathrm{Ext}^p_A(M,N)\,\}$$

with each $\mathrm{Tor}^A_p(-,-)$ and $\mathrm{Ext}^p_A(-,-)$ itself a graded vector space. Here $p$ is called the *homological degree* and we often write

$$\left[\,\mathrm{Tor}^A_p(M,N)\,\right]_r = \mathrm{Tor}^A_{p,r-p}(M,N) \text{ and } \left[\,\mathrm{Ext}^p_A(M,N)\,\right]^r = \mathrm{Ext}^{p,r-p}_A(M,N)\,.$$

For any left $A$-module $Q$, $\mathrm{Hom}(Q, \Bbbk)$ is the right $A$-module given by $(f \cdot a)(x) = f(a \cdot x)$, $f \in \mathrm{Hom}(Q, \Bbbk)$, $a \in A$ and $x \in Q$. Thus if $P_* \to M$ is an $A$-free resolution of $M$ we have the identification

$$\mathrm{Hom}_A(P_p, \mathrm{Hom}(Q, \Bbbk)) = \mathrm{Hom}(P_p \otimes_A Q, \Bbbk)\,.$$

Recall that we denote $\mathrm{Hom}(-, \Bbbk)$ by $(-)^{\#}$. Thus the identifications above yield the important relation

$$\mathrm{Ext}_A(M, Q^{\#}) = \left[\,\mathrm{Tor}^A(M, Q)\,\right]^{\#}\,. \tag{10.1}$$

**Definition.** If $A$ is a graded algebra then the *grade*, $\mathrm{grade}_A M$, of a right $A$-module $M$ is the least integer $k$ such that $\mathrm{Ext}_A^k(M, A) \neq 0$. If $\mathrm{Ext}_A(M, A) = 0$ we say $\mathrm{grade}_A M = \infty$.

The *depth* of an augmented graded algebra $A$, $\mathrm{depth}\, A$, is the grade of the trivial $A$-module $\Bbbk$.

The *global dimension* of $A$, $\mathrm{gldim}\, A$, is the greatest integer $k$ (or $\infty$) such that $\mathrm{Ext}_A^k(\Bbbk, -) \neq 0$.

Next, suppose given a graded Lie algebra $L$. A representation of $L$ in $M$ is the same as a $UL$-module structure in $M$. Thus we may abuse language and refer to $M$ as an $L$-module. If $M$ and $N$ are $L$-modules, then $M \otimes N$ becomes a $L$-module via the diagonal $UL \to UL \otimes UL$ extending $x \mapsto x \otimes 1 + 1 \otimes x$, $x \in L$. Similarly, $\mathrm{Hom}(M, N)$ is endowed with the $L$-structure given by

$$(x{\cdot}f)(m) = x{\cdot}f(m) + (-1)^{\deg x \cdot \deg f} f(x{\cdot}m)\,, \quad x \in L\,, m \in M, f \in \mathrm{Hom}(M, N)\,.$$

Now recall from §2.1 the Cartan-Chevalley-Eilenberg construction $(\wedge sL \otimes M, \partial)$ associated with any left $L$-module $M$. If $M = UL$ and the representation of $L$ in $UL$ is by left multiplication, then $H(\wedge sL \otimes UL, \partial) = \Bbbk$, as follows from the Poincaré-Birkhoff-Witt theorem ( [18], Proposition 22.3). Thus $(\wedge sL \otimes UL, \partial)$ is a resolution of $\Bbbk$ by free right $UL$-modules. It follows that for any right $L$-module, $M$, $(\wedge sL \otimes UL, \partial) \otimes M$ becomes, via the diagonal in $UL$, a complex of $L$-modules.

**Lemma 10.1.** *$(\wedge sL \otimes UL, \partial) \otimes M$ is a $UL$-free resolution of $M$.*

*proof:* The augmentation $(\wedge sL \otimes UL) \xrightarrow{\simeq} \Bbbk$ defines a quasi-isomorphism $(\wedge sL \otimes UL, \partial) \otimes M \xrightarrow{\simeq} M$. It remains to show that $\wedge^p sL \otimes UL \otimes M$ is $UL$ free, and for this it is sufficient to show that $UL \otimes M$ is $UL$ free. Consider the right $UL$-module $M \otimes UL$ and define a $UL$-linear map $\varphi : M \otimes UL \to$

$UL \otimes M$ by $\varphi : \Phi \otimes a \mapsto (1 \otimes \Phi) \cdot a$. Then recall ( [18], Proposition 21.2) that we may identify $\wedge L$ with $UL$ via the linear isomorphism $\wedge L \overset{\cong}{\to} UL$ given by $x_1 \wedge \cdots \wedge x_r \mapsto \sum (-1)^\sigma x_{\sigma(1)} \cdot \cdots \cdot x_{\sigma(r)}$ (Here the sum is over all permutations of $\{1, \cdots, r\}$.) Since $UL$ acts diagonally in $UL \otimes M$ it follows that for $a \in \wedge^r L$, $\varphi(\Phi \otimes a) - (-1)^{\deg \Phi \cdot \deg a} a \otimes \Phi \in \wedge^{<r} L \otimes M$, and hence $\varphi$ is an isomorphism.

□

As a consequence of Lemma 10.1 it follows that if $N$ is a left $UL$-module and $Q$ is a right $UL$-module, then

$$\mathrm{Tor}_*^{UL}(M,N) = H_*([C_*(L) \otimes UL \otimes M] \otimes_{UL} N) = H_*(C_*(L) \otimes M \otimes N)\,,$$

and

$$\mathrm{Ext}_{UL}^*(M,Q) = H(\mathrm{Hom}(C_*(L) \otimes M, Q)) = H(\mathrm{Hom}(C_*(L), \mathrm{Hom}(M,Q)))\,.$$

Note that setting

$$x \cdot m = -(-1)^{\deg x \cdot \deg m} m \cdot x\,, \qquad m \in M\,, x \in L\,,$$

determines a bijection between left and right representations of $L$ in $M$. Thus *we shall simply refer to $M$ as a $UL$-module* or, by abuse of language, as an $L$-module.

In particular a quick computation gives

$$\mathrm{Tor}_*^{UL}(M,N) = \mathrm{Tor}_*^{UL}(\Bbbk, M \otimes N)$$

and

$$\mathrm{Ext}_{UL}^*(M,N) = \mathrm{Ext}_{UL}^*(\Bbbk, \mathrm{Hom}(M,N))$$

with $UL$ acting diagonally from the left in $M \otimes N$, and in $\mathrm{Hom}(M,N)$.

Now suppose $E \subset L$ is a sub Lie algebra and write $L = E \oplus S$. This defines a multiplicative isomorphism $C_*(E) \otimes \wedge(sS) \overset{\cong}{\to} C_*(L)$. Moreover, the filtration $\mathcal{F}_p$ of $C_*(L) \otimes M \otimes N$ corresponding under this isomorphism to $\oplus_{i \le p} C_*(E) \otimes \wedge^i(sS) \otimes M \otimes N$ is independent of the choice of $S$ and is preserved by the differential. Here $E$ acts on $sS$ by the formula $x \cdot sy = s[x,y]$ and on $\wedge sS$ by derivations. Moreover, when $E$ is an ideal the same formula applies to the action of $L$ on $\wedge sS$. The corresponding first quadrant spectral sequence is called the *Hochschild-Serre spectral sequence* ([39]). It converges from

$$E^1_{p,q} = \mathrm{Tor}_q^{UE}(\wedge^p s(L/E) \otimes M, N) \Longrightarrow \mathrm{Tor}_{p+q}^{UL}(M,N)\,,$$

where $E$ acts by the adjoint representation in $L/E$.

Analogously there is a Hochschild-Serre spectral sequence converging from

$$E_1^{p,q} = \mathrm{Ext}^q_{UE}(\wedge^p s(L/E)\otimes M, N) \Longrightarrow \mathrm{Ext}^{p+q}_{UL}(M,N)\,.$$

When $E$ is an ideal $I$, a presentation of the spectral sequence is given in [18], §34. The adjoint action of $L$ in $\wedge sI$, and the action of $L$ in $M$ and $N$ gives a representation of $L$ in $\wedge sI\otimes M\otimes N = C_*(I)\otimes M\otimes N$ which commutes with the differential. The induced representation in $\mathrm{Tor}_*^{UI}(M,N)$ restricts to the trivial representation in $I$, and so makes each $\mathrm{Tor}_q^{UI}(M,N)$ into a $UL/I$-module. Since $\wedge^p sL/I$ is a trivial $UI$-module,

$$E^1_{p,q} = C_*(L/I)\otimes \mathrm{Tor}_q^{UI}(M,N)\,,$$

and

$$E^2_{p,q} = \mathrm{Tor}_p^{UL/I}(\Bbbk, \mathrm{Tor}_q^{UI}(M,N))\,.$$

The trivial case of $I\subset I$ also gives a spectral sequence $\widetilde{E}^i_{p,q}$ for which

$$\widetilde{E}^2_{p,q} = \begin{cases} \mathrm{Tor}_q^{UI}(M,N), & p=0\\ 0 & p>0\,. \end{cases}$$

Moreover the inclusion $I\subset L$ induces a morphism of spectral sequences. At the $E^2$-level this exhibits the map $\mathrm{Tor}_q^{UI}(M,N)\to \mathrm{Tor}_q^{UL}(M,N)$ as the composite

$$\begin{aligned} &\mathrm{Tor}_q^{UI}(M,N)\to \Bbbk\otimes_{UL/I}\mathrm{Tor}_q^{UI}(M,N)\\ &= \mathrm{Tor}_0^{UL/I}(\Bbbk,\mathrm{Tor}_q^{UI}(M,N)) = E^2_{0,q} \twoheadrightarrow E^\infty_{0,q} \rightarrowtail \mathrm{Tor}_q^{UL}(M,N)\,. \end{aligned} \tag{10.2}$$

For an ideal $I\subset L$ there is also a convergent Hochschild-Serre spectral sequence

$$\mathrm{Ext}^p_{UL/I}(\Bbbk, \mathrm{Ext}^q_{UI}(M,N)) \Longrightarrow \mathrm{Ext}^{p+q}_{UL}(M,N)\,.$$

Here, the map $\mathrm{Ext}^q_{UI}(M,N)\leftarrow \mathrm{Ext}^q_{UL}(M,N)$ decomposes as

$$\begin{aligned} &\mathrm{Ext}^q_{UI}(M,N)\leftarrow \mathrm{Hom}_{UL/I}(\Bbbk,\mathrm{Ext}^q_{UI}(M,N))\\ &= \mathrm{Ext}^0_{UL/I}(\Bbbk,\mathrm{Ext}^q_{UI}(M,N)) = E_2^{0,q} \leftarrowtail E_\infty^{0,q} \twoheadleftarrow \mathrm{Ext}^q_{UL}(M,N)\,. \end{aligned} \tag{10.3}$$

**Lemma 10.2.** *Suppose $I$ is an ideal in a graded Lie algebra $L$ and that $L=\Bbbk x\oplus I$ with $\deg x$ odd. Then for any right $L$-module $M$, $\mathrm{Ext}^*_{UI}(M,UL)$ is a free $U\,L/I$-module.*

*proof:* First recall that the right adjoint representation of $L$ in $C_*(I)$, together with the (right) representation of $L$ in $M$ and right multiplication by $UL$ in $UL$ define a right representation of $L$ in the complex $\mathrm{Hom}(C_*(I) \otimes M, UL)$ which induces the representation of $L$ in $\mathrm{Ext}_{UI}(M, UL)$ used to define $\mathrm{Ext}_{U\,L/I}(\Bbbk, \mathrm{Ext}_{UI}(M, UL))$. We denote this representation by

$$\alpha \otimes y \mapsto \alpha \wedge y\,, \qquad y \in L,\, \alpha \in \mathrm{Ext}_{UI}(M, UL)\,,$$

noting that it depends only on the image $\overline{y}$ of $y$ in $L/I$.

Next note that left multiplication by $UL$ in $UL$ is a map of right $UL$-modules and therefore makes $\mathrm{Ext}_{UI}(M, UL)$ into a left $UL$-module, denoted by

$$y \otimes \alpha \mapsto y \wedge \alpha\,, \qquad y \in L,\, \alpha \in \mathrm{Ext}_{UI}(M, UL)\,.$$

Finally, note that the right adjoint representation of $L$ in $UI$, together with the right adjoint representation of $L$ in $C_*(I)$ and the (right) representation of $L$ in $M$ define a right representation of $L$ in the complex $\mathrm{Hom}(C_*(I) \otimes M, UI)$ which induces a right representation of $L$ in $\mathrm{Ext}_{UI}(M, UI)$. This we denote by

$$\beta \otimes y \mapsto \beta \bullet y\,, \qquad y \in L, \beta \in \mathrm{Ext}_{UI}(M, UI)\,.$$

Thus, since the inclusion of $UI$ in $UL$ induces an inclusion of $\mathrm{Ext}_{UI}(M, UI)$ in $\mathrm{Ext}_{UI}(M, UL)$, a straightforward calculation gives

$$\beta \wedge y = \beta \bullet y + (-1)^{\deg \beta \cdot \deg y} y \wedge \beta\,, \qquad y \in L, \beta \in \mathrm{Ext}_{UI}(\Bbbk, UI)\,.$$

Now observe that the Poincaré Birkoff Witt Theorem implies that multiplication in $UL$ is an isomorphism $\wedge x \otimes UI \xrightarrow{\cong} UL$ of right $UI$-modules. Since $\dim \wedge x < \infty$ this gives an isomorphism

$$\mathrm{Hom}(C_*(I) \otimes M, UL) = \wedge x \otimes \mathrm{Hom}(C_*(I) \otimes M, UI)$$

and hence shows that the map $x \otimes \beta \mapsto x \wedge \beta$ induces an isomorphism

$$\wedge x \otimes \mathrm{Ext}_{UI}(M, UI) \xrightarrow{\cong} \mathrm{Ext}_{UI}(M, UL)\,.$$

The observations above then yield

$$\beta \wedge x = \beta \bullet x + (-1)^{\deg \beta} x \wedge \beta$$

so that

$$\mathrm{Ext}_{UI}(M, UL) = \mathrm{Ext}_{UI}(M, UI) \oplus \mathrm{Ext}_{UI}(M, UI) \wedge x\,.$$

But the map $- \wedge x$ depends only on $\overline{x} \in L/I$ and, in $U\,L/I$, $\overline{x}^2 = 0$. In other words, $\mathrm{Ext}_{UI}(M, UL)$ is a free $U\,L/I$-module. □

**Lemma 10.3.** *Suppose $M$ and $N$ are $UL$-modules, and $I$ is an ideal in $L$. If $N = N_{\geq 0}$ has finite type and $I$ acts trivially in $N$ then the $E_2$-term of the Hochschild-Serre spectral sequence for $Ext_{UL}(M, N)$ is given by*

$$E_2^{p,q} = Ext^p_{UL/I}\left(Tor_q^{UI}(M, \Bbbk), N\right)\,.$$

*proof*: We have to show that

$$\mathrm{Ext}^p_{UL/I}\left(\Bbbk, \mathrm{Ext}^q_{UI}(M,N)\right) = \mathrm{Ext}^p_{UL/I}\left(\mathrm{Tor}^{UI}_q(M,\Bbbk), N\right) .$$

Because of our hypotheses on $N$, it follows from (10.1) that

$$\mathrm{Ext}^q_{UI}(M,N) = \left(\mathrm{Tor}^{UI}_q(M, N^\#)\right)^\#$$

and so

$$\mathrm{Ext}^p_{UL/I}\left(\Bbbk, \mathrm{Ext}^q_{UI}(M,N)\right) = \left[\mathrm{Tor}^{UL/I}_p(\Bbbk, \mathrm{Tor}^{UI}_q(M, N^\#))\right]^\# .$$

On the other hand, because $N$ (and thus $N^\#$) are trivial $UI$-modules, we have $\mathrm{Tor}^{UI}_q(M, N^\#) = \mathrm{Tor}^{UI}_q(M,\Bbbk) \otimes N^\#$ as $UL/I$-modules. Thus

$$\mathrm{Tor}^{UL/I}_p\left(\Bbbk, \mathrm{Tor}^{UI}_q(M,\Bbbk)\otimes N^\#\right) = \mathrm{Tor}^{UL/I}_p\left(\mathrm{Tor}^{UI}_q(M,\Bbbk), N^\#\right) .$$

The Lemma follows from a second application of (10.1).

□

## 10.2 The depth of a minimal Sullivan algebra

Let $(\wedge V, d)$ be a minimal Sullivan algebra and recall (§2.1) that the associated quadratic Sullivan algebra $(\wedge V, d_1)$ is defined by $d_1 : V \to \wedge^2 V$ and $d - d_1 : V \to \wedge^{\geq 3} V$. Then by Corollary 3.4 (p. 105) the acyclic closure of $(\wedge V, d_1)$ can be written in the form $(\wedge V \otimes \wedge U, d_1)$ in which $d_1 : \wedge U \to V \otimes \wedge U$. Using the canonical linear isomorphism $\mathrm{Hom}_{\wedge V}(\wedge V \otimes M, -) = \mathrm{Hom}(M, -)$ we let $\mathrm{Hom}^p_{\wedge V}(\wedge V \otimes \wedge U, \wedge V) \subset \mathrm{Hom}_{\wedge V}(\wedge V \otimes \wedge U, \wedge V)$ correspond to the subspace $\mathrm{Hom}(\wedge U, \wedge^p V)$. Then a chain complex of graded vector spaces

$$\begin{aligned}\mathrm{Hom}^0_{\wedge V}(\wedge V \otimes \wedge U, \wedge V) \to \cdots &\to \mathrm{Hom}^p_{\wedge V}(\wedge V \otimes \wedge U, \wedge V)\\ &\to \mathrm{Hom}^{p+1}_{\wedge V}(\wedge V \otimes \wedge U, \wedge V) \to \cdots\end{aligned}$$

is defined by setting $d_1(f) = d_1 \circ f - (-1)^{\deg f} f \circ d_1$.

Now $(\wedge V \otimes \wedge U, d_1)$ is a $(\wedge V, d_1)$-semifree resolution of $\Bbbk$ and the homology of this complex is independent of the choice of bigraded semifree resolution. Thus we write

$$H^p(\mathrm{Hom}^*_{\wedge V}(\wedge V \otimes \wedge U, \wedge V), d_1) = \mathrm{Ext}^p_{(\wedge V, d_1)}(\Bbbk, (\wedge V, d_1))$$

and make the

**Definition.** The *depth* of the minimal Sullivan algebra $(\wedge V, d)$ is the least $p$ (or $\infty$) such that $\mathrm{Ext}^p_{(\wedge V, d_1)}(\Bbbk, (\wedge V, d_1)) \neq 0$.

Similarly, the wedge degree in $\wedge V$ defines a grading in $H(\wedge V, d_1)$, which we denote by $H^{p,*}(\wedge V, d_1)$ and we make the

**Definition.** The *global dimension* of $(\wedge V, d)$ is the largest $p$ (or $\infty$) such that $H^{p,*}(\wedge V, d_1) \neq 0$.

The next Theorem generalizes Theorem 35.13 in [18].

**Theorem 10.1.** *For any minimal Sullivan algebra* $(\wedge V, d)$,

$$\mathit{depth}\,(\wedge V, d) \leq \mathit{cat}\,(\wedge V, d) \leq \mathit{gldim}\,(\wedge V, d)\,.$$

*Moreover, if* $\mathit{depth}\,(\wedge V, d) = \mathit{cat}\,(\wedge V, d)$ *then* $\mathit{cat}\,(\wedge V, d) = \mathit{gldim}\,(\wedge V, d)$.

*proof*: As above let $(\wedge V, d_1)$ denote the quadratic Sullivan algebra for $(\wedge V, d)$. Suppose cat $(\wedge V, d) = m$. In Corollary 9.1 (p. 257) we constructed a $(\wedge V, d)$-semifree resolution

$$(\wedge V \otimes (k \oplus Z_0^{m,*}), d) \xrightarrow{\simeq} (\wedge V / \wedge^{>m} V, d)$$

which after filtering by the wedge degrees gives a bigraded quasi-isomorphism

$$(\wedge V \otimes (k \oplus Z_0^{m,*}), d_1) \xrightarrow{\simeq} (\wedge V / \wedge^{>m} V, d_1)\,.$$

It follows exactly as in the proof of Theorem 35.13 in [18] that it is sufficient to prove that if depth $(\wedge V, d) \geq m$ then

$$H^{p,*}(\mathrm{Hom}_{\wedge V}(\wedge V \otimes (Z_0^{m,*} \oplus k), \wedge V), d_1) = 0 \qquad \text{if } p < 0\,.$$

Following the idea in §35 of [18] we let

$$(\wedge V \otimes \wedge V \otimes \wedge U, d_1) \xrightarrow{\simeq} (\wedge V, d_1)$$

be a Sullivan model for the multiplication $(\wedge V, d_1) \otimes (\wedge V, d_1) \to (\wedge V, d_1)$. Since $\wedge V$ is not required to be simply connected, we need to rely on Proposition 3.9 (p. 103). According to Corollary 3.4 (p. 105) we may assume that $d_1 : \wedge U \to (V \otimes 1 \oplus 1 \otimes V) \otimes \wedge U$. Applying the augmentation $\varepsilon : (\wedge V, d_1) \to k$ to the middle term of $(\wedge V \otimes \wedge V \otimes \wedge U, d_1)$ and to $(\wedge V, d_1)$ gives a quasi-isomorphism $(\wedge V \otimes \wedge U, d_1) \xrightarrow{\simeq} k$, exhibiting this cdga as an acyclic closure of $(\wedge V, d_1)$.

In the same way, dividing by $\wedge^{>k} V$ yields quasi-isomorphisms

$$(\wedge V \otimes (\wedge V / \wedge^{>k} V) \otimes \wedge U, d_1) \xrightarrow{\simeq} (\wedge V / \wedge^{>k} V, d_1)$$

which for $k = m$ lifts to a quasi-isomorphism of bigraded $(\wedge V, d_1)$-modules,

$$(\wedge V \otimes (\wedge V / \wedge^{>m} V) \otimes \wedge U, d_1) \xrightarrow{\simeq} (\wedge V \otimes (k \oplus Z_0^{m,*}), d_1)\,.$$

Thus we only have to prove that if depth $(\wedge V, d) \geq m$, then for $p < 0$,

$$H^{p,*}(\mathrm{Hom}_{\wedge V}(\wedge V \otimes \wedge V/\wedge^{>m} V \otimes \wedge U, \wedge V), d_1) = 0\,. \tag{10.4}$$

Denote $(\mathrm{Hom}_{\wedge V}(\wedge V \otimes (\wedge V/\wedge^{\geq k} V) \otimes \wedge U, \wedge V), d_1)$ by $C(\wedge V/\wedge^{\geq k} V)$, and identify $\wedge^{\geq k} V/\wedge^{\geq k+1} V$ as the trivial $\wedge V$-module $\wedge^k V$. Then it follows by induction on $k$ from the short exact sequences

$$0 \to C(\wedge V/\wedge^{\geq k} V) \to C(\wedge V/\wedge^{\geq k+1} V) \to C(\wedge^k V) \to 0$$

that $H^{p,*}(C(\wedge V/\wedge^{>k} V)) = 0$, for $p < m - k$. This proves (10.4). □

Recall from Theorem 9.3 (p. 248) that if $(\wedge V, d) \to (\wedge Z, d)$ is a surjective morphism of minimal Sullivan algebra then cat $(\wedge V, d) \geq$ cat $(\wedge Z, d)$. Here we establish the analogous

**Proposition 10.1.** *Suppose $(\wedge W \otimes \wedge Z, d)$ is a decomposition of a minimal Sullivan algebra as a Sullivan extension $(\wedge W \otimes \wedge Z, d)$ of a minimal Sullivan algebra $(\wedge W, d)$. Then the fibre, $(\wedge Z, \overline{d})$ satisfies*

$$\textit{depth}\,(\wedge W \otimes \wedge Z, d) \geq \textit{depth}\,(\wedge Z, \overline{d})\,.$$

*proof*: The associated quadratic Sullivan algebra of $(\wedge W \otimes \wedge Z, d)$ decomposes as a Sullivan extension of quadratic Sullivan algebras $(\wedge W, d_1) \to (\wedge W \otimes \wedge Z, d_1) \to (\wedge Z, \overline{d}_1)$. Thus according to Proposition 3.13 (p. 112), the acyclic closure of $(\wedge W \otimes \wedge Z, d_1)$ may be decomposed as a $\Lambda$-extension

$$(\wedge W \otimes \wedge U_W, d_1) \to (\wedge W \otimes \wedge U_W \otimes \wedge Z \otimes \wedge U_Z, d_1)$$

in which $(\wedge W \otimes \wedge U_W, d_1)$ is the acyclic closure of $(\wedge W, d_1)$, the fibre $(\wedge Z \otimes \wedge U_Z, \overline{d}_1)$ is the acyclic closure of $(\wedge Z, \overline{d}_1)$ and $d_1 : \wedge U_Z \to (W \oplus Z) \otimes \wedge U_W \otimes \wedge U_Z$. Thus we may write

$$\begin{aligned}
&\mathrm{Ext}_{(\wedge W \otimes \wedge Z, d_1)}(\Bbbk, (\wedge W \otimes \wedge Z, d_1)) \\
&= H\left(\mathrm{Hom}^*_{\wedge W \otimes \wedge Z}(\wedge W \otimes \wedge Z \otimes \wedge U_W \otimes \wedge U_Z, \wedge W \otimes \wedge Z)\right) \\
&= H\left(\mathrm{Hom}^*_{\wedge W}(\wedge W \otimes \mathrm{Hom}^*_{\wedge Z}(\wedge Z \otimes \wedge U_W \otimes \wedge U_Z, \wedge W \otimes \wedge Z))\right)\,.
\end{aligned}$$

Filtering by the wedge degree of $\wedge W$ in the target space produces a first quadrant spectral sequence $E_*^{p,q}$, for which

$$E_1^{p,q} = \mathrm{Hom}^p_{\wedge W}\left(\wedge W \otimes \wedge U_W, \mathrm{Ext}^q_{(\wedge Z, \overline{d}_1)}(\Bbbk, (\wedge W, 0) \otimes (\wedge Z, \overline{d}_1))\right)\,.$$

If $\mathrm{Ext}^q_{(\wedge Z, \overline{d}_1)}(\Bbbk, (\wedge Z, \overline{d}_1)) = 0$, $q < m$ then it follows that

$$\mathrm{Ext}^r_{(\wedge W \otimes \wedge Z, d_1)}(\Bbbk, (\wedge W \otimes \wedge Z, d_1)) = 0\,, \qquad r < m\,.$$ □

**Example 1.** The depth of a minimal Sullivan algebra $(\wedge V, d)$ is equal to 0 if and only if $V$ is finite dimensional and is concentrated in even degrees.

Suppose first that $V$ is finite dimensional and concentrated in even degrees. In that case $\wedge U$ is an exterior algebra, $\wedge U = \wedge(u_1, \cdots, u_r)$. Then a cycle $f \in \mathrm{Hom}^0_{\wedge V}((\wedge V \otimes \wedge U, d), (\wedge V, d))$ is defined by $f(u_1 \cdots u_r) = 1$ and $f(\alpha) = 0$ if $\alpha \in \wedge^{<r} U$. This cannot be a boundary, because $d(U) \subset V \otimes \wedge U$.

Suppose now $\dim V = \infty$ and depth $(\wedge V, d) = 0$. Then there is a cycle, $f \in \mathrm{Hom}^0_{\wedge V}((\wedge V \otimes \wedge U, d_1), (\wedge V, d_1))$ and an $\alpha \in \wedge U$ with $f(\alpha) = 1$. Then there is a sub quadratic minimal Sullivan algebra, $(\wedge W, d_1) \subset (\wedge V, d_1)$ such that $\dim W < \infty$ and $d_1(\alpha) \in W \otimes \wedge U$. Since $\dim V = \infty$, for some $v \in V$, $v \notin W$ and $d_1 v \in \wedge^2 W$. Thus for some $u \in U$, $d_1 u - v \in W \otimes \wedge^+ U$, and so

$$f(d_1(u \cdot \alpha)) - v f(\alpha) \in W \otimes f(\wedge^+ U)\,.$$

Since $f$ is a 0-cycle, $f(\wedge^+ U) \subset \Bbbk$ and $f d_1 = \pm d_1 \circ f$. Thus $v \in d_1(\wedge V) + W$, which is impossible.

It remains to show that if $\dim V < \infty$ and depth $(\wedge V, d) = 0$, then $V = V^{\text{even}}$. But if not we may decompose $(\wedge V, d)$ as a minimal Sullivan extension $(\wedge V, d) = (\wedge W \otimes \wedge Z, d)$ in which $Z = \Bbbk z \oplus Z^{\text{even}}$ and $z$ has odd degree. But by Proposition 10.1, depth $(\wedge Z, \bar{d}) = 0$. Moreover, if $Z^{\text{even}} \neq 0$ then for some $v \in Z^{\text{even}}$, $dv = 0$ and again Proposition 10.1 applied to the Sullivan extension $(\wedge v, 0) \to (\wedge Z, d)$ implies that the fibre $(\wedge z \otimes \wedge(Z/v), d)$ has zero depth. Proceeding in this way we find that depth $(\wedge z, 0) = 0$ while a trivial calculation shows that depth $(\wedge z, 0) = 1$.

□

**Example 2.** Let $(\wedge V_1, d)$ and $(\wedge V_2, d)$ be two minimal Sullivan algebras with both $V_1$ and $V_2$ non-zero. Denote by $(\wedge V, d)$ the minimal Sullivan model of $(\wedge V_1, d) \oplus_{\Bbbk} (\wedge V_2, d)$. We prove that

$$\text{depth}\,(\wedge V, d) = 1\,.$$

Note first that we have a commutative diagram

$$\begin{array}{ccc}
(\wedge V_1, d) \otimes (\wedge V_2, d) & \longrightarrow & (\wedge V_1, d) \oplus_{\Bbbk} (\wedge V_2, d) \\
& \searrow & \uparrow \varphi \, \simeq \\
& & (\wedge V, d) = (\wedge V_1 \otimes \wedge V_2 \otimes \wedge W, d)\,.
\end{array}$$

Since the kernel of $\varphi$ is acyclic, $d(W)$ is contained in the ideal generated by $W$ and $V_1 \otimes V_2$. Now denote by $(\wedge V_1 \otimes \wedge U_1, d)$ and $(\wedge V_2 \otimes \wedge U_2, d)$ acyclic closures for $(\wedge V_1, d)$ and $(\wedge V_2, d)$. Then the acyclic closure of $(\wedge V, d)$ is

$$(\wedge V \otimes \wedge U, d) = ((\wedge V_1 \otimes \wedge U_1, d) \otimes (\wedge V_2 \otimes \wedge U_2, d) \otimes \wedge W \otimes \wedge U_W, d)\,.$$

Since the projection

$$\rho : (\wedge V \otimes \wedge U, d) \to (\wedge V_1 \otimes \wedge U_1, d) \oplus_{k\!\!k} (\wedge V_2 \otimes \wedge U_2, d)$$

is a quasi-isomorphism, its kernel is acyclic. Therefore $d(U_W)$ is contained in the ideal generated by $W$, $U_W, U_1 \otimes V_2, U_1 \otimes U_2, V_1 \otimes V_2$ and $V_1 \otimes U_2$.

To define a cycle of degree 1, we define a map $g \in \mathrm{Hom}^0_{(\wedge V_1,d)}((\wedge V_1 \otimes \wedge U_1, d), (\wedge V_1, d))$ by setting $g(1) = 1$ and $g(\wedge^+ U_1) = 0$. Then we define

$$f \in \mathrm{Hom}^1_{(\wedge V,d)}((\wedge V \otimes \wedge U, d), (\wedge V_1, d) \oplus_{k\!\!k} (\wedge V_2, d))$$

by setting $f(\alpha) = -gd(\alpha)$ for $\alpha \in \wedge U_1$ and $f(\alpha) = 0$ for $\alpha \in \wedge U_1 \otimes \wedge^+(U_2 \oplus U_W)$. Then $f$ is a 1-cycle. To show that $f$ is not a boundary let $u_1 \in U_1$ and $u_2 \in U_2$ satisfy $du_1 = v_1 \in V_1$ and $du_2 = v_2 \in V_2$. If $f = d(h)$ then for degree reason $h(1)$ must be a scalar in $k\!\!k$, and we have

$$-v_1 = dh(u_1) - h(1)v_1 \quad \text{ and } \; 0 = dh(u_2) - h(1)v_2\,.$$

The second equation gives $h(1) = 0$, which makes the first impossible. This shows that depth $(\wedge V, d) \leq 1$, and it follows from Example 1 that depth $(\wedge V, d) \neq 0$.

□

## 10.3 The depth of a Sullivan Lie algebra

Recall from §2.2 that a Sullivan Lie algebra is a graded Lie algebra $L = L_{\geq 0}$ satisfying

$$\dim L_0/[L_0, L_0] < \infty\,, \qquad \dim L_i < \infty\,,\; i \geq 1 \;\text{ and } \cap_n L^n = 0\,. \qquad (2.2)$$

Note that if $\dim L_i^p < \infty$ then the condition $\cap_n L_i^n = 0$ is equivalent to $[\underbrace{L_0, \cdots, L_0}_{n}, L_i] = 0$ for some $n$. In particular, if $L = L_{\geq 0}$ is a graded Lie algebra of finite type in which $L_0$ is nilpotent, then $L$ is a Sullivan Lie algebra if and only if $L_0$ acts nilpotently in each $L_i$.

Recall as well from §2.2 that if $L$ is a Sullivan Lie algebra then

$$(\wedge V, d_1) = \varinjlim C^*(L/L^n)$$

is its *associated quadratic Sullivan algebra.* In particular the natural morphism $\varinjlim_n C^*(L/L^n) \to C^*(L)$ induced by the surjections $L \to L/L^n$ may be interpreted as a morphism

$$\mu : (\wedge V, d_1) \to C^*(L)\,. \qquad (10.5)$$

As observed in the proof of Theorem 2.1 (p. 50), $V = \varinjlim s(L/L^n)^\#$, and $\mu$ restricts to the induced linear maps $V \to (sL)^\#$. Finally, again by Theorem 2.1, the pairing $V \times L \to \Bbbk$ induces an inclusion $L \to V^\#$ which extends to an isomorphism of $\widehat{L}$ onto the homotopy Lie algebra of $(\wedge V, d_1)$.

**Proposition 10.2.** *If $L$ is a Sullivan Lie algebra with associated quadratic Sullivan algebra $(\wedge V, d_1)$ then $\mu(L) : (\wedge V, d_1) \to C^*(L)$ is a quasi-isomorphism if and only if $\mu(L_0) : (\wedge V^1, d_1) \to C^*(L_0)$ is a quasi-isomorphism.*

*proof:* Since $L_{\geq 1}$ has finite type and $L_0$ acts locally nilpotently in $L_{\geq 1}$, it follows that $C^*(L_{\geq 1})$ is the Sullivan algebra $(\wedge V^{\geq 2}, d_1)$ and that the isomorphism $C_*(L) \cong C_*(L_0) \otimes C_*(L_{\geq 1})$ of graded vector spaces dualizes to exhibit $C^*(L)$ as the $\Lambda$-extension

$$C^*(L) = \left(C^*(L_0) \otimes \wedge V^{\geq 2}, d_1\right) .$$

Moreover, $\mu(L) = \mu(L_0) \otimes id : (\wedge V^1 \otimes \wedge V^2, d_1) \to (C^*(L_0) \otimes \wedge V^{\geq 2}, d_1)$ and the Proposition follows from Proposition 3.10 (p. 107).

□

Next recall from Theorem 3.2 (p. 110) that $(\wedge V, d_1)$ has a unique acyclic closure $(\wedge V \otimes \wedge U, d_1)$ in which, by Corollary 3.4 (p. 105) we may assume that

$$d_1 : 1 \otimes \wedge U \to V \otimes \wedge U .$$

In particular, the quotient differential in $\wedge U$ is zero. Thus (cf. §4.3) the Sullivan extension $(\wedge V \otimes \wedge U, d_1)$ determines a *holonomy representation*, $\theta$, of $L$ in $\wedge U$. If $x \in L$ and $a \in \wedge U$ then $\theta : L \to \mathrm{Der} \wedge U$ is given explicitly by

$$\theta(x)a = -\sum \langle v_i, x\rangle\, a_i ,$$

where $d_1(1 \otimes a) = \sum v_i \otimes a_i$.

Now recall that for any Lie algebra $I_L \subset UL$ is the ideal generated by $L$, and that $\widehat{UL} = \varprojlim_n UL/I_L^n$.

**Definition.** Suppose $L = L_{\geq 0}$ is any graded Lie algebra. Then,

(i) *depth* $L$ = least $p$ (or $\infty$) such that $\mathrm{Ext}^p_{UL}(\Bbbk, \widehat{UL}) \neq 0$.
(ii) *gldim* $L$ = gldim $UL$ = largest $p$ (or $\infty$) such that $\mathrm{Tor}_p^{UL}(\Bbbk, \Bbbk) \neq 0$.

In particular if $L = L_{\geq 1}$ then $\widehat{UL} = UL$, and the definition above coincides with that in ([18], p. 474).

**Proposition 10.3.** *If L is a Sullivan Lie algebra then, with the notation above,*

$$depth\, L = least\ p\ (or\ \infty)\ such\ that\ Tor_p^{UL}(\Bbbk, \wedge U) \neq 0\,.$$

*proof:* It follows from Lemma 2.12 (p. 61) with $\widehat{L} = \varprojlim_n L/L^n$ that $\widehat{UL} \xrightarrow{\cong} \widehat{U\widehat{L}}$. Since $\widehat{L}$ is the homotopy Lie algebra of $(\wedge V, d)$, Theorem 6.1 (p. 180) gives $\widehat{UL} \cong (\wedge U)^{\#}$ as $L$-modules. In this case

$$\left[\mathrm{Tor}^{UL}(\Bbbk, \wedge U)\right]^{\#} = \mathrm{Ext}_{UL}(\Bbbk, (\wedge U)^{\#}) = \mathrm{Ext}_{UL}(\Bbbk, \widehat{UL})\,.$$

□

**Theorem 10.2.** *Suppose L is a Sullivan Lie algebra. If μ is a quasi-isomorphism then*

$$depth\, L = depth\,(\wedge V, d_1) \quad and \quad gldim\, L = gldim\,(\wedge V, d_1)\,.$$

*proof:* The first equality will follow from the isomorphism

$$\left[\mathrm{Tor}_p^{UL}(\Bbbk, \wedge U)\right]^{\#} \cong \mathrm{Ext}^p_{(\wedge V, d_1)}(\Bbbk, (\wedge V, d_1)) \tag{10.6}$$

which we now establish. In fact, by the Poincaré-Birkoff-Witt Theorem ( [18], p. 286) $C_*(L, UL)$ is a $UL$-free resolution of $\Bbbk$, where $C_*(L, UL) = C_*(L) \otimes UL$ is the complex defined in §2.1. Since $C_*(L, \wedge U) = C_*(L, UL) \otimes_{UL} \wedge U$ it follows that

$$\left[\mathrm{Tor}^{UL}(\Bbbk, \wedge U)\right]^{\#} = H(C_*(L) \otimes \wedge U)^{\#}\,.$$

The natural isomorphism $(C_*(L) \otimes \wedge U)^{\#} \cong \mathrm{Hom}(\wedge U, C^*(L))$ then gives

$$\left[\mathrm{Tor}^{UL}(\Bbbk, \wedge U)\right]^{\#} = H(\mathrm{Hom}(\wedge U, C^*(L)))\,.$$

On the other hand, the canonical isomorphism

$$\mathrm{Hom}_{\wedge V}(\wedge V \otimes \wedge U, \wedge V) = \mathrm{Hom}(\wedge U, \wedge V)$$

identifies

$$\mathrm{Ext}_{(\wedge V, d_1)}(\Bbbk, (\wedge V, d_1)) = H(\mathrm{Hom}(\wedge U, \wedge V), d_1)\,.$$

Thus (10.5) will be proved if we show that composition with $\mu$:

$$\mu\circ : (\mathrm{Hom}(\wedge U, \wedge V), d_1) \to (\mathrm{Hom}(\wedge U, C^*(L)), d)$$

is a quasi-isomorphism.

The fact that $\mu\circ$ commutes with the differentials is a straightforward but tedious calculation from the definition in §2.1 of the differential $\partial$ in $C_*(L, \wedge U)$ and the fact that $\mu$ commutes with the differentials.

On the other hand, since a holonomy representation is locally nilpotent (cf. §4.3) there is an increasing filtration

$$0 = F_0 \subset F_1 \subset \cdots$$

of $\wedge U$ by $L$-submodules such that $\wedge U = \cup_k F_k$ and each $F_{k+1}/F_k$ is a trivial $L$-module. Thus the quasi-isomorphism $\mu$ induces quasi-isomorphisms of differential graded vector spaces

$$\mathrm{Hom}(F_{k+1}/F_k, \wedge V) \xrightarrow{\simeq} \mathrm{Hom}(F_{k+1}/F_k, C^*(L)) .$$

Hence, by a short exact sequence argument it induces quasi-isomorphisms

$$\mathrm{Hom}(F_k, \wedge V) \xrightarrow{\simeq} \mathrm{Hom}(F_k, C^*(L)) , \qquad k \geq 0 .$$

Now an easy inverse limit argument shows that $\mu\circ$ itself is a quasi-isomorphism, and the proof of (10.6) is complete.

Finally the equality $\mathrm{gldim}\, L = \mathrm{gldim}\, (\wedge V, d_1)$ is immediate from the fact that $H(\mu) : H^{p,*}(\wedge V, d_1) \xrightarrow{\cong} H^{p,*}(C^*(L))$.

□

## 10.4 Sub Lie algebras and ideals of a Sullivan Lie algebra

In this section we again consider Sullivan Lie algebras $L$, whose associated quadratic Sullivan algebra will always be denoted by $(\wedge V, d_1)$, and whose acyclic closure will be denoted by $(\wedge V \otimes \wedge U, d_1)$. As in §10.3 we may assume that $d_1 : \wedge U \to V \otimes \wedge U$. Finally, the holonomy representation of $L$ in $\wedge U$ will again be denoted by $x \mapsto \theta(x)$, $x \in L$, and we write

$$\theta(x)(\Phi) = x \cdot \Phi .$$

Now suppose $E \subset L$ is a sub Lie algebra. We set

$$K_E = \{ v \in V \mid \langle v, E \rangle = 0 \}$$

and let $Z_E$ be any vector space direct summand of $K_E$ in $V$: $V = K_E \oplus Z_E$, and

$$\wedge V = \wedge K_E \otimes \wedge Z_E .$$

**Lemma 10.4.** *With the hypotheses and notation above:*

*(i) If $E \subset L$ is a sub Lie algebra then $d_1 : K_E \to K_E \wedge V$.*
*(ii) If $I \subset L$ is an ideal then $\wedge K_I$ is preserved by $d_1$ and hence is a sub quadratic Sullivan algebra of $(\wedge V, d_1)$.*

*proof:* (i) By construction, the pairing $V \times L \to \Bbbk$ induces an injection $V \to L^\#$ (of degree $-1$). It follows that this induces an injection $Z_E \to E^\#$, which then extends to an injection $\wedge Z_E \to \wedge sE^\#$. Now let $w \in K_E$ and write $d_1 w = \Phi_1 + \Phi_2$ with $\Phi_1 \in K_E \wedge V$ and $\Phi_2 \in Z_E \wedge Z_E$. Then for $x, y \in E$,

$$0 = \langle w, [x, y]\rangle = \pm\langle \Phi_1, x, y\rangle \pm \langle \Phi_2, x, y\rangle \,.$$

But $\langle \Phi_1, x, y\rangle = 0$, and so $\langle \Phi_2, x, y\rangle = 0$ for all $x, y \in E$, which implies that $\Phi_2 = 0$.

(ii) If $w \in K_I$ then it follows from (i) that $d_1 w = \Psi_1 + \Psi_2$ with $\Psi_1 \in \wedge^2 K_I$ and $\Psi_2 \in K_I \otimes Z_I$. If $\Psi_2 \neq 0$ it follows from the fact that $Z_I \to I^\#$ and $K_I \to L^\#$ are injective maps that for some $x \in I$ and $y \in L$, $\langle \Psi_2, x, y\rangle \neq 0$. But $\langle \Psi_1, x, y\rangle = 0$ because, since $I$ is an ideal, for any $x \in I$ and $y \in L$,

$$\langle d_1 w, x, y\rangle = \pm\langle w, [x, y]\rangle = 0\,.$$

Thus $\Psi_2 = 0$ and $\wedge^2 K_I$ is preserved by $d_1$.

□

**Corollary 10.1.** *With the hypotheses and notation above*

*(i) $(\wedge Z, \overline{d}_1) := (\wedge V/(K_E \wedge \wedge V), \overline{d}_1)$ is a quadratic Sullivan algebra, and the quotient morphism*

$$\rho : (\wedge V, d_1) \twoheadrightarrow (\wedge Z, \overline{d}_1)$$

*identifies $Z_E$ with $Z$.*
*(ii) The homotopy Lie algebra of $(\wedge Z, \overline{d}_1)$, regarded as a sub Lie algebra under $\rho^\#$ of the homotopy Lie algebra of $(\wedge V, d_1)$, contains $E$ as a sub Lie algebra.*
*(iii) For an ideal $I$, $\rho$ is the fibre projection of a Sullivan extension*

$$(\wedge K_I, d_1) \to (\wedge K_I \otimes \wedge Z, d_1) \to (\wedge Z, \overline{d}_1)\,.$$

□

**Definition.** With the hypotheses and notation above, the *Sullivan closure* $\overline{E}$ of a sub Lie algebra $E \subset L$ is the sub Lie algebra $\overline{E} \supset E$ given by

$$\overline{E} := (sK_E)^\perp = \{\, x \in L \,|\, \langle K_E, x\rangle = 0\,\}\,.$$

In particular, $E$ is *Sullivan closed* if $\overline{E} = E$.

**Lemma 10.5.** *If $E$ is a sub Lie algebra of a Sullivan Lie algebra $L$, then*

*(i)* $\overline{E}$ *is Sullivan closed and* $K_E = K_{\overline{E}}$.
*(ii)* $\overline{E} = \cap_n(E + L^n)$, *and so $E$ is Sullivan closed if and only if* $E = \cap_n(L^n + E)$. *In particular, $E$ is closed if and only if $E_0$ is closed.*
*(iii) If $E$ is a Sullivan Lie algebra and $E^k$ is Sullivan closed, then for any $r$ there is some $n$ such that* $E_r^k \supset (E \cap L^n)_r$.
*(iv) If either* $E_0 \supset L_0^n$ *for some $n$, or* $\dim E_0 < \infty$, *then $E$ is Sullivan closed. If* $\dim L_0/E_0 < \infty$ *and $E$ is Sullivan closed then* $E_0 \supset L_0^n$ *for some $n$.*
*(v) Suppose that* $[L_0, E_0] \subset E_0$, $L_0^n \subset E_0$ *for some $n$, and that $L_0$ acts nilpotently in* $E_0/[E_0, E_0]$. *Then* $\dim E_0/[E_0, E_0] < \infty$ *and for each $k$ there is an $r$ such that* $L_0^r \subset E_0^k$.
*(vi) If $E$ is a Sullivan closed ideal then $L/E$ is a Sullivan Lie algebra.*

*proof*: (i) This is immediate from the definitions.

(ii) All these assertions follow from the equality $\overline{E} = \cap_n(E + L^n)$. To establish this, recall from §2.3 that $V = \cup_n V_n$ with $V_n = \left(s\,L/L^{n+2}\right)^{\#}$. Let $E(n)$ denote the image of $E$ in $L/L^{n+2}$ and let $K_n$ denote $K_E \cap V_n$. By definition, $K_n = E(n)^{\perp}$. Thus, since $L/L^{n+2}$ has finite type, $E(n) = K_n^{\perp}$. But $x \in E$ if and only if its image in each $L/L^{n+2}$ is in $K_n^{\perp}$. Thus $x \in \overline{E}$ if and only if $x \in E + L^n$ for all $n$.

(iii) Denote by $E[n]$ the image in $E/E^k$ of $E \cap L^n$. Because $E$ is a Sullivan Lie algebra, $E/E^k$ has finite type. It follows that for any $r$ there is some $n$ such that $E[n]_r = E[n+s]_r$, $s \geq 0$. Thus $E_r^k + (E \cap L^n)_r = E_r^k + (E \cap L^{n+s})_r$, $s \geq 0$. This implies that $E_r^k + (E \cap L^n)_r = \cap_s(E^k + L^s)_r = E_r^k$, because $E^k$ is Sullivan closed.

(iv) If $E_0 \supset L_0^n$ then it follows from (ii) that $E$ is Sullivan closed. Now suppose $\dim E_0 < \infty$, and that $y \in \cap_n(E_0 + L_0^n)$. Then for each $n \geq 1$, $y = x_n + z_n$ with $x_n \in E_0$ and $z_n \in L_0^n$. But since $\dim E_0 < \infty$, it follows that for some $k$, $E_0 \cap L_0^k = E_0 \cap L_0^{k+s}$, $s \geq 0$. Thus if $u \in E_0 \cap L_0^k$ then $u \in \cap_s L_0^{k+s} = 0$. This implies that for $n \geq k$, $z_{n+1} - z_n = x_n - x_{n+1} \in E_0 \cap L_0^k$, and so $z_{k+s} = z_k$, $s \geq 0$. This implies that $z_k \in \cap_s L^{k+s} = 0$, and so $y = x_k \in E_0$. It follows from (ii) that $E$ is Sullivan closed.

Finally, suppose $\dim L_0/E_0 < \infty$ and $E$ is Sullivan closed. The pairing $V \times L \to \Bbbk$ induces an injection $K_E \to (L/E)^{\#}$ and so, since $\dim L_0/E_0 < \infty$, $\dim (K_E)^1 < \infty$. Thus, with respect to the filtration $V_n^1 = (L_0/L_0^n)^{\#}$ of $V^1$, we have $K_E^1 \subset V_n^1$ for some $n$ and so $E_0 = (sK_E^1)_0^{\perp} \supset (sV_n^1)_0^{\perp} = L_0^n$.

(v) The Lie bracket induces surjections

$$E_0/[E_0, E_0] \otimes E_0^k/E_0^{k+1} \to E_0^{k+1}/E_0^{k+2}.$$

Thus it follows by induction on $k$ that $L_0$ acts nilpotently in each $E_0^k/E_0^{k+1}$. In particular, given $k$ there is some $\ell$ such that $[\underbrace{L_0, \cdots, L_0}_{\ell}, E_0^k] \subset E_0^{k+1}$. Now it follows, again by induction on $k$, that for each $k$ there is an $r$ for which $L_0^r \subset E_0^k$. In particular, for some $r$, $\dim E_0/[E_0, E_0] \leq \dim L_0/L_0^r < \infty$.

(vi) Since $L/[L, L] \to (L/E)/[L/E, L/E]$ is surjective, the quotient $(L/E)/[L/E, L/E]$ has finite type. Moreover, suppose $x \in L$ maps to $\overline{x} \in \cap_n (L/E)_0^n$. Then $x \in \cap_n (E + L^n)_0 = E_0$, because $E$ is Sullivan closed. Thus $\overline{x} = 0$ and $L/E$ is a Sullivan Lie algebra.

□

**Example**: An ideal that is not Sullivan closed.

Let $F$ be the Lie algebra with basis $y_1, y_2, y_3$ and Lie brackets $[y_i, y_j] = y_k$ where $i, j, k$ is a cyclic permutation of $1, 2, 3$. Denote by $\mathbb{L} = \mathbb{L}(x_1, x_2, x_3)$ the free Lie algebra generated by $x_1, x_2, x_3$ and define

$$\rho : \mathbb{L}(x_1, x_2, x_3) \to F$$

by $\rho x_i = y_i$ and denote $I = \operatorname{Ker} \rho$. Since $F = [F, F]$ the morphism $\rho$ restricts for every $n$ to a surjection $\mathbb{L}^n \to F$. Therefore $\mathbb{L} = I + \mathbb{L}^n$ for all $n$, and by Lemma 10.5(ii), $\overline{I} = \mathbb{L}$, and $I$ is not Sullivan closed.

**Proposition 10.4.** *Let $E$ and $I$ be respectively a sub Lie algebra and an ideal in a Sullivan Lie algebra $L$. Then, with the notation above,*

*(i) If $[E_0, E_0] \supset L_0^n$ for $n$ sufficiently large, then $E$ is Sullivan closed and a Sullivan Lie algebra.*

*(ii) Suppose $E_0 \supset L_0^n$ for some $n$. If $[L_0, E_0] \subset E_0$ and $L_0$ acts nilpotently in $E_0/[E_0, E_0]$, then $\dim E_0/[E_0, E_0] < \infty$ and each $E^k$ is Sullivan closed.*

*(iii) If $\dim E_0/[E_0, E_0] < \infty$ then the pairing $Z \times E \to \Bbbk$ induces a natural morphism from $(\wedge Z, \overline{d}_1)$ to the associated quadratic Sullivan algebra for $E$. If each $E^k$ is Sullivan closed this morphism is an isomorphism.*

*(iv) If $I$ is Sullivan closed then $L/I$ is a Sullivan Lie algebra and the pairing $K_I \times L/I \to \Bbbk$ induces an isomorphism from $(\wedge K_I, d_1)$ to the associated quadratic Sullivan algebra for $L/I$.*

*proof*: (i) It follows from Lemma 10.5(iv) that $E$ is Sullivan closed. Since $E \subset L$, $E_{\geq 1}$ has finite type and $\cap E^n \subset \cap L^n = 0$. To show that $E$ is a Sullivan Lie algebra it remains to show that $\dim E_0/[E_0, E_0] < \infty$. Since $[E_0, E_0] \supset L_0^n$, $E/(L^n \cap E) \to E/[E, E]$ is surjective in degree 0. But $E/(L^n \cap E)$ embeds in $L/L^n$, and it follows that $\dim E_0/[E_0, E_0] < \infty$.

(ii) This is immediate from Lemma 10.5(v) and Lemma 10.5(iv).

(iii) Again denote $\left(s\,L/L^{k+2}\right)^{\#}$ by $V_k$ and let $Z_k$ correspond under the isomorphism $Z_E \xrightarrow{\cong} Z$ to $Z_E \cap V_k$. Since $L/L^{k+2}$ has finite type it follows as in the proof of Lemma 10.5(ii) that the pairing $Z_E \times E \to \Bbbk$ induces isomorphisms $Z_k \xrightarrow{\cong} (E/E \cap L^{k+2})^{\#}$. These then extend to isomorphisms $(\wedge Z_k, \overline{d}_1) \xrightarrow{\cong} C^*(E/E \cap L^{k+2})$. Thus we obtain the natural morphism

$$(\wedge Z, \overline{d}_1) \xrightarrow{\cong} \varinjlim_k C^*(E/E \cap L^{k+2}) \to \varinjlim_k C^*(E/E^{k+2})$$

from $(\wedge Z, \overline{d}_1)$ to the associated Sullivan algebra for $E$.

Now suppose each $E^n$ is closed in $L$. Then by Lemma 10.5(iii), for sufficiently large $n$, $(E/E^{k+2})_{\leq p}$ embeds in the sequence

$$(E/E \cap L^n)_{\leq p} \to (E/E^{k+2})_{\leq p} \to (E/E \cap L^{k+2})_{\leq p}\,,$$

and these identify

$$\varinjlim_k C^*(E/E^{k+2})_{\leq p} = \varinjlim_k C^*(E/E \cap L^{k+2})_{\leq p} \cong ((\wedge Z)_{\leq p}, \overline{d}_1)\,.$$

Since this holds for each $p$, assertion (iii) is established.

(iv) It follows from Lemma 10.5(vi) that $L/I$ is a Sullivan Lie algebra. Now note that the short exact sequences

$$0 \to I/(L^{n+2} \cap I) \to L/L^{n+2} \to (L/I)\,/\,(L/I)^{n+2} \to 0$$

dualize to short exact sequences which fit in the row exact commutative diagram

$$\begin{array}{ccccccccc}
0 & \longrightarrow & \left[\frac{L/I}{(L/I)^{n+2}}\right]^{\#} & \longrightarrow & (L/L^{n+2})^{\#} & \longrightarrow & (I/L^{n+2} \cap I)^{\#} & \longrightarrow & 0 \\
 & & \uparrow \alpha & & \cong \uparrow \beta & & \cong \uparrow \gamma & & \\
0 & \longrightarrow & K_I \cap V_n & \longrightarrow & V_n & \longrightarrow & V_n/(K_I \cap V_n) & \longrightarrow & 0
\end{array}$$

in which $\beta$ is an isomorphism by the definition of $V_n = (sL/L^{n+1})^{\#}$ and $\gamma$ is an isomorphism because $I$ is Sullivan closed. It follows that $\alpha$ is also an isomorphism. In particular, the isomorphisms $\alpha$ extend to isomorphisms

$$\wedge(K_I \cap V_n) \xrightarrow{\cong} C^*((L/I)\,/\,(L/I)^{n+2})$$

and these commute with the differentials because in both cases the differential is defined via the Lie bracket. Passing to the limit completes the proof.

□

Suppose now that $E$ is a sub Lie algebra of a Sullivan Lie algebra $L$. Again, let $(\wedge V \otimes \wedge U, d_1)$ denote the acyclic closure of the quadratic Sullivan algebra $(\wedge V, d_1)$ associated with $L$. Thus restriction to $E$ of the holonomy representation of $L$ makes $\wedge U$ into a left $UE$-module. For the Hochschild-Serre spectral sequence applications in the next section we analyse the structure of $\wedge U$ as a $UE$-module.

For this, as above let $(\wedge Z, \overline{d}_1)$ be the quotient quadratic Sullivan algebra of $(\wedge V, d_1)$ obtained by dividing by $K_E \wedge \wedge V$. Thus, by Proposition 3.9 (p. 103), the surjection $\rho : (\wedge V, d_1) \to (\wedge Z, \overline{d}_1)$ extends to a quasi-isomorphism

$$\rho : (\wedge V \otimes \wedge S, d_1) \xrightarrow{\simeq} (\wedge Z, \overline{d}_1)$$

from a minimal $\Lambda$-extension of $(\wedge V, d_1)$.

Use $\rho$ to identify $Z$ with some arbitrary direct summand of $K_E$ in $V$. In view of Corollary 3.4 (p. 105), we may assume that

$$\rho(S) = 0 \quad \text{and} \quad d_1 : S \to (K_E \otimes \wedge S) \oplus (Z \otimes \wedge^+ S). \tag{10.7}$$

Since $\rho$ is a surjective quasi-isomorphism it has a right inverse $\lambda_Z : (\wedge Z, \overline{d}_1) \to (\wedge V \otimes \wedge S, d_1)$. Moreover, since $\rho$ maps $\wedge^k V \otimes \wedge S$ to $\wedge^k Z$ and since (by (10.7)) $d_1$ and $\overline{d}_1$ respectively map $\wedge^k V \otimes \wedge S$ and $\wedge^k Z$ to $\wedge^{k+1} V \otimes \wedge S$ and $\wedge^{k+1} Z$, it follows that $\lambda_Z$ may be chosen to map $Z$ to $V \otimes \wedge S$. Now the component of $\lambda_Z(Z)$ in $V \otimes 1$ is a direct summand of $K_E$ in $V$ and so we may and do suppose that

$$\lambda_Z(z) - 1 \otimes z \otimes 1 \in V \otimes \wedge^+ S, \qquad z \in Z. \tag{10.8}$$

Next, let $(\wedge Z \otimes \wedge U_Z, \overline{d}_1)$ denote an acyclic closure for $(\wedge Z, \overline{d}_1)$. Then use the morphism $\lambda_Z$ to construct the morphism,

$$(\wedge V, d_1) \to (\wedge V \otimes \wedge S) \otimes_{\wedge Z} (\wedge Z \otimes \wedge U_Z, d) = (\wedge V \otimes \wedge S \otimes \wedge U_Z, d_1) \tag{10.9}$$

which by Corollary 3.8 (p. 114) is the acyclic closure $(\wedge V \otimes \wedge U, d_1)$. Thus dividing $\wedge V \otimes \wedge S \otimes \wedge U_Z$ by the ideal $\wedge^+ V$ gives an isomorphism

$$\wedge S \otimes \wedge U_Z \cong \wedge U, \tag{10.10}$$

which identifies $S \oplus U_Z = U$.

Since $(\wedge V \otimes \wedge U, d_1)$ is an acyclic closure, $U$ is the increasing union of subspaces $0 = U(0) \subset \cdots \subset U(k) \subset \cdots$ such that for each $k \geq 1$

$$d_1(1 \otimes u) \in \wedge^+ V \otimes \wedge U(k-1)\,, \qquad u \in U(k)\,. \tag{10.11}$$

Now set

$$S(k) = S \cap U(k) \quad \text{ and } U_Z(k) = U_Z \cap U(k)\,.$$

Then $U(k) \subset S \oplus U_Z(k)$ and we have

$$\left.\begin{array}{ll} & d_1(1 \otimes u) \in \wedge^+ V \otimes \wedge S \otimes \wedge U_Z(k-1)\,, \qquad u \in U_Z(k) \\ \text{and} & \\ & d_1(1 \otimes u) \in \wedge^+ V \otimes \wedge^+ S(k-1)\,, \qquad u \in S(k). \end{array}\right\} \tag{10.12}$$

For convenience we use the isomorphism (10.10) to define $\widetilde{U}(k) \subset U$ by setting $S \oplus U(k) = \widetilde{U}(k)$.

Recall now from Corollary 10.1 (p. 280) that the inclusion of the homotopy Lie algebra, $F$, of $(\wedge Z, \overline{d}_1)$ in the homotopy Lie algebra of $(\wedge V, d_1)$ identifies $E$ as a sub Lie algebra of $F$. In particular, the holonomy representation of $F$ in $\wedge U_Z$ restricts to a (holonomy) representation, denoted by $x \mapsto \overline{\theta}_Z(x)$ of $E$. On the other hand, the Sullivan extensions $(\wedge V, d_1) \to (\wedge V \otimes \wedge S, d_1)$ and $(\wedge V, d_1) \to (\wedge V \otimes \wedge U, d_1)$ determine holonomy representations, denoted by $x \mapsto \overline{\theta}(x)$, of $L$ in $\wedge S$ and in $\wedge U$, with $\wedge S$ a sub $L$-module of $\wedge U$. These also restrict to representations of $E$.

**Lemma 10.6.** *Let $E$ be a sub Lie algebra of a Sullivan Lie algebra $L$. Then, with the notation above,*

- *(i) $\wedge^+ S$ and $\wedge^+ S \otimes \wedge U_Z$ are $E$-modules.*
- *(ii) The surjection $\rho_Z : \wedge U \to \wedge U_Z$, with kernel $\wedge^+ S \otimes \wedge U_Z$, is a morphism of $E$-modules.*
- *(iii) $\wedge U_Z$ is the increasing union of subspaces $F_k$ such that each $\wedge S \otimes F_k$ is an $L$-module and $\wedge S \otimes (F_{k+1}/F_k)$ is the tensor product of the $L$-module $\wedge S$ with the trivial action in $(F_{k+1}/F_k)$.*

*proof*: (i) By (10.7), $d_1 : S \to (K_E \otimes \wedge S) \oplus (Z \otimes \wedge^+ S)$. It follows from the definition (§4.3) of the holonomy representation that $E \cdot S \subset \wedge^+ S$. Since $E$ acts by derivations, $\wedge^+ S$ and $\wedge^+ S \otimes \wedge U_Z$ are preserved by $E$.

(ii) This is an immediate consequence of (i).

(iii) This follows from (10.12).

□

We now modify the isomorphism (10.10) to obtain an isomorphism compatible with the holonomy representation of $E$.

**Proposition 10.5.** *Suppose $E$ is a sub Lie algebra of a Sullivan Lie algebra $L$ with associated quadratic Sullivan algebra $(\wedge V, d_1)$. Then, with the notation above, the inclusion $\wedge S \to \wedge U$ extends to an isomorphism of graded algebras*

$$\xi : \wedge U \xrightarrow{\cong} \wedge S \otimes \wedge U_Z$$

*restricting to isomorphisms $\wedge \widetilde{U}(k) \xrightarrow{\cong} \wedge S \otimes \wedge U_Z(k)$, $k \geq 0$, and satisfying*

$$\left[id \otimes \overline{\theta}_Z(x)\right] \circ \xi = \xi \circ \overline{\theta}(x)\,, \qquad x \in E\,.$$

*proof:* Recall from §6.1 the diagonal homomorphism, $\Delta : \wedge U \to \wedge U \otimes \wedge U$, and use (10.10) to define $\rho_S : \wedge U \to \wedge S$ as the surjection extending $id_{\wedge S}$ and with kernel $\wedge S \otimes \wedge^+ U_Z$. Then set

$$\xi = (\rho_S \otimes \rho_Z) \circ \Delta : \wedge U \to \wedge S \otimes \wedge U_Z\,.$$

From Proposition 6.4 (p. 181) we have $\Delta \circ \overline{\theta}(y) = (id \otimes \overline{\theta}(y)) \circ \Delta$, $y \in L$. Thus for $x \in E$,

$$\xi \circ \overline{\theta}(x) = (\rho_S \otimes \rho_Z) \circ (id \otimes \overline{\theta}(x)) \circ \Delta = id \otimes \overline{\theta}_Z(x) \circ \xi\,.$$

Next recall from Lemma 6.2 (p. 173) that

$$\Delta = (\varepsilon_{\wedge V} \otimes id) \circ \sigma' : \wedge U \to \wedge U \otimes \wedge U\,,$$

where $\sigma' : \wedge U \to (\wedge V \otimes \wedge U \otimes \wedge U, d_1)$ is a right inverse for $\varepsilon_{\wedge V} \otimes \varepsilon_{\wedge U} \otimes id$. We show by induction on $k$ that we may choose $\sigma'$ so that

$$\left.\begin{array}{l} \sigma'(u) - 1 \otimes 1 \otimes u \in \wedge V \otimes \wedge U \otimes \wedge U(k-1)\,, \quad u \in U(k) \\ \text{and} \\ \sigma'(u) - 1 \otimes 1 \otimes u \in \wedge V \otimes \wedge U \otimes \wedge S(k-1)\,, \quad u \in S(k)\,. \end{array}\right\} \quad (10.13)$$

In fact, by (10.11), if $u \in U(k)$ then $d_1(1 \otimes 1 \otimes u)$ is a cycle in $\wedge V \otimes \wedge U \otimes \wedge U(k-1)$ mapping to zero under the quasi-isomorphism $\varepsilon_{\wedge V} \otimes \varepsilon_{\wedge U} \otimes id$. Thus for some $\Omega \in \wedge V \otimes \wedge U \otimes \wedge U(k-1)$, we have

$$d_1(1 \otimes 1 \otimes u - \Omega) = 0\,.$$

By hypothesis, $\sigma'$ is defined in $\wedge U(k-1)$ and we extend it to $U(k)$ by setting $\sigma'(u) = 1 \otimes 1 \otimes u - \Omega + \sigma\rho\Omega$.

On the other hand, if $u \in S(k)$ then by (10.12) $d_1(1 \otimes 1 \otimes u)$ is a cycle in $\wedge^+ V \otimes \wedge U \wedge S(k-1)$, and $\sigma' : \wedge S(k-1) \to \wedge V \otimes \wedge U \otimes \wedge S(k-1)$

is defined by induction. Thus the same argument as above establishes the second assertion of (10.13).

Now suppose $\Phi \in \wedge S$. Then by Proposition 6.2(i) (p. 174) $(id \otimes \varepsilon_{\wedge U})\Delta\Phi = \Phi$. But by (10.13), $\Delta\Phi \in \wedge U \otimes \wedge S$ and so

$$(\rho_S \otimes \rho_Z)(\Delta\Phi) = \rho_S(id \otimes \varepsilon_{\wedge U})(\Delta\Phi) = \rho_S\Phi = \Phi\,.$$

Thus $\xi$ extends the inclusion of $\wedge S$.

On the other hand, if $\Psi \in \wedge U_Z(k)$ then by (10.13)

$$\Delta\Psi - 1 \otimes \Psi \in \wedge U \otimes (\wedge S \otimes \wedge U_Z(k-1))\,.$$

Thus

$$(\rho_S \otimes \rho_Z)\Delta\Psi - 1 \otimes \Psi \in \wedge S \otimes \wedge U_Z(k-1)\,.$$

Set $\Gamma = (\rho_S \otimes \rho_Z) \circ \Delta(\Psi) - 1 \otimes \Psi$. Then, since $\Delta$ is an algebra homomorphism,

$$(\rho_S \otimes \rho_Z) \circ \Delta(\Phi \otimes \Psi) - \Phi \otimes \Psi = (\rho_S \otimes \rho_Z \circ \Delta)(\Phi) \wedge (\rho_S \otimes \rho_Z \circ \Delta)(\Psi) - \Phi \otimes \Psi$$

$$= \Phi \wedge \Gamma \in \wedge S \otimes \wedge U_Z(k-1)\,.$$

It follows that $\xi$ is an isomorphism, and in fact that it restricts to isomorphisms

$$\wedge S \otimes \wedge U_Z(k) \xrightarrow{\cong} \wedge S \otimes \wedge U_Z(k),, \qquad k \geq 0\,.$$

□

**Corollary 10.2.** *$\wedge S$ is a trivial $E$-module.*

## 10.5 Depth and relative depth

*In this section we consider a fixed Sullivan Lie algebra $L$. The associated quadratic Sullivan algebra is denoted $(\wedge V, d_1)$ and its acyclic closure is denoted $(\wedge V \otimes \wedge U, d_1)$.*

Recall now from §10.3 that depth $L$ is the least $p$ (or $\infty$) such that $\mathrm{Tor}_p^{UL}(\Bbbk, \wedge U) \neq 0$.

**Definition.** Let $E \subset L$ be a sub Lie algebra. Then the *depth of $E$ with respect to $L$*, $\mathrm{depth}_L E$, is the least $p$ (or $\infty$) such that $\mathrm{Tor}_p^{UE}(\Bbbk, \wedge U) \neq 0$.

**Remark.** Note that in principle $\mathrm{depth}_L E$ might not be an invariant of $E$, as it depends on the inclusion $E \subset L$.

However, recall from §10.4 that if $E$ is a Sullivan Lie algebra then it determines a decomposition $\wedge K_E \otimes \wedge Z$ of $\wedge V$, an induced surjection $\rho : (\wedge V, d_1) \to (\wedge Z, \overline{d}_1)$, and by Proposition 10.4(iii), (p. 282) a natural morphism from $(\wedge Z, \overline{d}_1)$ to the associated quadratic Sullivan algebra $(\wedge W, d_1)$ for $E$.

**Theorem 10.3.** *Suppose a sub Lie algebra $E$ of the Sullivan Lie algebra $L$ is itself a Sullivan Lie algebra. If the morphism $(\wedge Z, \overline{d}_1) \to (\wedge W, d_1)$ is a quasi-isomorphism then*

$$depth_L E = depth\, E\,.$$

*proof:* As specified at the start of this section, $(\wedge V \otimes \wedge U, d_1)$ denotes the acyclic closure of the quadratic Sullivan algebra associated with $L$. Moreover, by Proposition 10.5 (p. 286) there is an isomorphism $\wedge U = \wedge S \otimes \wedge U_Z$ which identifies the holonomy representation of $E$ in $\wedge U$ with the holonomy representation in $\wedge U_Z$ tensored with the trivial representation in $\wedge S$. Finally, since $(\wedge Z, \overline{d}_1) \to (\wedge W, d_1)$ is a quasi-isomorphism of minimal Sullivan algebras, it is an isomorphism. Thus Proposition 10.4(iii) (p. 282) identifies $(\wedge Z \otimes \wedge U_Z, \overline{d}_1)$ as the acyclic closure of $(\wedge W, d_1)$.

It follows that

$$\begin{aligned} \mathrm{depth}_L E &= \text{least } p \text{ such that } \mathrm{Tor}_p^{UE}(\Bbbk, \wedge U) \neq 0 \\ &= \text{least } p \text{ such that } \mathrm{Tor}_p^{UE}(\Bbbk, \wedge U_Z) \neq 0 \\ &= \mathrm{depth}\, E\,. \end{aligned}$$

□

**Corollary 10.3.** *Suppose $E$ is a sub Lie algebra of the Sullivan Lie algebra $L$, and that $E$ is itself a Sullivan Lie algebra. Then*

*(i) If each $E_0^n$ is Sullivan closed in $L_0$ then $depth_L E = depth\, E$.*
*(ii) If $dim\, E_0 < \infty$ then $depth_L E = depth\, E$.*
*(iii) Suppose $E_0 \supset L_0^n$, some $n$, that $[L_0, E_0] \subset E_0$, and that $L_0$ acts nilpotently in $E_0/[E_0, E_0]$. Then $depth_L E = depth\, E$.*

*proof:* It follows from Proposition 10.4(iii) (p. 282) that $(\wedge Z, \overline{d}) \xrightarrow{\cong} (\wedge W, d_1)$ and so (i) follows from the Theorem. Now (ii) is a special case of (i), and it follows from Proposition 10.4(ii) that (iii) $\Longrightarrow$ (i).

□

**Example.** Let $L$ be the homotopy Lie algebra of a Sullivan 1-algebra $(\wedge V, d_1)$ with $\dim H^1(\wedge V, d) < \infty$. Suppose $\dim L = \infty$ and denote by $E$ the sub Lie algebra generated by a subspace mapping isomorphically to $L/[L,L]$. Then $K_E = 0$, so that $E$ is not Sullivan closed and $\overline{E} = L$.

On the other hand, it follows from Theorem 2.1 (p. 50) that the inclusion of $E$ in $L$ induces isomorphisms $E/E^n \xrightarrow{\cong} L/L^n$. Thus the natural morphism $(\wedge V, d_1) \to \varinjlim_n C^*(E/E^n)$ is an isomorphism. Since $E$ is (trivially) a Sullivan Lie algebra it follows from Theorem 10.3 that

$$\operatorname{depth} E = \operatorname{depth}_L E\,.$$

The remaining results in this section rely heavily on Hochschild-Serre spectral sequences, and in particular on Lemma 10.9 below. Suppose $E$ is any sub Lie algebra of a graded Lie algebra $F = F_{\geq 0}$, and that $M$ is an $E$-module. Then recall from §2.1 the definition of the complex $C_*(E,M) = (\wedge sE \otimes M, \partial)$. It follows from Corollary 10.1 (p. 280) that $H(C_*(E,M)) = \operatorname{Tor}_*^{UE}(\mathbb{k}, M)$ with $\operatorname{Tor}_p^{UE}(\mathbb{k}, M)$ corresponding to the cycles in $\wedge^p sE \otimes M$.

Now suppose $E$ is an ideal in $F$ and that the representation of $E$ in $M$ is the restriction of a representation of $F$. Then this representation, together with the adjoint representation of $F$ in $E$, defines a representation of $F$ in $C_*(E,M)$ and hence a representation of $F$ in $\operatorname{Tor}_*^{UE}(\mathbb{k}, M)$.

**Lemma 10.7.** *With the notation and hypotheses above*

- *(i) The inclusion $sE \to \wedge sE$ induces an isomorphism $s(E/[E,E]) \xrightarrow{\cong} Tor_1^{UE}(\mathbb{k}, \mathbb{k})$. If $E$ is an ideal in $F$ this is an isomorphism of $F$-modules.*
- *(ii) Suppose $E$ is an ideal in $F$. If the representation of $F$ in $Tor_*^{UE}(\mathbb{k}, \mathbb{k})$ is locally nilpotent and if the representation of $E$ in $M$ extends to a locally nilpotent representation of $F$, then the representation of $F$ in $Tor_*^{UE}(\mathbb{k}, M)$ is locally nilpotent.*

*proof:* (i) This is immediate from the definition of $\partial$.

(ii) By hypothesis $M$ is the direct limit of sub $F$-modules $M(k) \subset M(k+1)$ such that each $M(k+1)/M(k)$ is a trivial $F$-module. Induction on $k$ via the short exact sequences

$$0 \to C_*(E, M(k)) \to C_*(E, M(k+1)) \to C_*\left(E, M(k+1)/M(k)\right)$$

shows that each $\operatorname{Tor}_*^{UE}(\mathbb{k}, M(k))$ is a locally nilpotent $F$-module. Since $\operatorname{Tor}_*^{UE}(\mathbb{k}, -)$ commutes with direct limits, (ii) follows. □

**Proposition 10.6.** *Let $E \subset F$ be sub Lie algebras of the Sullivan Lie algebra $L$. If the adjoint action of $E$ in $F/E$ is locally nilpotent then $depth_L\, E \leq depth_L\, F$.*

**Corollary 10.4.** *If $E$ is an ideal in $F$, or if $dim\, F/E < \infty$ and $E$ is Sullivan closed in $L$, then $depth_L E \leq depth_L F$. In particular, if $E$ and $F$ satisfy the hypotheses of Theorem 10.3 then*

$$depth\, E \leq depth\, F\,.$$

*proof:* If $E$ is an ideal in $F$ the action of $E$ in $F/E$ is trivial. If $\dim F/E < \infty$ and the action of $E$ in $F/E$ is not locally nilpotent, then for some $0 \neq \overline{x} \in F/E$, and some $k$, $\overline{x} \in [\underbrace{E, E, \cdots, E}_{n}, F/E]$, $n \geq k$. Thus if $x \in F$ maps to $\overline{x}$ then $x \in \cap_n(E + L^n)$ and so by Lemma 10.5(ii), $E$ is not Sullivan closed.

□

*proof of Proposition 10.6:* Recall from §10.1 that for any right $UE$-module $M$ and left $UE$-module $N$,

$$\mathrm{Tor}^{UE}(M, N) = \mathrm{Tor}^{UE}(l\!k, M \otimes N)$$

where $E$ acts from the left in $M$ via $x \cdot m = -(-1)^{\deg x \cdot \deg m} m \cdot x$, $m \in M$, $x \in E$. Thus the Hochschild-Serre spectral sequence corresponding to $E \subset F$ converges to $\mathrm{Tor}_*^{UF}(l\!k, \wedge U)$ from

$$E^1_{p,q} = \mathrm{Tor}_q^{UE}(l\!k, \wedge^p s(F/E) \otimes \wedge U)\,.$$

Since $E$ acts locally nilpotently in $F/E$, $\wedge s(F/E)$ is the increasing union of $UE$-submodules $M(k)$ where relative quotients $M(k+1)/M(k)$ are trivial $UE$-modules. It follows exactly as in the proof of Lemma 10.7(ii) that $\mathrm{Tor}_q^{UE}(l\!k, \wedge s(F/E) \otimes \wedge U) = 0$, $q < \mathrm{depth}_L\, E$. Thus $E^1_{p,q} = 0$, $q < \mathrm{depth}_L\, E$, and so $E^\infty_{p,q} = 0$ if $p + q < \mathrm{depth}_L E$. Hence $\mathrm{Tor}_r^{UF}(l\!k, \wedge U) = 0$, $r < \mathrm{depth}_L E$.

□

**Proposition 10.7.** *If $L$ is a Sullivan Lie algebra and $depth\, L = 0$, then $dim\, L < \infty$.*

*proof:* Let $(\wedge V, d_1)$ be the associated quadratic Sullivan algebra for $L$ and let $(\wedge V \otimes \wedge U, d_1)$ be its acyclic closure. Then $V$ has a basis $(v_\ell)_{\ell \geq 1}$ such that $dv_1 = 0$ and $dv_{\ell+1} \in \wedge(v_1, \cdots, v_\ell)$. Furthermore $U$ has a basis $(u_\ell)_{\ell \geq 1}$ such that $du_1 = v_1$ and $du_{\ell+1} - v_{\ell+1} \in \wedge(v_1, \cdots, v_\ell) \otimes \wedge U$. Finally, there are vectors $(x_\ell)_{\ell \geq 1}$ in $L$ such that $\langle v_i, x_j \rangle = \delta_{ij}$. It follows from the definition of the holonomy representation that for $i \geq j$, $x_i \cdot u_j = -\delta_{ij}$.

Now suppose $\dim L = \infty$. Any $\Phi \in \wedge U$ is an element in some $\wedge(u_1, \cdots, u_r)$. Since $L$ acts by derivations in $\wedge U$ it follows that $x_{r+1} \cdot (u_{r+1} \wedge \Phi) = -\Phi$, and hence $\mathrm{Tor}_0^{UL}(\Bbbk, \wedge U) = \Bbbk \otimes_{UL} \wedge U = 0$. Thus $\mathrm{depth}\, L \geq 1$.

□

**Proposition 10.8.** *Suppose $L$ decomposes as the direct sum of two ideals $I$ and $J$. Then $I$ and $J$ are Sullivan closed Sullivan Lie algebras, and*

$$\mathit{depth}\, L = \mathit{depth}\, I + \mathit{depth}\, J\,.$$

*Moreover, if the map $Tor^{UI}(\Bbbk, \wedge U) \to Tor^{UL}(\Bbbk, \wedge U)$ is non-zero then $\mathit{depth}\, J = 0$.*

*proof*: In this case $L/L^n = I/I^n \oplus J/J^n$. In particular, each $I/I^n$ and $J/J^n$ have finite type, as do $I_{\geq 1}$ and $J_{\geq 1}$. Thus $I$ and $J$ are Sullivan Lie algebras. Moreover $C^*(L/L^n) = C^*(I/I^n) \otimes C^*(J/J^n)$ and so, if $(\wedge V_I, d_1)$ and $(\wedge V_J, d_1)$ are the quadratic Sullivan algebras associated to $I$ and $J$, then

$$(\wedge V, d_1) = (\wedge V_I, d_1) \otimes (\wedge V_J, d_1)\,.$$

In particular, $I$ and $J$ are Sullivan closed in $L$. Furthermore, denoting the corresponding acyclic closures by $(\wedge V_I \otimes \wedge U_I, d_1)$ and $(\wedge V_J \otimes \wedge U_J, d_1)$, we have $\wedge U = \wedge U_I \otimes \wedge U_J$. Thus,

$$\mathrm{Tor}^{UL}(\Bbbk, \wedge U) = \mathrm{Tor}^{UI}(\Bbbk, \wedge U_I) \otimes \mathrm{Tor}^{UJ}(\Bbbk, \wedge U_J)\,,$$

and it follows that $\mathrm{depth}\, L = \mathrm{depth}\, I +$ $\mathrm{depth}\, J$. Finally, since $\mathrm{Tor}^{UI}(\Bbbk, \wedge U) \to \mathrm{Tor}^{UL}(\Bbbk, \wedge U)$ factors through $\mathrm{Tor}^{UI}(\Bbbk, \wedge U_I) \otimes \mathrm{Tor}_0^{UJ}(\Bbbk, \wedge U_J)$, it follows that $\mathrm{Tor}_0^{UJ}(\Bbbk, \wedge U_J) \neq 0$.

□

**Proposition 10.9.** *Suppose $M$ and $N$ are $L$-modules. Then for $0 \neq \alpha \in Tor^{UL}(M, N)$ there is a finitely generated sub Lie algebra $E \subset L$ and there are finitely generated $E$-modules $Q_M$ and $Q_N$ such that $\alpha$ is in the image of $Tor^{UE}(Q_M, Q_N)$.*

*proof*: This follows because Tor commutes with direct limits in $E$, $Q_M$ and $Q_N$. □

Next, suppose $E$ is a sub Lie algebra of the Sullivan Lie algebra $L$, with corresponding decomposition $\wedge U = \wedge S \otimes \wedge U_Z$ as described in (10.10), p. 284.

**Proposition 10.10.** *For any $L$-module, $M$, if $Tor_q^{UL}(M, \wedge U) \neq 0$ then $Tor_q^{UL}(M, \wedge S) \neq 0$.*

*proof*: Recall from Lemma 10.6 (p. 285) that $\wedge U$ is the direct limit of sub $L$-modules $F_k$ of the form $\wedge S \otimes F_k$ in which $\wedge S \otimes F_{k+1}/F_k$ is the tensor product of the holonomy representation of $L$ in $\wedge S$ with the trivial representation. Thus

$$\mathrm{Tor}_q^{UL}(M, \wedge U) = \varinjlim_k \mathrm{Tor}_q^{UL}(M, \wedge S \otimes F_k)\,.$$

Suppose $\mathrm{Tor}_q^{UL}(M, \wedge S) = 0$. Then it follows by induction on $k$ via a short exact sequence argument that $\mathrm{Tor}_q^{UL}(M, \wedge S \otimes F_k) = 0$ for all $k$, and hence that $\mathrm{Tor}_q^{UL}(M, \wedge U) = 0$.

□

**Proposition 10.11.** *Suppose $J$ and $I = \Bbbk x \oplus J$ are ideals in the Sullivan Lie algebras $L$ and that $J \supset L_0^n$, some $n$. If $x$ acts locally nilpotently in $Tor_*^{UJ}(\Bbbk, \Bbbk)$ then*

$$depth_L I = \begin{cases} depth_L J, & \text{if } deg\, x \text{ is odd} \\ depth_L J + 1, & \text{if } deg\, x \text{ is even.} \end{cases}$$

*proof:* The universal enveloping algebra of $\Bbbk x$ is just $\wedge x$, and we shall deduce the Proposition from the Hochschild-Serre spectral sequence,

$$E^2_{p,q} = \mathrm{Tor}_p^{\wedge x}(\Bbbk, \mathrm{Tor}_q^{UJ}(\Bbbk, \wedge U)) \Longrightarrow \mathrm{Tor}_{p+q}^{UI}(\Bbbk, \wedge U)\,.$$

For this, first observe that a general $\wedge x$-module $M$ is defined by a single linear map $\varphi : a \mapsto x \cdot a$, $a \in M$. It is an easy and classical fact that

$$\mathrm{Tor}_0^{\wedge x}(\Bbbk, M) = \mathrm{Coker}\, \varphi\,,$$

and, if $\deg x$ is even,

$$\mathrm{Tor}_1^{\wedge x}(\Bbbk, M) = \ker \varphi\,.$$

We apply this when $M$ is the $\wedge x$-module $\mathrm{Tor}_{d_J}^{UJ}(\Bbbk, \wedge U)$ and $d_J = \mathrm{depth}_L J$.

Recall as before that the holonomy representation of $L$ in $\wedge U$ is locally nilpotent. Since $x$ acts locally nilpotently in $\mathrm{Tor}_*^{UJ}(\Bbbk, \Bbbk)$, Lemma 10.7(ii)

(p. 289) asserts that $x$ acts locally nilpotently in $\mathrm{Tor}_{d_J}^{UJ}(\Bbbk, \wedge U)$. In particular, $\ker\varphi \neq 0$.

Next, recall that $(\wedge V, d_1)$ is the quadratic Sullivan algebra for $L$ and that $K_I = \{v \in V \,|\, \langle v, I\rangle = 0\}$. Since $J \supset L_0^n$ it follows that $K_J = \Bbbk w \oplus K_I$ for some $w$ such that $\langle w, x\rangle = 1$. In particular, $(\wedge K_I, d_1)$ is a sub Sullivan algebra and $d_1 w \in \wedge^2 K_I$. Thus the acyclic closure of $(\wedge K_J, d_1)$ has the form $(\wedge K_J \otimes \wedge S \otimes \wedge u, d_1)$ in which $(\wedge K_I \otimes \wedge S, d_1)$ is the acyclic closure of $(\wedge K_I, d_1)$ and $d_1 u - w \in \wedge^+ K_I \otimes \wedge S$. Then, as in (10.10), p. 284, the acyclic closure for $(\wedge V, d_1)$ has the form $(\wedge V \otimes \wedge S \otimes \wedge u \otimes \wedge U_Z, d_1)$.

In particular, the holonomy representation of $J$ in $\wedge U$ satisfies $y \cdot u = 0$, $y \in J$. Since $J$ acts by derivations this implies that for $y \in J$ and $\Phi \in \wedge U$,

$$y \cdot (\Phi \wedge u) = (y \cdot \Phi) \wedge u\,.$$

Thus right multiplication by $1 \otimes u$ in $C_*(J) \otimes \wedge U$ is a map of chain complexes, and so induces a map $\alpha \mapsto \alpha \cdot u$ in $\mathrm{Tor}_*^{UJ}(\Bbbk, \wedge U)$.

Moreover, since $\langle K_I, x\rangle = 0$, it follows that the holonomy representation of $I$ in $\wedge U$ satisfies $x \cdot u = -1$. Thus

$$x \cdot (\alpha \cdot u) = (-1)^{\deg x \cdot \deg \alpha + 1} \alpha + (x \cdot \alpha) \cdot u\,, \qquad \alpha \in \mathrm{Tor}_*^{UJ}(\Bbbk, \wedge U)\,.$$

Now suppose $\deg x$ is even, so that $\wedge x$ is the polynomial algebra $\Bbbk[x]$. Since $x$ acts locally nilpotently in $\mathrm{Tor}_*^{UJ}(\Bbbk, \wedge U)$, this implies that

$$\mathrm{Tor}_*^{UJ}(\Bbbk, \wedge U) = x \cdot \mathrm{Tor}_*^{UJ}(\Bbbk, \wedge U)\,,$$

and so $\mathrm{Coker}\,\varphi = 0$. Since $\ker\varphi \neq 0$, the Proposition is proved in this case.

Finally, suppose $\deg x$ is odd. We only have to show that in this case $\mathrm{Coker}\,\varphi \neq 0$. But if $\mathrm{Coker}\,\varphi = 0$ then every $\alpha \in \mathrm{Tor}_{d_J}^{UJ}(\Bbbk, \wedge U)$ has the form $x \cdot \beta$, and so $x \cdot \alpha = x \cdot (x \cdot \beta) = 0$. Choose $\alpha \neq 0$ in $\mathrm{Tor}_{d_J}^{UJ}(\Bbbk, \wedge U)$. Then the equation above gives $x \cdot (\alpha \cdot u) = \pm\alpha \neq 0$, and it follows that $\mathrm{Coker}\varphi \neq 0$, which completes the proof in this case.

□

**Theorem 10.4.** *Suppose $r$ elements $x_{ij}$ of even degree in a Sullivan Lie algebra $L$ satisfy*

- *(i) For each $j$ the $x_{ij} \in L^j$ and represent linearly independent elements in $L^j/L^{j+1}$.*
- *(ii) Each $ad\, x_{ij}$ is locally nilpotent.*

*Then $r \leq depth\, L$.*

**Corollary 10.5.** *For some $n$, and for each non-zero element $x \in L^n$, there is a $y \in L$ such that $(ad\, x)^k y \neq 0$, $k \geq 0$.*

**Corollary 10.6.** *If $N \subset L_{even}$ is the subspace spanned by the elements acting locally nilpotently in $L$ via the adjoint representation, then $dim\, N \leq depth\, L$.*

*proof of Theorem 10.4.* First note that for some $N$ the $x_{ij} \in L_{<N}$ for all $i, j$. Now let $r_j$ denote the number of elements $x_{ij}$ and define ideals $I(i,j)$ by

$$I(i,j) = \oplus_{\ell \geq i} k x_{i,j} \oplus L^{j+1} \oplus L_{\geq N}\,, \qquad i \leq r_j$$

and

$$I(r_i + 1, j) = L^{j+1} \oplus L_{\geq N}\,.$$

Then it follows from Proposition 10.10 and Corollary 10.4 (p. 290) that

$$\text{depth}_L(L^{j+1} + L_{\geq N}) \leq \text{depth}_L I(1,j) + r_i \leq \text{depth}_L(L^j + L_{\geq N})\,.$$

This implies that for some $n$, $\text{depth}_L(L^n + L_{\geq N}) + \sum r_i \leq \text{depth}\, L$, and since $\sum r_i = r$, the Theorem follows.

□

Recall again that $(\wedge V \otimes \wedge U, d_1)$ is the acyclic closure of the quadratic Sullivan algebra $(\wedge V, d_1)$ associated with the Sullivan Lie algebra, $L$.

**Theorem 10.5.** *Suppose $I$ is an ideal in a Sullivan Lie algebra $L$ for which $I_0 \supset L_0^n$ for some $n$ and $dim\, L/I < \infty$. If $L$ acts locally nilpotently in $Tor_*^{UI}(k, k)$, then*

$$depth\, L = dim\,(L/I)_{even} + depth_L I\,.$$

*If also $L_0$ acts nilpotently in $I_0/[I_0, I_0]$ then*

$$depth\, L = dim\,(L/I)_{even} + depth\, I\,.$$

**Corollary 10.7.** *Suppose $L$ is a Sullivan Lie algebra and $I_0 \subset L_0$ is a finite dimensional ideal. If $depth\, L < \infty$ then*

$$dim\, I_0 + depth\, L_{\geq 1} = depth\,(I_0 \oplus L_{\geq 1}) \leq depth\, L\,.$$

*proof:* Set $J = I_0 \oplus L_{\geq 1}$. Then $J$ is a Sullivan closed ideal and it follows from Corollary 10.3(ii) (p. 288) and Corollary 10.4 (p. 290) that

$$\text{depth}\, J = \text{depth}_L J \leq \text{depth}\, L\,.$$

On the other hand, $L_{\geq 1}$ is an ideal in $J$, and since $\text{depth}\, J \leq \infty$, and $J$ is trivially a Sullivan Lie algebra, the pair $(J, L_{\geq 1})$ satisfies the conditions of Theorem 10.5. □

*proof of Theorem 10.5*: By Corollary 10.1 (p. 280) the associated Sullivan algebra $(\wedge V, d_1)$ decomposes as a Sullivan extension $(\wedge K_I \otimes \wedge Z, d_1)$. Since $I_0 \supset L_0^n$, $I$ is Sullivan closed and the induced pairing $K_I \wedge L/I \to \Bbbk$ is a non-degenerate scalar product. Thus $\dim K_I < \infty$ and there is a basis $w_1, \cdots, w_s$ of $K_I$ in which $d_1 w_1 = 0$ and $d_1 w_{i+1} \in \wedge(w_1, \cdots, w_i)$.

Define ideals $J(k) \subset L$ by setting

$$J(k) = \{ y \in L \mid \langle w_i, y \rangle = 0 \,, \quad 1 \le i \le k \} \,, \qquad k \ge 1 \,,$$

and $J(k-1) = J(k) \oplus \Bbbk x_k$ with $\langle w_k, x_k \rangle = 1$. Since $I$ is Sullivan closed, $I = J(k)$.

However, clearly $L^{s+1} \subset I$ it follows that $L$ acts nilpotently in $J(k)/I$ and hence also acts nilpotently in $\wedge s(J(k)/I)$. Since by hypothesis $L$ acts locally nilpotently in $\mathrm{Tor}_*^{UI}(\Bbbk, \Bbbk)$ it follows from the Hochschild-Serre spectral sequence

$$E^1_{p,q} = \wedge^p s(J(k)/I) \otimes \mathrm{Tor}_q^{UI}(\Bbbk, \Bbbk) \Longrightarrow \mathrm{Tor}_{p+q}^{UJ(k)}(\Bbbk, \Bbbk)$$

that $L$ acts locally nilpotently in each $\mathrm{Tor}_*^{UJ(k)}(\Bbbk, \Bbbk)$. Thus the first assertion of Theorem follows from Proposition 10.11 (p. 292).

Finally, we show that if $L_0$ acts nilpotently in $I_0/[I_0, I_0]$ then $I$ satisfies the hypotheses of Theorem 10.3. Observe that it follows from the definition of the associated quadratic Sullivan algebra that $\langle K_I, L^{s+1} \rangle = 0$. Thus in this case Proposition 10.4(ii) asserts that $I$ is a Sullivan Lie algebra and that each $I^k$ is Sullivan closed. Now Proposition 10.4(iii) asserts that $(\wedge Z, \overline{d}_1)$ maps isomorphically to the quadratic Sullivan algebra associated with $I$. Thus $I$ satisfies the hypotheses of Theorem 10.3 and $\mathrm{depth}_L I = \mathrm{depth}\, I$.

□

## 10.6 The radical of a Sullivan Lie algebra

Recall that the *derived series* of a graded Lie algebra $E$ is the sequence of ideals $E^{(k)}$ defined by $E^{(1)} = E$ and $E^{(k+1)} = [E^{(k)}, E^{(k)}]$. The Lie algebra $E$ is *solvable* of *solv length* $n$ if $E^{(n)} \neq 0$ and $E^{(n+1)} = 0$, and the *radical*, $\mathrm{rad}\, E$, is the sum of the solvable ideals in $E$.

**Theorem 10.6.** *If a Sullivan Lie algebra, $L$, has finite depth then*

*(i) $\mathrm{rad}\, L$ is a finite dimensional ideal;*
*(ii) The adjoint representation of $L$ in $\mathrm{rad}\, L$ is nilpotent;*
*(iii) $\dim\, (\mathrm{rad}\, L)_{even} = \mathrm{depth}\ \mathrm{rad}\, L \le \mathrm{depth}\, L$.*

**Lemma 10.8.** *If $L$ is a Sullivan Lie algebra of finite depth and $J \subset L_0$ is an abelian sub Lie algebra for which $[L_0, J] \subset J$ then*

*(i) For some $n$, $J \cap L_0^n = 0$.*
*(ii) $\dim J < \infty$.*
*(iii) $L_0$ acts nilpotently in $J$ via the adjoint representation.*

*proof:* By Theorem 10.4 there is some $n$ such that for each $0 \neq x \in L_0^n$ there is a $y \in L$ with $(ad\, x)^k y \neq 0$, $k \geq 0$. Since $L_0$ acts nilpotently in each $L_r$, $r \geq 1$, it follows that $y \in L_0$. But if $x \in J \cap L_0^n$ then since $J$ is an abelian ideal in $L_0$, $[x, y] \in J$ and $[x, [x, y]] = 0$. It follows that $J \cap L_0^n = 0$. The remaining two assertions are immediate consequences.

□

*proof of Theorem 10.6.* First observe that if $I$ is an abelian ideal in $L$ then by Lemma 10.8, $\dim I_0 < \infty$, and so by Corollary 10.7, $\dim I_0 <$ depth $L$. On the other hand, $I_{\geq 1}$ is an abelian ideal in $L_{\geq 1}$ and by Theorem 10.3 (p. 288) depth $L_{\geq 1} \leq$ depth $L$.

Thus Corollary 10.4 (p. 290) and Corollary 10.7 (p. 294) give depth $I_{\geq 1} \leq$ depth $L_{\geq 1}$. Now if $I_{\geq 1}$ is infinite dimensional, then $I_{\geq 1}$ is the sum of $r$ infinite dimensional subspaces $I(j)$, with $r =$ depth $L_{\geq 1} + 1$. Since the $I(j)$ are ideals in the Lie algebra $I$, Proposition 10.8 (p. 291) gives

$$r > \operatorname{depth} I_{\geq 1} = \sum_{j=1}^{r} \operatorname{depth} I(j)\,.$$

Thus for some $j$, depth $I(j) = 0$. Now Proposition 10.7 (p. 290) asserts that $\dim I(j) < \infty$. This contradiction shows that $\dim I_{\geq 1} < \infty$ and so $\dim I < \infty$.

Suppose next that $I$ is a solvable ideal of solv length $n$ and assume by induction on $n$ that $[I, I]$ is finite dimensional. Then for some $n$, $[I, I]$ maps injectively to $L/L^n$ and so $I \cap L^n$ is abelian. In particular, by Lemma 10.8, $\dim I_0 < \infty$. On the other hand, for some $N$ $I_{\geq N}$ is an abelian ideal. Hence $I_{\geq 1}$, and therefore $I$, are finite dimensional.

Now, since $I$ is finite dimensional, by Corollary 10.3 (p. 288) and Corollary 10.4 (p. 290) we have depth $I \leq$ depth $L$. Thus Theorem 10.5, applied to the inclusion $0 \subset I$, gives

$$\dim I_{even} \leq \operatorname{depth} I \leq \operatorname{depth} L\,.$$

Since $I$ is any solvable ideal it follows that $\dim\, (\mathrm{rad}\, L)_{even} \leq$ depth $L$ and so for some $N$, $(\mathrm{rad}\, L)_{\geq N}$ is concentrated in odd degrees. Hence $(\mathrm{rad}\, L)_{\geq N}$ is an abelian ideal and therefore finite dimensional. Since dim $(\mathrm{rad}\, L)_{even} < \infty$, it follows that $\mathrm{rad}\, L$ is finite dimensional. In particular, for some $n$, $\mathrm{rad}\, L$ maps injectively to $L/L^n$ and hence $L$ acts nilpotently in $\mathrm{rad}\, L$.

Finally, Corollary 10.3 and Corollary 10.4 give depth $\mathrm{rad}\, L \leq \mathrm{depth}\, L$ and so Theorem 10.4 (p. 293) applied to the inclusion $0 \subset \mathrm{rad}\, L$ gives

$$\dim(\ \mathrm{rad}\, L)_{even} = \mathrm{depth}\ \mathrm{rad}\, L \leq \mathrm{depth}\, L\,.$$

□

**Corollary 10.8.** *If a Sullivan Lie algebra $L$ has finite depth, then*

*(i) $rad(L_0)$ is finite dimensional and the adjoint representation of $L_0$ in $rad\ (L_0)$ is nilpotent.*
*(ii) $rad(L_0) \supset (rad\, L)_0$ and $rad\ (L_{\geq 1}) = (rad\, L)_{\geq 1}$.*
*(iii) $dim\ (rad\, L)_{even} \leq dim\ rad\ (L_0) +\ dim\ (rad\, L_{\geq 1})_{even} \leq depth\, L$.*

*proof.* (i) We first observe that if $I \subset L_0$ is a solvable Lie algebra and if $[L_0, I] \subset I$ then $\dim I < \infty$. In fact, if $I$ is abelian this is Lemma 10.8, and in general it follows by induction on solv length $I$ exactly as in the proof of Theorem 10.6. Thus Corollary 10.7 gives $\dim I \leq \mathrm{depth}\, L$, and it follows, since $I$ is any solvable $L_0$-ideal, that

$$\dim\ \mathrm{rad}(L_0) \leq \mathrm{depth}\, L\,.$$

Since dim $\mathrm{rad}(L_0) < \infty$, its intersection with some $L_0^n$ is zero and so $L_0$ acts nilpotently in $\mathrm{rad}(L_0)$.

(ii) By Theorem 10.6, $(\mathrm{rad}\, L)_0$ is solvable, and clearly also an $L_0$-ideal. Therefore $(\mathrm{rad}\, L)_0 \subset \mathrm{rad}(L_0)$. Similarly, $\mathrm{rad}(L_{\geq 1}) \supset (\mathrm{rad}\, L)_{\geq 1}$. But if $x \in L_0$, $y \in L_{\geq 1}$ and $z \in \mathrm{rad}(L_{\geq 1})$, then

$$[y, [x, z]] = [[y, x], z] \pm [x, [y, z]] \in [L_{\geq 1}, \mathrm{rad}(L_{\geq 1}] + [x, \mathrm{rad}(L_{\geq 1})]\,.$$

It follows that $[x, \mathrm{rad}(L_{\geq 1})] + \mathrm{rad}(L_{\geq 1})$ is a finite dimensional ideal in $L_{\geq 1}$, and hence coincides with $\mathrm{rad}(L_{\geq 1})$. Thus $\mathrm{rad}(L_{\geq 1})$ is an ideal in $L$ and so $\mathrm{rad}(L_{\geq 1}) = (\ \mathrm{rad}\, L)_{\geq 1}$.

(iii) Corollary 10.3 and Corollary 10.4 give $\mathrm{depth}(\ \mathrm{rad}(L_0) \oplus L_{\geq 1}) \leq \mathrm{depth}\, L$. Corollary 10.7 then gives dim $\mathrm{rad}(L_0)+$ depth $L_{\geq 1} \leq \mathrm{depth}\, L$, and so

$$\dim(\mathrm{rad}\, L)_{even} \leq \dim\ \mathrm{rad}\ (L_0) + \dim\ \mathrm{rad}(L_{\geq 1})_{even} \leq \mathrm{depth}\, L\,.$$

□

**Corollary 10.9.** *Suppose $L$ is a Sullivan Lie algebra of finite depth.*

*(i) If $\dim L_0 = \infty$ then no $L_0^n$ is solvable*
*(ii) If $\dim L_{\geq 1} = \infty$ then no $L_0^n$ is solvable.*

*proof:* If $L_0^n$ (respectively $L_{\geq 1}^n$) is solvable then it is finite dimensional and so $L_0$ (respectively $L_{\geq 1}$) is finite dimensional.

□

**Proposition 10.12.** *Suppose $L$ is a Sullivan Lie algebra of finite depth.*

*(i) If $J \subset L_0$ is an $L_0$-ideal and for some $n$, $\dim [L_0^n, J] < \infty$ then $\dim J < \infty$.*
*(ii) If $I \subset L$ is an ideal and for some $n$, $\dim [L^n, I] < \infty$ then $\dim I < \infty$.*

*proof.* (i) As observed in Lemma 10.8, if $\dim J = \infty$ then for some $q \geq n$ there are elements $0 \neq x \in L_0^q \cap J$ and $y \in L_0$ such that $(ad\, x)^k y \neq 0$, $k \geq 0$. But $[x, y] \in J$ and so $[x, [x, y]] \in [L_0^n, J]$. Since $x$ acts nilpotently in $[L_0^n, J]$ this is impossible, and so $\dim J = \infty$.

(ii) This follows by the same argument as in (i).

□

## 10.7 Sullivan Lie algebras of finite type

*In this section we fix a Sullivan Lie algebra $L$ in which $\dim L_0 < \infty$. As in earlier sections $(\wedge V \otimes \wedge U, d_1)$ denotes the acyclic closure of the quadratic Sullivan algebra $(\wedge V, d_1)$ associated with $L$.*

**Proposition 10.13.** *With the hypotheses above:*

*(i) $V = sL^\#$ and $sL = V^\#$.*
*(ii) Every sub Lie algebra of $L$ is a Sullivan closed Sullivan Lie algebra.*
*(iii) $L$ is the homotopy Lie algebra of $(\wedge V, d_1)$.*
*(iv) The morphism $\mu : (\wedge V, d_1) \to C^*(L)$ defined on p. 276 is an isomorphism, dualizing to an isomorphism $\wedge sL \xrightarrow{\cong} (\wedge V)^\#$.*
*(v) If $I \subset L$ is an ideal then $\operatorname{depth} I \leq \operatorname{depth} L$.*

*proof:* Assertions (i)-(iv) are immediate from the fact that $L$ has finite type. Assertion (v) follows from Corollary 10.4 (p. 290) and since by Proposition 10.4 (p. 282) $I$ satisfies the hypotheses of Theorem 10.3. □

Because the holonomy representation of $L$ in $\wedge U$ is locally nilpotent (cf. §4.3), this extends to a representation of the completion $\widehat{UL} = \varprojlim UL/I_L^n$, where $I_L$ is the augmentation ideal of $L$.

Recall also from §10.1 (p. 268) that for any augmented graded algebra $A$, $\operatorname{depth} A =$ least $p$ of $\infty$ such that $\operatorname{Ext}_A^p(\Bbbk, A) \neq 0$.

**Proposition 10.14.** *With the notation and hypotheses above:*

(i) *depth* $(\wedge V, d_1) = \operatorname{depth} L$.

(ii) $Ext^p_{\widehat{UL}}(\Bbbk, \widehat{UL}) = \left( Tor_p^{UL}(\Bbbk, \wedge U)\right)^{\#}$.

(iii) *depth* $(\wedge V, d_1) = \operatorname{depth} \widehat{UL}$ *and gldim* $(\wedge V, d_1) = \operatorname{gldim} UL$.

For the proof of Proposition 10.14 we need

**Lemma 10.9.** *With the hypotheses above on* $L$, $\wedge sL \otimes \widehat{UL} = (\wedge sL \otimes UL) \otimes_{UL} \widehat{UL}$ *is a* $\widehat{UL}$*-free resolution of* $\Bbbk$:

$$H(\wedge sL \otimes \widehat{UL}, \partial) = \Bbbk\,.$$

*proof:* Recall from §6.2 the isomorphism $\eta_L : \widehat{UL} \to (\wedge U)^{\#}$. The canonical involution $\omega_L$ of $\widehat{UL}$ defined in §6.2 converts right multiplication to left multiplication. Thus the isomorphism $\omega_L \circ \eta_L^{-1}$, together with the identification $(\wedge V)^{\#} = \wedge sL$, defines an isomorphism

$$(\wedge V \otimes \wedge U, d_1)^{\#} \xrightarrow{\cong} (\wedge sL \otimes \widehat{UL}, \partial)\,.$$

A straightforward but tedious verification shows that this is compatible with the differentials, and the Lemma follows.

□

*proof of Proposition 10.14*: By Proposition 10.13(iii), $\mu : (\wedge V, d_1) \to C^*(L)$ is a quasi-isomorphism. Thus we may apply Theorem 10.2 (p. 278) and formula (10.6) (p. 278). The first gives (i).

To prove (ii) note that because of Lemma 10.9,

$$\operatorname{Ext}_{\widehat{UL}}(\Bbbk, -) = H\left(\operatorname{Hom}_{\widehat{UL}}(\wedge sL \otimes \widehat{UL}, -)\right)\,.$$

Thus from (10.6) and the isomorphism $\eta_L$ we obtain

$$\begin{aligned}\operatorname{Ext}_{\widehat{UL}}(\Bbbk, \widehat{UL}) &= H\left(\operatorname{Hom}(\wedge sL, \widehat{UL})\right) = H(\operatorname{Hom}(\wedge sL, (\wedge U)^{\#}) \\ &= [H(\wedge sL \otimes \wedge U)]^{\#} = \operatorname{Tor}^{UL}(\Bbbk, \wedge U)^{\#} \\ &= \operatorname{Ext}_{(\wedge V, d_1)}(\Bbbk, (\wedge V, d_1))\,.\end{aligned}$$

(iii) is an immediate consequence of (i) and Theorem 10.2.

□

**Proposition 10.15.** *Suppose $E$ is a sub Lie algebra, and $I$ is an ideal, in a Sullivan Lie algebra $L$ of finite type. Then*

*(i) $E$ is a Sullivan closed Sullivan Lie algebra.*
*(ii) $depth_L E = depth\, E$.*
*(iii) If $L$ acts locally nilpotently in $Tor_*^{UI}(\Bbbk, \Bbbk)$ and if $dim\, L/I < \infty$ and $depth\, L < \infty$ then*

$$depth\, L = dim\, (L/I)_{even} + depth\, I\,.$$

*proof:* (i) This is immediate because for some $n$, $L_0^n = 0$.

(ii) Since each $E^n$ is (trivially) Sullivan closed, Proposition 10.4(iii) asserts that $E$ satisfies the hypotheses of Theorem 10.3 (p. 288), and (ii) follows.

(iii) This follows immediately from Theorem 10.5 (p. 294).

□

**Corollary 10.10.** *Suppose $L = L_{\geq 0}$ is a Sullivan Lie algebra of finite type. Then*

$$depth\, L = dim\, L_0 + depth\, L_{\geq 1}\,.$$

*proof*: Apply Corollary 10.7 (p. 294) with $I = L_{\geq 1}$.

□

# Chapter 11

# Depth of a connected graded Lie algebra of finite type

A graded Lie algebra, $L$, is *connected* if $L = L_{\geq 1}$ and of *finite type* if each $\dim L_i < \infty$. In this chapter we consider *only* connected graded Lie algebras of finite type, a property which trivially holds for sub and quotient Lie algebras.

## 11.1 Summary of previous results

Let $L$ be a connected graded Lie algebra of finite type. Then $L$ is trivially pronilpotent since (§10.4) that is only a condition on $L_0$ if $L$ has finite type. Thus the invariants depth and gldim are defined for $L$ in §10.4, and the results of Chapter 10 can be expressed directly in terms of $L$, and its sub Lie algebras and ideals. We do this now followed by a brief explanation of the translation.

First recall, as described in §10.1, that every left module $M$ for a graded Lie algebra $L$ is naturally a right module and conversely; we simply refer to $M$ as an $L$-module.

Then note that for a connected graded Lie algebra $L$ of finite type, right multiplication in $UL$ makes $UL$ into a $UL$-module and that this dualizes to a representation in $(UL)^{\#}$. We call this the *standard representation* of $UL$ in $(UL)^{\#}$. Since $((UL)^{\#})^{\#} = UL$ by (10.1),

$$\mathrm{Ext}_{UL}(-, UL) = \left[\mathrm{Tor}^{UL}(-, (UL)^{\#})\right]^{\#} .$$

Finally, note that since $\widehat{UL} = \varinjlim_n UL/I_L^n$ (cf. p. 56) it follows that $UL = \widehat{UL}$.

**Proposition 11.1.** *Let $L$ be a connected graded Lie algebra of finite type. Then*

*(i) $depth\, L$ is the least $p$ (or $\infty$) such that $Ext^p_{UL}(\Bbbk, UL) \neq 0$.*
*(ii) $gldim\, L$ is the largest integer $p$ (or $\infty$) such that $Ext^p_{UL}(\Bbbk, \Bbbk) \neq 0$.*
*(iii) If $L$ is the homotopy Lie algebra of a minimal Sullivan algebra $(\wedge V, d)$ then $depth\, L = depth(\wedge V, d_1)$,*

$$depth\, L \leq cat(\wedge V, d) \leq gldim\, UL\,,$$

*and if $depth\, L = cat(\wedge V, d)$ then $depth\, L = gldim\, L$.*

**Proposition 11.2.** *Let $E$ and $I$ be respectively a sub Lie algebra and an ideal in a connected graded Lie algebra $L$ of finite type. Then*

*(i) $depth\, E$ is the least $p$ (or $\infty$) such that $Ext^p_{UE}(\Bbbk, UL) \neq 0$.*
*(ii) If the adjoint action of $E$ in $L/E$ is locally nilpotent then $depth\, E \leq depth\, L$.*
*(iii) $depth\, I \leq depth\, L$.*

**Proposition 11.3.** *If a connected graded Lie algebra, $L$, of finite type satisfies $depth\, L = 0$ then $L$ is finite dimensional, concentrated in odd degrees, and $Ext^k_{UL}(\Bbbk, UL) = 0$, $k \geq 1$.*

**Proposition 11.4.** *Let $L$ be a connected graded Lie algebra of finite type, and let $M$ and $N$ be $L$-modules. Then*

*(i) If $L$ is the direct sum of ideals $I$ and $J$ such that the restriction map $Ext_{UL}(\Bbbk, UL) \to Ext_{UI}(\Bbbk, UL)$ is non-zero, then $J$ is finite dimensional and concentrated in odd degrees.*
*(ii) If $0 \neq \alpha \in Ext_{UL}(M, N^{\#})$ then there is a finitely generated sub Lie algebra $E \subset L$ and finitely generated submodules $Q_M \subset M$ and $Q_N \subset N$ such that*

$$Ext_{UL}(M, N^{\#}) \to Ext_{UE}(Q_M, Q_N^{\#})$$

*is non-zero on $\alpha$.*
*(iii) If $I$ is an ideal in $L$ and $Ext^m_{UL}(M, UL) \neq 0$, then $Ext^m_{UL}(M, UL/I) \neq 0$.*

**Proposition 11.5.** *Suppose $L = L_{\geq 1}$ is a graded Lie algebra of finite type.*

*(i) If $L$ is the direct sum of ideals $I$ and $J$ then*

$$depth\, L = depth\, I + depth\, J\,.$$

*(ii) If $L$ is solvable and if $depth\, L < \infty$ then $dim\, L < \infty$ and $depth\, L = dim\, L_{even}$.*

**Theorem 11.1.** *Suppose $L = L_{\geq 1}$ is a graded Lie algebra of finite type. If depth $L = m$ then rad $L$ is a finite dimensional ideal in $L$ and*

$$\text{depth}\,\text{rad}\,L = \dim(\text{rad}\,L)_{even} \leq m\,.$$

*In particular, rad $L$ is nilpotent.*

A graded Lie algebra is *locally solvable* if every finitely generated sub Lie algebra is solvable.

**Corollary 11.1.** *A locally solvable graded Lie algebra $L = L_{\geq 1}$ of finite depth is finite dimensional, and satisfies $\dim L_{even} = \text{depth}\,L$.*

*proof:* Suppose depth $L = m$. By Proposition 11.4(ii) there is a finitely generated Lie algebra $F \subset L$ such that the restriction map

$$\text{Ext}^m_{UL}(\Bbbk, UL) \to \text{Ext}^m_{UF}(\Bbbk, UL)$$

is non-zero. In particular for every $x \in L$ there is a finitely generated Lie algebra $H$ containing $x$, with $F \subset H \subset L$. Since $F \subset H$, depth $H \leq m$. Since $H$ is solvable, by Theorem 11.1 $\dim H_{\text{even}} \leq m$. Therefore $\dim L_{\text{even}} \leq m$. Thus for some $q$, $L_{>q}$ is concentrated in odd degrees and is abelian. It follows that $L_{>q}$ is solvable and finite dimensional (Proposition 11.5). Thus $L$ is finite dimensional and so $L = \text{rad}\,L$. In particular, by Theorem 11.1, $\dim L_{\text{even}} = \text{depth}\,L$.

□

Now we describe briefly how the Propositions and Theorem above follow from the results of §10. The key fact is that a connected graded Lie algebra $L$ of finite type is the homotopy Lie algebra of its associated quadratic Sullivan algebra $(\wedge V, d_1) = C^*(L)$, as shown in Proposition 2.1 (p. 55); here $C^*(L)$ is the Cartan-Chevalley-Eilenberg construction defined in §2.1.

Let $(\wedge V \otimes \wedge U, d_1)$ denote the acyclic closure of $(\wedge V, d_1)$. Then, since $L$ is connected, Theorem 6.1 (p. 180) provides an isomorphism $UL \xrightarrow{\cong} (\wedge U)^{\#}$ which converts right multiplication in $UL$ to the dual of the holonomy representation of $L$ in $\wedge U$. Thus this isomorphism dualizes to an isomorphism

$$\wedge U \xrightarrow{\cong} (UL)^{\#}$$

of $UL$-modules, where $L$ acts in $\wedge U$ by the holonomy representation and in $(UL)^{\#}$ by the dual of right multiplication in $UL$.

With these identifications, Proposition 11.1 (i) follows immediately from the isomorphism above, and (ii) restates the definitions, p. 268. For assertion (iii) recall that by definition the quadratic part, $(\wedge V, d_1)$ of $(\wedge V, d)$ is $C^*(L)$, and so Theorem 10.2 (p. 278) gives

$$\operatorname{depth} L = \operatorname{depth}(\wedge V, d_1) = \operatorname{depth}(\wedge V, d) \text{ and}$$
$$\operatorname{gldim} L = \operatorname{gldim}(\wedge V, d_1) = \operatorname{gldim}(\wedge V, d).$$

Thus assertion (iii) restates Theorem 10.1 (p. 273).

For Proposition 11.2 note that since $(UL)^\# \cong \wedge U$, $\operatorname{Ext}^p_{UE}(\Bbbk, UL) = \left[\operatorname{Tor}^{UE}_p(\Bbbk, \wedge U)\right]^\#$. Thus assertion (i) translates to $\operatorname{depth} E = \operatorname{depth}_L E$, which is Proposition 10.15(ii) (p. 300). With this, assertions (ii) and (iii) follow from Proposition 10.6 (p. 290).

For Proposition 11.3 observe first that by Proposition 10.7 (p. 290), $\dim L < \infty$. Then it follows from Theorem 10.5 (p. 294) that $L$ is concentrated in odd degrees, and hence is an abelian Lie algebra. Thus $\operatorname{Tor}^{UL}_*(\Bbbk, UL)$ is the finite tensor product of spaces of the form $\operatorname{Tor}^{\wedge x}(\Bbbk, \wedge x)$ with $\deg x$ odd, and a straightforward calculation gives $\operatorname{Tor}^{UL}_{\geq 1}(\Bbbk, UL) = 0$.

In view of Proposition 11.3, Proposition 11.4(i) is the dual to Proposition 10.8 (p. 291) while Proposition 11.4(ii) is dual to Proposition 10.9 (p. 291). For Proposition 11.4(iii), recall from Proposition 10.4(iv) (p. 282) that the acyclic closure of $C^*(L/I)$ has the form $(C^*(L/I) \otimes \wedge S, d_1)$ in which $\wedge S$ is the $L/I$-module of Proposition 10.10 (p. 292). Thus assertion (iii) follows by dualizing Proposition 10.10. Next, in Proposition 11.5, assertion (i) is Proposition 10.8 (p. 291) and (ii) is a special case of Theorem 10.6 (p. 295). Finally, Theorem 11.1 is just Theorem 10.6.

## 11.2 Modules over an abelian Lie algebra

**Proposition 11.6.** *Suppose a connected graded Lie algebra, $L$, of finite type is abelian. If for some $L$-module, $M$, and some $m$, $Ext^m_{UL}(M, UL) \neq 0$, then for some $x \in M$ and some $n$ the map $U(L_{\geq n}) \to M$, $a \mapsto a \cdot x$, is injective.*

**Lemma 11.1.** *If $F$ is a connected graded Lie algebra of finite type and concentrated in odd degrees, and if $N$ is any $F$-module, then* $\operatorname{Ext}^p_{UF}(N, UF) = 0$, $p \geq 1$.

*proof:* Since $F$ is concentrated in odd degrees it is abelian. In view of Proposition 11.4(ii), it is sufficient to prove the Lemma when $F$ is finitely generated (and therefore finite dimensional). Let $v_1, \cdots, v_r$ be a basis of $F$. The short exact sequence $0 \to v_1N \to N \to N/v_1N \to 0$ makes it sufficient to prove the Lemma for $v_1N$ and $N/v_1N$. But $v_1^2 = \frac{1}{2}[v_1, v_1] = 0$ in $UF$ and so $v_1 \cdot v_1N = 0$. Thus we are reduced to the case of $F$-modules $N$ for which $v_1N = 0$. Iterating this argument for $v_2, \cdots, v_r$ reduces us to the case $F \cdot N = 0$. Thus $N = \oplus \Bbbk\, e_i$ is the direct sum of one dimensional $F$-modules.

But in this case

$$\mathrm{Ext}_{UF}(\oplus_i \Bbbk e_i, UF) = \prod_i \mathrm{Ext}_{UF}(\Bbbk e_i, UF)$$

and it follows from Proposition 11.3 that $\mathrm{Ext}^*_{UF}(\Bbbk, UF) = \mathrm{Ext}^0_{UF}(\Bbbk, UF)$. □

*proof of Proposition 11.6:* First note that because $L$ is abelian *all subspaces of L are ideals.*

Now if $m = 0$ then, because $\mathrm{Ext}^0_{UL} = \mathrm{Hom}_{UL}$, there is an $x \in M$ and an $f \in \mathrm{Hom}_{UL}(M, UL)$ such that $f(x) \neq 0$. Choose $n$ such that $f(x) \in (UL)_{<n}$. It follows from the Poincaré-Birkoff-Witt Theorem that multiplication $(UL)_{<n} \otimes UL_{\geq n} \to UL$ is an inclusion of right $UL_{\geq n}$-modules. Thus the Proposition follows in this case.

Suppose next by induction that the Proposition holds for $p < m$. It is clearly sufficient to prove the conclusion for some subquotient of $M$. Thus by Proposition 11.4(ii) we may suppose $M$ finitely generated. Then a simple exact sequence argument reduces to the case $M$ has the form $M = UL \cdot z$.

Set $E = L_{\text{even}}$ and $F = L_{\text{odd}}$, and let $N \subset M$ be the $L$-submodule of elements $y$ such that $UE \to UE \cdot y$ is not injective. Since $\mathrm{Ext}^m_{UL}(M, UL) \neq 0$, either $\mathrm{Ext}^m_{UL}(N, UL) \neq 0$ or $\mathrm{Ext}^m_{UL}(M/N, UL) \neq 0$.

Suppose $\mathrm{Ext}^m_{UL}(N, UL) \neq 0$. By Proposition 11.4(ii) there is a finitely generated $L$-submodule $P \subset N$ such that $\mathrm{Ext}^m_{UL}(P, UL) \neq 0$. Since it is sufficient to establish the conclusion of the Proposition for a subquotient $L$-module of $P$ we are reduced to the case $P = UL \cdot y$, and that $a \cdot y = 0$ for some $a \in UE$. Choose $k$ so that $a \in U(E_{<k})$ and write $I = E_{<k}$.

Since $a \cdot P = 0$ it follows that $\mathrm{Hom}_{UI}(P, UI) = 0$, as $UI$ has no zero divisors. Because $UI$ has finite type, $((UI)^\#)^\# = UI$. Thus

$$\mathrm{Hom}_{UI}(P, UI) = \mathrm{Ext}^0_{UI}(P, UI) = \left[\mathrm{Tor}_0^{UI}(P, (UI)^\#)\right]^\#$$

and so $\mathrm{Tor}_0^{UI}(P, UI)^\# = 0$.

Write $L = I \oplus J$. Then $J$ is an ideal, isomorphic to $L/I$, and the Hochschild-Serre spectral sequence converges to $\mathrm{Tor}^{UL}_{p+q}(P, UL)$, from

$$\mathrm{Tor}_p^{UJ}(\Bbbk, \mathrm{Tor}_q^{UI}(P, (UI)^\#) \otimes (UJ)^\#) = \mathrm{Tor}_p^{UJ}(\mathrm{Tor}_q^{UI}(P, (UI)^\#), (UJ)^\#)\,.$$

It follows via (10.1) that $\mathrm{Ext}^p_{UI}(\mathrm{Tor}^{UI}_q(P,(UI)^{\#}),UJ)\neq 0$ for some $p<m$.

By induction on $m$, for some $\alpha\in\mathrm{Tor}^{UI}_q(P,(UI)^{\#})$ and some $\ell\geq k$, the map $(UJ)_{\geq\ell}\to\mathrm{Tor}^{UI}_q(P,(UI)^{\#})$, $b\mapsto b\cdot\alpha$, is injective. But this map is induced by multiplication by $b$ in $P$. Since $UJ_{\geq\ell}=UL_{\geq\ell}$ the Proposition is proved in this case.

On the other hand, suppose $\mathrm{Ext}^m_{UL}(M/N,UL)\neq 0$. Since it is sufficient to establish the conclusion of the Proposition for $M/N$, we may restrict consideration to modules $M$ of the form $UL\cdot z$ and satisfying for each non-zero $y\in M$ that $UE\to UE\cdot y$ is injective.

If the Proposition fails here we may find an infinite sequence $k_0<k_1<\cdots$ and elements in $UL_{(k_ik_{i+1})}$ which map $z$ to 0. Write these elements as $a_i+b_i$ with $b_i\in UE$ and $a_i\in(UF)_+\cdot UE$. For each $i$ there is some $r$ for which $a_i^r=x_{j_1}\cdots x_{j_s}\Phi$ with $x_{j_\nu}\in F_{(k_i,k_{i+1})}$ and $\Phi\in UE$. It follows that $a_i^rbz=0$ and hence that $x_{j_1}\cdots x_{j_p}z=0$.

Since for any map $\varphi: M\to UF$ there is a $p$ such that $\varphi(z)\in UF_{\leq p}$ it follows that $\mathrm{Hom}_{UF}(M,UF)=0$. Now by Lemma 11.1, $\mathrm{Ext}_{UF}(M,UF)=0$. Since $UL$ is $UF$-free this implies that $\mathrm{Ext}_{UF}(M,UL)=0$.

Finally, the Hochschild-Serre spectral sequence

$$\mathrm{Ext}^q_{UL/F}(\Bbbk,\mathrm{Ext}^p_{UF}(M;UL))\Longrightarrow\mathrm{Ext}^{p+q}_{UL}(M,UL)$$

implies that $\mathrm{Ext}^*_{UL}(M,UL)=0$, a contradiction. □

Suppose $E$ is a sub Lie algebra of a connected Lie algebra $L$ of finite type. Denote by $UZ$ a subspace of $UE$ which maps isomorphically to $U(E/[E,E])$ and by $U(Z_{\geq n})$ the subspace of $UZ$ corresponding to $U(E/[E,E]_{\geq n})$.

**Proposition 11.7.** *If depth $L<\infty$ then for some $n$, some $s\leq$ depth $L$ and some element $z\in\wedge^s(s(L/E)\oplus[E,E])$ the adjoint representation of $UE$ in $\wedge^s((L/E)\otimes[E,E])$ restricts to an injection*

$$U(Z_{\geq n})\hookrightarrow U(Z_{\geq n})\cdot z\,.$$

*proof:* The Hochschild-Serre spectral sequence for the sub algebra $E$,

$$E_1^{p,q}=\mathrm{Ext}^q_{UE}(\wedge^p s(L/E),UL)\Rightarrow\mathrm{Ext}^{p+q}_{UL}(\Bbbk,UL)\,,$$

implies that for some $p$ and $q$, $\mathrm{Ext}^q_{UE}(\wedge^p s(L/E),UL)\neq 0$, with $p+q\leq$ depth $L$. As $UL$ is $UE$-free it follows that $\mathrm{Ext}^q_{UE}(\wedge^p s(L/E),UE)\neq 0$. Write $I=[E,E]$. By Proposition 11.4(iii) this implies that

$$\mathrm{Ext}^q_{UE}(\wedge^p s(L/E),U(E/I))\neq 0\,.$$

But by Lemma 10.3 there is a Hochschild-Serre spectral sequence

$$\mathrm{Ext}^r_{U(E/I)}(\mathrm{Tor}^{UI}_s(\wedge^p s(L/E), \Bbbk), U(E/I)) \Longrightarrow \mathrm{Ext}^{q=p+k}_{UE}(\wedge^p s(L/E), U(E/I))\,.$$

Thus the $E_2$-term is non-zero. Since $E/I$ is abelian, by Proposition 11.6 there are integers $n$ and $s$ and a $\gamma \in \mathrm{Tor}^{UI}_k(\wedge^p s(L/E), \Bbbk)$, such that $U((E/I)_{\geq n}) \to U((E/I)_{\geq n}) \cdot \gamma$ is injective.

Now, as observed in §10.1, $\mathrm{Tor}^{UI}_*(\Bbbk, \wedge^p s(L/E)) = \mathrm{Tor}^{UI}_*(\wedge^p s(L/E), \Bbbk)$ is the homology of a complex of $E$-modules of the form $\wedge(sI) \otimes \wedge^p s(L/E)$, where the representation of $E$ is the adjoint representation. In particular we may represent $\gamma$ by a cycle $z \in \wedge^k sI \otimes \wedge^p s(L/E)$, with $k \leq q$. Thus $z \in \wedge^s(sI \oplus L/E)$ with $s \leq \operatorname{depth} L$, and the map

$$U(Z_{\geq n}) \to U(Z_{\geq n}) \cdot z \quad \text{is injective}\,.$$

□

## 11.3 Weak depth

As usual, in this section $L$ denotes a connected graded Lie algebra of finite type.

**Definition.** (i) An $L$-module, $M$, is *weakly locally finite* if $M$ is the increasing union of finite dimensional subspaces $M(1) \subset M(2) \subset \cdots$ such that $M(k)$ is preserved by $L_{\leq k}$.

(ii) A sub-Lie algebra $E \subset L$ is *admissible* if the quotient $L/E$ is a weakly locally finite $E$-module for the adjoint representation.

Note that if $M$ and $N$ are weakly locally finite $L$-modules, so are $M \otimes N$, each $\wedge^q M$, and any sub-quotient module of $M$.

**Definition.** The *weak depth* of $L$, w-depth $L$, is the least $m$ (or $\infty$) such that for some weakly locally finite $L$-module, $M$, we have $\mathrm{Ext}^m_{UL}(M, UL) \neq 0$.

**Lemma 11.2.** *(i) If $E$ is an admissible sub Lie algebra of $L$, then w-depth $E \leq$ w-depth $L$. In particular for any ideal $I \subset L$, w-depth $I \leq$ w-depth $L$.*

*(ii) w-depth $L \leq$ depth $L$, and equality holds if $L$ is finitely generated.*

*proof:* Set $m =$ weak depth $L$ and let $M$ be a weakly locally finite $L$-module $M$ such that $\mathrm{Ext}^m_{UL}(M, UL) \neq 0$.

(i) Because $E$ is admissible in $L$, $L/E$ is a weakly locally finite $E$-module; hence so is $\wedge^p s(L/E) \otimes M$. Now it follows from the Hochschild-Serre spectral sequence that

$$E_1^{p,q} = \mathrm{Ext}^q_{UE}(\wedge^p s(L/E) \otimes M, UL) \neq 0\,, \qquad \text{some } p+q=m\,.$$

Thus w-depth $E \leq q \leq m$.

(ii) Since $\Bbbk$ is weakly locally finite, by definition w-depth $L \leq$ depth $L$. Since $L$ is finitely generated, $M$ is the union of an increasing sequence of finite dimensional modules $M(r)$, and each finitely generated submodule is contained in some $M(r)$. Therefore, by Proposition 11.4 for some $r$, $\mathrm{Ext}^m_{UL}(M(r), UL) \neq 0$. Now, since $M(r)$ is finite dimensional, a simple long exact sequence argument shows that $\mathrm{Ext}^m_{UL}(\Bbbk, UL) \neq 0$, i.e. depth $L \leq m$. □

**Definition.** The centralizer $Z$ of a subalgebra $E$ in a graded Lie algebra $L$ is the sub Lie algebra defined by

$$Z = \{x \in L \mid [x,a] = 0 \quad \text{for all } a \text{ in } E\}\,.$$

**Proposition 11.8.** *Let $E \subset L$ be a sub Lie algebra with centralizer $Z$. Suppose for some weakly locally finite $L$-module, $M$, and some $m$, that the restriction map*

$$\rho : Ext^m_{UL}(M, UL) \to Ext^m_{UE}(M, UL)$$

*is non-zero. Then $Z$ is finite dimensional, and dim $Z_{even} \leq m$.*

*proof:* In view of Proposition 11.4(ii) we may and do assume that $E$ is a finitely generated Lie algebra. Then, since by definition $M$ is an increasing union of finite dimensional $E$-modules $M(k)$, it follows from Proposition 11.4(ii) that for some $k$ the restriction map

$$\mathrm{Ext}^m_{UL}(M, UL) \to \mathrm{Ext}^m_{UE}(M(k), UL)$$

is non-zero. In particular since $UL$ is a free $UE$-module, $\mathrm{Ext}^m_{UE}(M(k), UE)$ is non-zero. Then, because dim $M(k) < \infty$, a short exact sequence argument shows that $\mathrm{Ext}^m_{UE}(\Bbbk, UE) \neq 0$. Thus depth $E \leq m$.

Let $Z$ be the centralizer of $E$ in $L$. Then $Z \cap E$ is an abelian ideal in $E$, and so it follows from Theorem 11.1 that dim $Z \cap E < \infty$. Choose $q$ so that $Z_{\geq q} \cap E = 0$ and set $F = Z_{\geq q}$. Now the restriction map $\mathrm{Ext}^m_{UL}(M, UL) \to \mathrm{Ext}^m_{UE}(M, UL)$ factors through $\mathrm{Ext}^m_{U(E\oplus F)}(M, UL)$, and since $UL$ is a free $U(E \oplus F)$-module, the restriction map

$$\rho : \mathrm{Ext}^m_{U(E\oplus F)}(M, U(E \oplus F)) \to \mathrm{Ext}^m_{UE}(M, U(E \oplus F))$$

is non-zero.

Finally, $E$ and $F$ are ideals in $E \oplus F$ and therefore it follows from Proposition 11.4(i) that $F$ is finite dimensional and concentrated in odd degrees. Thus $Z$ is finite dimensional, and $Z_{\geq q}$ is concentrated in odd degrees. But $Z_{<q}$ is an abelian ideal in $E$, and since depth $E \leq m$ it follows from Theorem 11.1 that $\dim Z_{\text{even}} = \dim (Z_{<q})_{\text{even}} \leq m$.

□

**Corollary 11.2.** *Let $Z$ be the centralizer of an ideal $I$ in a connected graded Lie algebra $L$ of finite type. If the restriction map $Ext_{UL}(\Bbbk, UL) \to Ext_{UI}(\Bbbk, UL)$ is non-zero then for some $q$, $Z_{\geq q}$ is finite dimensional and concentrated in odd degrees.*

**Corollary 11.3.** *Let $L$ be a connected graded Lie algebra of finite type and finite depth. Then there are elements $x_1, \cdots, x_r$ of odd degrees such that for all sufficiently large odd integers $n$, the map*

$$\sum_{i=1}^{r}[x_i, -] : L_n \to \bigoplus_{i=1}^{r} L_{n+\deg x_i}$$

*is injective.*

*proof*: Denote by $J$ the sub Lie algebra generated by $L_{\text{odd}}$. Then $J$ is an ideal, necessarily of finite depth by Proposition 11.2. By Proposition 11.4(ii) there is a sub Lie algebra $E$ generated by finitely many elements $x_1, \cdots, x_r$ in $J_{\text{odd}}$ such that the restriction map $\text{Ext}_{UJ}(\Bbbk, UJ) \to \text{Ext}_{UE}(\Bbbk, UJ)$ is non-zero. Therefore the centralizer $Z$ of $E$ in $J$ is finite dimensional. Choose $n_0$ such that $Z_{\geq n_0} = 0$. Then for each $n \geq n_0$ the map

$$\sum_{i=1}^{r}[x_i, -] : J_n \to \oplus_{i=1}^{r} L_{n+\deg x_i}$$

is injective. But for $n$ odd, $L_n = J_n$.

□

**Corollary 11.4.** *Let $L$ be a connected graded Lie algebra of finite type and finite depth. Then for some $r$ the centralizer of $L_{\leq r}$ is finite dimensional.*

*proof:* Choose $r$ so that for some $m$, $\text{Ext}^m_{UL}(\Bbbk, UL) \to \text{Ext}^m_{UE}(\Bbbk, UL)$ is non-zero, where $E$ is the sub Lie algebra generated by $L_{\leq r}$. Then $Z$ is the centralizer of $E$ and the Corollary is a special case of Proposition 11.8.

□

**Lemma 11.3.** *Let $E \subset L$ be a sub Lie algebra of a graded Lie algebra $L$ such that $L/E$ is finite dimensional and concentrated in odd degrees. Then for any $L$-module $M$, the restriction morphism*

$$Ext_{UL}^*(M, UL) \to Ext_{UE}^*(M, UL)$$

*is injective.*

*proof:* It is clearly sufficient to consider the case $L = E \oplus \Bbbk x$ with $x$ of odd degree. In this case $E$ is an ideal in $L$ and the quotient Lie algebra $L/E$ has the exterior algebra $\wedge x$ as its universal enveloping algebra. It follows from Lemma 10.2 (p. 270) that $\mathrm{Ext}_{UE}^*(M, UL)$ is the free $\wedge x$-module $\mathrm{Ext}_{UE}^*(M, UE) \otimes \wedge x$. Thus in the Hochschild-Serre spectral sequence, converging to $\mathrm{Ext}_{UL}^*(M, UL)$ we have

$$E_2^{p,q} = \mathrm{Ext}_{\wedge x}^p(\Bbbk, \mathrm{Ext}_{UE}^q(M, UL)) = 0\,, \qquad p \geq 1\,.$$

Hence

$$\mathrm{Ext}_{UL}^q(M, UL) = E_2^{0,q} = \mathrm{Hom}_{\wedge x}(\Bbbk, \mathrm{Ext}_{UE}^q(M, UL))\,.$$

Since the restriction morphism $\mathrm{Ext}_{UL}^q(M, UL) \to \mathrm{Ext}_{UE}^q(M, UL)$ is the composition

$$\mathrm{Ext}_{UL}^q(M, UL) \xrightarrow{\cong} E_2^{0,q} \hookrightarrow \mathrm{Ext}_{UE}^q(M, UL)\,,$$

(cf. (10.3)) the injectivity of the restriction morphism follows.

□

**Proposition 11.9.** *Suppose $L$ is the union of an increasing sequence of sub Lie algebras $L(r)$. Assume further that $z_0, z_1, \cdots$ are elements in $L$ of even degree such that $L(r)$ contains $z_0, \cdots, z_{r-1}$ and for $r, i \geq 0$*

$$dim\,[z_{r+i}, L(r)] < \infty \qquad and \quad [z_{r+i}, L(r)]_{even} = 0\,.$$

*Then the weak depth of $L$ is infinite.*

*proof:* Suppose w-depth $E = m < \infty$, and let $F(r) \subset L(r)$ be the sub Lie algebra of elements that commute with each $z_{r+i}$, $0 \leq i \leq m$. Since $[z_{r+i}, L(r)]$ is finite dimensional and concentrated in odd degrees it follows that $L(r)/F(r)$ is also finite dimensional and concentrated in odd degrees. By Lemma 11.3, the restriction morphism

$$\mathrm{Ext}_{UL(r)}(M, UL) \to \mathrm{Ext}_{UF(r)}(M, UL)$$

is injective for any $L$-module $M$.

On the other hand, since w-depth $L = m$ there is a weakly locally finite $L$-module $M$ such that $\mathrm{Ext}^m_{UL}(M, UL) \neq 0$. Moreover, since $L = \cup_r L(r)$, for some $r$ the restriction morphism $\mathrm{Ext}^m_{UL}(M, UL) \to \mathrm{Ext}^m_{UL(r)}(M, UL)$ is non-zero. Hence also the composite

$$\mathrm{Ext}^m_{UL}(M, UL) \to \mathrm{Ext}^m_{UF(r)}(M, UL) \text{ is non-zero.}$$

Fix such an $r$ and set $F = F(r)$. Since for $i < j$, $[z_{r+j}, L(r+i+1)] \subset L_{\mathrm{odd}}$, it follows that for all $i, j$, $[z_{r+j}, z_{r+i}] = 0$. In particular, $Z = \oplus_{i=r}^{m+r} \Bbbk\, z_i$ is an abelian Lie algebra commuting with $F$. Because $F \to L$ factors as $F \to F + Z \to E$, the restriction map $\mathrm{Ext}^m_{UL}(M, UL) \to \mathrm{Ext}^m_{UF}(M, UL)$ factors through $\mathrm{Ext}^m_{U(F+Z)}(M, UL)$. Since $UL$ is a free $U(F+Z)$-module, it follows that w-depth $(F + Z) \leq m$. Since $Z$ is an ideal in $F + Z$, it follows from Lemma 11.2 that w-depth $Z \leq m$. But $Z$ is finitely generated and so Lemma 11.2 asserts that depth $Z \leq m$. On the other hand, since $Z$ is abelian and concentrated in even degrees, Theorem 11.1 asserts that depth $Z = m + 1$, and we have the desired contradiction.

□

# Chapter 12

# Trichotomy

*In this Chapter $X$ will always denote a connected CW complex, and the ground field will always be $\mathbb{Q}$. Consistent with the convention in §1.4, $H^*(-;\mathbb{Q})$ and $H_*(-;\mathbb{Q})$ will be denoted simply by $H(-)$ and $H_*(-)$. As in Chapter 6, $\widetilde{X}$ will denote the universal cover of $X$.*

Since $\pi_k(X)$ is abelian for $k \geq 2$ we may form the rational vector spaces $\pi_k(X) \otimes \mathbb{Q}$, and we make the

**Definition.** For $k \geq 2$ the rank of $\pi_k(X)$, denoted by $\mathrm{rk}\, \pi_k(X)$, is given by

$$\mathrm{rk}\, \pi_k(X) = \dim \pi_k(X) \otimes \mathbb{Q}\,.$$

Note that if $\pi_k(X)$ is finitely generated this coincides with the classical definition: the number of copies of $\mathbb{Z}$ in a direct sum decomposition of cyclic groups of $\pi_k(X)$.

The objective of this and the next Chapter is to establish the properties of the ranks of the homotopy groups of $X$ under either of the two conditions

$$cat\, X < \infty \qquad \text{or} \qquad \dim X < \infty\,.$$

## 12.1 Overview of results

First observe that for any connected CW complex $X$ there are exactly three, naturally exclusive possibilities:

$$\sum_{k=2}^{\infty} \mathrm{rk}\, \pi_k(X) < \infty \tag{12.1}$$

$$\text{Each } \mathrm{rk}\, \pi_k(X) < \infty \text{ but } \sum_{k \geq 2} \mathrm{rk}\, \pi_k(X) = \infty \tag{12.2}$$

$$\text{Some } \operatorname{rk} \pi_k(X) = \infty . \tag{12.3}$$

We will call $X$ *rationally elliptic* (resp. *rationally hyperbolic*, *rationally infinite*) if it satisfies (12.1) (resp. (12.2), (12.3)).

For instance, compact simply connected Lie groups and homogeneous spaces are rationally elliptic ([18], Proposition 15.16). Simply connected finite CW complexes with negative Euler characteristic are rationally hyperbolic ([18], Proposition 32.10). If $X$ is a simply connected finite CW complex with $H_+(X) \neq 0$, then $S^1 \vee X$ is rationally infinite, because its universal cover has the homotopy type of a wedge of infinitely many copies of $X$. By contrast, for $n \geq 2$, $S^n \vee X$ is rationally hyperbolic.

With the conventions above we state the main results:

**Theorem 12.1.** *If a rationally elliptic connected CW complex satisfies* $\operatorname{cat} X < \infty$ *then* $H(\widetilde{X})$ *is finite dimensional. Moreover, in this case*

$$\pi_k(X) \otimes \mathbb{Q} = 0, \quad k \geq 2N \quad \text{and} \quad \sum_{k \geq 2} \operatorname{rk} \pi_k(X) \leq 2 \operatorname{cat} X ,$$

*where* $N = \max\{n \mid H^n(\widetilde{X}) \neq 0\}$.

In the rationally hyperbolic case the dominant phenomenom for the $\operatorname{rk} \pi_k(X)$ is exponential growth. To make this precise we first recall the invariant *log index* : if $(x_n)$ is a sequence of non-negative numbers then

$$\text{log index}\,(x_k) = \limsup_k \frac{\log x_k}{k} ,$$

where as usual $\log 0 = -\infty$. If $S = \{S_k\}_{k \geq 0}$ and $V = \{V^k\}_{k \geq 0}$ are graded vector spaces of finite type we write

$$\text{log index}\, S = \limsup_k \frac{\log \dim S_k}{k} \qquad \text{and log index}\, V = \frac{\log \dim V^k}{k} .$$

In particular, for a rationally hyperbolic connected CW complex $X$ we write

$$\alpha_k(X) = \frac{\log \operatorname{rk} \pi_k(X)}{k} \qquad \text{and} \quad \alpha(X) = \limsup_k \alpha_k(X) .$$

Thus

$$\alpha(X) = \text{log index}\,(\operatorname{rk} \pi_k(X)\,) = \text{log index}\, \pi_{\geq 2}(X) \otimes \mathbb{Q} .$$

**Proposition 12.1.** *Suppose $X$ is rationally hyperbolic simply connected CW complex. Then log index$H(X) \leq \alpha(X)$ and log index$H(X) < \infty$ if and only if $\alpha(X) < \infty$.*

**Theorem 12.2.** *If a rationally hyperbolic connected CW complex $X$ satisfies* $\operatorname{cat} X < \infty$ *then* $\alpha(X) > 0$. *If, in addition,* $\alpha(X) < \infty$ *then for some fixed $R$,*

$$\lim_{n\to\infty} \max_{n\le k\le n+R} \alpha_k(X) = \alpha(X)\,.$$

**Corollary 12.1.** *For each $\varepsilon > 0$ there is a $K$ such that for $n \ge K$,*

$$e^{(\alpha(X)-\varepsilon)n} \le \max_{n\le k\le n+R} rk\,\pi_k(X) \le e^{(\alpha(X)+\varepsilon)n}\,.$$

**Remark.** The crux of Theorem 12.2 is that $\alpha(X) < \infty$ and that the limit exists. It is immediate that the limit is $\alpha(X)$.

When $X$ is finite dimensional Theorem 12.2 can be strengthened in two ways:

**Theorem 12.3.** *Suppose $X$ is an $N$-dimensional rationally hyperbolic connected CW complex. Then for $n \ge 0$ there is some $k \in [n+2, n+N]$ such that* $\pi_k(X) \otimes \mathbb{Q} \neq 0$.

**Theorem 12.4.** *Suppose $X$ is a connected rationally hyperbolic CW complex and both* $\operatorname{cat} X$ *and $\alpha(X)$ are finite.*

*If* $\log \operatorname{index} H(\widetilde{X}) < \alpha_X$ *then there are positive constants $\beta, \gamma$, and a positive integer $R$ with the following property : for some $K$, and for all $n \ge K$,*

$$\alpha(X) - \gamma\frac{\log n}{n} \le \max_{n+2\le k\le n+R} \alpha_k(X) \le \alpha(X) + \frac{\beta}{n}\,.$$

**Corollary 12.2.** *For all $n \ge K$,*

$$\frac{1}{n^\gamma} e^{\alpha(X)n} \le \max_{n+2\le k\le n+R} rk\,\pi_k(X) \le e^\beta e^{\alpha(X)n}\,.$$

**Theorem 12.5.** *Suppose $X$ is an $N$-dimensional rationally hyperbolic connected CW complex. Then there are constants $\beta, \gamma > 0$, depending only on $H(\widetilde{X})$, and with the following property : for some $K$, and for all $n \ge K$,*

$$\alpha(X) - \gamma\frac{\log n}{n} \le \max_{n+2\le k\le n+N} \alpha_k(X) \le \alpha(X) + \frac{\beta}{n}\,.$$

**Corollary 12.3.** *For all $n \ge K$,*

$$\frac{1}{n^\gamma} e^{\alpha(X)n} \le \max_{n+2\le k\le n+N} rk\,\pi_k(X) \le e^\beta e^{\alpha(X)n}\,.$$

In the rationally infinite case we have

**Theorem 12.6.** *Suppose* $(\wedge V, d)$ *is the minimal Sullivan model of a rationally infinite simply connected CW complex,* $X$*. If* $cat_0 X < \infty$ *then for some* $K$ *and* $r$*, and for all* $n \geq K$*,*

$$dim\, V^k = \infty\,, \qquad some\ \ k \in [n, n+r]\,.$$

**Theorem 12.7.** *Suppose* $X$ *is a rationally infinite connected CW complex. If* $cat\, X < \infty$ *then for some* $K$ *and* $R$*, and for all* $n \geq K$*,*

$$dim\, \pi_k(X) \otimes \mathbb{Q} = \infty\,, \qquad some\ \ k \in [n, n+R]\,.$$

**Theorem 12.8.** *Suppose* $X$ *is an* $N$*-dimensional rationally infinite connected CW complex. Then for each* $n \geq 0$ *there is some* $k \in [n+2, n+N]$ *such that* $dim\, \pi_k(X) \otimes \mathbb{Q} = \infty$*.*

The proofs of Theorems 12.1 - 12.8 are presented as follows. Theorem 12.1 follows easily from the simply connected case proved as Theorem 32.15 in [18]; the extension to the non-simply connected case is in §12.2. Proposition 12.1, Theorems 12.2, 12.4, and 12.5 are reduced in §12.3 to the analogous Proposition 12.2, Theorems 12.8, 12.9, and 12.10 for minimal Sullivan algebras. These are proved in Chapter 13. Theorem 12.3 is proved in §12.3 and Theorems 12.5 and 12.6 are proved in §12.4. Finally, Theorem 12.7 is proved in §12.5.

Throughout this and the next Chapter we shall rely on the following

**Lemma 12.1.** *Suppose* $X$ *is a connected CW complex.*

*(i)* $cat\, X \leq dim\, X$.
*(ii)* $cat_0 \widetilde{X} \leq cat\, \widetilde{X} \leq cat\, X$.
*(iii) If* $rk\, \pi_k(X) < \infty$ *for all* $k \geq 2$ *then* $H(\widetilde{X})$ *has finite type. In this case a minimal Sullivan model* $\varphi : (\wedge V, d) \to A_{PL}(\widetilde{X})$ *(cf. (1.12), p. 36) induces an isomorphism*

$$\pi_k(\varphi) : \pi_k(\widetilde{X}) \otimes \mathbb{Q} \to V^{\#}\,.$$

*(iv) If* $dim\, X = N$ *for some* $N < \infty$ *and if* $rk \pi_k(X) < \infty$ *for all* $k \geq 2$*, then* $dim\, H(\widetilde{X}) < \infty$*.*

*proof:* (i) An $N$-dimensional CW complex is an $N$-cone (defined in §9.1). Thus by Theorem 9.1 (p. 245), $\mathrm{cat}\, X \leq N$.

(ii) Let $\rho : \widetilde{X} \to X$ be the covering projection and suppose $\mathrm{cat}\, X = m$. Then there is an open cover, $U_0, \cdots, U_m$, of $X$ such that each $U_i$ is contractible in $X$. Thus the inclusions of $\rho^{-1}(U_i)$ in $\widetilde{X}$ are homotopic to maps $\rho^{-1}(U_i) \to \rho^{-1}(*)$ where $*$ is the base point of $X$. But $\rho^{-1}(*)$ is a discrete subspace of the path connected space $\widetilde{X}$, and so the inclusion of $\rho^{-1}(*)$ in $\widetilde{X}$ is homotopic to a constant map. Thus $\mathrm{cat}\, \widetilde{X} \leq m$. The inequality $\mathrm{cat}_0\widetilde{X} \leq \mathrm{cat}\, \widetilde{X}$ is Theorem 9.2, p. 247.

(iii) Since $\pi_*(\widetilde{X}) \otimes \mathbb{Q}$ has finite type it follows, as observed in Remark 1, p. 208 of [18], that $H(\widetilde{X})$ has finite type. Thus Theorem 1.6 (p. 43) asserts that $\pi_k(\varphi)$ is an isomorphism.

(iv) $\widetilde{X}$ is also $N$-dimensional and so $H^{>N}(\widetilde{X}) = 0$. Thus (iv) follows from (iii).

□

## 12.2 The rationally elliptic case

**Theorem 12.1.** *If a rationally elliptic CW complex satisfies* $\mathit{cat}\, X < \infty$ *then* $H(\widetilde{X})$ *is finite dimensional. Moreover, in this case*

$$\pi_k(X) \otimes \mathbb{Q} = 0\,, \quad k \geq 2N \qquad \textit{and} \quad \sum_{k\geq 2} rk\, \pi_k(X) \leq 2\, \mathit{cat} X\,,$$

*where* $N = \max\{n \mid H^n(\widetilde{X}) \neq 0\}$.

*proof:* Since $\pi_{\geq 2}(X) \otimes \mathbb{Q}$ is finite dimensional it follows from Lemma 12.1(iii) (p. 316) that $H(\widetilde{X})$ has finite type and that a minimal Sullivan model $(\wedge V, d)$ for $\widetilde{X}$ satisfies $\dim V < \infty$.

Moreover, again by Lemma 12.1 (p. 316), $\mathrm{cat}(\wedge V, d) \leq \mathrm{cat}\, X < \infty$. Thus $(\wedge V, d)$ is a retract of a Sullivan algebra whose homology is isomorphic to $H(\wedge V / \wedge^{>m} V, d)$, where $m = \mathrm{cat}(\wedge V, d)$. In particular, $H(\widetilde{X}) \cong H(\wedge V, d)$ is finite dimensional.

Now Theorem 12.1 follows from Theorem 32.15 in [18], applied to $\widetilde{X}$.

□

## 12.3 The rationally hyperbolic case

In view of Lemma 12.1, Proposition 12.1, and Theorems 12.2, 12.4 and 12.5 follow respectively from Theorems 12.9, 12.10 and 12.11 below, applied to the minimal Sullivan model of $\widetilde{X}$. These will be proved in Chapter 13.

**Proposition 12.2.** *Let $L$ be the homotopy Lie algebra of a minimal Sullivan algebra $(\wedge V, d)$ in which $V = V^{\geq 2}$ has finite type and infinite dimension. Then $\log\operatorname{index} H(\wedge V, d) \leq \log\operatorname{index} L$ and if $\log\operatorname{index} H(\wedge V, d) < \infty$ so is $\log\operatorname{index} L$.*

**Theorem 12.9.** *Let $L$ be the homotopy Lie algebra of a minimal Sullivan algebra $(\wedge V, d)$ for which $V = V^{\geq 2}$ has finite type and infinite dimension. If $\operatorname{cat}(\wedge V, d) < \infty$ then $\log\operatorname{index} L > 0$.*

*Moreover, if, in addition, $\log\operatorname{index} H(\wedge V, d) < \infty$ then for some fixed $R$,*

$$\lim_{n\to\infty}\left(\max_{n\leq k\leq n+R}\frac{\log \dim L_k}{k}\right) = \log\operatorname{index} L < \infty\,.$$

**Theorem 12.10.** *Let $L$ be the homotopy Lie algebra of a minimal Sullivan algebra $(\wedge V, d)$ for which $V = V^{\geq 2}$ has finite type and infinite dimension. If $\operatorname{cat}(\wedge V, d) < \infty$ and $\log\operatorname{index} H(\wedge V, d) < \log\operatorname{index} L$, then there are constants $R, \beta, \gamma > 0$ with the following property: for some $K$ and all $n \geq K$,*

$$\log\operatorname{index} L - \gamma\,\frac{\log n}{n} \leq \max_{n\leq k\leq n+R}\left(\frac{\log \dim L_k}{k}\right) \leq \log\operatorname{index} L + \frac{\beta}{n}\,.$$

**Theorem 12.11.** *Let $L$ be the homotopy Lie algebra of a minimal Sullivan algebra $(\wedge V, d)$ for which $V = V^{\geq 2}$ has finite type and infinite dimension. If $H^{>N}(\wedge V, d) = 0$ then there are constants $\beta, \gamma > 0$, depending only on $H(\wedge V, d)$, and with the following property: for some $K$ and all $n \geq K$,*

$$\log\operatorname{index} L - \gamma\,\frac{\log n}{n} \leq \max_{n+2\leq k\leq n+N}\left(\frac{\log \dim L_k}{k}\right) \leq \log\operatorname{index} L + \frac{\beta}{n}\,.$$

## 12.4 The gap theorem

**Theorem 12.3.** *Suppose $X$ is an $N$-dimensional rationally hyperbolic connected CW complex. Then for each $n \geq 0$ there is some $k \in [n+2, n+N]$ such that $\pi_k(X) \otimes \mathbb{Q} \neq 0$.*

*proof:* The covering space $\widetilde{X}$ is also an $N$-dimensional rationally hyperbolic connected CW complex. Thus by Lemma 12.1 (p. 316) $H(\widetilde{X}) = H^{\leq N}(\widetilde{X})$ is finite dimensional. Thus the minimal Sullivan model, $(\wedge V, d)$, for $\widetilde{X}$

satisfies $H(\wedge V, d) = H^{\leq N}(\wedge V, d)$ is finite dimensional and so $V = V^{\geq 2}$ has finite type. The proof of Theorem 33.3 in [18] now shows that for each $n$ there is some $k \in [n+2, n+N]$ such that $V^k \neq 0$. This gives Theorem 12.3.

□

## 12.5 Rationally infinite spaces of finite category

**Theorem 12.6** *Suppose $(\wedge V, d)$ is the minimal Sullivan model of a rationally infinite simply connected CW complex, $X$. If $cat\, X < \infty$ then for some $K$ and $R$, and for all $n \geq K$,*

$$dim\, V^k = \infty\,, \qquad some\;\; k \in [n, n+R]\,.$$

*proof:* We proceed in three steps and prove successively:

**Step One**: $\dim V^k = \infty$, some $k$.

**Step Two**: $\dim V^k = \infty$ for infinitely many $k$.

**Step Three**: proof of the Theorem.

**Step One**: Suppose dim $V^k < \infty$ for all $k$. Then $H(X) = H(\wedge V, d)$ has finite type and so Theorem 1.6 (p. 43) shows that $\pi_k(X) \otimes \mathbb{Q} \cong V$ has finite type, contradicting our hypothesis. Thus $\dim V^k = \infty$ for some $k$.

**Step Two**: Suppose $\dim V^k = \infty$ for only finitely many $k$. Then for some $N$ we have

$$\dim V^N = \infty \quad \text{and} \quad \dim V^k < \infty\,, \qquad k > N\,.$$

Divide the Sullivan algebra by $V^{<N}$ and by a subspace of $V^N$ to obtain a quotient Sullivan algebra $(\wedge W^N \otimes \wedge V^{>N}, \overline{d})$ in which $\dim W^N$ is finite, but may be arbitrary large. If cat $X = m < \infty$ then Theorem 9.2 (p. 247), and the Mapping Theorem 9.3 (p. 248) give

$$\text{cat}(\wedge W^N \otimes \wedge V^{>N}, \overline{d}) \leq \text{cat}(\wedge V, d) \leq m\,.$$

Since $\overline{d}(W^N) = 0$ it follows by Proposition 9.3 (p. 250) that $\wedge^{m+1} W^N \subset \text{Im}\, \overline{d}$.

Now for $v \in V^{>N}$ write $\overline{d}v = \varphi(v) + \psi(v)$ with $\varphi(v) \in \wedge^{\geq 2} W^N$ and $\psi(v) \in \wedge W^N \otimes \wedge^+ V^{>N}$. Since $\wedge^{m+1} W^N \subset \text{Im}\, \overline{d}$ it follows that $\wedge^{m+1} W^N$

is contained in the ideal generated by $\varphi(V^{[N+1,(m+1)N-1]})$. Thus setting $p = \dim V^{[N+1,(m+1)N-1]}$ we have

$$\dim \wedge^{m+1} W^N \leq p \dim \wedge^{m-1} W^N .$$

But as $\dim W^N \to \infty$ so does $\dim \wedge^{m+1} W^N / \dim \wedge^{m-1} W^N$, and we have again arrived at a contradiction. Thus $\dim V^k = \infty$ for infinitely many $k$.

**Step Three**: Here we suppose the Theorem itself is false, and deduce a contradiction. Again denote $\mathrm{cat}(\wedge V, d)$ by $m$. Since the Theorem is false, we may find an infinite sequence $N_1 < N_2 < \cdots$ with the following two properties:

$$\dim V^{N_i} = \infty \quad \text{ and } \dim V^j < \infty , \quad N_i + 1 \leq j \leq N_i + i + 1 .$$

We now construct a quotient Sullivan algebra $\varphi : (\wedge V, d) \to (\wedge W, \overline{d})$ such that $W$ has finite type, and such that

$$\begin{cases} \varphi : V^s \xrightarrow{\cong} W^s , & \text{if } \dim V^s < \infty . \\ \dim W^{N_i} \geq (m+1)i \displaystyle\sum_{j=N_i+1}^{N_i+i+1} \dim V^j , & i \geq 1 . \end{cases} \tag{12.4}$$

The construction is by an obvious induction: If $\varphi$ is constructed in $\wedge V^{\leq s}$ it defines a surjection $(\wedge V, d) \to (\wedge W^{\leq s} \otimes \wedge V^{>s}, \overline{d})$. If $\dim V^{s+1} < \infty$ then extend $\varphi$ by the identity in $V^{s+1}$. If $\dim V^{s+1}$ is $\infty$ then, since $(\wedge W^{\leq s})$ has finite type, there are subspaces $U^{s+1} \subset V^{s+1}$ of finite codimension for which $\overline{d} U^{s+1} = 0$. Extend $\varphi$ by dividing by some $U^{s+1}$, where $U^{s+1}$ is chosen so that $W^{s+1}$ satisfies (12.4) if $s+1 = N_i$, some $i$. Then by the Mapping Theorem 9.3 (p. 248),

$$\mathrm{cat}\,(\wedge W, \overline{d}) \leq m .$$

The rest of the proof proceeds by an analysis of the homotopy Lie algebra, $L$, of $(\wedge W, \overline{d})$. Since $W$ has finite type, so does $L$. Thus setting $\mathrm{depth}\, L = q$ we have from Theorem 10.1 (p. 273) and Theorem 10.2 (p. 278) that

$$q = \mathrm{depth}\, L \leq \mathrm{cat}(\wedge W, \overline{d}) \leq m .$$

Then, from (12.4), we obtain

$$\dim L_{N_i - 1} \geq (q+1)i \sum_{j=N_i}^{N_i+i} \dim L_j , \qquad i \geq 1 . \tag{12.5}$$

Next observe that for any sub Lie algebra $F \subset L$ there is an isomorphism $UL = UF \otimes C$ of $UF$-modules. We shall use without further reference the consequence:

$$\mathrm{Tor}_k^{UF}(\mathbb{Q}, UF) \neq 0 \Leftrightarrow \mathrm{Tor}_k^{UF}(\mathbb{Q}, UL) \neq 0 .$$

In particular, recall from Proposition 10.7 (p. 290) that there is a sub Lie algebra $E \subset L$ generated by finitely many elements $x_1, \cdots, x_t$ such that the inclusion induces a non-zero map

$$\mathrm{Tor}_q^{UE}(\mathbb{Q}, UE) \to \mathrm{Tor}_q^{UL}(\mathbb{Q}, UL) .$$

Thus $\operatorname{depth} E \leq q$. Moreover, if $\operatorname{depth} E = 0$ then by Proposition 11.3 (p. 302) $\mathrm{Tor}_k^{UE}(\mathbb{Q}, UL) = 0$, $k \geq 1$. Thus it would follow that $\operatorname{depth} L = 0$ and so, by that Proposition, that $\dim L < \infty$. Since by Step Two this is not the case, we have

$$0 < \operatorname{depth} E \leq q .$$

Now choose any $i > t \sum_{j=1}^{t} \deg x_j$. Then

$$\bigoplus_{j=1}^{t} ad_L(x_j) : L_{N_i - 1} \longrightarrow \bigoplus_{j=1}^{t} L_{N_i - 1 + \deg x_j}$$

maps $L_{N_i-1}$ into a vector space of dimension at most $t \sum_{j=N_i}^{N_i+i} \dim L_j$. It follows from (12.5) that for $Z = \cap_{j=1}^{t} \ker\, ad_L(x_j)$ we have the inequality

$$\dim(Z \cap L_{N_i - 1}) \geq qi \sum_{j=N_i}^{N_i+i} \dim L_j . \tag{12.6}$$

Moreover, since $Z$ commutes with $E$, the vector space $Z \cap E$ is a central ideal in $E$. Therefore Theorem 11.1 (p. 303) gives $\dim Z \cap E < \infty$, and so for some $M$, $Z \cap E_{\geq M} = 0$.

Denote $L(1) = L_{>M} \cap Z$. On the one hand, the composite

$$\mathrm{Tor}_q^{UE}(\mathbb{Q}, UL) \to \mathrm{Tor}_q^{U(E \oplus L(1))}(\mathbb{Q}, UL) \to \mathrm{Tor}_q^{UL}(\mathbb{Q}, UL)$$

is non-zero, and so $\operatorname{depth}(E \oplus L(1)) \leq q$. Moreover, $U(E \oplus L(1))$ is the tensor product of the algebras $UE$ and $UL(1)$, and thus

$$\mathrm{Tor}_k^{U(E \oplus L(1))}(\mathbb{Q}, U(E \oplus L(1))) = \bigoplus_{j+\ell=k} \mathrm{Tor}_j^{UE}(\mathbb{Q}, UE) \otimes \mathrm{Tor}_\ell^{UL(1)}(\mathbb{Q}, UL(1)) .$$

Thus $q \geq \operatorname{depth}(E \oplus L(1)) = \operatorname{depth} E + \operatorname{depth} L(1)$. Since $\operatorname{depth} E > 0$ this yields

$$\operatorname{depth} L(1) \leq q - 1 .$$

On the other hand, for $i$ sufficiently large ($N_i > M$ and $i \geq t\sum_{j=1}^{t} \deg x_j$), (12.6) yields

$$\dim L(1)_{N_i-1} \geq qi \sum_{j=N_i}^{N_i+i} \dim L(1)_j \, .$$

Since depth $L(1) < q$ we may now iterate the process above to construct an infinite sequence of infinite dimensional Lie algebras

$$L = L(0 \supset L(1) \supset \cdots$$

with strictly decreasing depths. This is obviously impossible, and so the assumption that the Theorem fails produces a contradiction. This completes the proof.

□

**Theorem 12.7.** *Suppose $X$ is a rationally infinite connected CW complex. If $\mathrm{cat}\, X < \infty$ then for some $K$ and $R$, and for all $n \geq K$,*

$$\dim \pi_k(X) \otimes \mathbb{Q} = \infty \, , \qquad \text{some } k \in [n, n+R] \, .$$

The Theorem will follow via the Postnikov-type decomposition Lemma below from a simple adaptation of the proof of Theorem 12.5.

**Lemma 12.2.** *Let $Y$ be a simply connected topological space such that for some $n$,*

$$\dim \pi_i(Y) \otimes \mathbb{Q} < \infty \, , \quad i < n \quad \text{and } \dim \pi_n(Y) \otimes \mathbb{Q} = \infty \, .$$

*Then for any $M \geq 0$ there is a fibration $p : Z \to Y$ such that*

*(i) $\pi_i(p) \otimes \mathbb{Q}$ is an isomorphism for $i \neq n$ and is injective for $i = n$.*
*(ii) $M < \dim \pi_n(Z) \otimes \mathbb{Q} < \infty$.*

*proof:* Adjoin cells of dimension $\geq n$ to $Y$ to construct a map $f : Y \to B$ such that $\pi_i(f)$ is an isomorphism for $i < n$ and $\pi_{\geq n}(B) = 0$. By Lemma 12.1(iv), $H_*(B)$ has finite type. Now, as in §3.2, convert $f$ into a fibration for which the fibre is $(n-1)$-connected and has infinite dimensional rational homotopy in degree $n$. Then a simple Serre spectral sequence argument applied to this fibration shows that the Hurewicz homomorphism

$$hur : \pi_n(Y) \otimes \mathbb{Q} \to H_n(Y)$$

has infinite dimensional image and finite dimensional kernel.

Next, fix a weak homotopy equivalence $S \xrightarrow{\simeq} Y$ from a CW complex, and let $S_{n-1}$ be the $(n-1)$-skeleton of $S$. Then $H_n(S) \to H_n(S/S_{n-1})$ is injective, and it follows that the map

$$\pi_n(S) \otimes \mathbb{Q} \to \pi_n(S/S_{n-1}) \otimes \mathbb{Q}$$

has infinite dimensional image and finite dimensional kernel.

Write the image of this map as $I(1) \oplus I(2)$, with $M < \dim I(1) < \infty$, and form the composite

$$\sigma : \pi_n(S) \to \pi_n(S) \otimes \mathbb{Q} \to I(1) \oplus I(2) \to I(2) .$$

According to Proposition 4.20 in [18] there is a continuous map $g : S \to K(I(2), n)$ such that $\pi_n(g) = \sigma$. (Here $K(I(2), n)$ denotes the Eilenberg-MacLane space.)

Convert $g$ into a fibration with fibre $F$. Then the corresponding map $h : F \to S$ satisfies:

(i) $\pi_i(h) \otimes \mathbb{Q}$ is an isomorphism, $i \neq n$, and $\pi_n(h) \otimes \mathbb{Q}$ is injective.
(ii) $M < \dim \pi_n(F) \otimes \mathbb{Q} < \infty$.

Finally, convert the composite $F \xrightarrow{h} S \xrightarrow{\simeq} Y$ into a fibration $p : Z \to Y$, with the desired properties.

□

*proof of Theorem 12.6*: Since by Lemma 12.1 $\operatorname{cat} \widetilde{X} \leq \operatorname{cat} X$, we may without loss of generality suppose $X$ is simply connected. We show first that $\dim \pi_k(X) \otimes \mathbb{Q} = \infty$ for infinitely many $k$.

If not, there is a maximum $n$ such that $\dim \pi_n(X) \otimes \mathbb{Q} = \infty$. Repeated application of Lemma 12.2 then provides a map $h : Y \to X$ such that $\pi_i(h) \otimes \mathbb{Q}$ is injective, $\dim \pi_i(Y) \otimes \mathbb{Q} < \infty$, $i \neq n$, and $\dim \pi_n(Y) \otimes \mathbb{Q} = \infty$.

Then for each $M$, Lemma 12.2 provides maps $f_M : Z_M \to Y$ such that

(i) $\pi_*(f_M) \otimes \mathbb{Q}$ is injective and $\pi_i(f_M)$ is an isomorphism for $i \neq n$.
(ii) $M < \dim \pi_n(Z_M) \otimes \mathbb{Q} < \infty$.

In particular, $\pi_*(Z_M) \otimes \mathbb{Q}$ has finite type and so Lemma 12.1, together with Theorem 1.6 (p. 43) shows that $Z_M$ has a minimal Sullivan model $(\wedge V_M, d)$ in which $\dim V_M^i = \dim \pi_i(Y) \otimes \mathbb{Q}$ if $i \neq n$ and $M < \dim V_M^n < \infty$.

Moreover, since $\pi_*(h \circ f_M)$ is injective it follows from Theorem 25.6 in [18] and Lemma 12.1 that

$$\operatorname{cat}(\wedge V_M, d) = \operatorname{cat}_0(Z_M) \leq \operatorname{cat}_0 X \leq \operatorname{cat} X .$$

Now the argument in the proof of Step Two of Theorem 12.5 shows that this is impossible. It follows that $\dim \pi_k(X) \otimes \mathbb{Q} = \infty$ for infinitely many $k$.

Finally, suppose Theorem 12.6 false, and let $m = \operatorname{cat} X$. Then we may find an infinite sequence $n_1 < n_2 < \cdots$ with the following two properties:

$$\dim \pi_{n_i}(X) \otimes \mathbb{Q} = \infty \quad \text{and} \dim \pi_j(X) \otimes \mathbb{Q} < \infty \,, \quad n_i + 1 \leq j \leq n_i + i + 1 \,.$$

Repeated application of Lemma 12.2 then yields an infinite sequence of fibrations

$$\to F_i \to F_{i-1} \to \cdots \to F_1 \to X$$

such that $\pi_j(F_i) \otimes \mathbb{Q} \to \pi_j(F_{i-1}) \otimes \mathbb{Q}$ is injective, and an isomorphism if $j < n_i$, and

$$\infty > \dim \pi_{n_i}(F_i) \otimes \mathbb{Q} \geq (m+1)i \sum_{j=n_i+1}^{n_i+i+1} \dim \pi_j(X) \otimes \mathbb{Q} \,.$$

Now according to Theorem 9.7 in [18] each $F_i$ admits a rational homology equivalence into a rational space $(F_i)_{\mathbb{Q}}$. The maps $F_i \to F_{i-1}$ induce maps $(F_i)_{\mathbb{Q}} \to (F_{i-1})_{\mathbb{Q}}$, which we may also convert to fibrations. Thus we may suppose that the $F_i$ are themselves rational spaces, and that the groups $\pi_*(F_i) = \pi_*(F_i) \otimes \mathbb{Q}$ satisfy the condition above.

Set $F = \varprojlim F_i$. By construction, $\pi_j(F_i) \xrightarrow{\cong} \pi_j(F_{i-1})$ if $j < n_i$, and so $\pi_j(F) = \varprojlim \pi_j(F_i) \xrightarrow{\cong} \pi_j(F_i)$ if $j < n_i$. Moreover, it follows from the conditions above that $\pi_*(F) = \pi_*(F) \otimes \mathbb{Q}$ has finite type. Thus, by Lemma 12.1 and Theorem 12.5, the minimal Sullivan model $(\wedge V, d)$ of $F$ satisfies

$$\dim W^{n_i} \geq (m+1)i \sum_{j=n_i+1}^{n_i+i+1} \dim W^j \,, \quad i \geq 1 \,.$$

It also follows that $\pi_*(F) \otimes \mathbb{Q} \to \pi_*(X) \otimes \mathbb{Q}$ is injective, and so again by Theorem 25.6 in [18],

$$\operatorname{cat}(\wedge W, d) = \operatorname{cat}_0(F) \leq \operatorname{cat}_0 X \leq m \,.$$

Thus the homotopy Lie algebra of $(\wedge W, d)$ satisfies (12.5) (p. 320) and the proof of Theorem 12.5 shows that this is impossible. This completes the proof of Theorem 12.6.

□

## 12.6 Rationally infinite CW complexes of finite dimension

**Theorem 12.8.** *Suppose $X$ is an $N$-dimensional rationally infinite connected CW complex. Then for each $n \geq 0$ there is some $k \in [n+2, n+N]$ such that*

$$dim\, \pi_k(X) \otimes \mathbb{Q} = \infty\,.$$

*proof:* All spaces in this proof will be based with base points denoted by $*$, and all maps and homotopies will be base point-preserving; for simplicity of notation a based space $(Y, *)$ will usually be denoted simply by $Y$. Moreover, in view of Lemma 12.1 it is sufficient to prove the Theorem for the universal cover $\widetilde{X}$. Thus we may assume $X$ is simply connected.

The proof of the Theorem will be in three steps

**Step One**: The groups $G_q(S, Y)$, $q \geq 2$.

**Step Two**: The exact sequence

$$G_q(S \cup_h D^{k+1}, Y) \to G_q(S, Y) \to \pi_{q+k}(Y)\,.$$

**Step Three**: Completion of the proof.

**Step One**: *The groups $G_q(S, Y)$, $q \geq 2$.*

Here we generalize the subgroups $G_q(-) \subset \pi_q(-)$ introduced by Gottlieb in [24]. For this, fix a continuous map

$$\sigma : S \to Y$$

from a connected CW complex pointed by a 0-cell. Then a *relative Gottlieb map* is a pair $(\varphi, f) : (S^q \times S, S^q \times \{*\}) \to Y$ such that $\varphi$ restricts to $f \vee \sigma$ in $S^q \vee S$. We say $(\varphi_1, f_1) \sim (\varphi_2, f_2)$ if there is a homotopy of pairs restricting to the constant homotopy in $* \times S$, and denote the set of homotopy classes by $G_q(S, Y)$.

Addition in $G_q(S, Y)$ is defined as follows. Denote the equator sphere in $S^q$ by $S^{q-1}$ and let $\Delta : S^q \to S^q/S^{q-1} = S^q \vee S^q$ be the quotient map. Using the equality $(S^q \vee S^q) \times S = (S^q \times S) \cup_{* \times S} (S^q \times S)$ we suppose $(\varphi_1, f_1)$ and $(\varphi_2, f_2)$ are relative Gottlieb maps and set

$$\varphi_1 + \varphi_2 = (\varphi_1 \cup \varphi_2) \circ (\Delta \times id_S) : S^q \times S \to Y\,.$$

Then $\varphi_1 + \varphi_2$ restricts to $f_1 + f_2$ in $S^q$ and we set $(\varphi_1, f_1) + (\varphi_2, f_2) = (\varphi_1 + \varphi_2, f_1 + f_2)$. This construction clearly passes to homotopy classes, and thereby makes $G_q(S, Y)$ into an abelian group. By construction, the map $(\varphi, f) \mapsto f$ defines a homomorphism,

$$\alpha : G_q(S, Y) \to \pi_q(Y) . \tag{12.7}$$

**Step Two**: *The exact sequence*

$$G_q(S \cup_h D^{k+1}, Y) \to G_q(S, Y) \to \pi_{q+k}(Y) .$$

Suppose now that the map $\sigma$ of Step One is the restriction of a map

$$\sigma : S \cup_h D^{k+1} \to Y ,$$

where $S \cup_h D^{k+1}$ is the CW complex obtained by attaching a $(k+1)$-cell to $S$ via a map $h : S^k \to S$. (As usual, for any $r \geq 1$, $D^r$ denotes the disk with boundary $S^{r-1}$.) Denote $S \cup_h D^{k+1}$ by $T$.

Then a relative Gottlieb map $(\varphi, f)$ for $S$ and $Y$ defines the map

$$\varphi \cup \sigma : (S^q \times S) \cup_{* \times S} T \to Y .$$

On the other hand, recall the classical identification

$$(D^{q+k+1}, S^{q+k}) = (D^q \times D^{k+1}, D^q \times S^k \cup_{S^{q-1} \times S^k} S^{q-1} \times D^{k+1}).$$

Let $\rho : D^q \to D^q/S^{q-1} = S^q$ be the quotient map. Thus we may define

$$\psi = (\varphi \circ (\rho \times h) \cup (* \times \sigma)) : S^{q+k} \to Y ,$$

where $*$ denotes the constant map of $S^{q-1}$ on the base point of $S$. The homotopy class of this map depends only on the homotopy class of $(\varphi, f)$. Thus the correspondence $\varphi \mapsto \psi$ gives a set map

$$G_q(S, Y) \to \pi_{q+k}(Y) . \tag{12.8}$$

On the other hand, restriction from $S^q \times T$ to $S^q \times S$ defines a homomorphism $G_q(T, Y) \to G_q(S, Y)$ and we complete this section by showing that (12.8) is a homomorphism with kernel the image of $G_q(T, Y)$.

To show that (12.8) is a homomorphism we define the additive structure in $\pi_{q+k}(T)$ by the map $S^{q+k} \to S^{q+k}/S^{q+k-1} = S^{q+k} \vee S^{q+k}$ with equator sphere $S^{q+k-1}$ given by

$$S^{q+k-1} = D^{q-1} \times S^k \cup_{S^{q-2} \times S^k} S^{q-2} \times D^{k+1} .$$

That (12.8) is a homomorphism is then a straightforward verification from the construction of $\psi$.

Finally, $\psi$ is null homotopic precisely when it extends to $D^q \times D^{k+1}$ or, equivalently, when $\varphi \circ \sigma$ extends to $((S^q \times S) \cup_{* \times S} T) \cup_\psi (D^q \times S^{k+1})$. But a second trivial verification gives

$$((S^q \times S) \cup_{* \times S} T) \cup_\psi (D^q \times D^{k+1}) = S^q \times T .$$

It follows that the kernel of (12.8) is the image of $G_q(T, Y)$.

**Step Three.** *Completion of the proof of Theorem 12.8.*

As observed at the start of the proof we may assume $X$ is a simply connected $N$-dimensional CW complex, pointed by a zero cell. We establish

**Lemma 12.3.** *Suppose for some $n \geq 0$ that*

$$dim\, \pi_k(X) \otimes \mathbb{Q} < \infty\,, \quad k \in [n+2, n+N]\,.$$

*Then $dim\, \pi_{n+1}(X) \otimes \mathbb{Q} < \infty$.*

*proof:* We suppose the Lemma fails for some $n$, and deduce a contradiction. For this we construct by induction an increasing sequence of finite sub complexes of $X$,

$$* = S_0 \subset S_1 \subset \cdots \subset X$$

and maps

$$\varphi_i : S^{n+1} \times S_i \to S_{i+1}$$

such that the maps $f_i = \varphi_i|_{S^{n+1} \times *}$ represent linearly independent elements of $\pi_{n+1}(X) \otimes \mathbb{Q}$.

By hypothesis $\dim \pi_{n+1}(X) \otimes \mathbb{Q} = \infty$ and so we may find $f_0 : S^{n+1} \to X$. Suppose $\varphi_i, S_i$ are constructed, and let $S_{i+1} \subset X$ be any finite sub complex containing $\varphi_i(S^{n+1} \times S_i)$. Then (cf. (12.7)) we have the commutative diagram

$$\begin{array}{ccc} G_{n+1}(S_{i+1}, X) & \subset & G_{n+1}(S_i, X) \\ & \searrow \quad \swarrow & \\ & \pi_{n+1}(X)\,. & \end{array}$$

It follows from Step Two by induction on the number of relative cells in $(S_{i+1}, S_i)$ that $G_{n+1}(S_{i+1}, X) \otimes \mathbb{Q}$ has finite codimension in $G_{n+1}(S_i, X)$. This will also be true for each $G_{n+1}(S_j, X) \otimes \mathbb{Q} \subset G_{n+1}(S_{j-1}, X) \otimes \mathbb{Q}$, $j \leq i$. In particular, since $G_{n+1}(*, X) \otimes \mathbb{Q} = \pi_{n+1}(X) \otimes \mathbb{Q}$, $G_{n+1}(S_{i+1}, X) \otimes \mathbb{Q}$ has finite codimension in $\pi_{n+1}(X) \otimes \mathbb{Q}$. Choose $(\varphi_{i+1}, f_{i+1}) \in G_{n+1}(S_{i+1}, X)$ so that $[f_{i+1}], [f_i], \cdots, [f_0]$ are linearly independent in $\pi_{n+1}(X) \otimes \mathbb{Q}$.

Finally, define $\gamma : S^{n+1} \times \cdots \times S^{n+1}$ ($N+1$ factors) $\to X$ by composing

$$\gamma : \underbrace{S^{n+1} \times \cdots \times S^{n+1}}_{N+1 \text{ factors}} \stackrel{id \times f_N}{\longrightarrow} \underbrace{S^{n+1} \times \cdots \times S^{n+1}}_{N \text{ factors}} \times S_{N-1}$$

$$\stackrel{id \times \varphi_{N-1}}{\longrightarrow} \underbrace{S^{n-1} \times \cdots \times S^{n-1}}_{N-1 \text{ factors}} \times S^{N-2} \to \cdots \to X\,.$$

By construction, $\pi_{n+1}(\gamma) \otimes \mathbb{Q}$ is injective. Moreover, as described at the start of the proof of Lemma 12.2, there is a fibration

$$F \to Y \to B$$

in which $\pi_{\geq n+1}(B) = 0$, $\pi_j(Y) \xrightarrow{\cong} \pi_j(B)$ for $j < n+1$, and $Y \simeq X$. Since $S^{n+1} \times \cdots \times S^{n+1}$ is $n$-connected the map $\gamma' : S^{n+1} \times \cdots \times S^{n+1} \to Y$ corresponding to $\gamma$ is homotopic to a map

$$\gamma'' : S^{n+1} \times \cdots \times S^{n+1} \to F$$

for which $\pi_{n+1}(\gamma'') \otimes \mathbb{Q}$ is injective.

But $\pi_*(F) \to \pi_*(Y) \xrightarrow{\cong} \pi_*(X)$ is injective, and so Theorem 25.6 in [18] and Theorem 9.2 (p. 247) give $\mathrm{cat}_0 F \leq \mathrm{cat}_0 X \leq N$. Now the Hurewicz homomorphism identifies $\pi_{n+1}(\gamma'') \otimes \mathbb{Q}$ with $H_{n+1}(\gamma'')$ and so $H_{n+1}(\gamma'')$ is injective. Denote by $\beta_i \in H_{n+1}(S^{n+1} \times \cdots \times S^{n+1})$ the image of $[S^{n+1}]$ under the inclusion of the $i^{th}$ sphere. Since $H_{n+1}(\gamma'')$ is injective there are classes $\omega_i \in H^{n+1}(F)$ such that

$$\langle H^{n+1}(\gamma'')\omega_i, \beta_j \rangle = \delta_{ij} \, .$$

In particular $\omega_0 \bullet \cdots \bullet \omega_N \neq 0$, contradicting the fact that $\mathrm{cat}_0 F \leq N$. This completes the proof of the Lemma.

□

It follows from Lemma 12.3 that if $\dim \pi_k(X) \otimes \mathbb{Q} < \infty$, $k \in [n+2, n+N]$ then $\dim \pi_k(X) \otimes \mathbb{Q} < \infty$ for $2 \leq k \leq N$. As described in the proof of Lemma 12.2 (p. 322) we may, by adjoining cells of dimension $\geq N+2$ to $X$, construct a CW complex $Z$ such that $\pi_{\leq N}(Z) = \pi_{\leq N}(X)$ and $\pi_{>N}(Z) = 0$. Convert the map $X \to Z$ to a fibration and use a Serre spectral argument to see that $H_{\leq N}(X) \xrightarrow{\cong} H_{\leq N}(Z)$.

Since $\pi_*(Z)$ has finite type so, by Theorem 1.6 (p. 43) and Lemma 12.2 (p. 322), does $H_*(Z)$. Thus $H_{\leq N}(X)$ has finite type. But $X$ is an $N$-dimensional CW complex. Thus $H_*(X) = H_{\leq N}(X)$ has finite type and Theorem 1.6 (p. 43) asserts that $\pi_*(X) \otimes \mathbb{Q}$ has finite type. This contradicts the hypothesis of Theorem 12.7 and thereby completes the proof.

□

# Chapter 13

# Exponential growth

Recall from §12.1 that for any graded vector spaces, $S = \{S_k\}_{k\geq 0}$ and $V = \{V^k\}_{k\geq 1}$, we write

$$\log \text{index}\, S = \limsup_k \frac{\log \dim\, S_k}{k} \qquad \text{and } \log \text{index}\, V = \limsup_k \frac{\log \dim\, V^k}{k}\,.$$

In this Chapter we consider minimal Sullivan algebra $(\wedge V, d)$ with homotopy Lie algebra $L$, in which $V = V^{\geq 2}$ has finite type but is infinite dimensional. Our purpose is to prove (Proposition 12.2)

$$\log \text{index}\, H(\wedge V, d) \leq \log \text{index}\, L\,,$$

and

$$\log \text{index}\, H(\wedge V, d) < \infty \Leftrightarrow \log \text{index}\, L < \infty\,,$$

and the following three Theorems stated in §12.3. Proposition 12.2 follows from Proposition 13.2 immediately below and the rest of the Chapter is devoted to the proofs of the Theorems. As explained in Chapter 12, these establish a very strict form of exponential growth for the sequence $(\dim L_n)$ and thus for the ranks of the homotopy groups of CW complexes.

**Theorem 12.9.** *Let $L$ be the homotopy Lie algebra of a minimal Sullivan algebra $(\wedge V, d)$ for which $V = V^{\geq 2}$ has finite type and infinite dimension. If cat$(\wedge V, d) < \infty$ then log index$\,L > 0$.*

*Moreover, if, in addition, log index$\,H(\wedge V, d) < \infty$ then for some fixed $R$,*

$$\lim_{n\to\infty}\left(\max_{n\leq k\leq n+R} \frac{\log dim\, L_k}{k}\right) = log\ index\, L < \infty\,.$$

**Theorem 12.10.** *Let $L$ be the homotopy Lie algebra of a minimal Sullivan algebra $(\wedge V, d)$ for which $V = V^{\geq 2}$ has finite type and infinite dimension. If $cat(\wedge V, d) < \infty$ and $log\ index H(\wedge V, d) < log\ index L$, then there are constants $R, \beta, \gamma > 0$ with the following property: for some $K$ and all $n \geq K$,*

$$log\ index L - \gamma \frac{\log n}{n} \leq \max_{n \leq k \leq n+R} \left( \frac{\log dim\ L_k}{k} \right) \leq log\ index L + \frac{\beta}{n}.$$

**Theorem 12.11.** *Let $L$ be the homotopy Lie algebra of a minimal Sullivan algebra $(\wedge V, d)$ for which $V = V^{\geq 2}$ has finite type and infinite dimension. If $H^{>N}(\wedge V, d) = 0$ then there are constants $\beta, \gamma > 0$, depending only on $H(\wedge V, d)$, and with the following property: for some $K$ and all $n \geq K$,*

$$log\ index L - \gamma \frac{\log n}{n} \leq \max_{n+2 \leq k \leq n+N} \left( \frac{\log dim\ L_k}{k} \right) \leq log\ index L + \frac{\beta}{n}.$$

All three Theorems establish exponential growth detected in intervals of a fixed length. Essentially the only difference is that in the first one the error is just some positive "$\varepsilon$" decreasing to zero, while in the second and third it can be established as $\beta/n$ and $\gamma/\log n$. Additionally in the third Theorem, where $H^{>N}(\wedge V, d) = 0$, the length of the interval is $N - 2$ and $\beta$ and $\gamma$ depend only on $H(\wedge V, d)$.

Consequently the proofs of the three Theorems can and will be presented as a single proof except that the technique for estimating the error will vary, depending on which set of hypotheses are being considered. The strategy for the proof is then as follows:

§13.1 contains the proof of Proposition 12.2.

§13.2 establishes essential technical lemmas about the growth of a universal enveloping algebra.

§13.3 detects exponential growth in intervals of the form $[k, k+(m+1)k]$ where $m = \text{cat}\,(\wedge V, d)$.

§13.4 establishes the basis for the error estimates $\beta/n$ and $\gamma/\log n$ in the second and third Theorems.

§13.5 detects exponential growth in intervals whose length grows, but in a more constrained way.

§13.6 detects exponential growth in intervals of constant length, and shows that if $H^{>N}(\wedge V, d) = 0$ then the exponential growth occurs in intervals of length $N - 2$.

## 13.1 The invariant log index

When log index $V > 0$ we say that $V$ *grows at most exponentially.* Recall also (§1.1) that the Hilbert series of a graded vector space $V$ of finite type is the formal power series $\sum_k \dim V_k z^k$.

**Remarks.** For graded vector spaces, $V$, of finite type:

(i) $\log \operatorname{index} V = -\infty$ if and only if $\dim V < \infty$. Otherwise $\log \operatorname{index} V \geq 0$.

(ii) $\log \operatorname{index} V = \alpha > 0$ means that :

  (a) For each $\varepsilon > 0$, there is some $K$ such that for $k \geq K$, $\dim V^k \leq e^{k(\alpha+\varepsilon)}$

  (b) there is an infinite sequence $k_i$ such that $\log\,(\dim V^{k_i})^{1/k_i} \to \alpha$.

(iii) In particular, $\alpha = \log \operatorname{index} V < \infty$ if and only if $e^{-\alpha}$ is the radius of convergence of the Hilbert series $\sum \dim V^k z^k$.

We begin by recalling the well-known result of Babenko:

**Proposition 13.1.** ([3]) *Let $L = L_{\geq 1}$ be a graded Lie algebra of finite type. If $\log \operatorname{index} L > 0$, then $L$ and $UL$ have the same log index.*

*proof:* Since $L \subset UL$, $\log \operatorname{index} L \leq \log \operatorname{index} UL$. Thus we need only consider the case $\log \operatorname{index} L < \infty$. Set $a_n = \dim L_n$. Then the Hilbert series of $UL$ is given by

$$\sum_p b_p z^p = \frac{\prod_n (1 + z^{2n+1})^{a_{2n+1}}}{\prod_n (1 - z^{2n})^{a_{2n}}}.$$

Denote by $R$ the radius of convergence of the series $\sum_n a_n z^n$. It follows from Remark (iii) that $R < 1$. Then for $z < R$, by the mean value theorem, we have

$$\frac{z}{1+R} \leq \log(1+z) \leq z \leq \log(1-z)^{-1} \leq \frac{z}{1-R}.$$

Therefore

$$\log(\sum_p b_p z^p) \leq \sum_{k\ odd} a_k z^k + \sum_{k\ even} \frac{a_k z^k}{1-R} \leq \sum_{k=1}^{\infty} a_k \frac{z^k}{1-R}.$$

This shows that the radius of convergence of $UL$ is greater than or equal to $R$, and since $UL$ contains $L$, they are equal.

□

Babenko's result has the following important consequence:

**Proposition 13.2.** *Let $L$ be the homotopy Lie algebra of a minimal Sullivan algebra $(\wedge V, d)$ in which $V = V^{\geq 2}$ has finite type. Then*

$$\text{log index}\, H(\wedge V, d) \leq \text{log index}\, L \leq 2 \sup_{i\geq 2} \frac{\log \dim H^i(\wedge V, d)}{i-1}\,.$$

*In particular,*

$$\text{log index}\, H(\wedge V, d) < \infty \quad \Leftrightarrow \quad \text{log index}\, L < \infty\,.$$

*proof:* Since $\wedge V$ is the universal enveloping algebra of $V$, regarded as an abelian Lie algebra, it follows from Proposition 13.1 (p. 331) that log index $\wedge V$ = log index $V$ = log index $L$ = log index $UL$. The first inequality is immediate.

For the second, we may suppose log index $H(\wedge V, d) < \infty$, in which case

$$\sup_{i\geq 2} \frac{\log \dim H^i(\wedge V, d)}{i-1} < \infty\,.$$

Thus $\omega = \sup_{i\geq 2} \dim \left(H^i(\wedge V, d)\right)^{\frac{1}{i-1}} < \infty$. We show now that

$$\dim\, (UL)_k \leq (2\omega)^k\,, \qquad k \geq 1\,. \tag{13.1}$$

Let $(\wedge V \otimes \wedge U, d)$ be the acyclic closure (§3.6) of $(\wedge V, d)$. By Proposition 3.12 (p. 111) $\dim U^i = \dim V^{i+1} = \dim L_i$. By the Poincaré Birkhoff Witt theorem, $UL \cong \wedge U$ as graded vector spaces and so $\dim (UL)_k = \dim\, (\wedge U)^k$, $k \geq 0$. Thus it suffices to prove (13.1) with $(UL)_k$ replaced by $(\wedge U)^k$.

Consider the short exact sequence

$$0 \to \wedge^+ V \otimes \wedge U \to \wedge V \otimes \wedge U \to \wedge U \to 0\,.$$

Since $H(\wedge V \otimes \wedge U, d) = \Bbbk$ and the quotient differential in $\wedge U$ is zero, it follows that $d$ induces an injection of degree one: $\wedge U \to H(\wedge^+ V \otimes \wedge U, d)$. From the spectral sequence $H^+(\wedge V, d) \otimes \wedge U \Rightarrow H(\wedge^+ V \otimes \wedge U, d)$ we then obtain

$$\dim(\wedge U)^k \leq \sum_{i=2}^{k+1} \dim H^i(\wedge V, d) \dim\, (\wedge U)^{k+1-i}\,, \qquad k \geq 1\,.$$

But when $k = 1$, $\dim\, (\wedge U)^1 = \dim U^1 = \dim H^2(\wedge V, d) \leq \omega$. Thus, supposing by induction that (13.1) holds for $k < \ell$, we find

$$\dim\, (\wedge U)^\ell \leq \sum_{i=2}^{\ell+1} \omega^{i-1} (2\omega)^{\ell+1-i} \leq (2\omega)^\ell\,.$$

This establishes (13.1). By definition, $\log\omega = \sup_{i\geq 2}\frac{\log\dim H^i(\wedge V,d)}{i-1}$, and so

$$\log \operatorname{index} L = \log \operatorname{index} UL \leq 2\log\omega \leq 2\sup_{i\geq 2}\frac{\log\dim H^i(\wedge V,d)}{i-1}\,.$$

The final assertion of the Proposition is immediate because for any graded vector space $S = S^{\geq 1}$ of finite type, if $\log \operatorname{index} S < \infty$ so is

$$\sup_{i\geq 2}\frac{\log\ \dim S_i}{i-1}\,.$$

□

## 13.2 Growth of graded Lie algebras

**Proposition 13.3.** *Let $L = L_{\geq 1}$ be a graded Lie algebra of finite type. Suppose an infinite sequence of integers $0 < r_0 < r_1 < \cdots$, and an infinite sequence of real numbers $(\lambda_i)_{i\geq 0}$, satisfy*

$$dim\, L_{r_i} \geq e^{\lambda_i r_i}\,,\ i\geq 0\,,\quad and\ \ \lambda_{i+1}\geq \lambda_i + 2\frac{\log(r_i+1)}{r_i}+\frac{\log 3}{r_{i+1}}\,,\quad i\geq 0\,.$$

*Then $L$ contains a sub Lie algebra $E$, generated by subspaces in degrees $r_i$, $i\geq 0$, such that*

$$\frac{1}{2}e^{\lambda_i r_i}\leq \dim\left(E/[E,E]\right)_{r_i}\leq e^{\lambda_i r_i}\,,\quad i\geq 0\,.$$

Before undertaking the proof of this Proposition we establish three necessary Lemmas.

**Lemma 13.1.** *For any integer $s\geq 1$ the coefficients in the power series expansion*

$$\frac{1-z}{1-z-z^s}=\sum_{k=0}^{\infty}a_k z^k$$

*satisfy $a_0 = 1$, $a_k = 0$ for $1<k<s$ and $a_k\leq (s+1)^{k/s}$, $k\geq s$.*

*proof:* Note that

$$\frac{1-z}{1-z-z^s}=\frac{1}{1-z^s/(1-z)}=1+\sum_{\ell=1}^{\infty}z^{s\ell}\left(\frac{1}{1-z}\right)^{\ell}$$

$$= 1 + \sum_{\ell=1}^{\infty} z^{s\ell} \sum_{j=0}^{\infty} \binom{\ell + j - 1}{\ell - 1} z^j .$$

In particular, $a_0 = 1$ and $a_k = 0$, $1 \leq k < s$.

Fix $k \geq s$ and let $q \geq 1$ and $i \in [0, s-1]$ be the unique integers such that $k = qs + i$. Then

$$a_k = \sum_{\ell=1}^{q} \binom{\ell + (q-\ell)s + i - 1}{\ell - 1} .$$

Write $m = \ell - 1$, so that

$$a_k = \sum_{m=0}^{q-1} \binom{m + (q-m-1)s + i}{m} .$$

But

$$m + (q-m-1)s + i \leq m + (q-m-1)s + s - 1 = (q-m)s + (m-1) .$$

Thus,

$$\begin{aligned}\binom{m+(q-m-1)s + i}{m} &\leq \frac{[(q-m)s+(m-1)] \cdots [(q-m)s+r] \cdots [(q-m)s]}{m!} \\ &\leq s^m \frac{(q-1)(q-2) \cdots (q-m+r) \cdots (q-m)}{m!} \\ &= s^m \binom{q-1}{m} .\end{aligned}$$

Hence for $k \geq s$

$$a_k \leq \sum_{m=0}^{q-1} s^m \binom{q-1}{m} = (1+s)^{q-1} \leq (1+s)^{k/s} .$$

□

The next Lemma shows that the growth of a graded Lie algebra $F$ and the growth of its space of indecomposable elements $F/[F,F]$ are very close when $F$ is sufficiently connected.

**Lemma 13.2.** *Suppose for some integer $s \geq 1$ and some $\beta \geq 0$ that a graded Lie algebra $F$ satisfies*

$$F_k = 0 \, , \; k < s \qquad \text{and } dim\,(F/[F,F])_k \leq e^{\beta k} \, , k \geq s .$$

*Then*

$$dim\,(UF)_k \leq e^{(\beta + \frac{1}{s}\log(s+1))k} , \qquad k \geq 0 .$$

*proof:* Let $W$ be a graded vector space satisfying $W_k = 0$, $k < s$, and suppose that $\dim W_k$ is the integral part of $e^{\beta k}$. The tensor algebra, $TW$, is the universal enveloping algebra of the free graded Lie algebra, $E$, generated by $W$: $TW = UE$. The respective Hilbert series $W(z)$ for $W$ and $UE(z)$ for $UE$ satisfy ($<<$ denotes coefficient-wise inequality)

$$W(z) << (e^\beta z)^s \sum_{k=0}^{\infty} (e^\beta z)^k = \frac{(e^\beta z)^s}{1 - e^\beta z}$$

and

$$UE(z) = \frac{1}{1 - W(z)} << \frac{1}{1 - \dfrac{(e^\beta z)^s}{1 - e^\beta z}} = \frac{1 - e^\beta z}{1 - e^\beta z - (e^\beta z)^s}\,.$$

Write $\dfrac{1-x}{1-x-x^s} = \sum_{k=0}^{\infty} a_k x^k$. Then $UE(z) << \sum_{k=0}^{\infty} a_k\, e^{\beta k} z^k$.

Since $W_k = 0$, $k < s$, it follows that $(UE)_k = 0$, $1 \le k < s$ and $a_k = 0$, $1 \le k < s$. On the other hand, Lemma 13.1 gives $a_k \le (s+1)^{k/s}$, $k \ge s$. Thus

$$\dim(UE)_k < e^{\frac{k}{s}\log(s+1) + \beta k} = e^{\left[\beta + \frac{\log(s+1)}{s}\right]k}, \qquad k \ge s\,.$$

It follows from the definition of $W$ that there is a surjective linear map $W \to F/[F,F]$. This extends to a surjective map of graded Lie algebras $E \to F$ and to a surjective map of universal enveloping algebras $UE \to UF$. The Lemma follows.

□

**Lemma 13.3.** *Suppose $E \subset L$ and $Y \subset L$ are respectively a sub Lie algebra and a graded subspace of a graded Lie algebra, $L = L_{\ge 1}$, of finite type. Suppose further that for some integer $s \ge 1$ and some $\beta > 0$ they satisfy*

$$\dim(UE)_k \le e^{\beta k}\,,\ k \ge 0\,, \qquad Y_k = 0\,,\ k < s\,, \qquad \text{and } \dim Y_k \le e^{\beta k}\,,\ k \ge s\,.$$

*Then the sub Lie algebra $G$ generated by $E$ and $Y$ satisfies*

$$\dim(UG)_k \le e^{\left(\beta + 3\frac{\log(s+1)}{s}\right)k}\,, \qquad k \ge 0\,.$$

*Moreover, if $Y = Y_s$ then*

$$\dim(UG)_k \le \begin{cases} e^{\beta k} & ,\ k < s \\ e^{\left(\beta + \frac{2\log(s+1)}{s}\right)k} & ,\ k \ge s\,. \end{cases}$$

*proof:* Denote the adjoint action of $UL$ in $L$ by "$\circ$" and note that for $j \geq s$

$$\dim(UE \circ Y)_j \leq \sum_i \dim(UE)_i \dim Y_{j-i} \leq je^{\beta j} \leq e^{\left(\beta + \frac{\log(s+1)}{s}\right)j},$$

because for $j \geq s \geq 1$, we have $\frac{\log j}{j} < \frac{\log(s+1)}{s}$.

It follows from Lemma 13.2 that the sub Lie algebra, $F$, generated by $UE \circ Y$ satisfies

$$\dim(UF)_k \leq e^{\left(\beta + \frac{2}{s}\log(s+1)\right)k}, \qquad k \geq 0.$$

Since $[E, F] \subset F$ it follows that $G = E + F$. Thus multiplication in $UG$ gives a surjective map $UE \otimes UF \to UG$, whence for $k \geq s \geq 1$,

$$\begin{aligned}\dim(UG)_k \leq \textstyle\sum_i \dim(UE)_i \dim(UF)_{k-i} &\leq k\, e^{\beta k}\, e^{\left[\frac{2}{s}\log(s+1)\right]k} \\ &\leq e^{\left(\beta + \frac{3}{s}\log(s+1)\right)k}.\end{aligned}$$

Now, if $k < s$, $\dim UG)_k = \dim\,(UE)_k \leq e^{\beta k} \leq e^{(\beta + \frac{3}{s}\log(s+1))k}$.

Finally, if $Y = Y_s$ then for $k \geq s$, $(UE \circ Y)_k = (UE)_{k-s} \circ Y_s$ and so $\dim(UE \circ Y)_k \leq e^{\beta k}$. Since $(UE \circ Y)_k = 0$ for $k < s$ we obtain from Lemma 13.2 that $\dim(UF)_k \leq e^{\left(\beta + \frac{\log(s+1)}{s}\right)k}$, $k \geq s$; clearly $(UF)_k = 0$ for $1 \leq k < s$. The same argument as above now gives the final assertion.

□

*proof of Proposition 13.3:* We construct a sequence of subspaces $W_{r_i} \subset L_{r_i}$, $i \geq 0$ such that

$$\frac{1}{2}\, e^{\lambda_i r_i} \leq \dim W_{r_i} \leq e^{\lambda_i r_i}, \qquad i \geq 0, \tag{13.2}$$

and such that the sub Lie algebra $E(i)$ generated by $W_{r_0}, \ldots, W_{r_i}$ satisfies

$$\dim[UE(i)]_k \leq e^{(\lambda_i + 2\frac{\log(r_i+1)}{r_i})k}, \quad k \geq 0 \tag{13.3}$$

and

$$E(i) = [E(i), E(i)] \oplus \left(\oplus_{j=0}^{i} W_{r_j}\right), \quad i \geq 0. \tag{13.4}$$

It is then immediate from (13.2) and (13.4) that $E = \cup_i E(i)$ satisfies the conditions of the Proposition.

The construction of the $W_{r_i}$ is by induction on $i$. By hypothesis on $L$ we may choose $W_{r_0}$ to satisfy (13.2). Apply Lemma 13.3 with $s = r_0$, $E = 0$ $Y = W_{r_0}$ and $\beta = \lambda_0$ to conclude that $UE(0)$ satisfies (13.3). In this case, (13.4) is immediate.

Suppose $W_{r_0}, \dots, W_{r_\ell}$ are constructed. Then by (13.3) and the hypothesis on the $\lambda_i$,

$$\dim[UE(\ell)]_{r_{\ell+1}} \le \frac{1}{3}\, e^{\lambda_{\ell+1} r_{\ell+1}} .$$

Thus, since $\dim L_{r_{\ell+1}} \ge e^{\lambda_{\ell+1} r_{\ell+1}}$, we may choose $W_{r_{\ell+1}} \subset L_{r_{\ell+1}}$ so that (13.2) holds for $i = \ell+1$ and so that $W_{r_{\ell+1}} \cap E(\ell)_{r_{\ell+1}} = 0$. Then (13.4) is immediate for $i = \ell+1$. Since $\lambda_\ell + 2\frac{\log(r_\ell+1)}{r_\ell} \le \lambda_{\ell+1}$, we may apply the final assertion of Lemma 13.3 with $Y = W_{r_{\ell+1}}$, $\beta = \lambda_{\ell+1}$ and $s = r_{\ell+1}$ to obtain (13.3) for $i = \ell+1$.

□

## 13.3 Weak exponential growth and critical degree

In this section $(\wedge V, d)$ is a minimal Sullivan algebra for which $V = V^{\ge 2}$ has finite type, $\sum \dim V^k = \infty$, and $\operatorname{cat}(\wedge V, d) = m < \infty$. Its homotopy Lie algebra will be denoted by $L$. If $r < s$, we write $L_{[r,s]} = \{L_i\}_{r \le i \le s}$ and $V^{[r,s]} = \{V^j\}_{r \le j \le s}$.

Our principal purpose is to prove

**Theorem 13.1.** *With the hypotheses above there is a sequence of integers $r_0 < r_1 < \cdots$ such that $r_i < r_{i+1} < (m+1)(r_i+1)$ and*

$$\textit{log index}\, L = \lim_i \frac{\log \textit{dim}\, L_{r_i}}{r_i} .$$

To prove the Theorem we require the three Propositions below.

**Proposition 13.4.** *With the hypotheses and notation above,*

*(i)* $L_{even}$ *is infinite dimensional.*
*(ii) There is an even integer $r_0$ such that for each non-zero element $x \in L_{\ge r_0}$ of even degree there is some $y \in L$ such that the iterated Lie brackets $[x,[x,\cdots[x,y]\cdots]]$ are all non-zero. Moreover $y$ may be chosen so that $y \notin [L,L]$.*
*(iii) There are integers $n_0$ and $r_0$ such that $L_{n_0+kr_0} \ne 0$, for all $k \ge 0$.*
*(iv) With $n_0$ and $r_0$ as in (iii), $\textit{dim}\, L_{[n_0+kr_0,\, n_0+2kr_0-1]} \ge k$.*
*(v) If $H^{>N}(\wedge V, d) = 0$, some $N < \infty$, then for each $k \ge 0$, $L_{[k(N-1),(k+1)(N-1)]} \ne 0$.*

*proof:* (i) By Proposition 11.1(iii) (p. 302), depth $L \leq m$. If dim $L_{even} < \infty$ then $L$ is solvable and hence finite dimensional by Theorem 11.1 (p. 303). Thus $L_{even}$ is infinite dimensional.

(ii) The first assertion is Corollary 11.1(ii) (p. 303). Write $L = S \oplus [L, L]$. If ad $x$ acts locally nilpotently on $S$ then it does so in $L$, because $S$ generates $L$. Thus we may choose $y \in S$.

(iii) follows directly from (ii).

(v) This follows from the proof of the Gap Theorem 12.3 (p. 315).

□

**Proposition 13.5.**

*(i) Suppose $r_0, n_0$ are as in Proposition 13.4(iii). Then for any $K$,*

$$dim\, V^j > K\,, \qquad some\ j \leq (m+1)[n_0 + 4r_0^2(m+1)^{m+3}K]\,.$$

*(ii) If $H^{>N}(\wedge V, d) = 0$ then for any $K$,*

$$dim\, V^j > K\,, \qquad some\ j \leq 4N^{N+5}K\,.$$

**Definition.** The *critical degree* for a simply connected minimal Sullivan algebra $(\wedge V, d)$ is the least integer $\sigma$ such that dim $V^{1+\sigma} > [2(m+1)]^{m+1}$.

**Corollary 13.1.** *The critical degree $\sigma$ for $(\wedge V, d)$ satisfies*

$$\sigma < (m+1)[n_0 + 2^{m+3}r_0^2(m+1)^{2m+4}]\,.$$

*If $H^{>N}(\wedge V, d) = 0$ then*

$$\sigma < 2^{N+2}N^{2N+5}\,.$$

The next lemma is the key technical step in establishing a weak form of exponential growth for the integers dim $L_j$.

**Lemma 13.4.** *Suppose for integers $0 < q < p \leq 2q$ that $dim\, V^{[q+1,p]} = s \geq 2m$, and set*

$$\lambda_i = \max_{i(q+1) \leq j \leq ip} dim\, V^{j-1}\,.$$

*Then*

$$\sum_{i=2}^{m+1} \frac{\lambda_i}{s^i}\,[i(p-q-1)+1] > \frac{1}{2(m+1)^{m+1}}\,.$$

*proof:* Division by the ideal generated by $V^{\leq q}$ defines a quotient Sullivan algebra $(\wedge V^{\geq q+1}, \overline{d})$. Moreover, the Mapping Theorem 9.3 (p. 248) gives

$$\operatorname{cat}(\wedge V^{\geq q+1}, \overline{d}) \leq m\,.$$

In particular, $V^{[q+1,p]} \subset \ker \overline{d}$, and hence

$$\wedge^{m+1} V^{[q+1,p]} \subset \operatorname{Im} \overline{d}\,.$$

Choose linear maps of degree 1, $\theta_i : V^{>p} \to \wedge^i V^{[q+1,p]}$, $i \geq 2$, such that $\overline{d} - \sum_i \theta_i : V^{>p} \to V^{>p} \wedge (\wedge V^{\geq q+1})$. Then

$$\wedge^{m+1} V^{[q+1,p]} \subset \sum_{i=2}^{m+1} \theta_i(V^{>p}) \wedge \left(\wedge^{m+1-i} V^{[q+1,p]}\right)\,.$$

Since $\wedge^i V^{[q+1,p]}$ is concentrated in degrees $[i(q+1), ip]$ it follows that

$$\wedge^{m+1} V^{[q+1,p]} \subset \sum_{i=2}^{m+1} \theta_i(V^{[i(q+1)-1, ip-1]}) \wedge \left(\wedge^{m+1-i} V^{[q+1,p]}\right)\,.$$

Next, since $s \geq m+1$,

$$\dim \wedge^{m+1} V^{[q+1,p]} \geq \binom{s}{m+1} \quad \text{and} \dim \wedge^{m+1-i} V^{[q+1,p]} \leq s^{m+1-i}\,.$$

It follows that

$$\sum_{i=2}^{m+1} \dim V^{[i(q+1)-1, ip-1]} s^{m+1-i} \geq \binom{s}{m+1}\,,$$

and so

$$\sum_{i=2}^{m+1} \lambda_i \,(i(p-q-1)+1) s^{m+1-i} \geq \binom{s}{m+1}\,.$$

On the other hand, induction on $m$ gives the inequality $(m+1)! \leq 2\left(\frac{m+1}{2}\right)^{m+1}$. Thus, since $s \geq 2m$, we have

$$\binom{s}{m+1} > \frac{(s/2)^{m+1} 2^m}{(m+1)^{m+1}} = \frac{s^{m+1}}{2(m+1)^{m+1}}\,.$$

The Lemma follows from these inequalities.

□

*proof of Proposition 13.5*: Set $q = n_0 + kr_0$ and $p = n_0 + 2kr_0 - 1$, with $k \geq 2m$. Then by Proposition 13.4(iv),

$$s = \dim V^{[q+1,p]} \geq k \geq 2m\,.$$

Now set $\lambda = \max \dim V^j$, $2q+1 \le j \le (m+1)p$. Thus, in the notation of Lemma 13.4, $\lambda_i \le \lambda$, $2 \le i \le m+1$. Since $1/s^i \le 1/s^2$, Lemma 13.4 gives

$$m\frac{\lambda}{s^2}\left[(m+1)(p-q)\right] > \frac{1}{2(m+1)^{m+1}} .$$

It follows that

$$\lambda > \frac{s^2}{2(p-q)m(m+1)^{m+2}}$$

and so, since $s \ge k$

$$\lambda > \frac{k}{2r_0(m+1)^{m+3}} . \tag{13.5}$$

But by definition, $\lambda = \dim V^j$ for some $j \le (m+1)p$, and thus (i) follows from (13.5).

(ii) Suppose $H^{>N}(\wedge V, d) = 0$. Thus $(\wedge V, d)$ is quasi-isomorphic to a commutative cochain algebra $(A, d)$ such that $A^+ = A^{[2,N]}$. It follows that $(A^+)^k = 0$ if $k > N/2$ and so it follows from Proposition 9.1(i) (p. 246) that $m \le N/2$. Moreover, in Proposition 13.4(iv) we may, by Proposition 13.4(v), take $n_0 = 0$ and $r_0 = N-1$ in applying part (i) of this Proposition. When $N \ge 4$, since $m+1 \le N$ (ii) is immediate from (i). When $N = 2$ or 3 we have $r_0 = 1$ or 2 and $m = 1$, and so (ii) follows from (i) by a direct calculation.

□

**Proposition 13.6.** *(Weak exponential growth) Suppose for some $q$ (e.g. $q$ = critical degree) that $\dim V^{q+1} > [2(m+1)]^{m+1}$. Then $q$ extends to an infinite sequence $q = q_0 < q_1 < \cdots$ such that for each $i \ge 0$, $q_{i+1}+1 = \ell_i(q_i+1)-1$ with $2 \le \ell_i \le m+1$, and such that*

$$\dim L_{q_i} > [2(m+1)]^{m+1}, \qquad \dim L_{q_i} \ge \left[\frac{1}{2(m+1)}\right]^{m+1} \left(\dim L_{q_{i-1}}\right)^{\ell_i - 1},$$

*and*

$$\dim L_{q_i} \ge \left(\left[\frac{1}{2(m+1)}\right]^{m+1} \dim L_q\right)^{\frac{q_i+1}{q+1}} .$$

*proof:* By hypotheses, $\dim V^{q_0+1} > [2(m+1)]^{m+1}$. Now suppose the $q_i$ are constructed for $i \le k$ and set $\dim V^{q_k+1} = s$. By construction, $s \ge$

$[2(m+1)]^{m+1} \geq 2m$, and so we may apply Lemma 13.4 (p. 338) with $q = q_k$ and $p = q_k + 1$. Since then $p - q - 1 = 0$ we obtain

$$\sum_{i=2}^{m+1} \frac{\lambda_i}{s^i} > \frac{1}{2(m+1)^{m+1}},$$

where $\lambda_i = \dim V^{i(q_k+1)-1}$.

In particular, some $\lambda_j$ must satisfy $\lambda_j > \frac{s^j}{2(m+1)^{m+2}}$. We set $\ell_k = j$ and $q_{k+1} = \ell_k(q_k + 1) - 2$. Then

$$\begin{aligned}\dim V^{q_{k+1}+1} &> \frac{(\dim V^{q_k+1})^{\ell_k}}{2(m+1)^{m+2}} \geq \left(\frac{1}{2(m+1)}\right)^{m+1} \left(\dim V^{q_k+1}\right)^{\ell_k} \\ &\geq \left[\frac{1}{2(m+1)}\right]^{m+1} [2(m+1)]^{2(m+1)},\end{aligned}$$

which proves the first two inequalities.

Now set

$$a = \left(\frac{1}{2(m+1)}\right)^{m+1},$$

and $n = 1 + \ell_k + \cdots + \ell_2 \cdots \ell_k$. Then it follows from the second inequality that $\dim V^{q_{k+1}+1} \geq a^n \cdot (\dim V^{q+1})^{\ell_1 \ell_2 \cdots \ell_k}$. Since each $\ell_i \geq 2$, $n/(\ell_1 \ell_2 \cdots \ell_k) \leq \frac{1}{2^k} + \cdots + \frac{1}{2} < 1$. Thus

$$\dim V^{q_{k+1}+1} \geq \left(\left[\frac{1}{2(m+1)}\right]^{m+1} \dim V^{q+1}\right)^{\ell_1 \ell_2 \cdots \ell_k}.$$

But $\frac{q_{k+1}+1}{q+1} \leq \ell_1 \ell_2 \cdots \ell_k$ and this proves the third inequality.

□

**Corollary 13.2.** *Given a sequence $(q_i)$ as in the Proposition then for all $i > j \geq 0$,*

$$\frac{\log \mathit{dim}\, L_{q_i}}{q_i + 1} \geq \frac{\log \mathit{dim}\, L_{q_j}}{q_j + 1} - \frac{(m+1)\log 2(m+1)}{q_j + 1}.$$

*In particular, if $s > q_0$ then*

$$\max_{s+1 \leq i < (m+1)(s+1)} \frac{\log \mathit{dim}\, L_i}{i+1} \geq \frac{\log \mathit{dim}\, L_{q_0} - (m+1)\log 2(m+1)}{q_0 + 1}.$$

*proof:* The first inequality is immediate and the second follows because some $q_i \in (s, (m+1)(s+1))$.

□

**Corollary 13.3.**

$$\log \mathit{index}\, L \geq \sup_k \left(\frac{\log \mathit{dim}\, L_k}{k+1} - \frac{(m+1)\log 2(m+1)}{k+1}\right).$$

**Corollary 13.4.** *Log index* $L > 0$.

*proof:* Since $q_{i+1} = \ell_i(q_i+1) - 2 \le (m+1)(q_i+1) - 2 \le (m+2)q_i$, Proposition 13.6 implies that $\dim L_{q_i} > C^{q_i}$, where $C = (a \dim L_q)^{\frac{1}{q+1}}$. It follows that $\log \operatorname{index} L > 0$.

□

**Corollary 13.5.** *There is some* $K > 1$ *and some integer* $n$ *such that*

$$\sum_{i=0}^{k} \dim L_i \ge K^k\,, \qquad k \ge n\,.$$

*proof:* Set $n = q$, so that $\dim L_q > \frac{1}{a} = [2(m+1)]^{m+1}$, and set $C = (a \dim L_q)^{\frac{1}{q+1}}$, and $K = C^{\frac{1}{m+2}}$. Then

$$\dim L_{q_i} \ge C^{q_i+1}\,.$$

For an integer $k \ge n$ choose $i$ so that $q_i \le k < q_{i+1}$. Then

$$\sum_{i=1}^{k} \dim L_i \ge \dim L_{q_i} \ge C^{q_i+1} = \left(C^{\frac{1}{\ell_i}}\right)^{q_{i+1}+2} \ge K^k\,.$$

□

*proof of Theorem 13.1:* We construct the sequence $r_i$ as follows. Set $r_0 = \sigma$ and then, given $r_i$, let $r_{i+1} \in (r_i, (m+1)(r_i+1))$ be chosen so that for $r \in (r_i, (m+1)(r_i+1))$

$$\frac{\log \dim L_r - (m+1)\log(2(m+1))}{r} \le \frac{\log \dim L_{r_{i+1}} - (m+1)\log(2(m+1))}{r_{i+1}}\,.$$

It follows from Proposition 13.6 that for all $i$, $\dim L_{r_i} > (2(m+1))^{m+1}$.

On the other hand, for any $\varepsilon > 0$ there are infinitely many integers $p$ such that $\frac{\log \dim L_p}{p} > \log \operatorname{index} L - \varepsilon$. Each such $p$ must belong to an interval $(r_{i-1}, (m+1)(r_{i-1}+1))$, and thus

$$\frac{\log \dim L_p - (m+1)\log 2(m+1)}{p} \le \frac{\log \dim L_{r_i} - (m+1)\log 2(m+1)}{r_i}\,.$$

It follows that

$$\frac{\log \dim L_{r_i}}{r_i} > \log \operatorname{index} L - \varepsilon - (m+1)\log 2(m+1)\left(\frac{1}{r_{i-1}} - \frac{1}{r_i}\right)\,.$$

For $p$ (and therefore $r_i$) sufficiently large, $\frac{(m+1)\log 2(m+1)}{r_{i-1}} < \varepsilon$.

$$\frac{\log \dim L_{r_i}}{r_i} > \log \operatorname{index} L - 2\varepsilon\,.$$

On the other hand, Corollary 13.2 extends $r_i$ to a sequence of integers $r_i = q_0 < q_1 < \cdots$ such that $2q_j \leq q_{j+1} < (m+1)(q_j+1)$ and

$$\frac{\log \dim L_{q_j}}{q_j+1} \geq \frac{\log \dim L_{r_i}}{r_i+1} - \frac{(m+1)\log 2(m+1)}{r_i+1}.$$

For any $k \geq i$, $(r_k, (m+1)(r_k+1))$ must contain some $q_j$ and it follows as above that

$$\frac{\log \dim L_{r_{k+1}}}{r_{k+1}} \geq \frac{\log \dim L_{r_i}}{r_i+1} + (m+1)\log 2(m+1)\left[\frac{1}{r_{k+1}} - \frac{1}{q_j+1} - \frac{1}{r_i+1}\right]$$
$$\geq \log \operatorname{index} L - 4\varepsilon\,.$$

Thus since $\log \operatorname{index} L = \limsup \frac{\log \dim L_k}{k}$ it follows that

$$\frac{\log \dim L_{r_i}}{r_i} \longrightarrow \log \operatorname{index} L\,.$$

□

## 13.4 Approximation of log index $L$

In this section $(\wedge V, d)$ is a minimal Sullivan algebra satisfying the following conditions:

$$\begin{cases} V = V^{\geq 2} \ \text{ has finite type and } \sum_{i\geq 2} \dim V^i = \infty\,, \\ \operatorname{cat}(\wedge V, d) = m < \infty\,, \end{cases} \tag{13.6}$$

and we fix the following notation:

$$\begin{cases} L \text{ is the homotopy Lie algebra of } (\wedge V, d) \\ \sigma = \text{critical degree of } (\wedge V, d) \\ \alpha_L = \ \log \operatorname{index} L. \end{cases}$$

**Theorem 13.2.** *Let $L$ be the homotopy Lie algebra of a minimal Sullivan algebra, $(\wedge V, d)$, satisfying (13.6). Suppose $H^{>N}(\wedge V, d) = 0$ and $\dim H^i(\wedge V, d) \leq h$, $i \leq N$. Then there are positive constants $B_0$ and $C_0$, depending only on $N$ and $h$, and such that if*

$$\frac{\log r}{2\log Nh} \geq \max\{4, \sigma, m+1\}$$

*then*

$$\max_{r\leq i<(m+1)r} \frac{\log \dim L_i}{i} - \frac{B_0}{r} \leq \alpha_L \leq \max_{r\leq i<(m+1)r} \frac{\log \dim L_i}{i} + \frac{C_0}{\log r}\,.$$

**Remark.** Suppose $(C^*, \delta)$ is a finite dimensional rational cochain algebra connected by a chain of quasi-isomorphisms with $C^*(X;\mathbb{Q})$, where $X$ is a rationally hyperbolic, simply connected, $N$-dimensional CW complex (e.g. the reduced simplicial cochains on a simply connected polyhedron). Then a classical result of Adams-Hilton ( [1]) shows that the dual of the bar construction on $(C^*, \delta)$ provides a chain complex of finite type whose homology is $H_*(\Omega X;\mathbb{Q} = UL_X$.

Thus Theorem 13.2 shows that log index $L_X$ is *computable*: that is it can be computed up to predetermined accuracy by an a priori bounded set of computations.

In Theorem 13.3 below we extend Theorem 13.2 to a second situation, and consider the two Scenarios below together.

**Scenario A** : $\dim H(\wedge V, d) < \infty$, and
**Scenario B** : $\log \operatorname{index} H(\wedge V, d) < \log \operatorname{index} L$.

Essentially a single proof works for both Scenarios, and in the case of Scenario A, Theorem 13.3 reduces to Theorem 13.2. In both cases we will assume $\log \operatorname{index} L < \infty$, which by Proposition 13.2 is equivalent to $\log \operatorname{index} H(\wedge V, d) < \infty$.

To consider both Scenarios at the same time we need to establish notation which will have different meanings depending on which Scenario is under consideration. This is as follows, with the Scenario in question indicated by (A) or (B).

$$(A)\begin{cases} N = \max\{i \mid H^i(\wedge V, d) \neq 0\}, \quad \text{and } h = \max_{i\leq N} \dim H^i(\wedge V, d). \\ n = Nh. \\ s_* = \max\{5, \sigma, m+1\}. \end{cases}$$

$$(B)\begin{cases} \alpha_H = \log \operatorname{index} H(\wedge V, d), \quad \text{and } \Delta = \alpha_L - \alpha_H. \\ n \text{ is an integer such that } \log n \geq \dfrac{\log \dim L_i}{i}, \quad i \geq 1 \ \text{(Prop. 13.2)}. \\ s_* \text{ is a fixed integer such that } s_* \geq \max\{5, \sigma, m+1, n\}, \\ \sup_{i\geq s_*} \dfrac{\log \dim H^i(\wedge V, d)}{i} \leq \alpha_H + \dfrac{\Delta}{3}, \quad \text{and} \\ \dfrac{\log \dim L_{s_*} - (m+1)\log 2(m+1)}{s_*+1} \geq \alpha_L - \dfrac{\Delta}{3}. \end{cases}$$

**Lemma 13.5.** *In the case of Scenario A,*

*(i)* $m \leq N/2$.
*(ii)* $\dim L_k \leq (Nh)^k$, $k \geq 1$.
*(iii)* $0 < \alpha_L \leq \log(Nh)$.

*proof:* (i) is established in the proof of Proposition 13.5(ii) (p. 340). (ii) follows from Proposition 13.2 (p. 332) because $N \geq 2$ and $h \geq \omega$. Finally Corollary 13.3 gives $\alpha_L > 0$ and (ii) gives $\alpha_L \leq Nh$.

□

The following relationships are then immediate in either Scenario.

$$\left.\begin{array}{l} n \geq 2\,. \\ \dim L_k \leq n^k\,, \quad k \geq 1\,. \\ \alpha_L \leq \log n\,. \end{array}\right\} \tag{13.7}$$

**Theorem 13.3.** *Let $L$ be the homotopy Lie algebra of a minimal Sullivan algebra $(\wedge V, d)$ satisfying (13.6). If either $\dim H(\wedge V, d) < \infty$ or if log index $H(\wedge V, d) <$ log index $L < \infty$, then there are positive constants $B_0, C_0$ such that for any $r \geq n^{2s_*}$,*

$$\max_{r \leq i < (m+1)r} \frac{\log \dim L_i}{i} - \frac{B_0}{r} \leq \alpha_L \leq \max_{r \leq i < (m+1)r} \frac{\log \dim L_i}{i} + \frac{C_0}{\log r}\,.$$

*Moreover, in the case of Scenario A, the constants $B_0$ and $C_0$ may be chosen to depend only on $N$ and $h$.*

Before the proof of Theorem 13.3 we require three Lemmas.

**Lemma 13.6.** *In the case of Scenario B, and with the hypotheses and notation above, for any $s \geq s_*$ and $r > n^{2(s-1)}$,*

*(i)*
$$\alpha_L - \frac{\Delta}{3} \leq \max_{s < i < r} \frac{\log \dim L_i}{i}\,.$$

*(ii)*
$$\sup_{s \leq i} \frac{\log \dim H^i(\wedge V, d)}{i} \leq \max_{s < i < r} \frac{\log \dim L_i}{i} - \frac{\Delta}{3}\,.$$

*proof:* Since $s \geq s_* \geq \sigma$, by Corollary 13.2 (p. 341),

$$\max_{(s+1)\leq i<(m+1)(s+1)} \frac{\log\dim L_i}{i} \geq \frac{\log\dim L_{s_*} - (m+1)\log 2(m+1)}{s_*+1}$$
$$\geq \alpha_L - \frac{\Delta}{3}.$$

On the other hand, since $n \geq 2$ and $s \geq s_* \geq m+1$ it follows that $(m+1)(s+1) \leq n^{2(s-1)}$. This proves (i), and (ii) is immediate, since for $i \geq s$,

$$\frac{\log\dim H^i(\wedge V, d)}{i} \leq \alpha_H + \frac{\Delta}{3} \leq \alpha_L - \frac{2\Delta}{3} \leq \max_{s\leq i<r} \frac{\log\dim L_i}{i} - \frac{\Delta}{3}.$$

□

**Lemma 13.7.** *If $s \geq s_*$ and $r > n^{2(s-1)}$ then*

$$\alpha_L \leq \sup_{s<i} \frac{\log dim\, L_i}{i} < \max_{s<i<r} \frac{\log dim\, L_i}{i} + \frac{C_1}{\log r}$$

*where $C_1 > 0$ is a constant. In Scenario A, the constant $C_1$ may be chosen to depend only on $N$ and $h$.*

*proof:* Set

$$c = \max_{i\leq s} \dim L_i\,, \qquad \mu_r = \max_{s<i<r} \frac{\log\dim L_i}{i} \qquad \text{and } \mu_\infty = \sup_{s<i} \frac{\log\dim L_i}{i}.$$

We show first that for $k \in [r, 2r-1]$,

$$\dim L_k \leq q(k+2)^{cs} e^{\mu_r(k+1)+3\sqrt{k+1}}\,, \tag{13.8}$$

where $q = n$ in Scenario A and

$$q = \dim H^{<s}(\wedge V, d) + 1/(1-e^{-\Delta/3})$$

in Scenario B.

For this, let $(\wedge V^{\leq r} \otimes \wedge U^{<r}, D)$ be an acyclic closure (§3.6) of $(\wedge V^{\leq r}, d)$. By Proposition 3.12 (p. 111),

$$\dim U^i = \dim V^{i+1} = \dim L_i\,, \qquad i < r\,.$$

Applying $\wedge V \otimes_{\wedge V^{\leq r}} -$ we obtain the Sullivan algebra

$$(\wedge V \otimes \wedge U^{<r}, D) = (\wedge V^{\leq r} \otimes \wedge U^{<r} \otimes \wedge V^{\geq r+1}, D)\,.$$

Since $H(\wedge V^{\leq r} \otimes \wedge U^{<r}, D) = \Bbbk$, this yields a quasi-isomorphism

$$(\wedge V \otimes \wedge U^{<r}, D) \xrightarrow{\simeq} (\wedge V^{\geq r+1}, \overline{d})$$

where $(\wedge V^{\geq r+1}, \overline{d})$ is the quotient of $(\wedge V, d)$ by the ideal generated by $V^{\leq r}$. Moreover, since $(\wedge V^{\geq r+1}, \overline{d})$ is minimal, it follows that $V^{[r+1,2r]}$ embeds in $H(\wedge V^{\geq r+1}, \overline{d})$.

Now by definition $\dim U^{[1,s]} \leq cs$. Thus if $n_i$ is the number of monomials of degree $i$ in $cs$ variables, then $\dim(\wedge U^{[1,s]})^i \leq n_i$. But

$$n_i = \binom{i+cs-1}{i} = \binom{i+cs-1}{cs-1}$$

is also the number of monomials of degree $cs-1$ in $i+1$ variables, and so

$$\dim\,(\wedge U^{[1,s]})^i \leq n_i \leq (i+1)^{cs-1}\,.$$

On the other hand,

$$\dim(\wedge U^{[s+1,r-1]})^j \leq \dim(\otimes_{i=s+1}^{r-1} TU^i)^j \leq \rho(j)e^{\mu_r j}$$

where $\rho(j)$ is the number of partitions of $j$; i.e. the number of solutions of $\sum ip_i = j$. According to [64], $\rho(k) \leq e^{3\sqrt{k}}$. Thus,

$$\dim(\wedge U)^\ell \leq (\ell+1)^{cs}e^{\mu_r \ell}e^{3/\sqrt{\ell}}\,, \qquad \ell \geq 1\,.$$

Now filtering by the degree in $\wedge V$ gives a spectral sequence converging from $H(\wedge V, d) \otimes H(\wedge U^{<r}, \overline{d})$ to $H(\wedge V^{\geq r+1}, \overline{d})$. Thus for $k \in [r, 2r-1]$,

$$\begin{aligned}\dim L_k &= \dim V^{k+1} \leq \dim H^{k+1}(\wedge V \otimes \wedge U^{<r}, D)\\ &\leq \textstyle\sum_{i=0}^{k+1} \dim H^i(\wedge V, d)\dim\,(\wedge U^{<r})^{k+1-i}\,.\end{aligned}$$

In Scenario A, by (13.7) this yields

$$\begin{aligned}\dim L_k &\leq n \max_{\ell \leq k+1} \dim\,(\wedge U^{<r})^\ell\\ &< n(k+2)^{cs}e^{\mu_r(k+1)}e^{3\sqrt{k+1}}\,,\end{aligned}$$

which gives (13.8).

In Scenario B this yields

$$\begin{aligned}\dim L_k \leq{}& \dim H^{<s}(\wedge V, d)(k+2)^{cs}e^{\mu_r(k+1)}e^{3\sqrt{k+1}}\\ &+ \textstyle\sum_{i=s}^{k+1} \dim H^i(\wedge V, d)(k+2)^{cs}e^{\mu_r(k+1-i)}e^{3\sqrt{k+1}}\,.\end{aligned}$$

But by Lemma 13.6 for $i \geq s$:

$$\dim H^i(\wedge V, d) \leq e^{-i\Delta/3}e^{\mu_r i}\,.$$

Since

$$\sum_{i=s}^{k+1} e^{-i\Delta/3} \leq \frac{1}{1-e^{-\Delta/3}}\,,$$

substitution of this in the previous inequality gives (13.8).

Next recall from (13.7) that $\frac{\log \dim L_i}{i} \leq \log n$, $i \geq 1$. Thus from (13.8) we obtain

$$\frac{\log\ \dim L_k}{k} \leq \frac{\log q}{k} + cs\frac{\log(k+2)}{k} + \mu_r + \frac{\mu_r}{k} + 3\frac{\sqrt{k+1}}{k}, \qquad k \in [r, 2r-1] .$$

Since $\mu_r \leq \log n$ and $\log(k+2) \leq 2\log k$ for $k \geq 2$ and since $\frac{\log x}{x}$ is a decreasing function for $x \geq e$, it follows that for $k \in [r, 2r-1]$,

$$\frac{\log \dim L_k}{k} \leq \mu_r + \frac{\log n + \log q}{r} + cs\frac{2\log r}{r} + 3\sqrt{\frac{2}{r}} .$$

Therefore

$$\mu_{2r} - \mu_r \leq \frac{\log n + \log q}{r} + cs\frac{2\log r}{r} + 3\sqrt{\frac{2}{r}} .$$

Thus, because the previous inequalities hold with any $r' \geq r$, we may replace $r$ by $2^i r$ in this inequality, and sum over $i \geq 1$ to obtain

$$\begin{aligned}\mu_\infty - \mu_r \leq & \left(\sum_{i=0}^{\infty} \frac{1}{2^i}\right) \frac{\log n + \log q}{r} + \left(\sum_{i=0}^{\infty} \frac{\log 2^i}{2^i}\right) \frac{2cs\log r}{r} \\ & + \left(3\sum_{i=0}^{\infty} \sqrt{\frac{2}{2^i}}\right) \frac{1}{\sqrt{r}} .\end{aligned}$$

Finally, by (13.7) $c \leq n^s$. Since $\frac{\log x}{x}$ is a decreasing function for $x \geq e$ and $s \geq s_* \geq n$, it follows that $s \leq n^{s/2}$. Moreover, since $s \geq s_* \geq 5$, $\frac{s}{s-1} \leq 5/4$. Thus, because $r \geq n^{2(s-1)}$, we have

$$cs \leq n^{\frac{3s}{2}} \leq r^{15/16} .$$

But for some positive constant, $A$, independent of $(\wedge V, d)$, and for all $r \geq 1$,

$$\frac{r^{15/16}\log r}{r} < \frac{A}{\log r} .$$

Now substitution in the previous inequality gives the inequality of the Lemma.

□

**Lemma 13.8.** *Suppose integers $s$ and $r$ satisfy $s \geq s_*$ and $r \leq n^{2s}$. Then*

$$\max_{s<i<r} \frac{\log dim\, L_i}{i} \leq \max_{s<i<r} \frac{\log dim\, L_i}{i+1} + \frac{2(\log n)^2}{\log r} .$$

*proof*: By (13.7) and the hypothesis above we have that

$$\frac{\log \dim L_i}{i} \le \log n\,, \quad i \ge 1 \qquad \text{and} \quad s \ge \frac{\log r}{2\log n}\,.$$

It follows that for $i > s$,

$$\begin{aligned}\frac{\log \dim L_i}{i} &= \frac{i+1}{i}\frac{\log \dim L_i}{i+1} < \frac{s+1}{s}\frac{\log \dim L_i}{i+1} \\ &< \frac{\log \dim L_i}{i+1} + \frac{2(\log n)^2}{\log r}\,.\end{aligned} \tag{13.9}$$

□

*proof of Theorem 13.3*: Denote by $s$ the integer defined by

$$\frac{\log r}{2\log n} \le s < \frac{\log r}{2\log n} + 1\,. \tag{13.10}$$

Then $n^{2(s-1)} < r \le n^{2s}$ and, since by hypothesis $r \ge n^{2s_*}$, $s_* \le s$. Since $m+1 \le s_* \le s$ and since $n \ge 2$ and $s_* \ge 4$ it follows that $(s, (m+1)(s+1)) \subset (s, r)$.

On the other hand, by the definition of the critical degree, $\dim L_\sigma > (2(m+1))^{m+1}$. Since $s \ge \sigma$ it follows from Proposition 13.6 (p. 340) that for some $\ell \in (s, (m+1)(s+1))$ we have $\dim L_\ell \ge (2(m+1))^{m+1}$.

Now let $i_0 \in (s, r)$ be an integer for which

$$\frac{\log \dim L_{i_0} - (m+1)\log 2(m+1)}{i_0+1} = \max_{s<i<r} \frac{\log \dim L_i - (m+1)\log 2(m+1)}{i+1}\,.$$

Then $\log \dim L_{i_0} - (m+1)\log 2(m+1) > 0$. Thus it follows from Proposition 13.6 that $i_0$ extends to an infinite sequence $i_0 < i_1 < \cdots$ for which, by Corollary 13.2 (p. 341),

$$\frac{\log \dim L_{i_\nu}}{i_\nu + 1} \ge \frac{\log \dim L_{i_0} - (m+1)\log 2(m+1)}{i_0+1}\,, \qquad \nu \ge 0\,.$$

Since $i_\nu \le (m+1)(i_{\nu-1}+1) - 2$ it follows that some $i_k$ satisfies $r \le i_k < (m+1)r$. Thus

$$\begin{aligned}\max_{r \le i < (m+1)r} \frac{\log \dim L_i}{i+1} \ge \frac{\log \dim L_{i_k}}{i_k+1} &\ge \frac{\log \dim L_{i_0} - (m+1)\log 2(m+1)}{i_0+1} \\ &= \max_{s<i<r} \frac{\log \dim L_i - (m+1)\log 2(m+1)}{i+1} \\ &\ge \max_{s<i<r} \frac{\log \dim L_i}{i+1} - \frac{(m+1)\log 2(m+1)}{s}\,.\end{aligned}$$

Now, by (13.10), $\frac{1}{s} \leq \frac{2\log n}{\log r}$. Apply Lemma 13.8 to obtain

$$\max_{r\leq i<(m+1)r} \frac{\log \dim L_i}{i+1} \geq$$
$$\max_{s<i<r} \frac{\log \dim L_i}{i} - \frac{2(\log n)^2 + 2(m+1)\log n \log 2(m+1)}{\log r}.$$

Thus it follows from Lemma 13.7 that

$$\alpha_L < \max_{r\leq i<(m+1)r} \frac{\log \dim L_i}{i} + \frac{C_0}{\log r},$$

where in Scenario A, the constant $C_0$ depends only on $N$ and $h$.

Finally, for $r \leq i < (m+1)r$, we have $\frac{i+1}{i} \geq 1 + \frac{1}{r}$. Thus since $\frac{\log \dim L_i}{i} \leq \log n$, Corollary 13.3 (p. 341) gives

$$\begin{aligned} \alpha_L &\geq \max_{r\leq i<(m+1)r} \frac{\log \dim L_i}{i+1} - \frac{(m+1)\log 2(m+1)}{r} \\ &\geq \max_{r\leq i<(m+1)r} \frac{\log \dim L_i}{i} - \frac{\log n + (m+1)\log 2(m+1)}{r}, \end{aligned}$$

and in Scenario A, $\log n = \log Nh$.

This completes the proof of Theorem 13.3, and with it that of Theorem 13.2.

□

## 13.5 Moderate exponential growth

This section is devoted to the next stage of the proof of Theorems 12.8 and 12.9 recalled at the start of this Chapter. For this, as in §13.4, we consider a minimal Sullivan algebra $(\wedge V, d)$ satisfying

$$\left.\begin{array}{l} V = V^{\geq 2} \text{ and } V \text{ has finite type.} \\ \operatorname{cat}(\wedge V, d) \leq m < \infty\,. \\ \log \operatorname{index} H(\wedge V, d) < \infty\,. \end{array}\right\} \tag{13.11}$$

In addition, as in §13.3, we adopt the notation

$$\left.\begin{array}{l} L = \text{homotopy Lie algebra of } (\wedge V, d)\,, \\ \alpha_L = \log \operatorname{index} L\,, \\ \alpha_H = \log \operatorname{index} H(\wedge V, d)\,, \\ \text{and if } \dim H(\wedge V, d) < \infty, \text{ then} \\ N = \max\{i \mid H^i(\wedge V, d) \neq 0\} \text{ and } h = \max\limits_{i\leq N} \dim H^i(\wedge V, d). \end{array}\right\} \tag{13.12}$$

**Theorem 13.4.** *Let $L$ be the homotopy Lie algebra of a minimal Sullivan algebra $(\wedge V, d)$ satisfying (13.11).*

*(i) For any $\varepsilon > 0$ there is an infinite sequence $q_0 < q_1 < \cdots$ of even integers such that*

$$\lim_j \frac{q_{j+1}}{q_j} = 1\,, \qquad \textit{and} \quad \frac{\log dim\, L_{q_j}}{q_j} \geq \alpha_L - \varepsilon\,.$$

*(ii) Suppose $\log index\, H(\wedge V, d) < \log index\, L$. Then there is an infinite sequence of even integers $q_0 < q_1 < \cdots$ and there are positive constants $C$, $B$ and a constant $M$ such that*

$$q_{j+1} \leq q_j + \sqrt{q_j} + M\,, \quad \textit{and} \;\; \alpha_L - \frac{C}{\log q_j} \leq \frac{\log dim\, L_{q_j}}{q_j} \leq \alpha_L + \frac{B}{q_j}\,.$$

*(iii) If $dim\, H(\wedge V, d) < \infty$ the constants $C$ and $B$ may be chosen to depend only on $N$ and $h$.*

We first prove the weaker version (Proposition 13.7) of Theorem 13.4 in which the integers $q_j$ are not required to be even.

**Proposition 13.7.** *Let $L$ be the homotopy Lie algebra of a minimal Sullivan algebra $(\wedge V, d)$ satisfying (13.11). Then in each of the situations (i), (ii) and (iii) of Theorem 13.4 there are infinite sequences $(p_j)$ of integers satisfying the conditions of the Theorem, but with the constants $B$ and $C$ replaced by $B/2$ and $C/2$. Moreover the constant $M$ in (ii) may be taken to be $-1$.*

The core of the proof of Proposition 13.7 is contained in the next two Lemmas, where always (i), (ii) and (iii) refer to those separate cases in Theorem 13.4.

**Lemma 13.9.** *In the situation of Proposition 13.7:*

*(i) Suppose given $\varepsilon > 0$. Then there is a sub Lie algebra $E \subset L$ and there are infinite sequences $r_0 \leq r_1 \leq \cdots$ of integers and $\lambda_0 < \lambda_1 < \cdots$ of numbers, such that for each $i$,*

$$2r_i < r_{i+1} < 2(m+1)(r_i + 1)\,, \qquad \lambda_i > \alpha_L - \varepsilon/2$$

*and*

$$\frac{1}{2} e^{\lambda_i r_i} \leq dim\,(E/[E,E])_{r_i} \leq e^{\lambda_i r_i}\,, \qquad i \geq 0\,. \qquad (13.13)$$

*(ii) If $\alpha_H < \alpha_L$ then sequences $r_i$ and $\lambda_i$ may be chosen to satisfy (13.13), with $r_i < r_{i+1} < (m+1)(r_i+1)$ and $\lambda_i = \alpha_L - F/\log r_i$, where $F$ is a positive constant.*

*(iii) If $H^{>N}(\wedge V, d) = 0$ then the constant $F$ in (ii) may be chosen to depend only on $N$ and $h$.*

*proof :* (i) Theorem 13.1 (p. 337) gives an infinite sequence $p_0 < p_1 < \cdots$ such that $p_0 > m+1$, $p_{i+1} < (m+1)(p_i+1)$, and for all $i$,

$$\frac{\log \dim L_{p_i}}{p_i} > \alpha_L - \varepsilon/4 .$$

Define $r_0 = p_0$, and for each $i$, let $r_i$ be the least $p_j$ for which $p_j > 2r_{i-1}$. Then since $2r_{i-1} + 1 \le r_i < (m+1)(p_{j-1}+1)$ it follows that $r_i < 2(m+1)(r_{i-1}+1)$. Moreover, $r_i > 2^i r_0$, and it follows that there are positive constants $a, b$ independent of the $r_i$ such that

$$\sum_{i\ge 0}\left(\frac{2\log(r_i+1)}{r_i} + \frac{\log 3}{r_{i+1}}\right) < \sum_{i\ge 0}\frac{ia\log r_0}{2^i r_0} < b\frac{\log r_0}{r_0} .$$

Thus for $r_0$ sufficiently large,

$$\sum_{i\ge 0}\left(\frac{2\log(r_i+1)}{r_i} + \frac{\log 3}{r_{i+1}}\right) < \varepsilon/4 .$$

Set

$$\varepsilon_i = \varepsilon - \sum_{j=0}^{i-1}\left(\frac{2\log(r_j+1)}{r_j} + \frac{\log 3}{r_{j+1}}\right) \quad \text{and} \quad \lambda_i = \alpha_L - \varepsilon_i .$$

Then

$$\alpha_L - \varepsilon_i < \alpha_L - \varepsilon/4 < \frac{\log \dim L_{r_i}}{r_i} \quad \text{and} \quad \varepsilon_{i+1} \le \varepsilon_i - \frac{2\log(r_i+1)}{r_i} - \frac{\log 3}{r_{i+1}} .$$

Now (i) follows from Proposition 13.3 (p. 333), with $\lambda_i = \alpha_L - \varepsilon_i$.

(ii) Theorem 13.3 (p. 345) gives an infinite sequence $r_0 < r_1 < \cdots$ with $r_0 > 2(m+1)$, and a positive constant, $C_0$, such that for each $i$, $r_i < r_{i+1} < (m+1)(r_i+1)$, and

$$\frac{\log \dim L_{r_i}}{r_i} \ge \alpha_L - \frac{C_0}{\log(r_{i-1}+1)} \ge \alpha_L - \frac{C_0}{\log r_i - \log(m+1)} .$$

Since $r_0 > 2(m+1)$, $\log r_i - \log(m+1) \ge \frac{1}{2}\log r_i$ and so

$$\frac{\log \dim L_{r_i}}{r_i} \ge \alpha_L - \frac{2C_0}{\log r_i} .$$

Since $r_i < r_{i+1} < (m+1)(r_i+1) < 2(m+1)r_i$ we obtain

$$\frac{1}{\log r_i} - \frac{1}{\log r_{i+1}} \leq \frac{\log 2(m+1)}{\log r_i \log r_{i+1}} .$$

It follows that for $r_0$ sufficiently large, and all $i \geq 0$,

$$\frac{2C_0}{\log r_i} - \frac{2\log(r_i+1)}{r_i} - \frac{\log 3}{r_{i+1}} > \frac{2C_0}{\log r_{i+1}} .$$

Set $F = 2C_0$. Then we may apply Proposition 13.3 (p. 333) with $\lambda_i = \alpha_L - \frac{F}{\log r_i}$ to obtain the Lemma in this case.

(iii) This is immediate, because according to Theorem 13.3 (p. 345), if $H^{>N}(\wedge V, d) = 0$, then the constant $C_0$ in (ii) depends only on $N$ and $h$.

□

**Lemma 13.10.** *Suppose $E \subset L$ is a sub Lie algebra of a graded Lie algebra $L = L_{\geq 0}$ of finite type, and that $0 < r_0 < r_1 < \cdots$ and $0 < \lambda_0 < \lambda_1 < \cdots$ are sequences such that for all $i$*

$$\frac{1}{2}e^{\lambda_i r_i} \leq dim\,(E/[E,E])_{r_i} \leq e^{\lambda_i r_i} .$$

*If $depth\, L \leq m$ there are positive integers $q$ and $n_0$ such that for any positive integer $k$ and any $r_i \geq q$,*

$$\max_{kr_i+1 \leq p \leq kr_i+n_0} \frac{\log dim\, L_p}{p} \geq \frac{\lambda_i r_i}{r_i+n_0} - \frac{F_0 + F_1 \log k}{r_i} ,$$

*where $F_0$ and $F_1$ are positive constants depending only on $n_0, q, m$ and $dim\, L_i$, $i \leq n_0$.*

*proof:* Write $[E,E] = I$ and $W = sI \oplus sL/E$. Denote the adjoint representation of $UE$ in $\wedge W$ by $\circ$. Since $\mathrm{depth}\, L \leq m$, by Proposition 11.7 (p. 306) there is an integer $s \leq m$, an integer $n_0$, an element $z \in \wedge^s(W_{\leq n_0})$, and an integer $q$ such that the adjoint representation of $UE$ restricts to an injection

$$A \to A \circ z , \tag{13.14}$$

where $A$ is a subspace mapping isomorphically to $U\left((E/I)_{\geq q}\right)$. Without loss of generality we suppose $r_0 > q$.

Next, fix some $r_i$, and let $\{u_j\} \subset E_{r_i}$ represent an ordered basis of $(E/I)_{r_i}$. Further, for $\ell \geq 1$, denote by $A^\ell \subset UE$ the linear span of the elements $u_{j_1} \bullet \cdots \bullet u_{j_\ell}$ with $j_1 < \cdots < j_\ell$. Since the map (13.14) is injective, it follows that for any $\ell \geq 1$,

$$A^\ell \longrightarrow A^\ell \circ z \quad \text{is injective} .$$

Since for any $\ell$, $\ell'$, $A^{\ell+\ell'} \subset A^\ell \cdot A^{\ell'}$ (multiplication in $UE$), we have

$$A^\ell \circ z \subset \sum_{t=1}^{s} \sum_{\substack{\ell_1+\cdots+\ell_t=\ell \\ \ell_1\leq\cdots\leq\ell_t}} (A^{\ell_1} \circ W_{\leq n_0}) \cdots (A^{\ell_t} \circ W_{\leq n_0}) \wedge^{s-t} (W_{\leq n_0}) .$$

When $\ell = ks$ we have $\ell_t \geq \frac{ks}{t} \geq k$ (because $s \geq t$). It follows that

$$A^{\ell_t} \circ W_{\leq n_0} \subset A^{\ell_t - k} \circ A^k \circ W_{\leq n_0} .$$

Thus, since $A^{ks} \to A^{ks} \circ z$ is injective, we obtain with $K = \dim \wedge^{\leq m} W_{\leq n_0}$ that

$$\dim A^{ks} \leq K^s \sum_{t=1}^{s} \sum_{\substack{\ell_1+\ldots+\ell_t=ks \\ \ell_1\leq\cdots\leq\ell_t}} \dim A^{\ell_1} \cdots \dim A^{\ell_t - k} \dim (A^k \circ W_{\leq n_0}) .$$

We now estimate $\dim A^\ell$ as follows. For any positive integer $p$ and any $x \in \mathbb{R}$ with $x > p$ we write

$$\binom{x}{p} = \frac{x(x-1)\cdots(x-p+1)}{p!} .$$

We choose $q$ sufficiently large that $e^{\lambda_0 r_0} > 4$. It follows that for $\ell \geq 1$, $e^{\lambda_i r_i \ell} > 4\ell$. Thus, in view of the bounds on $\dim(E/[E,E])$ we obtain

$$\frac{e^{\lambda_i r_i \ell}}{4^\ell \ell!} \leq \binom{\frac{1}{2}e^{\lambda_i r_i}}{\ell} \leq \dim A^\ell \leq \binom{e^{\lambda_i r_i}}{\ell} \leq \frac{e^{\lambda_i r_i \ell}}{\ell!} .$$

Applying this to the inequality above yields

$$\frac{e^{\lambda_i r_i ks}}{4^{ks}(ks)!} \leq \frac{\rho K^s e^{\lambda_i r_i k(s-1)}}{[k(s-1)]!} \dim \left(A^k \circ W_{\leq n_0}\right) ,$$

where $\rho$ is the number of partitions of $ks$. Since by [64] $\rho \leq e^{3\sqrt{ks}}$ and since $s \leq m$ we find

$$\max_{kr_i+1\leq p\leq kr_i+n_0} \dim W_p \geq \frac{1}{n_0} \dim \left(A^k \circ W_{\leq n_0}\right)$$

$$\geq \frac{1}{n_0 K^m e^{3\sqrt{km}}} \frac{e^{\lambda_i r_i k}}{4^{km}(km)!} \frac{1}{[k(m-1)]!} .$$

Thus, since $(km)^{km} \geq (km)!$, it follows that

$$\max_{kr_i+1\leq p\leq kr_i+n_0} \frac{\log\dim W_p}{p} \geq \max_{kr_i+1\leq p\leq kr_i+n_0} \frac{\log\dim W_p}{kr_i + n_0}$$

$$\geq \frac{\lambda_i r_i - F_0 - F_1 \log k}{r_i + n_0/k} \geq \frac{\lambda_i r_i}{r_i + n_0} - \frac{F_0 + F_1 \log k}{r_i},$$

where $F_0$ and $F_1$ are constants independent of $(\wedge V, d)$, $k$ and $r_i$.

□

*proof of Proposition 13.7*: We use the notation as developed in the preceding two Lemmas, and suppose that the sequences $(r_i)$ and the integer $n_0$ in Lemma 13.10 are chosen such that $r_0 \geq n_0$, and that $(r_i)$ satisfies the hypotheses of both Lemmas.

(i) Fix $i \geq 1$, and let $\sigma_i$ be the integer for which

$$\sigma_i < \frac{r_{i+1}}{r_i} \leq \sigma_i + 1\,.$$

Then by Lemma 13.10 there is a finite sequence of integers

$$q_j \in [jr_i + 1, jr_i + n_0]\,, \qquad i \leq j \leq i\sigma_i\,,$$

such that

$$\frac{\log \dim L_{q_j}}{q_j} \geq \frac{\lambda_i r_i}{r_i + n_0} - \frac{F_0 + F_1 \log j}{r_i}\,.$$

By the hypotheses of Lemma 13.9,

$$2r_i < r_{i+1} < 2(m+1)(r_i + 1) < 4(m+1)r_i\,,$$

and thus $\sigma_i \leq 4(m+1)$ and $r_i > 2^i r_0$. Thus

$$\frac{\log j}{r_i} \leq \frac{\log 4(m+1) + \log r_i}{r_i}\,.$$

It follows that for some $i_0$, and for all $i \geq i_0$,

$$\frac{F_0 + F_1 \log j}{r_i} < \frac{\varepsilon}{4}\,.$$

On the other hand, by Lemma 13.9, for each $i$, $\lambda_i > \alpha_L - \varepsilon/2$. It follows that for $i \geq i_0$ and for $i \leq j \leq i\sigma_i$,

$$\frac{\log \dim L_{q_j}}{q_j} > (\alpha_L - \varepsilon/2)\left(\frac{r_i}{r_i + n_0}\right) - \frac{\varepsilon}{4} \geq \alpha_L - \varepsilon\,.$$

Note next that since $r_0 \geq n_0$,

$$q_j < q_{j+1} \qquad \text{and that} \quad \frac{q_{j+1}}{q_j} < 1 + \frac{2}{j}\,.$$

Moreover, if $q'_j$ is the sequence corresponding to $i+1$ then $q'_j$ is defined for $i+1 \le j \le (i+1)\sigma_{i+1}$. Since $\sigma_i r_i < r_{i+1} \le (\sigma_i + 1)r_i$ it follows that

$$q_{i\sigma_i} < q'_{i+1} \qquad \text{and} \qquad \frac{q'_{i+1}}{q_{i\sigma_i}} < \frac{(1+1/i)r_{i+1} + n_0}{r_{i+1} - 2}\,.$$

Finally, write $p_{ij} = q_j$, so that $p_{i+1,j} = q'_j$. Then concatenating these finite sequences for $i \ge i_0$ yields the desired sequence $(p_i)$.

$$\cdots < p_{i,i} < \cdots < p_{i,i\sigma_i} < p_{i+1,i+1} < \cdots$$

(ii) We show that there are positive constants $C, B$ such that for sufficiently large $r$

$$\alpha_L - \frac{C/2}{\log r} \le \max_{r<j<r+\sqrt{r}} \frac{\log \dim L_j}{j} \le \alpha_L + \frac{B/2}{r}\,. \tag{13.15}$$

Once (13.15) is established the sequence $p_j$ is constructed inductively by choosing $p_{j+1} \in (p_j, p_j + \sqrt{p_j})$ so that

$$\frac{\log \dim L_{p_{j+1}}}{p_{j+1}} = \max_{p_j<i<p_j+\sqrt{p_j}} \frac{\log \dim L_i}{i}\,.$$

Then

$$\alpha_L - \frac{C/2}{\log p_{j+1}} \le \frac{\log \dim L_{p_{j+1}}}{p_{j+1}} \le \alpha_L + \frac{B/2}{p_{j+1}}\,.$$

This establishes Proposition 13.7(ii).

It remains to prove (13.15), and we do so using the sequence $(r_i)$ of Lemma 13.9(ii), and assuming $r > r_1^2$ and $r_0 > 1 + 4(m+2)^2$.

Let $i$ be the largest integer such that $r_i + n_0 < \sqrt{r}$ and let $k$ be the least integer such that $kr_i \ge r$. Then

$$kr_i + n_0 = (k-1)r_i + r_i + n_0 < r + \sqrt{r}\,,$$

and so $[kr_i + 1, kr_i + n_0] \subset [r+1, r+\sqrt{r}-1]$.

Moreover, since $r_{i+1} \le (m+1)(r_i+1) < (m+2)r_i$ (Lemma 13.9(ii)) and since $n_0 \le r_0$ we have

$$\sqrt{r} \le r_{i+1} + n_0 \le 2r_{i+1} \le 2(m+2)r_i\,.$$

It follows that

$$(k-1)r_i < r \le 4(m+2)^2 r_i^2\,.$$

Therefore

$$k < 1 + 4(m+2)^2 r_i < (1 + 4(m+2)^2)r_i < r_i^2\,. \tag{13.16}$$

Altogether then we have that

$$\frac{\log k}{r_i} < \frac{1}{\log r_i}$$

for $i$ sufficiently large.

On the other hand, by Lemma 13.10,

$$\begin{aligned} \max_{r<j<r+\sqrt{r}} \frac{\log \dim L_j}{j} &\geq \max_{kr_i+1\leq j\leq kr_i+n_0} \frac{\log \dim L_j}{j} \\ &\geq \frac{\lambda_i r_i}{r_i+n_0} - \frac{F_0+F_1\log k}{r_i} . \end{aligned}$$

By Lemma 13.9, in this case $\lambda_i = \alpha_L - \frac{F}{\log r_i}$, since $r_i^2 < r \leq 4(m+1)^2 r_i^2$ it follows from (13.16) that for $r$ (and hence $r_i$) sufficiently large

$$\max_{r<j<r+\sqrt{r}} \frac{\log \dim L_j}{j} \geq \alpha_L - \frac{C/2}{\log r}$$

where $C$ is a constant depending only on the constant $F$ of Lemma 13.9.

Finally, since $r+\sqrt{r} < 2r < (m+1)r$ the right-hand side inequality of (13.15) is the left-hand inequality of Theorem 13.3 (p. 345), with $B = 2B_0$.

(iii) This is immediate from (ii) since the constants $B/2$ and $C/2$ in (13.15) are obtained from the constants $B_0$ of Theorem 13.3 and $F$ from Lemma 13.9(iii) by a modification by completely independent constants.

□

*proof of Theorem 13.4*: In each of the three cases (i), (ii) and (iii) Proposition 13.7 provides an infinite sequence $p_1 < p_2 < \cdots$ satisfying the conclusion of the Theorem except that some of the $p_i$ may be odd and we may use $\varepsilon/2$ instead of $\varepsilon$. If $\dim L_{\text{odd}} < \infty$ by starting at some $p_i$ we have the conclusion of the Theorem. Suppose then that $\dim L_{\text{odd}} = \infty$.

By Corollary 11.3 (p. 309) for some odd integer $\ell_0$ there are elements $y_1, \cdots, y_k$ in $L$ of odd degrees $n_1, \cdots, n_k \leq \ell_0$, and such that for $\ell$ odd and $\ell \geq \ell_0$, the map

$$\sum_{i=1}^{k} [y_i, -] : L_\ell \to \bigoplus_{i=1}^{k} L_{\ell+n_i}$$

is injective. It follows that there is some even integer $s(\ell) \in [\ell, \ell+\ell_0]$ such that

$$\dim L_{s(\ell)} \geq \frac{1}{k} \dim L_\ell .$$

In this case we may assume $p_0 > \ell_0$, and we construct a new sequence $q_0 < q_1 < \cdots$ of even integers such that for some $i_0$ the sequence $(q_i)_{i\geq i_0}$

satisfies the conclusion of the Theorem. For this suppose $q_1, \cdots, q_r$ are constructed and that for each $i \leq r$ there is some $j(i)$ such that

$$q_i = \begin{cases} p_{j(i)}, & \text{if } p_{j(i)} \text{ is even} \\ s(p_{j(i)}), & \text{if } p_{j(i)} \text{ is odd}. \end{cases}$$

Then let $j(r+1)$ be the least integer such that $p_{j(r+1)} > q_r$ and define $q_{r+1}$ by the formula above. It remains to show that the conditions of Theorem 13.4 are satisfied if $i$ is sufficiently large.

In the case of (i), fix $i$ and, for simplicity write $j = j(i+1)$. Then $p_j - 1 \leq q_i$, and hence

$$\frac{q_{i+1}}{q_i} \leq \frac{p_j + \ell_0}{p_{j-1}}.$$

It follows that $q_{i+1}/q_i \to 1$. Moreover, since $q_{i+1} \geq p_j$, for $i$ sufficiently large,

$$\frac{\log \dim L_{q_{i+1}}}{q_{i+1}} \geq \frac{\log \dim L_{p_j} - \log k}{p_j + \ell_0} \geq \alpha_L - \varepsilon.$$

Thus Theorem 13.4 is proved in this case.

In the case of (ii) and (iii), define the $j(i)$ as above, set $M = \ell_0$, fix $i$ and write $j = j(i+1)$. Then

$$q_{i+1} \leq p_j + \ell_0 < p_{j-1} + \sqrt{p_{j-1}} + \ell_0 \leq q_i + \sqrt{q_i} + \ell_0.$$

Moreover, for $i$ sufficiently large,

$$\frac{\log \dim L_{q_i}}{q_i} \geq \frac{\log \dim L_{p_{j(i)}} - \log k}{p_{j(i)} + \ell_0} \geq \alpha_L - \frac{C}{\log q_i},$$

where in the case of (iii) $C$ depends only on $N$ and $h$. This proves the left-hand inequality; the right-hand inequality follows from Theorem 13.3 (p. 345).

□

## 13.6 Exponential growth

In this section we complete the proof of the Theorems recalled at the start of the Chapter. As in §13.5 we consider a minimal Sullivan algebra $(\wedge V, d)$ satisfying

$$\left.\begin{array}{l} V = V^{\geq 2} \text{ and } V \text{ has finite type}, \\ \operatorname{cat}(\wedge V, d) \leq m < \infty, \\ \log \operatorname{index} H(\wedge V, d) < \infty, \end{array}\right\} \qquad (13.11)$$

and adopt the notation

$$\left.\begin{aligned} &L = \text{ homotopy Lie algebra of } (\wedge V, d)\,,\\ &\alpha_L = \log \operatorname{index} L\,,\\ &\alpha_H = \log \operatorname{index} H(\wedge V, d)\,,\\ &\text{and if } \dim H(\wedge V,d) < \infty \text{ then}\\ &N = \max\{i \mid H^i(\wedge V,d) \neq 0\} \text{ and } h = \max_{i\leq N} \dim H^i(\wedge V,d)\,. \end{aligned}\right\} \qquad (13.12)$$

**Theorem 13.5.** *Let $L$ be the homotopy Lie algebra of a minimal Sullivan algebra satisfying (13.11). Then*

*(i) There is a positive integer $r$ such that*

$$\lim_{n\to\infty} \max_{n\leq i\leq n+r} \frac{\log dim\, L_i}{i} = \alpha_L\,.$$

*(ii) If $\alpha_H < \alpha_L$ then there is a positive integer $r$ and there are positive constants $\gamma$ and $\beta$ such that for some $K$,*

$$\alpha_L - \frac{\gamma}{\log n} \leq \max_{n\leq i\leq n+r} \frac{\log dim\, L_i}{i} \leq \alpha_L + \frac{\beta}{n}\,, \qquad n \geq K\,.$$

*(iii) If $H^{>N}(\wedge V,d) = 0$ then the constants $\gamma$ and $\beta$ may be chosen to depend only on $N$ and $h$.*

**Remark.** Theorem 13.5 contains both Theorems 12.9 and 12.10. For Theorem 12.11 we will still need to show that in (iii) we may choose $r = N-2$. That will be accomplished after the proof of Theorem 13.5.

*proof of Theorem 13.5:* Since $\alpha_L = \limsup_i \frac{\log \dim L_i}{i}$, it follows that for any $\varepsilon > 0$ and any $r$ that

$$\max_{n\leq i\leq n+r} \frac{\log \dim L_i}{i} \leq \sup_{n\leq i} \frac{\log \dim L_i}{i} < \alpha_L + \varepsilon\,,$$

if $n$ is sufficiently large. On the other hand, in the case of (ii), for any $r$, if $n > r/(m+1)$, Theorem 13.3 (p. 345) gives

$$\max_{n\leq i\leq n+r} \frac{\log \dim L_i}{i} \leq \max_{n\leq i\leq (m+1)n} \frac{\log \dim L_i}{i} \leq \alpha_L + \frac{B}{n}$$

where $B$ is a constant. Moreover in the case of (iii) $B$ depends only on $N$ and $h$. Thus we may set $\beta = B$ and achieve the right-hand inequality of Theorem 13.5 in cases (ii) and (iii).

To complete the proof we apply Theorem 13.4 (p. 351) recalling the sequence $q_j$ and its properties and the constant $C$. Then we fix a positive $\varepsilon < \alpha_L/2$ and show that for some $R$, and for all $n$ sufficiently large:

$$\max_{n\le i\le n+R} \frac{\log \dim L_i}{i} \geq \begin{cases} \alpha_L - 2\varepsilon\,, & \text{in case (i)} \\ \alpha_L - \dfrac{2C}{\log n}\,, & \text{in case (ii)}. \end{cases} \tag{13.17}$$

Then we set $\gamma = 2C$, noting from Theorem 13.4 that in case (iii) $\gamma$ depends only on $N$ and $h$. We proceed by assuming (13.16) fails for all $R$, and deducing a contradiction.

For this we construct an infinite sequence $(z_r)_{r\geq 0}$, of elements of $L$ and an increasing sequence of sub Lie algebras $E(r)$ such that

$$\left.\begin{array}{l} \text{(1) For } r\geq 1\,,\ z_r\neq 0\,, \deg z_r \text{ is even, and } \deg z_{r+1} > \deg z_r\,. \\ \text{(2) } z_j\in E(r) \text{ if } j<r\,. \\ \text{(3) } [E(r), z_r] \text{ is finite dimensional and concentrated} \\ \quad\ \text{in odd degrees}\,. \\ \text{(4) } E = \cup E(r) \text{ contains subspaces of the form } L_{[k,k+\ell]} \\ \quad\ \text{with } \ell \text{ arbitrarily large.} \end{array}\right\} \tag{13.18}$$

Once this construction is complete it will follow from (13.17)(4) that, in the terminology of §11.3, $E$ is admissible in $L$. Thus by Lemma 11.2, w-depth $E \leq$ w-depth $L \leq$ depth $L < \infty$. But this contradicts Proposition 11.9 (p. 310), and so completes the proof of Theorem 13.5.

We shall construct the elements $z_r$ of even degree and the Lie algebras $E(r)$ together with generating subspaces $S(r) \subset E(r)$. Our construction will depend on Theorem 13.4 (p. 351), and in particular on the properties of the sequence $(q_i)$ established there. The construction will satisfy the conditions below, where as usual, "$\circ$" denotes the adjoint representation of $UL$ in $L$:

- For some $\ell_r$,

$$S(r) = S(r)_{\leq \ell_r}\,. \tag{13.19}$$

- For some $k$, and all $\ell \geq 1$,

$$\dim UE(r)_\ell \leq e^{\beta_r \ell}, \text{where } \beta_r \leq \begin{cases} \alpha_L - \varepsilon_r\,, & \text{case (i)} \\ \alpha_L - \dfrac{C/2}{\log \ell_r}\,, & \text{case (ii)}\,, \end{cases} \tag{13.20}$$

where $\varepsilon_1 > \varepsilon_2 > \cdots > \varepsilon_r > \cdots$ is a decreasing sequence with each $\varepsilon_r > 2\varepsilon$.

- For $r \geq 1$,

$$\dim [E(r-1), z_{r-1}] < \infty\,, \text{ and } [E(r-1), z_{r-1}]_{\text{even}} = 0. \quad (13.21)$$

The construction is by induction on $r$. For $r = 1$ we invoke the failure of (13.16) to find an arbitrarily large integer $\ell_1$ such that

$$3\frac{\log(\ell_1+1)}{\ell_1} + \frac{\log \dim L_{\ell_1}}{\ell_1} < \begin{cases} \alpha_L - 2\varepsilon\,, & \text{case (i)} \\ \alpha_L - \dfrac{C/2}{\log \ell_1}\,, & \text{case (ii).} \end{cases}$$

Now set $S(1) = L_{\ell_1}$ and

$$\varepsilon_1 = \alpha_L - \frac{3\log(\ell_1+1)}{\ell_1} - \frac{\log \dim L_1}{\ell_1}\,.$$

Then (13.19) is true by definition. By construction, $\varepsilon_1 > 2\varepsilon$, and Lemma 13.3 (p. 335), applied with

$$\beta = \frac{\log \dim L_{\ell_1}}{\ell_1} \qquad \text{and } s = \ell_1\,,$$

shows that (13.20) holds with $r = 1$. Set $E(0) = 0$ and $z_0 = 0$. Then (13.21) is also true by definition.

Now suppose the $z_i$ are constructed for $i < r$ and that the $S(i)$, $E(i)$ and $\varepsilon_i$ are constructed for $i \leq r$. Our first step is to construct $z_r$.

For this, fix an integer $p > \ell_r$ for which if $q \geq p$ then

$$\begin{cases} 3\dfrac{\log(q+1)}{q} < \varepsilon_r - 2\varepsilon\,, & \text{case (i)} \\ \dfrac{3\log(q+1)}{q} < \dfrac{C/2}{\log q} & \text{case (ii)}\,. \end{cases} \quad (13.22)$$

Then choose $q_j > p$ so that for $i \geq j$

$$\left.\begin{array}{ll} q_{i+1} < \left(1 + \dfrac{\varepsilon}{2\alpha_L}\right) q_i, & \text{case (i)} \\ \dfrac{2(\sqrt{q_i}+M)\,\alpha_L}{q_i} < \dfrac{C}{\log q_i}\,, & \text{case (ii)} \end{array}\right\} \quad (13.23)$$

where $M$ is the constant in Theorem 13.4.

Then, since (13.16) fails, there is an interval $[k_r, k_r + \ell_r]$ with $k_r$ arbitrarily large, and such that for $k_r \leq i \leq k_r + \ell_r$,

$$\frac{\log \dim L_i}{i} < \begin{cases} \alpha_L - \varepsilon_r\,, & \text{case (i)} \\ \alpha_L - \dfrac{2C}{\log k_r}\,. & \text{case (ii)} \end{cases} \quad (13.24)$$

In particular, we may assume $k_r > q_j$.

Finally, let $q_k$ be the largest integer in the sequence $(q_i)$ for which $q_k \leq k_r$. It follows from (13.24) and the properties of the $(q_i)$ in Theorem 13.4 that no $q_i$ can belong to $[k_r, k_r + \ell_r]$. Since $q_k$ is the largest $q_i$ such that $q_i \leq k_r$, we have $k_r + \ell_r < q_{k+1}$. Thus Theorem 13.4 with $q_k$ sufficiently large, together with (13.23), gives

$$q_k < k_r < k_r + \ell_r < q_{k+1} < \begin{cases} q_k\left(1 + \dfrac{\varepsilon}{2\alpha_L}\right), & \text{case (i)} \\ q_k + s\sqrt{q_k} + M\,. & \text{case (ii)} \end{cases} \tag{13.25}$$

In particular, in view of (13.23) and since $k_r$ may be arbitrarily large, we may assume that $\log k_r < 4/3 \log q_k$.

On the other hand, a straightforward calculation from (13.20) and (13.24) shows that

$$\begin{aligned} &\log\dim\,((UE(r))_{i-q_k} \otimes L_i) \\ &\quad \leq \begin{cases} (\alpha_L - 2\varepsilon_r)q_k & \text{case (i)}\,, \\ (\alpha_L - \dfrac{3C/2}{\log q_k})q_k & \text{case (ii)}\,. \end{cases} \end{aligned} \tag{13.26}$$

Moreover, by Theorem 13.4,

$$\frac{\log\dim L_{q_k}}{q_k} > \alpha_L - \begin{cases} \varepsilon\,, & \text{case (i)} \\ C/\log q_k\,, & \text{case (ii)}. \end{cases}$$

Combining this with (13.25) yields, after another straightforward calculation, that for $i \in [k_r, k_r + \ell_r]$

$$\dim\,(UE(r))_{i-q_k}\dim L_i \leq \begin{cases} e^{-\varepsilon q_k}\dim L_{q_k}\,, & \text{case (i)} \\ e^{-\frac{C}{2\log q_k}q_k}\dim L_{q_k}\,. & \text{case (ii)} \end{cases}$$

In either case it follows, since $q_k$ may be arbitrarily large, that

$$\sum_{i=k_r}^{k_r+\ell_r} \dim\,(UE(r))_{i-q_k}\dim L_i < \dim L_{q_k}\,. \tag{13.27}$$

Now, the adjoint representation restricts to a linear map $UE(r) \otimes L_{q_k} \to L$, which in turn may be regarded as a linear map $L_{q_k} \to \mathrm{Hom}_{q_k}(UE(r), L)$. For $k_r \leq i \leq k_r + \ell_r$ this restricts to linear maps

$$\theta_i : L_{q_k} \to \mathrm{Hom}((UE(r))_{i-q_k}, L_i)\,.$$

Thus from (13.27) we obtain

$$\cap_i \ker \theta_i \neq 0\,.$$

Pick $z \in L_{q_k}$ so that $\theta_i(z) = 0$, $k_r \leq i \leq k_r + \ell_r$. Then $(UE(r) \circ z) \cap L_{[k_r, k_r+\ell_r]} = 0$. But $E(r)$ is generated in degrees $\leq \ell_r$, and therefore $UE(r) \circ z \subset L_{<k_r}$ is a finite dimensional subspace. Let $z_r \in UE(r) \circ z$ be a non-zero element of maximal even degree. Then $UE(r) \circ z_r$ is finite dimensional and concentrated in odd degrees. In particular, $z_r$ satisfies (13.17 (3)).

Finally, set $S(r+1) = S(r) \oplus \Bbbk z_r \oplus L_{[k_r, k_r+\ell_r]}$, and let $E(r+1)$ be the Lie algebra generated by $S(r)$. Thus $\ell_{r+1} = k_r + \ell_r$. We will apply Lemma 13.3 (p. 335) with $E = E(r)$ and $Y = \Bbbk z_r \oplus L_{[k_r, k_r+\ell_r]}$. Since $k_r$ could be arbitrarily large, we may assume $2 \log \ell_r < \log k_r$. Thus if we set

$$\beta = \begin{cases} \alpha_L - \varepsilon_r & \text{case (i)} \\ \alpha_L - \dfrac{C}{\log k_r} & \text{case (ii)} \end{cases}$$

(13.19) and (13.23) will give

$$\dim\left(UE(r)\right)_k \leq e^{\beta k}\,, \quad k \geq 0\,, \qquad \text{and } \dim L_i \leq e^{\beta i}\,, \quad i \in [k_r, k_r + \ell_r]\,.$$

Now set $|z_r| = \deg z_r$. Thus Lemma 13.3 asserts that

$$\dim\left(UE(r+1)\right)_k \leq e^{\beta_{r+1} k}$$

with

$$\beta_{r+1} = \begin{cases} \alpha_L - \varepsilon_r + 3\dfrac{\log(|z_r|+1)}{|z_r|} & \text{case (i)} \\ \alpha_L - \dfrac{C}{\log k_r} + \dfrac{3\log(|z_r|+1)}{|z_r|} & \text{case (ii)}\,. \end{cases}$$

It is immediate in case (i) from (13.21) that

$$\varepsilon_{r+1} = \varepsilon_r - \frac{\log(|z_r|+1)}{|z_r|} < 2\varepsilon$$

and so (13.19) is satisfied in this case. In case (ii), because of (13.22), we may assume that $k_r + \ell_r < 2q_k \leq 2|z_r|$. Thus (13.21) yields for $k_r$ sufficiently large that

$$\begin{aligned} \beta_{r+1} &< \alpha_L - \frac{C}{\log k_r} + \frac{C/4}{\log |z_r|} \\ &< \alpha_L - \frac{C}{\log k_r} + \frac{C/4}{\log(\frac{1}{2}(k_r + \ell_r))} \\ &< \alpha_L - \frac{C}{\log k_r} + \frac{C/2}{\log(k_r + \ell_r)} \\ &< \alpha_L - \frac{C/2}{\log(k_r + \ell_r)}\,. \end{aligned}$$

This establishes (13.19) in case (ii), and so closes the induction and completes the proof of Theorem 13.5.

□

It remains for the proof of Theorems 12.9, 12.10 and 12.10 to show that if $H^{>N}(\wedge V, d) = 0$ then the length $R$ of the interval in Theorem 13.5 can be chosen to be $N-2$. We accomplish this in the next Theorem.

**Theorem 13.6.** *Let $L$ be the homotopy Lie algebra of a minimal Sullivan algebra satisfying (13.11). If $H^{>N}(\wedge V, d) = 0$, then Theorem 13.5 is satisfied with $r = N-2$, and with $\gamma$ replaced by $2\gamma$.*

*proof*: Let $r$ and $K$ be the integers and let $\gamma$ be the constant for which the inequalities of Theorem 13.5 are satisfied. When $r$ is replaced by $N-2$ the right-hand side inequality of Theorem 13.5(iii) is immediate. We shall show that the left-hand side inequality also holds for $n$ sufficiently large, at the cost of replacing $\gamma$ by $2\gamma$. If $r \le N-2$ there is nothing to prove, so we suppose $r > N-2$. By ( [18], Proposition 33.6), and by taking $K$ sufficiently large, we may suppose that for all $n \ge K$,

$$\sum_{i=n+1}^{n+N-1} \dim L_i \ge \frac{1}{Nh} \dim L_n \,.$$

Now suppose $n > K+r$. By Theorem 13.5 for some $\ell \in [n+N-r, n+N]$ we have

$$\frac{\log\ \dim L_\ell}{\ell} \ge \alpha_L - \frac{\gamma}{\log(n+N-r)} \,.$$

On the other hand, there is a sequence $\ell < \ell(1) < \cdots$ such that $\ell(j+1) \le \ell(j) + N - 1$ and $\dim L_{\ell(j+1)} \le \frac{1}{Nh} \dim L_{\ell(j)}$. It follows that there is an integer $i \ge \ell$ such that $i \in [n+2, n+N]$ and

$$\dim L_i \ge \left(\frac{1}{Nh}\right)^{r-N+2} \dim L_\ell \,.$$

Then

$$\begin{aligned} \frac{\log \dim L_i}{\ell} &\ge \frac{\log \dim L_\ell}{\ell} - \frac{(r-N+2)\log Nh}{\ell} \\ &\ge \alpha_L - \frac{\gamma}{\log(k+N-r)} - \frac{(r-N+2)\log Nh}{\ell} \,. \end{aligned}$$

Therefore, for $n$ sufficiently large,

$$\begin{aligned}
\tfrac{\log \dim L_i}{i} &\geq \frac{\ell}{i}\left(\alpha_L - \frac{\gamma}{\log(n+N-r)}\right) - \frac{(r-N+2)\log Nh}{i} \\
&\geq \frac{n+N}{n+2}\alpha_L - \frac{n+N}{n+2}\frac{\log(n+2)}{\log(n+N-r)}\frac{\gamma}{\log(n+2)} - \frac{(r-N+2)\log Nh}{n+2} \\
&> \alpha_L - \frac{2\gamma}{\log(n+2)}\,.
\end{aligned}$$

Since $i \in [n+2, n+N]$ this completes the proof.

□

# Chapter 14

# Structure of a graded Lie algebra of finite depth

## 14.1 Introduction

*In this chapter all graded vector spaces $V = \{V_i\}_{i\in\mathbb{Z}}$ have finite type, and will be referred to simply as "graded vector spaces". We consider graded Lie algebras $L = L_{\geq 1}$ of finite type, and since $L_0 = 0$. They are trivially Sullivan Lie algebras. In particular the results of Chapters 10 and 11 all apply. In particular depth $L$ is defined and, as observed in Proposition 11.1*

$$\text{depth}\, L = \text{ least } p \text{ (or } \infty) \text{ such that } \mathrm{Ext}^p_{UL}(\Bbbk, UL) \neq 0\,.$$

Our objective is to establish consequences of the hypothesis

$$\text{depth}\, L < \infty\,, \tag{14.1}$$

with particular attention to the growth of the integers $\dim L_i$ and to the structure of the ideals in $L$. This analysis relies on the following means of comparing the size of graded vector spaces of finite type.

**Definition:** If $W$ and $V$ are graded vector spaces, then $V \leqslant W$ if for some fixed positive $\lambda, q$ and $N$,

$$\dim V_k \leq \lambda \sum_{i=k}^{k+q} \dim W_i\,, \qquad k \geq N\,.$$

If $W$ is a subspace of $V$ and $V \leqslant W$, we say that $W$ is *full* in $V$.

It is immediate from the definition that an equivalence relation on the graded vector spaces of finite type is defined by

$$V \sim W \quad \Leftrightarrow \quad W \leqslant V \text{ and } V \leqslant W\,.$$

**Definition.** Two graded vector spaces $V$ and $W$ are *equivalent* if $V \sim W$. If $V$ and $W$ are subspaces of a graded Lie algebra $L$, then $V, W$ are *$L$-equivalent* ($V \sim_L W$) if for all ideals $K$ in $L$, $V \cap K \sim W \cap K$.

The main result of this chapter is

**Theorem 14.1.** *Let $L = L_{\geq 1}$ be a graded Lie algebra of finite type and finite depth $m$. Then the number $\nu(L)$ of $L$-equivalence classes of ideals in $L$ satisfies $\nu(L) \leq 2^m$ and equality holds if and only if $L \sim_L I(1) \oplus \cdots \oplus I(m)$ where the $I(i)$ are ideals of depth* 1.

The comparison of the size of graded vector spaces require one more definition.

**Definition:** A subspace $W \subset V$ is *$S$-large* in $V$ with respect to a graded vector space, $S$, if $V/W \leqslant S$.

The next lemma follows via elementary linear algebra

**Lemma 14.1.** *Let $S, U, V, W$ be graded vector spaces.*

*(i) If $S, W \subset V$, then $V/W \leqslant S \Rightarrow V \leqslant (W + S)$.*
*(ii) If $U \subset W \subset V$ and $W/U \leqslant S$ and $V/W \leqslant S$, then $V/U \leqslant S$.*
*(iii) The finite intersection of $S$-large subspaces $W(i)$ of $V$ is also $S$-large in $V$.*
*(iv) If $U, W$ are subspaces of $V$ then $U + W \sim U \oplus W$, where $U + W$ means the sum in $V$, and*

$$U \sim W \iff U \text{ and } W \text{ are full in } U + W\,.$$

*(iv) If $S(i) \sim T(i)$, $i = 1, \cdots, r$, are graded vector spaces, then*

$$\oplus_{i=1}^r S(i) \sim \oplus_{i=1}^r T(i)\,.$$

## 14.2 The spectrum

Recall from Chapter 12 (p. 314), that for any graded vector space $V = V_{\geq 0}$,

$$\log \operatorname{index} V = \limsup_n \frac{\log \dim V_n}{n}\,.$$

**Definition.** The *spectrum* of a graded Lie algebra $L = L_{\geq 1}$ of finite type is the set

$$\operatorname{spectrum} L = \{\log \operatorname{index} I \mid I \text{ is an infinite dimensional ideal of } L\}\,.$$

We prove

**Theorem 14.2.** *If $L = L_{\geq 1}$ a graded Lie algebra of finite type and finite depth $m$, then the spectrum of L contains at most $m$ elements.*

To begin we consider the following four properties for an ideal $I$ in a graded Lie algebra $L = L_{\geq 0}$ of finite type and finite depth.

(P1) $\operatorname{depth} L = \operatorname{depth} I$.
(P2) The restriction map $\rho_I : \operatorname{Ext}_{UL}(\Bbbk, UL) \to \operatorname{Ext}_{UI}(\Bbbk, UL)$ is non-zero.
(P3) $I$ is full in $L$.
(P4) log index $L$ = log index $I$.

**Proposition 14.1.** *Let $I$ be an ideal in a graded Lie algebra $L = L_{\geq 1}$ of finite type and finite depth. Then*

$$(P1) \Longrightarrow (P2) \Longrightarrow (P3) \Longrightarrow (P4).$$

*proof:* (P1) $\Rightarrow$ (P2). Let $m = \operatorname{depth} L$. Since $\operatorname{depth} I = \operatorname{depth} L$, in the Hochschild-Serre spectral sequence,

$$E_2^{p,q} = \operatorname{Ext}^p_{UL/I}(\Bbbk, \operatorname{Ext}^q_{UI}(\Bbbk, UL)) \Rightarrow \operatorname{Ext}^{p+q}_{UL}(\Bbbk, UL)\,,$$

$E_2^{p,q} = 0$ for $q < m$ and

$$\operatorname{Ext}^m_{UL}(\Bbbk, UL) \cong E_\infty^{0,m} \cong E_2^{0,m} = \left(\operatorname{Ext}^m_{UI}(\Bbbk, UL)\right)^{L/I}\,.$$

By (10.3), p. 270, the restriction map $\rho_I$ is injective in homological degree $m$.

(P2) $\Rightarrow$ (P3). By hypothesis there is an element $a \in \operatorname{Ext}_{UL}(\Bbbk, UL)$ with $\rho_I(a) \neq 0$. From Proposition 11.4(ii) applied to the element $\rho_I(a)$ it follows that $I$ contains a finitely generated Lie subalgebra $E$ such that the restriction map $\rho_E : \operatorname{Ext}_{UI}(\Bbbk, UL) \to \operatorname{Ext}_{UE}(\Bbbk, UL)$ is non-zero on $\rho_I(a)$. Thus it follows from Proposition 11.8 (p. 308), that the centralizer of $E$ in $L$ is finite dimensional.

Denote by $x_1, \cdots, x_r$ a set of generators of $E$. Then for $n$ sufficiently large the map

$$\sum_{i=1}^r [-, x_i] : L_n \to \bigoplus_{i=1}^r I_{n+\deg x_i}$$

is injective. This implies that $I$ is full in $L$.

(P3) $\Rightarrow$ (P4). It follows from (P3) that there are positive integers $q$, $\lambda$ and $N$ such that we have the following inequality (coefficientwise) of series

$$t^q(\sum_{n\geq N} \dim L_n\, t^n) << \lambda(\sum_{i=1}^{q} t^i)\,(\sum_{n\geq K} \dim I_n\, t^n)\,.$$

Therefore they have the same radius of convergence, and so $\log \operatorname{index} L = \log \operatorname{index} I$.

□

**Remark.** It follows from Proposition 14.1 that if an ideal $I$ has the same depth as $L$, then it also has the same log index. The converse is not true. Take for instance $L$ to be the direct sum of two free graded Lie algebras $L_1$ and $L_2$ on two generators of degree 2. Then $L$ and $L_1$ have the same log index, but $\operatorname{depth} L = 2$ and $\operatorname{depth} L_1 = 1$.

**Definition.** The *hyper radical* of a graded Lie algebra, $L = L_{\geq 1}$ of finite type, is the ideal $R$ which is the sum of the ideals $I \subset L$ satisfying

$$\log \operatorname{index} I < \log \operatorname{index} L\,.$$

By convention, $R = \{0\}$ if there is no infinite dimensional ideal $I$ of $L$ with $\log \operatorname{index} I < \log \operatorname{index} L$.

**Lemma 14.2.** *For any graded Lie algebra $L = L_{\geq 1}$, the ideal $I(x)$ generated by $x \in L$ satisfies*

$$x \in R \Longleftrightarrow \textit{log index} I(x) < \textit{ log index}.L.$$

*proof:* If $x \notin R$ then $I(x) \not\subset R$ and so by definition $\log \operatorname{index} I(x) = \log \operatorname{index} L$. Conversely, suppose $x \in R$. Then $x$ is a finite sum $x = \sum_{i=1}^{p} x_i$ where $x_i$ belongs to an ideal $I_i$ with $\log \operatorname{index} I_i < \log \operatorname{index} L$. Let $I$ be the sum of the $I_i$. Then $\log \operatorname{index} I = \max_i \log \operatorname{index} I_i < \log \operatorname{index} L$, and so $I \subset R$. Since $I(x) \subset I$, $\log \operatorname{index} I(x) < \log \operatorname{index} L$.

□

**Lemma 14.3.** *Suppose $L$ is a graded Lie algebra satisfying the hypotheses of Theorem 14.2. Then*

*(i) The restriction map $Ext_{UL}(\Bbbk, UL) \to Ext_{UR}(\Bbbk, UL)$ is zero.*
*(ii) $\operatorname{depth} R < \operatorname{depth} L$.*
*(iii) $\log \operatorname{index} R < \log \operatorname{index} L$.*

*proof:* (i) Suppose $\alpha \in \mathrm{Ext}^p_{UL}(\Bbbk, UL)$ restricts to a non-zero element in $\mathrm{Ext}^p_{UR}(\Bbbk, UL)$. This in turn would restrict to a non-zero element in $\mathrm{Ext}^p_{UE}(\Bbbk, UL)$, where $E$ is a finitely generated sub Lie algebra of $R$, (Proposition 11.4(ii), p. 302). Let $x_1, \cdots, x_r$ generate $E$ and let $I$ be the sum of the ideals $I(x_j) \subset L$ generated by the $x_j$. By Lemma 14.2, $\log\mathrm{index}\, I(x_j) < \log\mathrm{index}\, L$ and

$$\log\mathrm{index}\, I = \max \log\mathrm{index}\, I(x_i) < \log\mathrm{index}\, L\,.$$

On the other hand the following commutative diagram of restrictions

$$\begin{array}{ccc}
\mathrm{Ext}_{UL}(\Bbbk, UL) & \longrightarrow & \mathrm{Ext}_{UR}(\Bbbk, UL) \\
\downarrow & & \downarrow \\
\mathrm{Ext}_{UI}(\Bbbk, UL) & \longrightarrow & \mathrm{Ext}_{UE}(\Bbbk, UL)
\end{array}$$

shows that the restriction map $\mathrm{Ext}_{UL}(\Bbbk, UL) \to \mathrm{Ext}_{UI}(\Bbbk, UL)$ is non-zero. Then by Proposition 14.1 $\log\mathrm{index}\, I = \log\mathrm{index}\, L$, in contradiction to the inequality above.

(ii) This follows because otherwise, by Proposition 14.1, the restriction map would be non-zero.

(iii) As observed in the proof of (i), $R$ contains a sub Lie algebra $E$ generated by elements $x_1, \cdots, x_r$ such that the restriction map

$$\mathrm{Ext}_{UR}(\Bbbk, UL) \to \mathrm{Ext}_{UE}(\Bbbk, UL)$$

is non-zero. Let $I(x_i)$ be the ideal generated by $x_i$ and let $I$ be the sum of the $I(x_i)$. Then $\mathrm{Ext}_{UR}(\Bbbk, UL) \to \mathrm{Ext}_{UI}(\Bbbk, UL)$ is non-zero and so by Proposition 14.1, $\log\mathrm{index}\, R = \log\mathrm{index}\, I$. Since $x_i \in R$, by Lemma 14.2 $\log\mathrm{index}\, I(x_i) < \log\mathrm{index}\, L$. Therefore $\log\mathrm{index}\, I = \max \log\mathrm{index}\, I(x_i) < \log\mathrm{index}\, L$, and so $\log\mathrm{index}\, R = \log\mathrm{index}\, I < \log\mathrm{index}\, L$.

□

*proof of Theorem 14.2:* If $\mathrm{depth}\, L = 0$, then by Proposition 11.3 (p. 302) $L$ is finite dimensional and concentrated in odd degrees. Thus, since $L$ is infinite dimensional, $\mathrm{depth}\, L \geq 1$. Let $R$ be the hyper radical of $L$. If $R$ is finite dimensional then $\mathrm{spectrum}\, L$ consists of the single element $\log\mathrm{index}\, L$. If $R$ is infinite dimensional then

$$\mathrm{spectrum}\, L = \mathrm{spectrum}\, R \cup \{\log\mathrm{index}\, L\}\,.$$

Since $\mathrm{depth}\, R < \mathrm{depth}\, L$ the Theorem follows by induction on $\mathrm{depth}\, L$.

□

## 14.3 Minimal sub Lie algebras

*In this section $S = S_{\geq 0}$ denotes a fixed graded vector space of finite type.*

**Definition.** Let $L = L_{\geq 1}$ be a graded Lie algebra of finite type, and let $M = \{M_i\}_{i\in\mathbb{Z}}$ be a $\mathbb{Z}$-graded $L$-module.

(i) The *isotropy sub Lie algebra* $L_V$ at a subspace $V \subset M$ is defined by

$$L_V = \{x \in L \,|\, x \cdot V = 0\}\,.$$

(ii) The $L$-module $M$ is $S$-*light* if whenever $\dim V < \infty$, then $L/L_V \leqslant S$.

(iii) The graded Lie algebra $L$ is $S$-*minimal with respect to an ideal* $I$ if every sub Lie algebra $E$ such that $L/E \leqslant S$ and $I \subset E \subset L$ satisfies

$$\operatorname{depth} E \geq \operatorname{depth} L\,.$$

(iv) The Lie algebra $L$ is $S$-*minimal* if $L$ is $S$-minimal with respect to 0, i.e. if $\operatorname{depth} E \geq \operatorname{depth} L$ for all sub Lie algebras $E$ of $L$ with $L/E \leqslant S$.

**Remark.** It is immediate from the definition that for any ideal $I \subset L$ there is a sub Lie algebra $E$ containing $I$ such that $L/E \leqslant S$ and $E$ is $S$-minimal with respect to $I$.

Now recall from §10, p. 268, that the *grade* of an $L$-module $M$ is defined by

$$\operatorname{grade}_L M = \text{ least } p \text{ (or } \infty) \text{ such that } \operatorname{Ext}^p_{UL}(M, UL) \neq 0\,.$$

Since $L = L_{\geq 1}$ and has finite type it follows from (10.1) (p. 268) that

$$\operatorname{grade}_L M = \text{ least } p \text{ (or } \infty) \text{ such that } \operatorname{Tor}_p^{UL}(M, (UL)^{\#}) \neq 0\,.$$

In particular,

$$\operatorname{depth}_L = \text{least } p \text{ (or } \infty) \text{ such that } \operatorname{Tor}_p^{UL}(\mathbb{k}, (UL)^{\#}) \neq 0\,.$$

Our principal objective in this section is

**Theorem 14.3.** *Let $I$ be an ideal in a graded Lie algebra $L = L_{\geq 1}$ of finite type and finite depth. If $M = \{M_i\}_{i\in\mathbb{Z}}$ is a non-zero $S$-light $L$-module satisfying $I \cdot M = 0$, then*

*(i) $I$ extends to a sub Lie algebra $E \subset L$ such that $L/E \leq S$ and*

$$\mathit{depth}\, E \leq \mathit{grade}_L M\,.$$

*(ii) If $L$ is $S$-minimal with respect to $I$, then*

$$\mathit{depth}\, E = \mathit{depth}\, L = \mathit{grade}_L M\,.$$

Theorem 14.3 provides a principal ingredient for the proof of Theorem 14.1. However its proof requires the considerable preliminary technical material contained in the next three Lemmas.

**Lemma 14.4.** *Let $L = L_{\geq 1}$ be a graded Lie algebra of finite type, and let $M = \{M_i\}_{i\in\mathbb{Z}}$ be an $L$-module. Then*

*(i) The direct sums and tensor products of $S$-light $L$-modules are $S$-light.*
*(ii) Any subquotient of an $S$-light $L$-module is $S$-light.*
*(iii) The sum, $N$, of all $S$-light submodules $N(\alpha) \subset M$ is itself $S$-light.*
*(iv) If $M$ is an $S$-light $L$-module, then each $\wedge^q M$ is also $S$-light.*
*(v) Suppose $M = M_{\geq 0}$. If $M \subset S$ then $M$ is $S$-light. In particular, $M$ is $M$-light.*
*(vi) An $L$-module $M$ is $S$-light if and only if $L/L_v \leq S$ for all $v \in M$.*
*(vii) $M$ is $S$-light if and only if for some set $\{v_i\}$ of generators for $M$ as an $L$-module, each $L/L_{v_i} \leqslant S$.*

*proof:* (i) Any finite dimensional subspace of a direct sum or tensor product is contained in a finite direct sum or tensor product. Thus it is sufficient to prove this for the sum and tensor product of two $S$-light $L$-modules $M$ and $N$. If $V \subset M \oplus N$ is finite dimensional then $V \subset U \oplus Z$ where $U$ and $Z$ are finite dimensional subspaces of $M$ and $N$. Since $L_V \supset L_U \cap L_Z$ and the intersection of $S$-large subspaces is $S$-large (Lemma 14.1(iii), p. 368), $L/L_V \leqslant S$.

(ii) Let $N \subset M$ be a submodule. If $V$ is finite dimensional subspace of $N$, then it is a finite dimensional subspace of $M$ and the result follows. Now suppose $p : M \to Q$ is a surjection of $L$-modules, and denote by $s : Q \to M$ a $\Bbbk$-linear map such that $p \circ s = id_Q$. If $V$ is a finite dimensional subspace of $Q$, then $s(V)$ is a finite dimensional subspace of $M$ and $L_V \supset L_{s(V)}$. Again the result follows.

(iii) This follows from (i) and (ii) because $N$ is a quotient of the direct sum of the $N(\alpha)$.

(iv) follows from (i) and (ii).

(v) Let $V$ be a finite dimensional subspace of $M$ with basis $(v_i)_{1\leq i\leq N}$. Then the action of $L$ on the $v_i$ induces a linear injection

$$(L/L_V)_k \to \bigoplus_{i=1}^{N} M_{k+\deg v_i}\,.$$

Since $M = M_{\geq 0}$ it follows that $L/L_V \leqslant M \leqslant S$.

(vi) If $V \subset M$ is finite dimensional, with basis $v_1, \cdots, v_p$ then $L_V = \cap_i L_{v_i}$. Thus (vi) follows since the finite intersection of $S$-large subspaces is $S$-large.

(vii) We first show that if $L/L_v \leq S$ for some $v \in V$, then $L/L_{a\cdot u} \leq S$ for all $a \in UL$. In fact, because of (ii), it is sufficient to show this when $a = x_1 \cdots x_r$ ($x_i \in L$) and we proceed by induction on $r$.

Set $w = x_2 \cdots x_r \cdot v$ and set $T = \{y \in L \,|\, [y, x_1] \in L_w\}$. Then $[-, x_1] : L/T \to L/L_w$ is injective and so $L/T \leqslant S$. On the other hand, for $z \in L$ we have

$$z \cdot x_1 \cdots x_r \cdot v = z \cdot x_1 \cdot w = [z, x_1] \cdot w \pm x_1 \cdot z \cdot w$$

and so $L_{x_1\cdot w} \supset T \cap L_w$. Now the inequalities above yield

$$L/L_{x_1 w} \leqslant L/T + L/L_w \leqslant S\,.$$

□

**Lemma 14.5.** *Let $L = L_{\geq 1}$ be a graded Lie algebra of finite type.*

*(i) If $E \subset L$ is a sub Lie algebra and $L/E \leq S$, then the $L$-module $UL \otimes_{UE} \Bbbk$ is $S$-light.*

*(ii) If $L$ acts by derivations in a Lie algebra $E$, and if $L/L_{x_\alpha} \leqslant S$ for a set $\{x_\alpha\}$ of generators for the Lie algebra $E$, then $E$ is an $S$-light $L$-module.*

*proof :* (i) The vector space $UL \otimes_{UE} \Bbbk$ is generated as an $L$-module by the single element $v = 1 \otimes 1$. Since $L_v = E$ and $L/E \leq S$, (i) follows from Lemma 14.4(vii).

(ii) Let $W$ be the linear span of the $w_\alpha$. Then $UL \cdot W$ is an $S$-light $L$-module by Lemma 14.4(vii). The natural linear map $UL \cdot W \to E$ extends to an $L$-linear algebra surjection $T(UL \cdot W) \to UE$. But $T(UL \cdot W)$ is

$S$-light by Lemma 14.4(i), and hence $E$, as a subquotient of $T(UL \cdot W)$, is $S$-light by Lemma 14.4(ii). □

**Lemma 14.6.** *Let $E$ be a sub Lie algebra of a graded Lie algebra $L = L_{\geq 1}$ of finite type. Then*

$$\mathit{Ext}_{UL}(UL \otimes_{UE} \Bbbk, -) \cong \mathit{Ext}_{UE}(\Bbbk, -)\,.$$

*proof :* Let $P_* \to \Bbbk$ be a $UE$-free resolution. Since $UL$ is $UE$ free on the right, $UL \otimes_{UE} P_* \to UL \otimes_{UE} \Bbbk$ is a $UL$-free resolution. But

$$\mathrm{Hom}_{UL}(UL \otimes_{UE} P_*, -) = \mathrm{Hom}_{UE}(P_*, -)\,.$$

□

We are now ready for the proof of Theorem 14.3 :

**Theorem 14.3.** *Let $I$ be an ideal in a graded Lie algebra $L = L_{\geq 1}$ of finite type and finite depth. If $M = \{M_i\}_{i \geq 0}$ is a non-zero $S$-light $L$-module satisfying $I \cdot M = 0$, then*

*(i) $I$ extends to a sub Lie algebra $E \subset L$ such that $L/E \leq S$ and*

$$\mathit{depth}\, E \leq \mathit{grade}_L M\,.$$

*(ii) If $L$ is $S$-minimal with respect to $I$, then*

$$\mathit{depth}\, L = \mathit{depth}\, E = \mathit{grade}_L M\,.$$

*proof:* (i) If $\mathrm{grade}_L M = \infty$ there is nothing to prove. Suppose $\mathrm{grade}_L M = m < \infty$. Then (Proposition 10.7, p. 290) for some finitely generated submodule $N \subset M$, the restriction $\mathrm{Ext}^m_{UL}(M, UL) \to \mathrm{Ext}^m_{UL}(N, UL)$ is non-zero. Denote by $v_1, \ldots, v_r$ a set of generators of $N$. Applying $\mathrm{Ext}_{UL}(-, UL)$ to the short exact sequence $0 \to UL \cdot v_1 \to N \to N/(UL \cdot v_1) \to 0$ yields a long exact sequence, and in particular the short exact sequence

$$\mathrm{Ext}^m_{UL}(UL \cdot v_1, UL) \leftarrow \mathrm{Ext}^m_{UL}(N, UL) \leftarrow \mathrm{Ext}^m_{UL}(N/(UL \cdot v_1), UL)\,.$$

It follows by induction on $r$ that there exists a subquotient module of $N$ of the form $UL \cdot v$ and for which $\mathrm{Ext}^m_{UL}(UL \cdot v, UL) \neq 0$. Moreover, as a subquotient of $M$, $UL \cdot v$ is $S$-light (Lemma 14.4(ii)).

Consider the short exact sequence of $L$-modules

$$0 \to K \to UL \otimes_{UL_v} \Bbbk \to UL \cdot v \to 0\,. \tag{14.2}$$

Since $UL \cdot v$ is $S$-light, $L/L_v \leqslant S$. Moreover, $UL \otimes_{UL_v} \Bbbk$ is generated by $w = 1 \otimes 1$ and $L_w = L_v$. Thus $L/L_w \leq S$ and so $UL \otimes_{UL_v} \Bbbk$ and $K$ are also $S$-light (Lemmas 14.4(vii) and 14.4(ii)). Note also that since $UL \cdot v$ is a subquotient of $M$, $I \cdot UL \cdot v = 0$. In particular, $I \subset L_v$ and, since $I$ is an ideal, it follows that $I \cdot (UL \otimes_{UL_v} \Bbbk) = 0$. Hence $I \cdot K = 0$.

On the other hand, since $\mathrm{Ext}^m_{UL}(UL \cdot v, UL) \neq 0$, from the exact sequence (14.2), we deduce that either

$$\mathrm{Ext}^{m-1}_{UL}(K, UL) \neq 0 \quad \text{ or } \mathrm{Ext}^m_{UL}(UL \otimes_{UL_v} \Bbbk, UL) \neq 0\,.$$

In the first case the Theorem follows by induction on $m$. In the second case by Lemma 14.6,

$$\mathrm{Ext}^m_{UL}(UL \otimes_{UL_v} \Bbbk, UL) \cong \mathrm{Ext}^m_{UL_v}(\Bbbk, UL)\,.$$

Thus $\mathrm{depth}\, L_v \leq m$. Set $E = L_v$ in this case.

(ii) Since $L$ is $S$-minimal with respect to $I$ it is sufficient to show that $\mathrm{depth}\, L = \mathrm{grade}_L M$. Let $E \subset L$ be a sub Lie algebra satisfying the conclusion of (i). It follows from the Hochschild-Serre spectral sequence,

$$\mathrm{Tor}^{UE}_p(\wedge^q sL/E, (UL)^\#) \Rightarrow \mathrm{Tor}^{UL}_{p+q}(\Bbbk, (UL)^\#)\,,$$

(§10.1, p. 269) that there exist $p, q$ with $p + q = \mathrm{depth}\, L$, and such that

$$\mathrm{grade}_E \wedge^q sL/E \leq p\,.$$

Since $L/E$ is an $S$-light $E$-module so is $\wedge^q s\,L/E$ (Lemma 14.4(iv)). Since $I \cdot L/E = 0$, (i) with $M$ and $L$ replaced by $\wedge^q sL/E$ and $E$, gives a sub Lie algebra $G \subset E$ satisfying $I \subset G$, $E/G \leq S$, and

$$\mathrm{depth}\, G \leq \mathrm{grade}_E \wedge^q sL/E \leq \mathrm{depth}\, L - q\,.$$

Since $L$ is $S$-minimal with respect to $I$, $\mathrm{depth}\, L \leq \mathrm{depth}\, G$; i.e. $p + q \leq p$. Thus $q = 0$, $\mathrm{depth}\, L = p$, and $\mathrm{grade}_E(\wedge^q s\,L/E) = \mathrm{grade}_E(\Bbbk) = \mathrm{depth}\, E$. This gives $\mathrm{depth}\, E \leq p = \mathrm{depth}\, L$. But $L$ was $S$-minimal with respect to $I$, so that $\mathrm{depth}\, L \leq \mathrm{depth}\, E$. This gives $\mathrm{depth}\, L = \mathrm{depth}\, E$ for any sub Lie algebra $E \subset L$ satisfying the conclusion of (i).

Next, let $M_+ = \{M_i\}_{i \geq 0}$ and set $N = M/M_+$; both $M_+$ and $N$ are $S$-light $L$-modules. If $M_+ \neq 0$, we can find a short exact sequence of $L$-modules of the form

$$0 \to K \to M_+ \to \Bbbk x \to 0\,.$$

If $K = 0$ then $M_+ = \Bbbk x$ and $\mathrm{grade}_L(M_+) = \mathrm{depth}\, L$. Otherwise, by (i) there is a sub Lie algebra $E \subset L$ satisfying $I \subset E$ and $\mathrm{depth}\, E \leq \mathrm{grade}_L K$. Since $L$ is $S$-minimal with respect to $I$, $\mathrm{depth}\, L \leq \mathrm{depth}\, E \leq \mathrm{grade}_L K$. Thus, if $p = \mathrm{depth}\, L$, the sequences

$$\mathrm{Ext}^i_{UL}(K, UL) \leftarrow \mathrm{Ext}^i_{UL}(M_+, UL) \leftarrow \mathrm{Ext}^i_{UL}(\Bbbk x, UL) \leftarrow 0\,, \qquad i \leq p$$

are exact. This implies that $\text{grade}_L M_+ = \text{depth}\, L$ in this case as well, which establishes (ii) if $N = 0$.

Next, suppose $N \neq 0$. Since $N$ is concentrated in negative degrees, and since $(UL)^\#$ is also concentrated in negative degrees, it follows that $(N \otimes (UL)^\#)^\# = N^\# \otimes UL$ as $L$-modules with diagonal action.

On the other hand $\text{Tor}_*^{UL}(N, (UL)^\#) = \text{Tor}_*^{UL}(\Bbbk, N \otimes (UL)^\#)$, and dualizing gives $\text{Ext}_{UL}^*(N, UL) = \text{Ext}_{UL}^*(\Bbbk, N^\# \otimes UL)$. Since $N^\# \otimes UL$ is a free $UL$-module (diagonal action) this shows that $\text{grade}_L N = \text{depth}\, L$. In particular this establishes (ii) if $M_+ = 0$.

Finally, suppose $M_+ \neq 0$ and $N \neq 0$. Since $\text{depth}\, L = \text{grade}_L M_+ = \text{grade}_L N = p$, the short exact sequence

$$0 \to M_+ \to M \to N \to 0$$

and the consequent exact sequences

$$\text{Ext}_{UL}^i(M_+, UL) \leftarrow \text{Ext}_{UL}^i(M, UL) \leftarrow \text{Ext}_{UL}^i(N, UL) \leftarrow 0\,, \qquad i \leq p$$

imply that $\text{grade}_L M = p = \text{depth}\, L$. □

## 14.4 The weak complements of an ideal

**Definition.** Let $I$ be an ideal in a graded Lie algebra $L = L_{\geq 1}$ of finite type and finite depth. A *weak complement* for $I$ in $L$ is a sub Lie algebra $E \subset L$ such that $E \cap I = 0$, $L/E \leqslant I$ (which implies that $E \oplus I$ is full in $L$), and $E$ is $I$-minimal.

**Theorem 14.4.** *Suppose $I$ is an ideal in a graded Lie algebra $L = L_{\geq 1}$ of finite type and finite depth. Then*

*(i) $I$ has a weak complement.*
*(ii) If $E$ is any weak complement for $I$ in $L$, then*

$$\text{depth}\, E + \text{depth}\, I = \text{depth}(E \oplus I) \leq \text{depth}\, L\,.$$

*proof* (i). Since $\text{depth}\, I \leq \text{depth}\, L < \infty$, (Proposition 11.2, p. 302), by Proposition 10.7 (p. 290) there are elements $x_1, \dots, x_r \in I$ such that the sub Lie algebra, $G$, generated by the $x_i$ satisfies $\text{Ext}_{UI}^*(\Bbbk, UL) \to \text{Ext}_{UG}^*(\Bbbk, UL)$ is non-zero. This implies that $A = \{\, y \in I \,|\, [y, x_i] = 0\,, 1 \leq i \leq r \,\}$ is a finite dimensional sub Lie algebra (Proposition 11.8, p. 308). Choose $n$ so that $A$ is concentrated in degrees $< n$ and set

$$F = \{\, y \in L_{\geq n} \,|\, [y, x_i] = 0\,, 1 \leq i \leq r \,\}\,.$$

Evidently $F \cap I = 0$.

On the other hand, $F$ is the kernel of the linear map $L_{\geq n} \to I \oplus \ldots \oplus I$ given by $x \mapsto ([x, x_1], \ldots, [x, x_r])$. Thus

$$\dim L_k / F_k \leq \sum_{i=1}^{r} \dim I_{k+\deg x_i} \,.$$

It follows that $L/F \leqslant I$. Now let $E$ be a sub Lie algebra of minimum depth among the sub Lie algebras $E \subset F$ satisfying $L/E \leqslant I$.

(ii) Since $L/E \leqslant I$, $L/(E \oplus I) \leqslant I$, and so by Lemma 14.4(v) (p. 373), $L/(E \oplus I)$ is an $I$-light $E$-module. Thus (ii) follows from Proposition 14.2 below.

□

**Proposition 14.2.** *Let $E$ and $I$ be respectively a sub Lie algebra and an ideal in a graded Lie algebra $L = L_{\geq 1}$ of finite type. If $E \cap I = 0$, then for any graded vector space $S = S_{\geq 0}$, of finite type,*

*(i) If $E$ is $S$-minimal and $I$ is an $S$-light $E$-module (adjoint representation), then $E \oplus I$ is $S$-minimal with respect to $I$, and*

$$\mathit{depth}\,(E \oplus I) = \mathit{depth}\, E + \mathit{depth}\, I \,.$$

*(ii) If, moreover, $L/(E \oplus I)$ is an $S$-light $E$-module, then*

$$\mathit{depth}\,(E \oplus I) \leq \mathit{depth}\, L \,.$$

*proof:* (i) Use the inclusions $E, I \to (E \oplus I)$ and multiplication in $U(E \oplus I)$ to write $U(E \oplus I) = UI \otimes UE$. Then for $x \in E$, $a \in UI$, $b \in UE$, we have

$$x \cdot (a \otimes b) = (\mathrm{ad}\, x)a \otimes b + (-1)^{\deg a \deg x} a \otimes x \cdot b \,.$$

It follows that $\mathrm{Tor}^{UI}(\mathbb{k}, U(E \oplus I)^{\#}) = \mathrm{Tor}^{UI}(\mathbb{k}, (UI)^{\#} \otimes (UE)^{\#})$ as $E$-modules. Thus the Hochschild-Serre spectral sequence converges from

$$E^2_{p,q} = \mathrm{Tor}_p^{UE}(\mathrm{Tor}_q^{UI}(\mathbb{k}, (UI)^{\#}), (UE)^{\#}) \quad \text{to} \quad \mathrm{Tor}_{p+q}^{U(E \oplus I)}(\mathbb{k}, (U(E \oplus I))^{\#}) \,.$$

Now since $I$ is an $S$-light $E$-module, so is each $\wedge^q sI \otimes (UI)^{\#}$, and hence so are the subquotients $\mathrm{Tor}_q^{UI}(\mathbb{k}, (UI)^{\#})$. Since $E$ is $S$-minimal we may apply Theorem 14.3(ii) to the Lie algebra $E$, the ideal 0, and the $E$-modules $\mathrm{Tor}_q^{UL}(\mathbb{k}, (UI)^{\#})$. It follows from that Theorem that if $\mathrm{Tor}_q^{UI}(\mathbb{k}, (UI)^{\#}) \neq 0$ then $\mathrm{depth}\, E = \mathrm{grade}_E \mathrm{Tor}_q^{UI}(\mathbb{k}, (UI)^{\#})$. Hence $E^2_{p,q} = 0$ for $q < \mathrm{depth}\, I$ or for $p < \mathrm{depth}\, E$, and $E^2_{p,q} \neq 0$ when $q = \mathrm{depth}\, I$ and $p = \mathrm{depth}\, E$. A standard corner argument now shows that $\mathrm{depth}\,(E \oplus I) = \mathrm{depth}\, E + \mathrm{depth}\, I$.

Finally, we show that $E \oplus I$ is $S$-minimal with respect to $I$. In fact, suppose $G \subset E \oplus I$, is a sub Lie algebra containing $I$ and satisfying $(E \oplus I)/G \leqslant S$. Then $G = F \oplus I$ where $F \subset E$ is a sub Lie algebra and $E/F \leqslant S$. Form the Hochschild-Serre spectral sequence

$$E^2_{p,q} = \mathrm{Tor}_p^{UF}(\mathrm{Tor}_q^{UI}(\Bbbk, (UI)^{\#}), (UF)^{\#}) \Rightarrow \mathrm{Tor}_{p+q}^{U(F\oplus I)}(\Bbbk, (U(F \oplus I))^{\#}).$$

It follows from this spectral sequence that for some $q \geq \mathrm{depth}\, I$,

$$\mathrm{grade}_F \mathrm{Tor}_q^{UI}(\Bbbk, (UI)^{\#}) \leq \mathrm{depth}(F \oplus I) - q.$$

But according to Theorem 14.3(i), applied with $L$, $I$ and $M$ replaced by $E$, 0 and $\mathrm{Tor}_q^{UI}(\Bbbk, (UI)^{\#})$, there is a sub Lie algebra $E' \subset F$ such that $F/E' \leqslant S$ and $\mathrm{depth}\, E' \leq \mathrm{grade}_F \mathrm{Tor}_q^{UI}(\Bbbk, (UI)^{\#})$. Thus

$$\mathrm{depth}\, E' \leq \mathrm{depth}(F \oplus I) - q \leq \mathrm{depth}(F \oplus I) - \mathrm{depth}\, I.$$

Since $E/F \leqslant S$, $E/E' \leqslant S$. But, because $E$ is $S$-minimal, $\mathrm{depth}\, E \leq \mathrm{depth}\, E'$. Thus $\mathrm{depth}\, E \leq \mathrm{depth}\,(F \oplus I) - \mathrm{depth}\, I$, and so

$$\mathrm{depth}\,(E \oplus I) = \mathrm{depth}\, E + \mathrm{depth}\, I \leq \mathrm{depth}(F \oplus I).$$

(ii) Consider the Hochschild-Serre spectral sequence converging from

$$E_1^{p,q} = \mathrm{Ext}^p_{U(E\oplus I)}(\wedge^q sL/(E \oplus I), UL) \quad \text{to} \quad \mathrm{Ext}^{p+q}_{UL}(\Bbbk, UL).$$

Since $L/(E \oplus I)$ is an $S$-light $E$-module annihilated by $I$, it is also an $S$-light $E \oplus I$-module. Thus each $\wedge^p sL/(E \oplus I)$ is an $S$-light $(E \oplus I)$-module annihilated by $I$. Since $E \oplus I$ is $S$-minimal with respect to $I$, Theorem 14.3(ii) asserts that either $\wedge^q sL/(E \oplus I) = 0$, or else

$$\mathrm{depth}\,(E \oplus I) = \mathrm{grade}_{E\oplus I}(\wedge^q sL/(E \oplus I)).$$

Since $\mathrm{Ext}^p_{U(E\oplus I)}(\wedge^q sL/(E\oplus I), UL) \neq 0$ for some $p+q = \mathrm{depth}\, L$, assertion (ii) follows.

□

**Proposition 14.3.** *Let $J$ and $K$ be ideals in a graded Lie algebra $L = L_{\geq 1}$ of finite type and finite depth. Then there is a weak complement, $E$, for $J \cap K$ in $K$ that is also a weak complement for $J$ in $J + K$.*

*proof:* By definition $E$ is $J \cap K$-minimal and $K/E \leqslant J \cap K$. Now $E \cap J = (E \cap K) \cap J = 0$. Moreover, $K/E$ maps onto $(K + J)/(E \oplus J)$ and so $(K + J)/(E \oplus J) \leqslant J \cap K$. Thus

$$(K + J)/E \leqslant J \cap K + J \leqslant J.$$

Finally, since $E$ is $J \cap K$-minimal it is trivially $J$-minimal and hence is a weak complement for $J$ in $K + J$.

□

**Corollary 14.1.** *With the notation and hypotheses of Proposition 14.3, if* $\operatorname{depth} J = \operatorname{depth}(J + K)$ *then* $J \cap K$ *is full in* $J$.

*proof:* By Proposition 14.3 there is a weak complement, $E$, for $J \cap K$ in $J$ that is also a weak complement for $J$ in $J + K$. Thus by Theorem 14.4,

$$\operatorname{depth} E + \operatorname{depth} J = \operatorname{depth}(E \oplus J) \leq \operatorname{depth}(J + K)\,.$$

It follows that $\operatorname{depth} E = 0$. This implies (Proposition 11.3, p. 302) that $E$ is finite dimensional. Since $E \oplus (J \cap K)$ is full in $J$ it follows that $J \cap K$ is full in $J$.

□

Now let $R$ be the hyper radical of a graded Lie algebra $= L_{\geq 1} L$ of finite type and finite depth. Then $R$ has a weak complement $E$ in $L$. In particular $E \cap R = 0$ and $E \oplus R$ is full in $L$. By Lemma 14.3 (p. 370) $\log\operatorname{index} R < \log\operatorname{index} L$. This implies that $\log\operatorname{index} E = \log\operatorname{index} L$, while $\operatorname{depth} E + \operatorname{depth} R \leq \operatorname{depth} L$ by Theorem 14.4.

## 14.5 L-equivalence

**Notation.** Recall from §14.1 that two graded vector spaces $V$ and $W$ are *equivalent* ($V \sim W$) if $V \leqslant W$ and $W \leqslant V$. Thus, as recalled there, an equivalence relation (*L-equivalence*) on the ideals of a graded Lie algebra, $L = L_{\geq 1}$ of finite type is defined by:

$$I \sim_L J \qquad \Leftrightarrow \qquad \text{for all ideals } K \subset L,\ I \cap K \sim J \cap K\,.$$

The set of L-equivalence classes of ideals in $L$ will be denoted by $\mathcal{L}$.

If $I$ is an ideal in $L$ its L-equivalence class will be denoted by $[I]$. Finally, the number (possibly $\infty$) of L-equivalence classes of ideals is denoted by $\nu(L)$.

Our principal objective here is to establish the main Theorem of the Chapter:

**Theorem 14.1.** *Let $L = L_{\geq 1}$ be a graded Lie algebra of finite type and finite depth $m$. Then the number $\nu(L)$ of $L$-equivalence classes of ideals in $L$ satisfies $\nu(L) \leq 2^m$ and equality holds if and only if $L \sim_L I(1) \oplus \cdots \oplus I(m)$ where the $I(i)$ are ideals of depth 1.*

Theorem 14.1 is included in Theorem 14.5 below, following the necessary preliminary materials.

**Proposition 14.4.** *Let $L = L_{\geq 1}$ be a graded Lie algebra of finite type and finite depth.*

*(i) The structure of a distributive lattice in $\mathcal{L}$ with initial element $[0]$ and maximum element $[L]$ is defined by*

$$[J] \leq [I] \Longleftrightarrow I \cap J \sim_L J\,, \quad [I] \vee [J] = [I+J] \quad \text{and } [I] \wedge [J] = [I \cap J]\,.$$

*(ii) If $I$ and $J$ are ideals in $L$ then any chain of strict inequalities in $\mathcal{L}$ of the form*

$$[J] < [I(1)] < \cdots [I(r)] = [I]$$

*satisfies $r \leq \operatorname{depth} I - \operatorname{depth} J$. In particular, the length of such a chain is at most $\operatorname{depth} L$.*

*(iii) Maximal chains have the same length.*

**Definition.** The maximal length $r$ in the chain above is called the *height* of $[I]$ over $[J]$. When $[J] = [0]$, $r$ is called the *height* of $[I]$ and denoted by $ht[I]$. In particular $ht[L]$ is the *height* of $L$.

**Remark.** It follows from Proposition 14.4 that the height of $[I]$ over $[J]$ is just $ht[I] - ht[J]$.

The proof of Proposition 14.4 requires two Lemmas.

**Lemma 14.7.** *With the hypotheses of Proposition 14.4, if $I, J, K$ are ideals in $L$, then*

*(i) $I \cap J \sim [I, J]$.*
*(ii) $(I + J) \cap K \sim I \cap K + J \cap K$.*

*proof:* (i) By Theorem 14.4 there is a weak complement $E$ for $[I, J]$ in $I \cap J$. Since $E \oplus [I, J]$ is full in $I \cap J$ it follows that $E \oplus [I, J] \sim I \cap J$. But

$[E,E] \subset E \cap [I,J] = 0$ and so $E$ is an abelian Lie algebra of finite depth. Thus (Proposition 11.3, p. 302) $\dim E < \infty$ and so $[I,J] \sim I \cap J$.

(ii) $(I+J) \cap K \sim [I+J, K] = [I,K] + [J,K] \sim I \cap K + J \cap K$.

□

**Lemma 14.8.** *With the hypotheses of Proposition 14.4 let $I, J$ be ideals in $L$. Then*

*(i)* $I \sim_L J \Leftrightarrow I \sim_L (I+J) \sim_L J$.
*(ii)* $I \sim_L J \Leftrightarrow I \sim_L (I \cap J) \sim_L J$.
*(iii) If* $J \subset I$ *and* $\operatorname{depth} J = \operatorname{depth} I$, *then* $J \sim_L I$.
*(iv) If* $I(i) \sim_L J(i)$ *are pairs of L-equivalent ideals in L, then*

$$I(1) + \cdots + I(r) \sim_L J(1) + \cdots + J(r)\,.$$

*(v) For any ideal K,* $(I+J) \cap K \sim_L I \cap K + J \cap K$.
*(vi) If* $I \sim_L J$ *and K is any ideal in L then* $I \cap K \sim_L J \cap K$.

*proof:* (i) We need only show that $I \sim_L J \Rightarrow I \sim_L (I+J)$. If $K$ is any ideal in $L$, then $J \cap K \leqslant I \cap K$ and so $I \cap K + J \cap K \leqslant I \cap K$. It follows that $I \cap K \sim I \cap K + J \cap K$, and so by Lemma 14.7,

$$I \cap K \sim I \cap K + I \cap K \sim I \cap K + J \cap K \sim (I+J) \cap K\,.$$

Thus $I \sim_L I + J$.

(ii) We need only prove that $I \sim_L J \Rightarrow I \sim_L I \cap J$. Again let $K$ be an ideal in $L$. Then

$$(I \cap J) \cap K = I \cap (J \cap K) \sim J \cap (J \cap K) = J \cap K\,.$$

Thus $I \cap J \sim_L J$.

(iii) Let $K$ be an ideal in $L$. Since $J \subset J + (I \cap K) \subset I$ we have by Proposition 10.5 (p. 286) that

$$\operatorname{depth} J \leq \operatorname{depth}(J + (I \cap K)) \leq \operatorname{depth} I\,.$$

Thus $\operatorname{depth} J = \operatorname{depth}(J + (I \cap K))$. It follows from Corollary 14.1 (p. 380) that $J \cap (I \cap K)$ is full in $I \cap K$. But $I \cap J = J$ and so $J \cap K$ is full in $I \cap K$. Thus $J \cap K \sim I \cap K$ for all $K$; i.e. $J \sim_L I$.

(iv) We need only show that if $I \sim_L J$ and $H$ is an ideal in $L$, then $I + H \sim_L J + H$. But for any ideal $K$ we have by Lemma 14.7 and Lemma 14.1(iv) (p. 368) that

$$\begin{aligned} &K \cap (I+H) \sim (K \cap I) + (K \cap H) \sim K \cap I \oplus K \cap H \\ &\sim K \cap J \oplus K \cap H \sim (K \cap J) + (K \cap H) \sim K \cap (J+H)\,. \end{aligned}$$

Thus $I + H \sim_L J + H$.

(v) For any ideal $H \subset L$ we have by Lemma 14.7 that

$$(I + J) \cap K \cap H \sim (I \cap K \cap H) + (J \cap K \cap H) \sim [(I \cap K) + (J \cap K)] \cap H.$$

Thus $(I + J) \cap K \sim_L I \cap K \, + \, J \cap K$.

(vi) For any ideal $H$, $(I \cap K) \cap H = I \cap (K \cap H) \sim J \cap (K \cap H) = (J \cap K) \cap H$.

□

*proof of Proposition 14.4:* (i) If follows from Lemma 14.8(vi) that the condition $I \cap J \sim_L I$ depends only on $[I]$ and $[J]$; thus the partial order is well defined. Clearly $[0]$ and $[L]$ are initial and terminal elements. It follows from Lemma 14.8(iv) and Lemma 14.8(vi) that $[I] \vee [J]$ and $[I] \wedge [J]$ only depend on $[I]$ and $[J]$. Similarly it follows from Lemma 14.8 ((vi) and (v)) that if $[H] \leqslant [I]$, $[J] \leqslant [K]$ then $[H] \leqslant [I \cap J]$ and $[I + J] \leqslant$[K]. Thus $\mathcal{L}$ is a lattice and Lemma 14.8(v) shows that the lattice is distributive.

(ii) This follows from Lemma 14.8(iii) which asserts that if $[J] < [I]$ then $\operatorname{depth} J < \operatorname{depth} I$.

(iii) This is a standard fact about distributive lattices ( [25]).

□

**Theorem 14.5.** *Let $L = L_{\geq 1}$ be a graded Lie algebra of finite type, finite depth $m$, and height $r$. Then*

*(i) $\operatorname{depth} L \geq \operatorname{height} L$.*
*(ii) $\nu(L) \leq 2^r$.*
*(iii) $\nu(L) = 2^r$ if and only if $L \sim_L I(1) \oplus \cdots \oplus I(r)$ where the $I(i)$ are infinite dimensional ideals. In this case $I(i)$ has height 1.*
*(iv) If $\nu(L) = 2^m$, then $\operatorname{depth} L = \operatorname{height} L$ and the $I(i)$ are infinite dimensional ideals of depth* 1.

For the proof of Theorem 14.5 we require the following notation and Lemma.

**Notation.** If $I \subset L$ is an ideal then $\nu_L(I)$ is the number of elements of $\mathcal{L}$ represented by ideals $J$ in $L$ with $J \subset I$. (Note that this may be different from $\nu(I)$ since an ideal in the Lie algebra $I$ may not be an ideal in $L$.)

**Lemma 14.9.** *With the hypotheses of Theorem 14.5,*

*(i) If $I \subset J$ are ideals in $L$ then $\nu_L(I) \le \nu_L(J)$.*
*(ii) If $I$ and $J$ are ideals in $L$ and $I \sim_L J$ then $\nu_L(I) = \nu_L(J)$. In particular, $\nu_L[I]$ is well defined.*
*(iii) If $I$ is the direct sum of ideals $J$ and $K$ in $L$, then $\nu_L(I) = \nu_L(J)\nu_L(K)$.*

*proof :* (i) This is immediate from the definition.

(ii) Since $I \sim_L (I \cap J)$ (Lemma 14.8(ii)), if $H$ is an ideal in $L$ contained in $I$ it follows that $H = (H \cap I) \sim_L (H \cap I \cap J)$ (Lemma 14.8(vi)). Thus the set of L-equivalence classes of ideals in $I$ coincides with the set of L-equivalence classes of ideals in $I \cap J$, and so $\nu_L(I) = \nu_L(I \cap J) = \nu_L(J)$.

(iii) Any ideal $H$ in $L$ contained in $I$ satisfies $H \sim_L (H \cap J) \oplus (H \cap K)$ (Lemma 14.8(v)). Suppose $G$ is another ideal contained in $I$ such that $G \cap J \sim_L H \cap J$ and $G \cap K \sim_L H \cap K$. Then $G \sim_L (G \cap J) \oplus (G \cap K) \sim_L (H \cap J) \oplus (H \cap K) \sim_L H$. It follows that $\nu_L(I) = \nu_L(J)\nu_L(K)$.

□

*proof of Theorem 14.5*: (i) This is Proposition 14.4(ii).

(ii) Let $0 < [J(1)] < \ldots < [J(r)] = [L]$ be a maximal chain of strict inclusions in $\mathcal{L}$, and let $\mathcal{L}(k)$ denote the subset of $\mathcal{L}$ of elements $[J] \le [J(k)]$. Then, for $1 \le k \le r$, let $[K] \in \mathcal{L}$ be an element of minimum height satisfying the two conditions:

$$[K] \le [J(k)] \qquad \text{and} \qquad [K] \not\le [J(k-1)]\,.$$

We shall show that the map $\varphi(k) : \mathcal{L}(k-1) \times \mathbb{Z}_2 \to \mathcal{L}(k)$ given by

$$([J], 0) \mapsto [J] \quad \text{and} \quad ([J], 1) \mapsto [J] \vee [K]$$

is a surjection.

Since the chain is maximal, the conditions above imply that $[J(k-1)] \vee [K] = [J(k)]$. Thus for any $[J] \in \mathcal{L}(k)$ we have $[J] = ([J] \wedge [J(k-1)]) \vee ([J] \wedge [K])$. If $[J] \wedge [K] \not\le [J(k-1)]$ then it too satisfies the conditions above. Since $[J] \cap [K] \leqslant [K]$, it has height $\le ht[K]$. But $ht[K]$ was a minimum; thus $[J] \wedge [K] = [K]$ in this case. It follows that $\varphi(k)$ is indeed a surjection. In particular, $\nu_L[J(k)] \le 2\nu_L[J(k-1)]$ and so $\nu_L(L) \le 2^r \nu_L(0) = 2^r$.

(iii) Suppose first that $\nu_L = 2^r$. Then each $\varphi(k)$ is a bijection. But clearly

$$\varphi(k)([0], 1) = [K] = ([J(k-1)] \wedge [K]) \vee [K] = \varphi(k)([J(k-1)] \wedge [K], 1)\,.$$

Thus $J(k-1)\cap K \sim 0$, and hence is concentrated in degrees $< n$, some $n$. Now set $I(k) = J(k)\cap K_{\geq n}$. Since $K_{\geq n} \sim_L K$, $I(k) \sim_L K$. Thus $[J(k)] = [J(k-1)\oplus I(k)]$. This also implies that $I(k)$ is not L-equivalent to zero; i.e. $\dim I(k)$ is infinite. Iterating this construction gives $[L] = [I(1)\oplus\cdots\oplus I(r)]$.

Conversely, suppose $L \sim_L I(1)\oplus\cdots\oplus I(r)$, where each $I(i)$ is an infinite dimensional ideal. Then $[0] < [I(i)]$, each $i$, and so $\nu_L[I(i)] \geq 2$. By Lemma 14.9(iii), and Theorem 14.5(ii), $2^r \geq \nu_L = \prod_{i=1}^r \nu_L[I(i)] \geq 2^r$. Thus $\nu_L = 2^r$ and $\nu_L[I(i)] = 2$, $1\leq i\leq r$. In particular, each $I(i)$ has height 1.

(iv) If $\nu_L = 2^m$ we must have $m = r$. Since the $I(i)$ are infinite dimensional, $\operatorname{depth} I(i) \geq 1$ and because $m = \operatorname{depth} L \geq \sum_{i=1}^m \operatorname{depth} I(i)$, it follows that $\operatorname{depth} I(i) = 1$, each $i$. □

**Corollary 14.2.** *Let $L$ be as in Theorem 14.5. If $L \sim_L I(1)\oplus\cdots\oplus I(r)$, where the $I(i)$ are infinite dimensional ideals of height* 1, *then $ht[L] = r$ and every element $[I] \in \mathcal{L}$ of height $s \geq 1$ is uniquely of the form $[I] = [I_{i_1}]\vee\ldots\vee[I_{i_s}]$.*

*proof:* It is a trivial consequence of the distributive law that

$$[I_{i_1}]\vee\ldots\vee[I_{i_s}] = [I_{j_1}]\vee\ldots\vee[I_{j_q}]$$

if and only if $s = q$ and $\{i_1,\cdots,i_s\} = \{j_1,\cdots,j_q\}$. Thus the elements of the form $[I_{i_1}]\vee\ldots\vee[I_{i_s}]$, $1\leq s\leq r$ are $2^r - 1$ distinct elements of $\mathcal{L}$, and so $\nu_L \geq 2^r$.

On the other hand, because each $I(i)$ has height 1, it is also an immediate consequence of the distributive law that $0 < [I(1)] < \cdots < [I(1)]\vee\ldots\vee[I(k)] < \ldots < [L]$ is a chain of maximum length, so that $ht[L] = r$ and $\nu_L \leq 2^r$. Thus $\nu_L = 2^r$ and $\mathcal{L} = \{[0], [I_{i_1}]\vee\ldots\vee[I_{i_s}]\}$. Clearly $ht([I_{i_1}]\vee\ldots\vee[I_{i_s}]) = s$.

□

**Remark.** Theorem 14.5 and its Corollary show that the graded Lie algebras $L = L_{\geq 1}$ of finite type and finite depth, and satisfying $ht[L] = r$ and $\nu_L = 2^r$, are the analogues in this setting of the classical semi-simple Lie algebras. Note that this includes as a special case the graded Lie algebras $L$ with $\operatorname{depth} L = m$ and $\nu_L = 2^m$.

It is an outstanding problem as to whether the homotopy Lie algebra of a simply connected rationally hyperbolic CW complex contains a free Lie algebra on two generators. However we have:

**Theorem 14.6.** *Suppose $L = L_{\geq 1}$ is a graded Lie algebra of finite type and finite depth. If*

$$2ht[L] \leq depth\, L - 1$$

*then $L$ contains a free Lie algebra on two generators.*

**Lemma 14.10.** *Suppose $J \subset I$ are ideals in a graded Lie algebra $L = L_{\geq 1}$ of finite type and finite depth. If $[J] < [I]$ and $depth\, J + 1 = depth\, I$, then $I$ contains an infinite dimensional sub Lie algebra of depth 1.*

*proof:* Since $[J] < [I]$ there is an ideal $H \subset L$ such that $J \cap H$ is not full in $I \cap H$. Set $K = I \cap H$; then $J \cap K$ is not full in $K$. Thus a weak complement, $E$, for $J \cap K$ in $K$ is infinite dimensional.

Next note that since $J \subset J + K \subset I$, either depth $J$ = depth $(J + K)$ or depth$(J + K)$ = depth $I$. The first equality would imply $J \sim_L (J + K)$ (Lemma 14.8). Intersection with $K$ then gives $J \cap K \sim_L K$, which is impossible because $J \cap K$ is not full in $K$. Thus

$$\text{depth}(J + K) = \text{ depth}\, J + 1\,.$$

But since $E$ may be chosen to also be a weak complement for $J$ in $J+K$ (Proposition 14.3), Theorem 14.4 yields

$$\text{depth}\, E + \text{depth}\, J \leq \text{depth}(J + K) = \text{depth}\, J + 1\,.$$

This gives depth $E \leq 1$. But $E$ is infinite dimensional and thus depth $E = 1$. □

*proof of Theorem 14.6*: Let $r = ht[L]$ and let $0 < [I(1) < \ldots < [I(r)] = [L]$ be a chain of strict inclusions in $\mathcal{L}$. We may assume $I(1) \subset \cdots \subset I(r)$, and then it follows from Lemma 14.8 that $0 <$ depth $I(1) < \ldots <$ depth $I(r)$. In view of our hypothesis, either depth $I(1) = 1$ or for some $i$, depth $I(i+1) =$ depth $I(i) + 1$. Lemma 14.10 then implies that $I(i+1)$ contains an infinite dimensional sub Lie algebra of depth 1. Finally according to [19] each infinite dimensional sub Lie algebra of depth 1 contains a free Lie algebra on two generators. □

**Example.** Let $L_X$ be the homotopy Lie algebra of the space

$$X = S^3_a \vee S^3_b \vee S^5_z \cup_{[a,z]} e^8 \cup_{[a,[b,z]]} e^{10} \cup_{[b,[b,z]]} e^{10}\,.$$

Then $L_X$ has depth 2 and the lattice $\mathcal{L}$ has exactly three elements.

The Sullivan minimal model of $X$ is quasi-isomorphic to the differential graded algebra $(A, d) = (\wedge(x, y, z, t)/(xy, tz), d)$ where $\deg x = \deg y = 3$,

$\deg z = 5$, $\deg t = 7$, $dx = dy = dz = 0$, $d(t) = yz$. The algebra $(A, d)$ is a semifree $(\wedge(x, y)/(xy), 0)$-module. This gives a rational fibration

$$F = S^5 \vee S^7 \to X \to B = S^3 \vee S^3 .$$

The ideal $L_F$ has not the same log index as $L_X$, and so is neither L-equivalent to $L_X$ nor to 0. The exact sequence $0 \to L_F \to L_X \to L_B \to 0$ implies at once that

$$[0] < [L_F] < [L_X]$$

are the only elements of $\mathcal{L}$. In particular $L_F$ is the hyper radical of $L_X$.

□

## 14.6 The odd part of a graded Lie algebra

**Theorem 14.7.** *Let $L = L_{\geq 1}$ be a graded Lie algebra of finite type and finite depth. Then the sub Lie algebra $L_{even}$ is full in $L$. In particular log index $L_{even}$ = log index $L$.*

*proof:* Let $I$ be the sub Lie algebra generated by $L_{\text{odd}}$; $I$ is clearly an ideal in $L$ and hence has finite depth. Choose $x_1, \ldots, x_n$ of odd degrees $e_1 \leq \ldots \leq e_n$ that generate a sub Lie algebra $F$ for which $\mathrm{Ext}_*^{UI}(\Bbbk, UI) \to \mathrm{Ext}_*^{UF}(\Bbbk, UI)$ is non-zero (Proposition 10.7, p. 290). The centralizer of the $x_i$ in $I$ is therefore finite dimensional (Proposition 11.8, p. 308), which implies that for some $N$ the linear map $x \mapsto ([x, x_1], \ldots, [x, x_n])$ is an injection $I_k \to I_{k+e_1} \oplus \cdots \oplus I_{k+e_n}$, $k \geq N$. Since the $e_i$ are odd, it follows that for $k \geq N$,

$$\dim(I_{\text{odd}})_k \leq \sum_{j=k+e_1}^{k+e_n} \dim(I_{\text{even}})_j \quad \text{and} \quad \dim(I_{\text{even}})_k \leq \sum_{j=k+e_1}^{k+e_n} \dim(I_{\text{odd}})_j ,$$

which implies that both $I_{\text{odd}}$ and $I_{\text{even}}$ are full in $I$.

Finally, let $E$ be a weak complement for $I$ in $L$. Then $E \subset L_{\text{even}}$ and $E \oplus I$ is full in $L$. Since $I_{\text{even}}$ is full in $I$ it follows that $E \oplus I_{\text{even}}$ is full in $L$ and so $L_{\text{even}}$ is full in $L$. □

**Remark.** The log index of $L_{odd}$ can be strictly lower than the log index of $L$. Consider for instance the direct sum, $L = \mathbb{L}(x, y) \oplus \mathbb{L}(u, v)$, of two free Lie algebras with $x, y$ in degree 2 and $u, v$ in degree 5. Then log index $L = \sqrt{2}$ and log index $L_{odd} = \sqrt[5]{2}$.

# Chapter 15

# Weight decompositions of a Sullivan Lie algebra

## 15.1 Weight decompositions

Recall from §7.4 that a *weight decomposition* of a graded Lie algebra $L = L_{\geq 0}$ is a decomposition

$$L = \oplus_{p\geq 1} L(p)$$

for which each $[L(p), L(q)] \subset L(p+q)$. Recall also that the lower central filtration $(L^n)_{n\geq 1}$ of a graded Lie algebra $L$ is defined by $L^1 = L$ and $L^{n+1} = [L, L^n]$. In particular, $L^n \subset \oplus_{p\geq n} L(p)$, and there follows

**Lemma 15.1.** *Suppose $L = L_{\geq 0}$ is a graded Lie algebra equipped with a weight decomposition. If*

$$dim\, L_0/[L_0, L_0] < \infty \qquad and\ dim\, L_i < \infty\,, \quad i \geq 1\,, \tag{15.1}$$

*then $L$ is a Sullivan Lie algebra.*

**Example 1.** *The associated weighted Lie algebra.*

Suppose $L$ is a Sullivan Lie algebra (but not necessarily equipped with a weight decomposition). Then as observed in Corollary 2.5 (p. 70) and recalled in §7.4, a weighted Lie algebra $\mathcal{L}$ is defined by $\mathcal{L}(p) = L^p/L^{p+1}$ with Lie bracket induced from the Lie bracket $[-,-] : L^p \times L^q \to L^{p+q}$. This is called the *associated weighted Lie algebra* for $L$.

By definition, $[\mathcal{L}(1), \mathcal{L}(p)] = \mathcal{L}(p+1)$. Thus

$$\mathcal{L}/[\mathcal{L},\mathcal{L}] = \mathcal{L}(1) \quad \text{and } \mathcal{L}_0/[\mathcal{L}_0,\mathcal{L}_0] = \mathcal{L}(1)_0 = L_0/[L_0, L_0]\,.$$

In particular $\mathcal{L}$ is a Sullivan Lie algebra.

On the other hand, suppose $G$ is a group. Then as observed in §7.4 the classical identity of Hall ( [45], Theorem 5.3) shows that the commutator map $[\,,\,] : (a,b) \mapsto aba^{-1}b^{-1}$ induces linear maps

$$[\,,\,] : (G^n/G^{n+1} \otimes \mathbb{Q}) \otimes (G^p/G^{p+1} \otimes \mathbb{Q}) \to (G^{n+p}/G^{n+p+1} \otimes \mathbb{Q}) \tag{15.2}$$

which makes the direct sum $\mathcal{L}(G)\otimes\mathbb{Q} := \oplus_p(G^p/G^{p+1})\otimes\mathbb{Q}$ into a weighted Lie algebra. Moreover, according to Theorem 7.7 (p. 217), if $\dim G/[G,G]\otimes\mathbb{Q} < \infty$ and if $\mathcal{L}$ is the weighted Lie algebra for the homotopy Lie algebra of $BG$ then $\mathcal{L}_0 \cong \mathcal{L}(G)\otimes\mathbb{Q}$.

**Example 2.** Weight decomposition of $\mathbb{L}(x_1,\cdots,x_k)/(y_1,\cdots,y_\ell)$.

Suppose $\mathbb{L} = \mathbb{L}(x_1,\cdots,x_k)$ is the free graded Lie algebra on generators $x_i$ (cf [18], p. 289). Then $U\mathbb{L}$ is the tensor algebra on the subspace $X = \oplus_i \Bbbk\, x_i$ and it follows that

$$\mathbb{L} = \oplus_{p\geq 1}\mathbb{L}(p)\,, \qquad \text{with } \mathbb{L}(p) = \mathbb{L}\cap T^p(X)\,,$$

i.e. $\mathbb{L}(p)$ is the linear span of the iterated brackets $[x_{i_1},\cdots,x_{i_p}]$. This is clearly a weight decomposition of $\mathbb{L}$.

Now, suppose $y_1,\cdots,y_\ell \in \mathbb{L}$ satisfy $y_j \in \mathbb{L}(p_j)$, $1\leq j\leq \ell$. Then the ideal, $I$, generated by the $y_j$ has a weight decomposition given by $I(p) = I\cap\mathbb{L}(p)$. Hence a weight decomposition in the quotient

$$\mathbb{L}(x_1,\cdots,x_k)/(y_1,\cdots,y_\ell) = \mathbb{L}/I$$

is given by $(\mathbb{L}/I)\,(p) = \mathbb{L}(p)/I(p)$.

□

**Example 3.** Let $F = F_{\geq 2}$ be a graded Lie algebra concentrated in even degrees and define a new graded Lie algebra $L = L_0$ by setting

$$L = \bigoplus_p F_{2p}\,,$$

but with the same Lie bracket. A weight decomposition for $L$ is then given by $L(p) = F_{2p}$. If $\dim F/[F,F] < \infty$ it follows that $\dim L/[L,L] < \infty$ and so $L$ is a Sullivan algebra. In this case $F$ has finite type, and hence is the homotopy Lie algebra of a quadratic Sullivan algebra $(\wedge W, d_1)$ with $W = \{W^{2p+1}\}$ and $W^{2p+1} = sF_{2p}^{\#}$. Thus the associated quadratic Sullivan algebra for $L$ is given by $(\wedge V, d_1)$ with $V = V^1 = \oplus_p W^{2p+1}$ and with the differential coinciding with that in $\wedge W$.

In particular, if $\dim H(\wedge W, d_1) < \infty$ then $\dim H(\wedge V, d_1) < \infty$.

□

Recall from §2.1 the Cartan-Chevalley-Eilenberg constructions $C_*(L)$ and $C^*(L)$ for a graded Lie algebra $L = L_{\geq 0}$. If $L$ is a Sullivan Lie algebra then (§2.3) its associated quadratic Sullivan algebra, $(\wedge V, d_1)$ is defined by

$$(\wedge V, d_1) = \varinjlim_n C^*(L/L^n)\,.$$

Moreover in this case both depth $(\wedge V, d_1)$ and depth $L$ are defined, respectively in §10.2, p. 251, and §10.3, p. 256. As noted there, in general these two invariants may not coincide.

However, consider the case of a quadratic Sullivan algebra $(\wedge V, d_1)$ which is both the associated quadratic Sullivan algebra for a minimal Sullivan algebra $(\wedge V, d)$ and also for a Sullivan Lie algebra $L$.

**Theorem 15.1.** *With the hypotheses and notation above suppose $L_0$ has a weight decomposition. If $cat(\wedge V, d) < \infty$ and either $H(\wedge V, d)$ or $H(C_*(L_0))$ has finite type, then*

$$depth\, L = depth\,(\wedge V, d) \leq cat\,(\wedge V, d)\,.$$

Before proving Theorem 15.1 we require some preliminary material. Recall the definitions $C_*(L) = (\wedge sL, \partial)$ and $C^*(L) = (\oplus_q(\wedge^q sL)^\#, \partial^\#)$. Here the "wedge degree" induces direct decompositions,

$$H(C_*(L)) = \oplus_q H_q(C_*(L)) \quad \text{ and } H(C^*(L)) = \oplus_q H^q(C^*(L))\,,$$

and $q$ is called the *homological degree.* The homological degree is usually different from the "degree" $r$, and in fact

$$H_q(C_*(L)) = \{\, H_q(C_*(L))_r\}_{r\geq 0} \quad \text{ and } H^q(C^*(L)) = \{H^q(C^*(L))^r\}_{r\geq 0}\,.$$

Recall now from (10.5) that the surjections $L \to L/L^n$ induce a morphism

$$\mu : \varinjlim_n C^*(L/L^n) \to C^*(L)$$

of cochain algebras. The main step in proving Theorem 15.1 is

**Proposition 15.1.** *Suppose $L$ is a Sullivan Lie algebra.*

*(i) If $L$ has a weight decomposition then the natural map*

$$C_*(L) \to \varprojlim_n C_*(L/L^n)$$

*is a quasi-isomorphism.*

*(ii) If $L$ has a weight decomposition and if either $H(C_*(L))$ or $H(\wedge V, d_1)$ is a graded vector space of finite type then*

$$\mu : \varinjlim_n C^*(L/L^n) \to C^*(L)$$

*is a quasi-isomorphism.*

*proof:* (i) The weight decomposition $L = \oplus_{p\geq 1} L(p)$ induces the direct decomposition $\wedge^q sL = \oplus_p(\wedge^q sL)(p)$ where $(\wedge^q sL)(p)$ is the linear span of the subspaces $sL(p_1)\wedge\cdots\wedge sL(p_q)$ with $p_1+\cdots+p_q = p$. Evidently $(\wedge sL)(p) = \oplus_q(\wedge^q sL)(p)$ is preserved by $\partial$, and so this induces direct decompositions

$$H_q(C_*(L)) = \bigoplus_{p\geq 1} H_q\left(C_*(L)(p)\right).$$

Now a simple induction shows that

$$L^n = \bigoplus_{p\geq n} L(p)\cap L^n\,, \tag{15.3}$$

so that this defines a weight decomposition of $L^n$. The Hochschild-Serre spectral sequence

$$E^2_{r,s} = \mathrm{Tor}_r^{U\,L/L^n}(\Bbbk, \mathrm{Tor}_s^{UL^n}(\Bbbk,\Bbbk)) \Longrightarrow \mathrm{Tor}^{UL}_{r+s}(\Bbbk,\Bbbk)$$

is then the direct sum of the spectral sequences for the different weights $p$. For $p<n$ it follows from (15.3) that

$$E^2_{r,s}(p) = \begin{cases} 0 & s\geq 1 \\ \mathrm{Tor}_r^{UL_n}(\Bbbk,\Bbbk)(p) & s=0\,, \end{cases}$$

and thus that

$$\mathrm{Tor}^{UL}(\Bbbk,\Bbbk)(p) \xrightarrow{\cong} \mathrm{Tor}^{U\,L/L^n}(\Bbbk,\Bbbk)(p)\,.$$

But this can be restated as

$$H(C_*(L))(p) \xrightarrow{\cong} H(C_*(L/L^n))(p)\,, \qquad n>p\,,$$

which proves (i).

(ii) Suppose $H(C_*(L))$ has finite type. Then it follows that $H_q(C_*(L)) = (H^q(C^*(L)))^\#$. Thus the dual of $H(\mu)$ is a linear map

$$H(\mu)^\# : \varprojlim_n H(C_*(L/L^n)) \longleftarrow H(C_*(L))$$

and this is clearly the isomorphism of (i). In particular, $H(\mu)$ is an isomorphism.

On the other hand, suppose $H(\wedge V, d_1)$ has finite type. Since $L$ is a Sullivan Lie algebra each $L/L^n$ has finite type (Lemma 2.5, p. 54). It follows that each $(\wedge^q sL/L^n)^\# = \wedge^q\,(sL/L^n)^\#$, and has finite type. Hence each $C_q(L/L^n) \xrightarrow{\cong} \left(\wedge^q(sL/L^n)^\#\right)^\#$. It follows that

$$\varprojlim_n C_q(L/L^n) \xrightarrow{\cong} \varprojlim_n \left(\wedge^q(sL/L^n)^\#\right)^\# = \left(\varinjlim_n C^q(L/L^n)\right)^\#.$$

But $(\wedge V, d_1) = \varinjlim_n C^*(L/L^n)$. Therefore $H^q(\varinjlim_n C^*(L/L^n))$ has finite type and thus so does its dual. Now it follows from (i) that each $H_q(C_*(L))$ has finite type, and the previous assertion applies.

□

*proof of Theorem 15.1:* Because $L$ is a Sullivan Lie algebra, $L_{\geq 1}$ has finite type, and so $C^*(L_{\geq 1})$ is the associated quadratic Sullivan algebra $(\wedge Z, \overline{d}_1)$. In this case $\mu$ is just the identity. Moreover, because $L_0$ acts nilpotently in each $L_j$, $j \geq 1$, it follows that the exact sequence

$$0 \leftarrow L_0 \leftarrow L \leftarrow L_{\geq 1} \leftarrow 0$$

induces a decomposition of $C^*(L)$ as the Sullivan extension

$$C^*(L_0) \to (C^*(L_0) \otimes \wedge Z, d_1) \to (\wedge Z, \overline{d_1}) .$$

If $H(C^*(L_0))$ has finite type then by Proposition 15.1, $\mu : (\wedge V^1, d) \to C^*(L_0)$ is a quasi-isomorphism. On the other hand, suppose $H(\wedge V, d)$ has finite type. Since by hypothesis the quadratic Sullivan algebra for the Sullivan Lie algebra $L$ is $(\wedge V, d_1)$, it follows that $\dim V^i < \infty$, $i \geq 2$. Hence by Lemma 3.2 (p. 101), $H(\wedge V^1, d)$ has finite type. Now Proposition 15.1 asserts that $\mu : (\wedge V^1, d_1) \to C^*(L_0)$ is a quasi-isomorphism. Thus in either case it follows from Proposition 10.2 (p. 277) that $\mu$ is a quasi-isomorphism.

Now Theorem 10.2 (p. 278) gives $\operatorname{depth} L \leq \operatorname{depth}(\wedge V, d_1)$. But by definition, $\operatorname{depth}(\wedge V, d_1) = \operatorname{depth}(\wedge V, d)$ and Theorem 10.1 (p. 273) gives $\operatorname{depth}(\wedge V, d) \leq \operatorname{cat}(\wedge V, d)$.

□

## 15.2 Exponential growth of $L$

Recall that for any graded vector space $S$ of finite type,

$$\log \operatorname{index} S = \limsup_n \frac{\log \dim S_n}{n} .$$

Recall also from Example 1, that associated with any graded Lie algebra $L = L_{\geq 0}$ is the weighted graded Lie algebra $\mathcal{L} = \oplus_n L^n/L^{n+1}$.

**Theorem 15.2.** *Suppose $L$ is an infinite dimensional Sullivan Lie algebra, concentrated in even degrees. If the associated weighted Lie algebra, $\mathcal{L}$, satisfies $\dim H(C_*(\mathcal{L})) < \infty$, then*

*(i)* $\alpha_{\mathcal{L}} = \limsup_n \frac{\log \dim L^n/L^{n+1}}{n}$ *satisfies*

$$0 < \alpha_{\mathcal{L}} < \infty .$$

*(ii) there are constants $b, c$ and $K$ such that for $n \geq K$,*

$$\frac{1}{n^c} e^{n\alpha_{\mathcal{L}}} \leq \dim L^n/L^{n+1} \leq e^b e^{n\alpha_{\mathcal{L}}} .$$

*proof:* As stated in Example 1, $\mathcal{L} = \oplus \mathcal{L}(n)$ is a Sullivan Lie algebra. Let $(\wedge V, d_1)$ be the quadratic Sullivan algebra associated with $\mathcal{L}$. Then, since $\dim H(C_*(\mathcal{L})) < \infty$, it follows from Proposition 15.1 that

$$\mu : (\wedge V, d_1) \to C^*(\mathcal{L})$$

is a quasi-isomorphism. Hence $H(\wedge V, d_1)$ is finite dimensional.

Now $V$ is the direct sum of the subspaces $V(n) = [\, s\mathcal{L}(n)\,]^{\#}$ and

$$d_1 : V(n) \to +_{i+j=n} V(i) \wedge V(j) . \tag{15.4}$$

Moreover, since $L$ is concentrated in even degrees, $V$ is concentrated in odd degrees. Thus a new quadratic Sullivan algebra $(\wedge W, d_1)$, is defined as follows. First, set $W = \{W^{2p+1}\}$ with $W^{2p+1} = V(p)$. Since $V$ is concentrated in odd degrees this extends to an isomorphism of algebras $\wedge W = \wedge V$ (which may not preserve degrees). Finally, use this isomorphism to identify the differential in $\wedge V$ with a differential $d_1$ in $\wedge W$. Then it is immediate from (15.4) that $(\wedge W, d_1)$ is a quadratic Sullivan algebra.

Moreover, since $H(\wedge W, d_1) = H(\wedge V, d_1) = H(C^*(\mathcal{L}))$ is finite dimensional, we may apply Theorem 12.11 (p. 318).

Let $\alpha$ denote the log index of the homotopy Lie algebra $F$ of $(\wedge W, d_1)$. Then Theorem 12.11 gives positive constants $\beta$, $\gamma$, $N$ and $K$ such that for $q \geq K$,

$$e^{\alpha q - \gamma \log q} \leq \max_{q+2 \leq k \leq q+N} \dim F_k \leq e^{\alpha q + \beta} .$$

But $F$ is concentrated in even degrees and

$$\dim F_{2k} = \dim W^{2k+1} = \dim \mathcal{L}(k) = \dim L^k/L^{k+1} .$$

In particular it follows that $\log \operatorname{index} F = \frac{1}{2}\alpha_{\mathcal{L}}$. Thus the inequality above, with $q = 2p$, yields

$$e^{p\alpha_{\mathcal{L}} - \gamma \log 2p} \leq \max_{p+1 \leq k \leq p+N/2} \dim L^k/L^{k+1} \leq e^{p\alpha_{\mathcal{L}} + \beta} \qquad p \geq K .$$

The inequality on the right (with $n = p + 1$) gives

$$\dim L^{n+1}/L^{n+2} \leq e^{(n+1)\alpha_{\mathcal{L}}+\beta}, \qquad n \geq K.$$

On the other hand, since $\dim H(C_*(\mathcal{L})) < \infty$, $\dim L/[L,L] < \infty$. Let $x_1, \cdots, x_r \in L$ represent a basis of $L/[L,L]$. Then

$$\sum ad\, x_i : \oplus_{i=1}^r L^n/L^{n+1} \to L^{n+1}/L^{n+2}$$

is surjective, and it follows that for all $k$, $\dim L^k/L^{k+1} \leq r\left(\dim L^{k-1}/L^k\right)$. In particular,

$$\max_{n+2\leq k\leq n+N} \dim L^k/L^{k+1} \leq r^N \dim L^n/L^{n+1}.$$

Therefore for $n \geq K$,

$$\frac{1}{r^N} e^{n\alpha_{\mathcal{L}}-\gamma \log n} \leq \dim L^n/L^{n+1}.$$

The inequality on the left of the Theorem follows.

□

## 15.3 The fundamental Lie algebra of 1-formal Sullivan algebra

Recall from §2.10 that a minimal Sullivan algebra $(\wedge V, d)$ is *formal* if there is a quasi-isomorphism of the form $(\wedge V, d) \xrightarrow{\simeq} (H, 0)$, and 1-*formal* if there is a morphism $\varphi : (\wedge V, d) \to (H, 0)$ such that $H^1(\varphi)$ is an isomorphism and $H^2(\varphi)$ is injective. A topological space is *formal* (or 1-*formal*) if its minimal Sullivan model is formal (or 1-formal).

Now, according to Proposition 2.7 (p. 88) if a Sullivan algebra $(\wedge V, d)$ is 1-formal, then $V^1$ admits a direct sum decomposition $V^1 = \oplus_{n\geq 0} M_n$ such that $V_n^1 = \oplus_{i\leq n} M_i$ is the filtration defined in §2.2 (p. 48), and such that $d : M_n \to +_{i+j=n-1} M_i \wedge M_j$.

Suppose $(\wedge V, d)$ is a 1-formal minimal Sullivan algebra, and define a direct sum decomposition $V^1 = \oplus_{n\geq 1} V^1(n)$ by setting $V^1(n) = M_{n-1}$. Then setting $(\wedge^q V^1)(n)$ to be the linear span of the subspaces $V^1(n_1) \wedge \cdots \wedge V^1(n_q)$ with $n_1 + \cdots + n_q = n$ gives a direct sum decomposition

$$(\wedge V^1, d_1) = \bigoplus_{n\geq 1} ((\wedge V^1)(n), d_1).$$

In particular, as follows from Corollary 2.9 (p. 89), this gives a direct sum decomposition of $H(\wedge V^1, d)$ with $H^1(\wedge V, d) = H(\wedge V^1, d)(1)$ and $H^2(\wedge V, d) = H(\wedge V^1, d)(2)$.

This decomposition of $V^1$ then defines the weighted Lie algebra $\mathcal{L} = \mathcal{L}_0$ given by

$$\mathcal{L} = \bigoplus_{n\geq 1} \mathcal{L}(n)\,, \qquad \text{with } \mathcal{L}(n) = V^1(n)^\#\,, \tag{15.5}$$

with Lie bracket dual to the differential in $(\wedge V^1, d_1)$.

Thus $\mathcal{L}$ is a sub Lie algebra of the fundamental Lie algebra $L_0$ of $(\wedge V, d_1)$ defined in §2.1, and $\mathcal{L}$ is also the associated weighted Lie algebra for $L_0$ defined in §15.1. In particular, $\mathcal{L}^n = \oplus_{p\geq n}\mathcal{L}(p)$ and $L_0 = \varprojlim_n \mathcal{L}/\mathcal{L}^n$.

**Theorem 15.3.** *Suppose $(\wedge V, d)$ is a 1-formal minimal Sullivan algebra and that both $H(\wedge V, d)$ and $V^{\geq 2}$ have finite type. If $cat(\wedge V, d) < \infty$ then*

*(i) $dim\, H(\wedge V^1, d) < \infty$.*

*(ii) The fundamental Lie algebra $L_0$ of $(\wedge V, d)$ and its associated weighted Lie algebra $\mathcal{L}$ are Sullivan Lie algebras. Moreover,*

$$depth\,\mathcal{L} = depth\,(\wedge V^1, d) \leq cat\,(\wedge V^1, d) < \infty\,.$$

*(iii) $rad\,\mathcal{L} = rad\,L_0$ and*

$$dim\ rad\,\mathcal{L} \leq depth\,\mathcal{L}\,.$$

*(iv) There are constants $b, c, \alpha_{\mathcal{L}}$ and $K$ such that $0 < \alpha_{\mathcal{L}} < \infty$ and*

$$\frac{1}{n^c} e^{n\alpha_{\mathcal{L}}} \leq dim\,\mathcal{L}(n) = dim\, L_0^n/L_0^{n+1} \leq e^b e^{n\alpha_{\mathcal{L}}}\,, \qquad n \geq K\,.$$

*proof:* (i) Let $\mathrm{cat}\,(\wedge V, d) = m$. Then by Proposition 9.6 (p. 259), $\mathrm{cat}\,(\wedge V^1, d) < \infty$ and for some $N$, $H^i(\wedge V, d_1) = 0$, $i \geq N$. On the other hand, by Lemma 3.2 (p. 101), $H(\wedge V^1, d)$ has finite type, and thus $\dim H(\wedge V^1, d) < \infty$.

(ii) The first assertion is in Example 1 of §15.1. Moreover, since $\mathcal{L}$ has a weight decomposition and $(\wedge V^1, d)$ is its associated quadratic Sullivan algebra we may, in view of (i), apply Theorem 15.1 with $\mathcal{L} = L$ and $(\wedge V^1, d) = (\wedge V, d)$ to obtain the second assertion.

(iii) In view of (ii) the inequality $\dim\ \mathrm{rad}\,\mathcal{L} \leq \mathrm{depth}\,\mathcal{L}$ follows from Theorem 10.6 (p. 295). Since $\dim\ \mathrm{rad}\,\mathcal{L} < \infty$, $\mathrm{rad}\,\mathcal{L} \subset \oplus_{i\leq r}\mathcal{L}(i)$ for some $r$. Now any element of $L_0$ may be written as an infinite sum $x = \sum_{p\geq 1} x_p$ with $x_p \in \mathcal{L}(p)$. For each $p$, $[x_p, \mathrm{rad}\,\mathcal{L}] \subset \mathrm{rad}\,\mathcal{L} \subset \oplus_{i\leq r}\mathcal{L}(i)$. It follows that $[x_p, \mathrm{rad}\,\mathcal{L}] = 0$ for $p > r$ and so $[x, \mathrm{rad}\,\mathcal{L}] \subset \mathrm{rad}\,\mathcal{L}$. Thus $\mathrm{rad}\,\mathcal{L}$ is a solvable ideal in $\mathcal{L}_0$ and so $\mathrm{rad}\,\mathcal{L} \subset \mathrm{rad}\,L_0$.

On the other hand, let $I \subset L_0$ be a solvable ideal and let $I(p)$ be the image of $I \cap L_0^p$ in $\mathcal{L}(p)$. If $x \in I(p)$ and $y \in \mathcal{L}(q)$ then $[y, x] \subset I(p+q)$, and it follows that $\oplus_p I(p)$ is an ideal in $\mathcal{L}$. Moreover, if $x \in I(p)$ and $y \in I(q)$ then there are elements $x' \in L_0^{p+1}$ and $y' \in L_0^{q+1}$ such that $x+x'$ and $y+y'$ are in $I$. Thus if $I$ is abelian it follows that $[x,y]=0$ and so $\oplus_p I(p)$ is an abelian ideal in $\mathcal{L}$ and hence contained in $\mathrm{rad}\,\mathcal{L}$.

Now set $J = I + \mathrm{rad}\,\mathcal{L}$ and for $x \in J$ write $x = \sum x_p$ with $x_p \in \mathcal{L}(p)$. If $x_1, \cdots, x_q \in \mathrm{rad}\,\mathcal{L}$ then $\sum_{i>q} x_i \in J$, and so $x_{q+1} \in I(q+1) + (\mathrm{rad}\,\mathcal{L})(q+1) \subset \mathrm{rad}\,\mathcal{L}$. Thus each $x_i \in \mathrm{rad}\,\mathcal{L}$, and since $\dim \mathrm{rad}\,\mathcal{L} < \infty$ it follows that for some $q$, $x_i = 0$, $i > q$. It follows that $x \in \mathrm{rad}\,\mathcal{L}$ and $I \subset \mathrm{rad}\,\mathcal{L}$. An easy induction now shows that any solvable ideal in $L_0$ is contained in $\mathrm{rad}\,\mathcal{L}$ and hence $\mathrm{rad}\,L_0 = \mathrm{rad}\,\mathcal{L}$.

(iv) It follows from Proposition 15.1(ii) that $(\wedge V^1, d) \xrightarrow{\cong} C^*(\mathcal{L})$. Then by (i) $\dim H(C_*(\mathcal{L})) < \infty$, and (iv) follows from Theorem 15.2.

□

**Corollary 15.1.** *If $G$ is a discrete group whose classifying space $BG$ is a 1-formal Sullivan space for which $cat(BG) < \infty$, then the Lie algebra $\mathcal{L}(G) = \oplus_p G^p/G^{p+1}$ of Example 1 (p. 389) satisfies*

*(i) $depth(\mathcal{L}(G) \otimes \mathbb{Q}) \le cat(BG)$,*
*(ii) $dim\ rad(\mathcal{L}(G) \otimes \mathbb{Q}) \le depth(\mathcal{L}(G) \otimes \mathbb{Q}) \le cat(BG)$.*
*(iii) There are constants $b, c, \alpha_G$ and $K$ such that $0 < \alpha_G < \infty$ and*

$$\frac{1}{n^c} e^{n\alpha_G} \le dim\ \left(G^n/G^{n+1}\right) \otimes \mathbb{Q} \le e^b e^{n\alpha_G}\,, \qquad k \ge K\,.$$

*proof:* By Theorem 7.2 (p. 197), $BG$ has a minimal Sullivan model of the form $(\wedge V^1, d)$ and $H(\wedge V^1, d)$ has finite type. Moreover, by Theorem 9.2 (p. 247), $\mathrm{cat}\,(\wedge V^1, d) \le \mathrm{cat}\,(BG)$. Thus the Corollary follows from Theorem 15.3.

□

Next, again consider a 1-formal minimal Sullivan algebra $(\wedge V, d)$, and suppose $\dim H^1(\wedge V^1, d) < \infty$. Denote by $\mathcal{L}$ the weighted Lie algebra defined in (15.5). Then it follows from Theorem 2.1 (p. 50) that $\mathcal{L}(1) \cong \mathcal{L}/[\mathcal{L},\mathcal{L}]$ is finite dimensional, and that $\mathcal{L}^n = \oplus_{i\ge n}\mathcal{L}(i)$. In particular, $\mathcal{L}$ is a Sullivan Lie algebra, $(\wedge V^1, d)$ is the associated quadratic Sullivan algebra, and $\mathcal{L}$ is the associated weighted Lie algebra (cf. Example 1) for the fundamental Lie algebra of $(\wedge V, d)$.

Denote by $\mathbb{L}$ the free Lie algebra on $\mathcal{L}(1)$. Then the identity of $\mathcal{L}(1)$ extends to a surjective morphism of weighted Lie algebras

$$\rho : \mathbb{L} \to \mathcal{L}\,,$$

which is an isomorphism in weight 1.

**Definition.** ([11]) The *Chen Lie algebra* of $(\wedge V, d)$ is the quotient Lie algebra

$$h(\wedge V, d) := \mathbb{L}/K\,,$$

where $K$ is the ideal generated by $\ker \rho(2) : \mathbb{L}(2) \to \mathcal{L}(2)$. If $(\wedge V, d)$ is the minimal Sullivan model of a path connected topological space $X$ then the Chen Lie algebra is called the *Chen Lie algebra* of $X$ and is denoted by $h(X)$.

**Theorem 15.4.** *Suppose $(\wedge V, d)$ is a 1-formal minimal Sullivan algebra and that $\dim H^1(\wedge V, d) < \infty$. Then, with the notation above, the morphism $\rho' : \mathbb{L}/K \xrightarrow{\cong} \mathcal{L}$ induced by $\rho$ is an isomorphism.*

*proof:* Denote $\mathbb{L}/K$ by $E$ and by $(\wedge Z, d)$ the quadratic Sullivan 1-algebra associated to the Sullivan Lie algebra $E$. The morphism $\rho'$ induces a morphism of Sullivan algebras

$$\varphi : (\wedge Z, d) \leftarrow (\wedge V^1, d)\,,$$

with the restriction of $\varphi$ to $V^1$ the dual of $\rho'$. Since $\rho'$ is by definition surjective, $\varphi : V^1 \to Z$ is injective. In particular, $\varphi$ defines a Sullivan extension

$$(\wedge Z, d) = (\wedge V^1 \otimes \wedge W, d)\,.$$

Now suppose $W \neq 0$. Then for some non-zero element $w \in W$, $dw$ is a cycle in $\wedge^2 V^1$. But since $(\wedge V^1, d)$ is 1-formal, by Proposition 2.7 (p. 88) every homology class in $H^2(\wedge V^1, d)$ is represented by a cycle in $\wedge^2 V_0^1$. Thus we may assume $dw \in \wedge^2 V_0^1$, and so $w \in Z_1$.

On the other hand, $Z_1 = (\mathbb{L}/K)(2)^\# \oplus (\mathbb{L}/K)(1)^\#$ and by construction $\rho' : (\mathbb{L}/K)(2) \oplus (L/K)(1) \xrightarrow{\cong} \mathcal{L}(2) \oplus \mathcal{L}(1)$. Thus $\varphi : V_1 \xrightarrow{\cong} Z_1$ and it follows that $w \in V_1$, which is a contradiction. Thus $\varphi : \wedge V \xrightarrow{\cong} \wedge Z$ and $\rho'$ is an isomorphism.

□

**Example.** Let $(\wedge W, d)$ be the minimal Sullivan model of an oriented Riemann surface of genus $\geq 2$. According to Proposition 8.5 (p. 240) $(\wedge W, d)$ is a formal Sullivan 1-algebra. Thus as observed above, its homotopy Lie algebra is the completion of its associated weighted Lie algebra $\mathcal{L} = \oplus_{n \geq 1} L(n)$. Moreover it follows from Theorem 15.4 and §8.5 that

$$\mathcal{L} = \mathbb{L}(a_1, b_1, \cdots, a_g, b_g)/\sum [a_i, b_i]\,,$$

a result which first appears in Sullivan [62].

□

Next, let $E$ be the associated weighted Lie algebra for $L$. The gradation $E = \oplus_{n\geq 1}E(n)$ extends to a gradation of the enveloping algebra $UE$ and we denote by $UE(t)$ the Hilbert series

$$UE(t) = \sum_{n=0}^{\infty} \dim (UE)(n)t^n .$$

**Proposition 15.2.** *With the above notation, denote $b_q = \dim H^q(\wedge V, d)$. Then*

$$UE(t) = \frac{1}{\sum_q (-1)^q\, b_q t^q} .$$

*proof:* Since $(\wedge V, d)$ is formal, we have a quasi-isomorphism $\varphi : (\wedge V, d) \to (H, 0)$, and $V$ is equipped with a lower gradation $V = \oplus_{p\geq 0}V_{(p)}$ such that $V_{(p)} = (E(p+1))^{\#}$. This induces a lower gradation $\wedge V = \oplus_p(\wedge V)_{(p)}$ and $d : V_{(p)} \to (\wedge^2 V)_{(p-1)}$. Denote by $(\wedge V \otimes \wedge U, d)$ the acyclic closure of $(\wedge V, d)$. The quasi-isomorphism $\varphi$ induces a quasi-isomorphism

$$\varphi \otimes 1 : (\wedge V \otimes \wedge U, d) \to (H \otimes \wedge U, d) .$$

It follows from the construction of the acyclic closure (p. 111) that the lower gradation of $V$ induces one in $\wedge U$ for which $d : U_{(p)} \to (\wedge V \otimes \wedge U)_{(p-1)}$ and such that the linear part of the differential induces an isomorphism $U_{(p)} \to V_{(p-1)}$. Therefore we have isomorphisms of graded vector spaces $(\wedge U)_{(p)} \cong UE(p)$ and $UE(t) = \sum_{p\geq 0} \dim (\wedge U)_{(p)} t^p$.

Because $(\wedge V, d)$ is formal, it follows from Proposition 2.7 (p. 88) that $\wedge V_{(0)}$ generates $H$. Thus $H^q = H^q_{(0)}$, $(H \otimes \wedge U)^q_{(p)} = H^q \otimes (\wedge U)_{(p)}$ and $d(U_{(p)}) \subset H^1 \otimes (\wedge U)_{(p-1)}$. Thus for each $n \geq 0$ the direct sum

$$\bigoplus_{p+q=n} H^q \otimes (\wedge U)_{(p)}$$

is a subcomplex.

Now since $H^n(\wedge V \otimes \wedge U, d) = 0$ if $n > 0$, we have

$$\sum_{p+q=n} (-1)^q \dim\, (H \otimes \wedge U)^q_{(p)} = \begin{cases} 1, & \text{if } n = 0 \\ 0, & \text{if } n > 0 . \end{cases}$$

Now remark that

$$\left(\sum_q (-1)^q b_q t^q\right) \cdot \left(\sum_p \dim(\wedge U)_{(p)} t^p\right) = \sum_{p,q} (-1)^q \dim(H \otimes \wedge U)^q_{(p)} t^{p+q} = 1 .$$

□

**Example.** There are many examples of 1-formal Sullivan 1-algebras, and a detailed study of 1-formality is given in the survey of Papadima and Suciu ( [54]). For instance, when $G$ is an Artin group, then $K(G,1)$ is 1-formal. In [37], Kohno proves that the complement $X$ in $\mathbb{C}^n$ of complex projective hypersurfaces is 1-formal. More precisely, if $X$ is the complement of a family $\{H_1, H_2, \cdots, H_r\}$ of hyperplanes in $\mathbb{C}^n$ the Chen Lie algebra is generated by the $\{x_1, x_2, \cdots, x_r\}$ with relations

$$\left[x_i, x_{j_1} + x_{j_2} + \cdots + x_{j_s}\right] = 0\,,$$

where $j_1 < j_2 < \cdots < j_s$ are such that $\operatorname{codim} H_{j_1} \cap H_{j_2} \cap \cdots H_{j_s} = 2$. □

# Chapter 16

# Problems

**Problem 1. Cohomology of the spatial realization.** If $(\wedge V, d)$ is the minimal Sullivan model of a path connected space $X$ and if $|\wedge V, d|$ is a Sullivan space, then Theorem 7.8 shows that the natural map $\mu_X : X \to |\wedge V, d)|$ is a rational homology equivalence. Is the converse true? In this case must $X$ be a Sullivan space?

**Problem 2. Model and 1-model.** Let $(\wedge V, d)$ be a minimal Sullivan algebra for which $\dim H^1(\wedge V, d) < \infty$ and $\dim V^i < \infty$, $i \geq 2$. If $\operatorname{cat}(\wedge V, d) < \infty$ then Proposition 9.6 and Theorem 9.3 assert that $\operatorname{cat}(\wedge V^1, d) < \infty$ and $\operatorname{cat}(\wedge V^{\geq 2}, \overline{d}) < \infty$. Is the converse true?
More generally, for any minimal Sullivan extension $(\wedge W \otimes \wedge Z, d)$, if $\operatorname{cat}(\wedge W, d) < \infty$ and $\operatorname{cat}(\wedge Z, \overline{d}) < \infty$, is $\operatorname{cat}(\wedge W \otimes \wedge Z, d) < \infty$?

**Problem 3. Exponential growth in homotopy.** Chapters 12 and 13 establish exponential growth of the ranks of the homotopy groups of a simply connected rationally hyperbolic space. Moreover, Chapter 15 shows that, under some hypotheses, for the fundamental Lie algebra $L_0$ of a minimal Sullivan algebra, the sequence $\dim L_0^n / L_0^{n+1}$ grows exponentially.

- Suppose $L$ is the rational homotopy Lie algebra of a simply connected rationally hyperbolic space and that $\dim L/[L, L]$ is finite dimensional. Does the sequence $\dim L^n / L^{n+1}$ grow exponentially?
- Suppose for a group $G$ that $BG$ is a Sullivan space with infinite dimensional fundamental Lie algebra $L_0$. Does the sequence $\dim L_0^n / L_0^{n+1}$ grow exponentially?
- Let $L$ be a graded Lie algebra with depth $L < \infty$ and $\dim L/[L, L] < \infty$. Does the sequence $L^n / L^{n+1}$ grow faster than any polynomial?

**Problem 4. Structure of the homotopy Lie algebra.** Chapter 14 establishes general structure theorems for the lattice of $L$-equivalence classes of ideals in a graded Lie algebra $L = L_{\geq 1}$ of finite depth. This suggests the questions:

- Is there an analogous lattice for a Lie algebra $L_0$ of finite depth, with analogous structure theorems?
- Are there analogous results for $\mathbb{Q}[\pi_1(X)] \otimes UL_{\geq 1}$ in general or in the case of Sullivan spaces?
- Let $L$ be the homotopy Lie algebra of a simply connected rationally hyperbolic space. If $\operatorname{depth} L < \infty$, does it follows that $\operatorname{depth} L_{even} < \infty$?
- In Chapter 14 it is shown that log index $[L, L] =$ log index $L$. What conditions imply that log index $L/[L, L] <$ log index $L$? Is this always the case if $L$ is the homotopy Lie algebra of a simply connected rationally hyperbolic finite CW complex?
- Suppose $E$ is an abelian sub Lie algebra of the homotopy Lie algebra $L$ of a simply connected space of finite category. If $\dim L = \infty$, is it always true that log index $E <$ log index $L$?

**Problem 5. Sullivan spaces.**

- Determine a large class of Sullivan spaces of the form $BG$.
- If $X$ is a connected finite CW complex such that $\pi_i(X) = 0$ if $i \geq 2$, is $X$ a Sullivan space?
- If $X$ has a single 2-cell and no cell of higher degree, must $X$ be 1-formal?
- If $G$ is the quotient of a free group by one relation, is $BG$ a Sullivan space?

**Problem 6. Depth and category.** According to Chapter 10, $\operatorname{cat}(\wedge V, d) \geq \operatorname{depth}(\wedge V, d_1)$. Moreover if $\mu : (\wedge V, d_1) \to C^*(L)$ is a quasi-isomorphism, then $\operatorname{depth}(\wedge V, d_1) = \operatorname{depth} L$. Are there other conditions which imply $\operatorname{depth} L = \operatorname{depth}(\wedge V, d_1)$?

**Problem 7. Rationalization.** Compare the rationalization $X_{\mathbb{Q}} = |\wedge V, d|$ with the Bousfield-Kan construction $\mathbb{Q}_\infty(X)$ ( [7]).

**Problem 8.** Determine some algebraic conditions on a pronilpotent Lie algebra $L$ which imply an isomorphism

$$\mathrm{Ext}^{UL}(\Bbbk, \Bbbk) \cong \mathrm{Ext}^{\Bbbk[G_L]}(\Bbbk, \Bbbk)\,.$$

**Problem 9.**

- Determine some algebraic conditions on a graded Lie algebra that imply that the natural morphism

$$\varinjlim_n C^*(L/L^n) \to C^*(L)$$

  is a quasi-isomorphism.
- Determine conditions on a group $G$ which imply that

$$\varinjlim_n C^*((G/G^n); \mathbb{Q}) \to C^*(G; \mathbb{Q})$$

  is a quasi-isomorphism.

**Problem 10.** Suppose $L$ is a Sullivan Lie algebra concentrated in degree zero.

- When is there a subspace $S \subset L$ mapping isomorphically to $L/[L, L]$ and generating a sub Lie algebra $E \subset L$ isomorphic to the associated weighted Lie algebra of Chapter 15?
- Is there a weaker condition on $S$ for which

$$\varinjlim_n C^*(E/E^n) \xrightarrow{\simeq} C^*(E)?$$

**Problem 11.** If $L$ is a graded Lie algebra then each $L_k$ is filtered by the subspaces $L_k(p) = \{\, x \in L_k \mid [\underbrace{L_0, \cdots, L_0}_{p+1}, x] = 0 \,\}$. To what extent can the results about Sullivan Lie algebras be extended to Lie algebras in which $\cap_n L_0^n = 0$ and $L_0/[L_0, L_0]$ and each $L_k(p+1)/L_k(p)$ are finite dimensional.

**Problem 12.** Suppose $L$ is an infinite dimensional Sullivan Lie algebra concentrated in degree zero. Must it be true that if $\operatorname{depth} L < \infty$ then for some $n$ $\dim L^n/[L^n, L^n] = \infty$?

**Problem 13. Local and weak local nilpotence**

- According to Theorem 10.3 (p. 288) if $L$ is a Sullivan Lie algebra of depth $m$ then the elements $x \in L_{even}$ such that $ad\,x$ is locally nilpotent span a vector space of dimension $\leq m$. Do the elements $x \in L_{odd}$ for which $ad\,x$ is locally nilpotent span a finite dimensional vector space?
- If $L$ is a Sullivan Lie algebra an element $x \in L$ is *weakly locally nilpotent* if for some sequence $n(k)$, $(ad\,x)^k : L \to L^{n(k)}$ and $\lim_{k\to\infty} k/n(k) = 0$. If depth $L < \infty$ can every $x \in L$ be weakly locally nilpotent?

**Problem 14.** It is known ([43]) that a minimal simply connected Sullivan algebra whose homology is a Poincaré duality algebra is the model of a commutative cochain algebra which itself is a Poincaré duality algebra. Does this remain true in the non-simply connected case?

# Bibliography


[1] J.F. Adams and P.J. Hilton, *On the chain algebra of a loop space*, Comment. Math. Helvetici **30** (1956), 305-330.

[2] J. Amoros, *On the Malcev completion of Kahler groups*, Comment. Math. Helv. **71** (1996), 192-212

[3] I.K. Babenko, *Analytical properties of Poincaré series of a loop space*, Mat. Zametski **27** (1980) 751-764.

[4] H. Baues and J.M. Lemaire, *Minimal models in homotopy theory*, Math. Ann. **225** (1977), 219-242.

[5] J. Birman and T. Brendle, *Handbook of geometric topology*, Elsevier (2005).

[6] A.K. Bousfiled and V.K. Gugenheim, *On PL de Rham theory and rational homotopy type*, Mem. Amer. Math. Soc. **179** (1976).

[7] A.K. Bousfield and D.M. Kan, *Homotopy limits, completions and localizations*, Lecture Notes in Math. 304, Springer (1972).

[8] E. Brieskorn, *Sur les groupes de tresses (d'après Arnold)*, Séminaire Bourbaki 1971/72, Lecture Notes in Mathematics, 317, Springer-Verlag (1973).

[9] H. Cartan and S.Eilenberg, *Homological algebra*, Princeton University Press (1956).

[10] B. Cenkl and R. Porter, *Malcev's completion of a group and differential forms*, J. Differential Geometry **15** (1980), 531-542.

[11] K.T. Chen, *Extension of $C^\infty$-function algebra by integrals and Malcev completion of $\pi_1$*, Advances in Math. **23** (1977), 181-210.

[12] O. Cornea, G. Lupton, J. Oprea and D. Tanré, *Lusternik-Schnirelmann category*, Mathematical Surveys and Monographs 103, AMS (2003).

[13] P. Deligne, P. Griffiths, J. Morgan and D. Sullivan, *Real homotopy theory of Kähler manifolds*, Invent. Math. **29** (1975), 245-274.

[14] G. Denham and A. Suciu, *On the homotopy Lie algebra of an arrangement*, Michigan Math. J. **54** (2006), 319-340.

[15] M. Falk, *The minimal model of the complement of an arrangement of hyperplanes*, Trans. Amer. Math. Soc. **309** (1988), 543-556.

[16] M. Falk and R. Randell, *The lower central series of a fiber-type arrangement*, Invent. Math. **82** (1985), 77-88.

[17] Y. Félix, *La dichotomie elliptique-hyperbolique en homotopie rationnelle*,

Astérisque 176, SMF (1989).
[18] Y. Félix, S. Halperin and J.-C. Thomas, *Rational Homotopy Theory*, Graduate Texts in Mathematics **205**, Springer-Verlag (2001).
[19] Y. Félix, S. Halperin and J.-C. Thomas, *An asymptotic formula for the ranks of the homotopy groups of finite complex*, Expositiones Math. **25** (2007), 67-76.
[20] Y. Félix, S. Halperin and J.-C. Thomas, *The ranks of the homotopy groups of a finite complex*, Can. J. Math. **65** (2013), 82-119.
[21] Y. Félix, J. Oprea and D. Tanré, *Algebraic models in geometry*, Oxford Graduate Texts in Math. 17 (2008).
[22] W. Fulton, *Algebraic curves*, Addison-Wesley (1969).
[23] T. Ganea, *Lusternik-Schnirelmann category and strong category*, Illinois J. Math. **11** (1967), 417-427.
[24] D.H. Gotllieb, *Evaluation subgroups of homotopy groups*, Amer. J. Math. **91** (1969), 729-756.
[25] G. Grätzer, *General lattice theory*, Birkhäuser (1998).
[26] P. Griffiths and J. Morgan, *Rational homotopy theory and differential forms*, Second Edition, Progress in Mathematics 19, Birkhäuser (2013).
[27] P.-P. Grivel, *Formes différentielles et suites spectrales*, Ann. Inst. Fourier **29** (1979), 17-37.
[28] S. Halperin, *Lectures on Minimal Models*, Mémoires de la Société Mathématique de France, **230** (1983).
[29] S. Halperin and J.-M. Lemaire, *Suites inertes dans les algèbres de Lie graduées*, Math. Scand. **61** (1987), 39-67.
[30] S. Halperin and J. Stasheff, *Obstructions to homotopy equivalences*, Adv. in Math. **32** (1979), 233-279.
[31] K. Hasegawa, *Minimal models of nilmanifolds*, Proc. Amer. Math. Soc. **106** (1989), 65-71.
[32] A. Hatcher, *Algebraic topology*, Cambridge University Press (2002).
[33] K. Hess, *A proof of Ganea conjecture for rational spaces*, Topology **30** (1991), 205-214.
[34] K. Hess, *Rational homotopy theory: a brief introduction*, Interactions between homotopy theory and algebra, 175-202, Contemp. Math. **436**, Amer. Math. Soc. (2007).
[35] P. Hilton, G. Mislin and J. Roitberg, *Localization of nilpotent groups and spaces*, North-Holland (1975).
[36] I. James, *On category in the sense of Lusternik-Schnirelmann*, Topology **17** (1978), 331-348.
[37] T. Kohno, *On the holonomy Lie algebra and the nilpotent completion of the fundamental group of the complement of hypersurfaces*, Nagoya Math. J. **92** (1983), 21-37.
[38] T. Kohno, *Série des Poincaré-Koszul associée aux groupes de tresses pures*, Invent. Math. **82** (1985), 57-75.
[39] J.-L. Koszul, *Homologie et cohomologie des algèbres de Lie*, Bull. Soc. Math. France **78** (1950), 65-127.
[40] A.L. Kurosh, *Theory of groups, II*, Chelsea Publishing (1956).

[41] P. Lambrechts, *Analytic properties of Poincaré series of spaces*, Topology **37** (1998), 1363-1370.
[42] M. Lazard, *Sur les groupes nilpotents et les anneaux de Lie*, Ann. Ec. Norm. Sup. **71** (1954), 101-190.
[43] J.M. Lemaire and F. Sigrist, *Sur les invariants d'homotopie rationnelle liés à la LS catégorie*, Comment. Math. Helv. **56** (1981), 103-122.
[44] MacLane, *Homology*, Springer-Verlag (1975).
[45] W. Magnus, A. Karrass and D. Solitar, *Combinatorial group theory*, Interscience Publishers (1966).
[46] A.L. Malcev, *Nilpotent groups without torsion*, Jzv. Akad. Nauk. SSSR, Math. **13** (1949), 201-212.
[47] P. May, *Simplicial objects in algebraic topology*, Chicago Univ. Press (1967).
[48] P. May, *Classifying spaces and fibrations*, Mem. Amer. Math. Soc. **155** (1975).
[49] J. Milnor and J.C. Moore, *On the structure of Hopf algebras*, Annals Math. **81** (1965), 211-264.
[50] J. Morgan,*The algebraic topology of smooth algebraic varieties*, Inst. Hautes Etudes Sci Publ. Math. **148** (1970), 137-204.
[51] J. Neisendorfer and L. Taylor, *Lie algebras, coalgebras and rational homotopy theory for nilpotent spaces*, Pacific J. Math. **74** (1978), 429-460.
[52] P. Orlik, *Seifert Manifold*, Lecture Notes in Mathematics, **231** Springer-Verlag (1972).
[53] P.Orlik and L.Solomon, *Combinatorics and topology of complements of hyperplanes*, Invent. Math. **56** (1980), 167-189.
[54] S. Papadima and A. Suciu, *Geometric and algebraic aspects of 1-formality*, Bull. Math. Soc. Sci. Math. Roumanie **52** (2009), 355-375.
[55] S. Papadima and A. Suciu, *Chen Lie algebras*, Internat. Math. Res. Notices **21** (2004), 1057-1086.
[56] S. Papadima and A. Suciu, *Homotopy Lie algebras, lower central series and the Koszul property*, Geom. Topol. **8** (2004), 1079-1125.
[57] S. Papadima and S. Yuzvinsky, *On rational* $K(\pi, 1)$ *spaces and Koszul algebras*, J. Pure Appl. Alg. **144** (1999), 157-167.
[58] D. Quillen, *Rational homotopy theory*, Ann. Math. **90** (1969), 205-295.
[59] H.K. Schenck and A. Suciu, *Lower central series and free resolutions of hyperplane arrangements*, Trans. Amer. Math. Soc. **354** (2002), 3409-3433.
[60] J.P. Serre, *Lie algebras and Lie groups*, Benjamin Inc. (1965).
[61] D. Sullivan, *Differential forms and the topology of manifolds*, Proceedings of the International Conference on Manifolds and Related Topics in Topology, Tokyo, 1973, University of Tokyo Press (1975), 37-49.
[62] D. Sullivan, *Infinitesimal computations in topology*, Publ. IHES **47** (1977), 269-331.
[63] G.H Toomer, *Lusternik-Schnirelmann category and the Moore spectral sequence*, Math. Z. **138** (1974), 123-143.
[64] J.H. van Lint and R.M. Wilson, *A course in combinatorics*, Cambridge University Press, second edition (2001).
[65] C. Weibel, *An introduction to homological algebra*, Cambridge University

Press (1995).
[66] G. Whitehead, *Elements of homotopy theory*, Graduate Text in Math. **61**, Springer-Verlag (1978).
[67] J.H.C. Whitehead, *A certain exact sequence*, Ann. Math. **52** (1951), 51-110.

# Index

www.ingramcontent.com/pod-product-compliance
Ingram Content Group UK Ltd.
Pitfield, Milton Keynes, MK11 3LW, UK
UKHW020403150926
8624IPUK00004B/312